S0-BCN-221

Measurement and Detection of Radiation

McGRAW-HILL SERIES IN NUCLEAR ENGINEERING

THOMAS H. PIGFORD Consulting Editor

Ash *Nuclear Reactor Kinetics*
Benedict, Pigford, and Levi *Nuclear Chemical Engineering*
Bonilla *Nuclear Engineering*
Ellis *Nuclear Technology for Engineers*
El-Wakil *Nuclear Power Engineering*
Glower *Experimental Reactor Analysis and Radiation Measurements*
Hoisington *Nucleonics Fundamentals*
Knief *Nuclear Energy Technology*
Meghreblian and Holmes *Reactor Analysis*
Price *Nuclear Radiation Detection*
Schultz *Control of Nuclear Reactors and Power Plants*
Stephenson *Introduction to Nuclear Engineering*
Tsoulfanidis *Measurement and Detection of Radiation*

Measurement and Detection of Radiation

Nicholas Tsoulfanidis
University of Missouri-Rolla

HEMISPHERE PUBLISHING CORPORATION *Washington New York London*

McGRAW-HILL BOOK COMPANY *New York St. Louis San Francisco Auckland Bogotá Hamburg Johannesburg London Madrid Mexico Montreal New Delhi Panama Paris São Paulo Singapore Sydney Tokyo Toronto*

211012

539.7
T 882

Figures 3.3 and 4.22 are from "The Atomic Nucleus" by R. D. Evans. Copyright © 1972 by McGraw-Hill. Used with the permission of McGraw-Hill Book Company.

Figures 7.19, 10.24, 10.34, 10.37, and 12.22 are reproduced from *Instruments for Research and Applied Sciences* by permission of EG&G ORTEC, Oak Ridge, Tennessee.

MEASUREMENT AND DETECTION OF RADIATION

Copyright © 1983 by Hemisphere Publishing Corporation. All rights reserved. Printed in the United States of America. Except as permitted under the United States Copyright Act of 1976, no part of this publication may be reproduced or distributed in any form or by any means, or stored in a data base or retrieval system, without the prior written permission of the publisher.

1 2 3 4 5 6 7 8 9 0 BCBC 8 9 8 7 6 5 4 3 2

This book was set in Press Roman by Hemisphere Publishing Corporation. The editors were Christine Flint and Elizabeth Dugger; the designer was Sharon Martin DePass; the production supervisor was Miriam Gonzalez; and the typesetter was Wayne Hutchins. BookCrafters, Inc., was printer and binder.

Library of Congress Cataloging in Publication Data

Tsoulfanidis, Nicholas, date.
Measurement and detection of radiation.

(McGraw-Hill series in nuclear engineering)
Bibliography: p.
Includes index.
1. Radiation—Measurement. 2. Nuclear counters.
I. Title. II. Series.
QC795.42.T78 1981 539.7'7 81-2940
ISBN 0-07-065397-6 AACR2

To Zizeta

contents

3. REVIEW OF ATOMIC AND NUCLEAR PHYSICS *75*

4. ENERGY LOSS AND PENETRATION OF RADIATION THROUGH MATTER *115*

5. GAS–FILLED DETECTORS *165*

6. SCINTILLATION DETECTORS *195*

■ preface

The material in this book, which is the result of a 10-year experience obtained in teaching courses related to radiation measurements at the University of Missouri-Rolla, is intended to provide an introductory text on the subject. It includes not only what I believe the beginner ought to be taught but also some of the background material that people involved in radiation measurements should have. The subject matter is addressed to upper-level undergraduates and first-year graduate students. It is assumed that the students have had courses in calculus and differential equations and in basic atomic and nuclear physics. The book should be useful to students in nuclear, mechanical, and electrical engineering, physics, chemistry (for radiochemistry), nuclear medicine, and health physics; to engineers and scientists in laboratories using radiation sources; and to personnel in nuclear power plants.

The structure and the contents of the book are such that the person who masters the material will be able to

1 Select the proper detector given the energy and type of particle to be counted and the purpose of the measurement.
2 Analyze the results of counting experiments, i.e., calculate errors, smooth results, unfold energy spectra, fit results with a function, etc.
3 Perform radiation measurements following proper health physics procedures.

The first chapter defines the energy range of the different types of radiation for which instruments and methods of measurement are considered; it gives a brief discussion of errors that emphasizes their importance; and, finally, it presents a very general description of the components of a counting system. This last part of the chapter is necessary because a course on radiation measurements involves laboratory work, and for this reason the students should be familiar from the very beginning with the general features and functions of radiation instruments.

The second chapter addresses the very important subject of errors. Since all experimental results have errors, and results reported without their corresponding errors are meaningless, this chapter is fundamental for a book such as this one. Further discussion of errors caused by the analysis of the results is presented in Chap. 11.

Chapters 3 and 4 constitute a quick review of material that should have been covered in previous courses. My experience has been that students need this review of atomic and nuclear physics and of penetration of radiation through matter. These two chapters can be omitted if the instructor feels that the students know the subject.

Chapters 5–7 describe the different types of radiation detectors. Full chapters have been devoted to gas-filled counters, scintillation detectors, and semiconductor detectors. Detectors with "special" functions are discussed in Chap. 17.

The subject of relative and absolute measurements is presented in Chap. 8. The solid angle (geometry factor) between source and detector and effects due to the source and the detector, such as efficiency, backscattering, and source self-absorption are all discussed in detail.

Chapter 9 is an introduction to spectroscopy. It introduces and defines the concepts used in the next four chapters. Chapter 10 discusses the features of the electronic components of a counting system that are important in spectroscopy. Its objective is not to make the reader an expert in electronics but to show how the characteristics of the instruments may influence the measurements.

Chapter 11 presents methods of analysis of experimental data. Methods of curve fitting, of interpolation, and of least-squares fitting are discussed concisely but clearly. A general discussion of folding, unfolding, and data smoothing, which are necessary tools in analysis of spectroscopic measurements, occupies the second half of this chapter. Special methods of unfolding for photons, charged particles and neutrons are further discussed in Chaps. 12 through 14, which also cover spectroscopy. Individual chapters are devoted to photons, charged particles, and neutrons. All the factors that affect spectroscopic measurements and the methods of analysis of the results are discussed in detail.

Chapter 15 is devoted to activation analysis, a field with wide-ranging applications. Health physics is discussed in Chap. 16. I feel that every person who handles radiation should know at least something about the effects of radiation, radiation units, and regulations related to radiation protection. This chapter may be omitted if the reader has already studied the subject.

Chapter 17 deals with special detectors and spectrometers that have found applications in many different fields but do not fit in any of the previous chapters. Examples are the self-powered detectors, which may be gamma or neutron detectors, fission track detectors, thermoluminescent dosimeters, photographic emulsions, and others.

The problems at the end of each chapter should help the student understand the concepts presented in the text. They are arranged not according to difficulty but in the order of presentation of the material needed for their solution.

The appendixes at the end of the book provide useful information to the reader.

I use the SI (metric) units with the exception of some well-established nonmetric units, which, it seems, are here to stay. Examples are MeV, keV, and eV for energy; the barn for cross sections; the curie, and the rem. These units are given in parentheses along with their SI counterparts.

Writing a book is a tremendous undertaking, a task too big for any single person. I was fortunate to have been helped by many individuals, and it gives me great pleasure to recognize them here. First and foremost, I thank all the former students who struggled through my typed notes when they took the radiation measurements course at the University of Missouri-Rolla. Their numerous critical comments are deeply appreciated. I thank my colleagues, Dr. D. Ray Edwards for his continuous support, Dr. G. E. Mueller for his many useful suggestions, and Drs. A. E. Bolon and T. J. Dolan for many helpful discussions over the last 10 years. I also thank Dr. R. H. Johnson of Purdue University for reviewing certain chapters. I especially thank my dear friend Professor B. W. Wehring of the University of Illinois for numerous lengthy discussions following his detailed critical review of most of the chapters. I am grateful to Mrs. Susan Elizagary for expertly typing most of the manuscript and to Mrs. Betty Volosin for helping in the final stages of typing.

No single word or expression of appreciation can adequately reflect my gratitude to my wife Zizeta for her moral support and understanding during the last three painstaking years, and to my children Steve and Lena for providing pleasant and comforting distraction.

Nicholas Tsoulfanidis

Measurement and Detection of Radiation

1

introduction to radiation measurements

1.1 WHAT IS MEANT BY RADIATION?

The word *radiation* was used until about 1900 to describe electromagnetic waves. Around the turn of the century, electrons, x-rays, and natural radioactivity were discovered and were also included under the umbrella of the term radiation. The newly discovered radiation showed characteristics of particles, in contrast to the electromagnetic radiation, which was treated as a wave. In the 1920s, DeBroglie developed his theory of the duality of matter, which was soon afterward proved correct by electron diffraction experiments, and the distinction between particles and waves ceased to be important. Today, radiation refers to the whole electromagnetic spectrum as well as to all the atomic and subatomic particles that have been discovered.

One of the many ways in which different types of radiation are grouped together is in terms of ionizing and nonionizing radiation. The word *ionizing* refers to the ability of the radiation to ionize a gas through which it passes.

Nonionizing radiation is electromagnetic radiation with wavelength λ of about 1.0 nm or longer. That part of the electromagnetic spectrum includes radiowaves, microwaves, visible light ($\lambda = 770$ nm to 390 nm), and ultraviolet light ($\lambda = 390$ nm to 1 nm).

Ionizing radiation includes the rest of the electromagnetic spectrum (x-rays, $\lambda \approx 1.0$–0.005 nm) and γ rays with wavelength shorter than that of x-rays. It also includes all the atomic and subatomic particles, such as electrons, positrons, protons, alphas, neutrons, heavy ions, and mesons.

The material in this text refers only to ionizing radiation. Specifically, it deals with detection instruments and methods, experimental techniques, and analysis of results for radiation in the energy range shown in Table 1.1. Particles with energies listed in Table 1.1 are encountered around nuclear reactors, around installations involving production or use of natural or manufactured radioisotopes, and also around low-energy accelerators. Not included in Table 1.1 are cosmic rays and particles produced by high-energy accelerators (GeV energy range).

1.2 STATISTICAL NATURE OF RADIATION EMISSION

Radiation emission is nothing more than release of energy by a system as it moves from one state to another.

According to classical physics, exchange or release of energy takes place on a continuous basis; i.e., any amount of energy, no matter how small, may be exchanged as long as the exchange is consistent with conservation laws. The fate of a system is

TABLE 1.1 Maximum Energy Calculated

Particle	Energy (MeV)
α	20
β	10
γ	20
n	15
Heavy ions	100

exactly determined if initial conditions and forces acting upon it are given. One may say that classical physics prescribed a "deterministic" view of the world.

Quantum theory changed all that. According to quantum theory, energy can be exchanged only in discrete amounts when a system moves from one state to another. The fact that conservation laws are satisfied is a necessary but not a sufficient condition for the change of a system. The fate of the system is not determined exactly if initial conditions and forces are known. One can only talk about the probability that the system will do something or do nothing. Thus, with the introduction of quantum theory, the study of the physical world changed from "deterministic" to "probabilistic."

The emission of atomic and nuclear radiation obeys the rules of quantum theory. As a result of this, one can only talk about the probability that a reaction will take place or that a particle will be emitted. If one attempts to measure the number of particles emitted by a nuclear reaction, that number is not constant in time; it has a statistical uncertainty because of the probabilistic nature of the phenomenon under study.

Consider a radioactive source emitting electrons and assume that one attempts to measure the number of electrons per unit time emitted by the source. For every atom of the source there is a probability, not a certainty, that an electron will be emitted during the next unit of time. One can never measure the "exact" number. The number of particles emitted per unit time is different for successive units of time. Therefore, one can only determine the average number of particles emitted. That average, like any average, carries with it an uncertainty, an error. The determination of this error is an integral part of any radiation measurement.

1.3 THE ERRORS AND ACCURACY AND PRECISION OF MEASUREMENTS

A measurement is an attempt to determine the value of a certain parameter or quantity. Anyone attempting a measurement should keep in mind the following two axioms regarding the result of the measurement:

Axiom 1 **No measurement yields a result without an error.**

Axiom 2 **The result of a measurement is almost worthless unless the error associated with that result is also reported.**

The term *error* is used to define the following concept:

Error = (measured quantity) − (true value)

or Error = estimated uncertainty of the measured quantity

Related to the error of a measurement are the terms *accuracy* and *precision.* The dictionary gives essentially the same meaning for both accuracy and precision, but in experimental work they have different meanings.

The accuracy of an experiment tells us how close the result of the measurement is to the true value of the measured quantity. The precision of an experiment is a measure of the exactness of the result. As an example, consider the measurement of the speed of light, which from theory is known to be 3×10^8 m/s. Assume that a measurement gave the result 2.9998×10^8 m/s. The difference between these two numbers is an estimate of the accuracy of the measurement. On the other hand, the precision of the measurement is related to the number of significant figures† representing the result. The number 2.9998×10^8 indicates that that the result has been determined to be between 2.9997 and 2.9999 or, equivalently, that it is known to 1 part in 30,000 (1/29998).

If the measurement is repeated and the new result is 2.9999×10^8 m/s, the accuracy changed but not the precision. If, on the other hand, the result of the measurement is 2.99985×10^8 m/s, both precision and accuracy have changed.

Every experimenter should consider accuracy and precision simultaneously. It would be a waste of effort to try to improve the precision of a measurement if it is known that the result is inaccurate. On the other hand, it is almost useless to try to achieve very high accuracy if the precision of the measurement is low.

Limitations in the accuracy and precision of measurements result from many causes. Among the most important are:

1 Incorrectly calibrated instruments.
2 Algebraic or reading errors of the observer.
3 Uncontrolled changes in environmental conditions, such as temperature, pressure, and humidity.
4 Inability to construct arbitrarily small measuring meter-sticks, rods, pointers, clocks, apertures, lenses, etc.
5 A natural limit of sensitivity for any real measuring instrument detecting individual effects of atoms, electrons, molecules, and protons.
6 Imperfect method of measurement in most cases.
7 Unknown exact initial state of the system. Or, even if the initial state is known, it is impossible to follow the evolution of the system. For example, to determine the state of a gas in a container one should know the exact position and velocity of every molecule at $t = 0$. Even if this is known, how practical is it to follow 10^{20} atoms or molecules moving in a box?
8 Statistical nature of some processes, e.g., radioactive decay. There is a probability

†As an example of the number of significant figures, each of the following numbers has five significant figures: 2.9998, 29998, 20009, .0029998, 2.9880×10^8.

that an atom of a radioactive isotope will decay in the next 10 s, and this is as much information as one can report on this matter. The probability can be calculated, but it is still a probability never a certainty.

1.4 TYPES OF ERRORS

There are many types of errors, but they are usually grouped into two broad categories, systematic and random.

Systematic (or determinate) errors are those that affect all the results in the same way. The experimenter should make every reasonable effort to minimize or, better yet, eliminate systematic errors. Examples of such errors are

1 Errors from badly calibrated instruments
2 Personal errors (algebraic, wrong readings, etc.)
3 Imperfect technique

Random (or indeterminate) errors can either decrease or increase the results of a measurement, but in a nonreproducible way. Most of the random errors cannot be eliminated. They can be reduced, however, by either improving the experimental apparatus, improving the technique, and/or repeating the experiment many times. Examples of random errors are

1 Errors resulting from experimental apparatus (reading of instruments, electronic noise, etc.)
2 Errors from uncontrolled change in condition such as V, T, P
3 Probabilistic nature of the phenomenon under study

The determination of error associated with the measurement is a very important task. It is probably as important as the measurement. Technical journals and scientific reports never report results of experiments without the error corresponding to these results. A measurement reported without an error is almost worthless. The study of errors has great practical applications for the work of scientists and engineers.

This text does not give a complete theory of error. Only the fundamentals needed for a basic understanding of the statistical analysis of errors are presented. The objective is to present methods that provide an estimate of the error of a certain measurement or a series of measurements and procedures that minimize the error.

1.5 NUCLEAR INSTRUMENTATION

1.5.1 Introduction

This section is addressed to the person who has not seen or used radiation instruments.† Its purpose is to present a general description of the physical appearance and operation of the basic components of a radiation counting system. Every component is treated like a "black box," i.e., input and output are discussed without any details

†The term *radiation instruments* refers to instruments used for the detection of ionizing radiation as explained in Sec. 1.1.

about how the output is obtained. Details about the construction and operation of individual units are given in later chapters. Detectors are discussed in Chaps. 5 through 7 and the rest of the electronics in Chap. 10.

Counting systems are classified into two types, according to the method of operation:

1 Pulse-type systems. The output consists of voltage pulses, one pulse per particle detected.
2 Current-type systems. The output is an average value, resulting from the detection of many particles.

A basic pulse-type system consists of the instruments shown in Fig. 1.1. The function of each component is discussed in later sections of this chapter.

A current-type system (e.g., an electrometer or a rate meter) is simpler than the pulse-type system. Such systems are discussed in Chap. 10. The remainder of this chapter concerns only pulse-type counting systems.

1.5.2 The Detector

The function of the detector is to produce a signal for every particle entering in it. Every detector works by using some interaction of particles with matter. Following is a list of the most common detector types.

1 Gas-filled counters (ionization, proportional, GM counters)
2 Scintillation detectors
3 Semiconductor detectors
4 Spark chambers (used with high energy particles)
5 Bubble chambers (used with high energy particles)
6 Photographic emulsions
7 Thermoluminescent dosimeters (TLDs)
8 Cerenkov counters
9 Self-powered neutron or gamma detectors

The signal at the output of most detectors is a voltage pulse, such as the one shown in Fig. 1.2. For others, the signal may be a change in color (emulsions) or some trace that can be photographed (bubble or spark chambers)

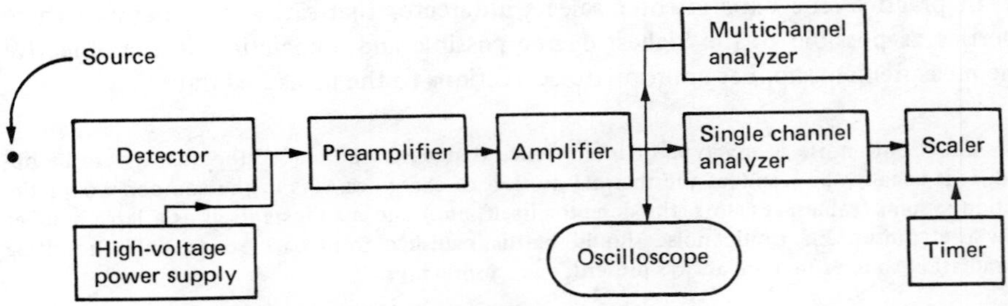

FIG. 1.1 A basic pulse-type detection system.

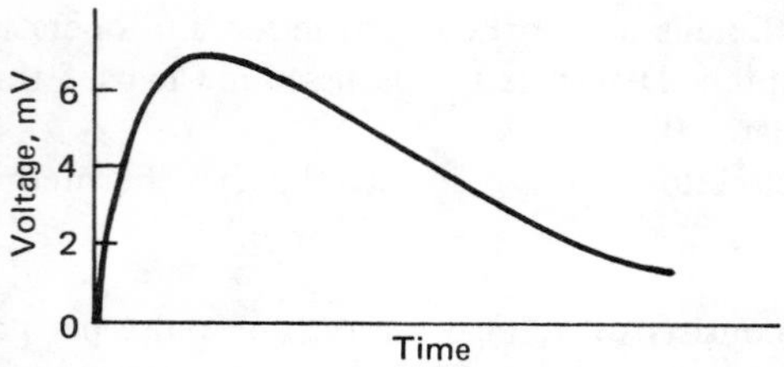

FIG. 1.2 A typical pulse-type detector signal.

The ideal pulse-type counter should satisfy the following requirements:

1 Every particle entering the detector should produce a pulse at the exit of the counter, which is higher than the electronic noise† level of the unit that accepts it (usually this unit is the preamplifier). In such a case, every particle entering the detector will be detected and the detector efficiency, defined as the ratio of the number of particles detected to the number of particles entering the counter, will be equal to 100 percent (for more details on efficiency, see Chap. 8).
2 The duration of the pulse should be short, so that particles coming in one after the other in quick succession produce separate pulses. The duration of the pulse is a measure of the dead time of the counter (see Sec. 2.21) and may result in loss of counts in the case of high counting rates.
3 If the energy of the particle is to be measured, the height of the pulse should have some known fixed relationship to the energy of the particle. To achieve this, it is important that the size of the counter is such that the particle deposits all its energy (or a known fraction) in it.
4 If two or more particles deposit the same energy in the detector, the corresponding pulses should have the same height. This requirement is expressed in terms of the energy resolution of the detector (see Chap. 9). Good energy resolution is extremely important if the radiation field consists of particles with different energies and the objective of the measurement is to identify (resolve) these energies. Figure 1.3 shows an example of good and bad energy resolution.

There is no detector that satisfies all these requirements. Few detectors have 100 percent efficiency. In practice, it is not feasible for gamma and neutron detectors to have all the energy of the particle deposited in the counter. Because of statistical effects there is no detector with ideal energy resolution. What should one do?

In practice, the experimenter selects a detector that satisfies as many of these properties as possible to the highest degree possible and, depending on the objective of the measurement, applies appropriate corrections to the measured data.

†Electronic noise is any type of interference that tends to "mask" the quantity to be observed. It is usually the result of the thermal motion of charge carriers in the components of the detection system (cables, resistors, the detector itself, etc.) and manifests itself as a large number of low-level pulses. Electronic noise should be distinguished from background pulses resulting from radiation sources that are always present, e.g., cosmic rays.

1.5.3 The NIM Concept

Most of the commercially available instruments that are used in radiation measurements conform to the standards on nuclear instrument modules (NIM) developed by the United States Atomic Energy Commission (now the Nuclear Regulatory Commission).[1] The objective of the NIM standard is the design of commercial modules that are interchangeable physically and electrically. The electrical interchangeability is confined to the supply of power to the modules and in general does not cover the design of the internal circuits.

The size of the smallest, called a single-width, NIM is 0.203 m × 0.02 m (8 in × $\frac{3}{4}$ in). Multiple-width NIMs are also made. The standard NIM bin will accommodate 12 single-width NIMs or any combination of them having the same total equivalent width. Figure 1.4 is a photograph of the front and back sides of a commercial standard bin. Figure 1.5 is a photograph of the bin filled with NIMs of different widths, made by different manufacturers.

1.5.4 The High-Voltage Power Supply

The *high-voltage power supply* (HVPS) provides a positive or negative voltage necessary for the operation of the detector. Most detectors need positive high voltage (HV). Typical HVs for common detectors are given in Table 1.2. The HVPS is constructed in such a way that the HV at the output changes very little even though the input voltage (110 V, ac) may fluctuate.

A typical commercial HVPS is shown in Fig. 1.6. The front panel has an indicator light that shows whether the unit is on or off and, if it is on, whether the output is positive or negative voltage. There are two knobs for voltage adjustment, one for coarse changes of 500-V intervals, the other for changes of 0.1 V. The output is at the rear of the unit.

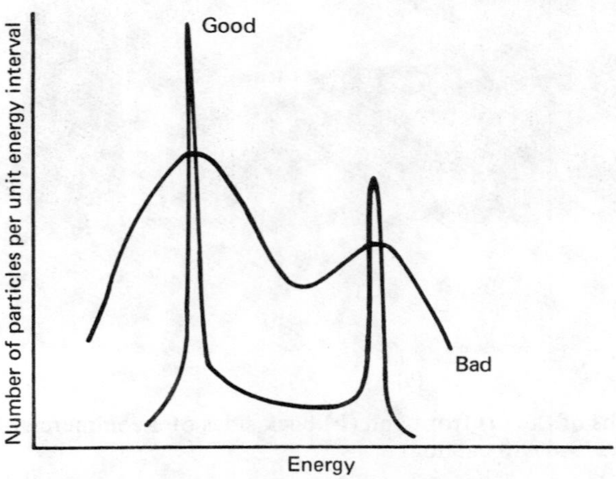

FIG. 1-3 Good and bad energy resolution.

(a)

(b)

FIG. 1.4 Photographs of the (*a*) front and (*b*) back sides of a commercial NIM bin (from Canberra 1979–1980 catalog).

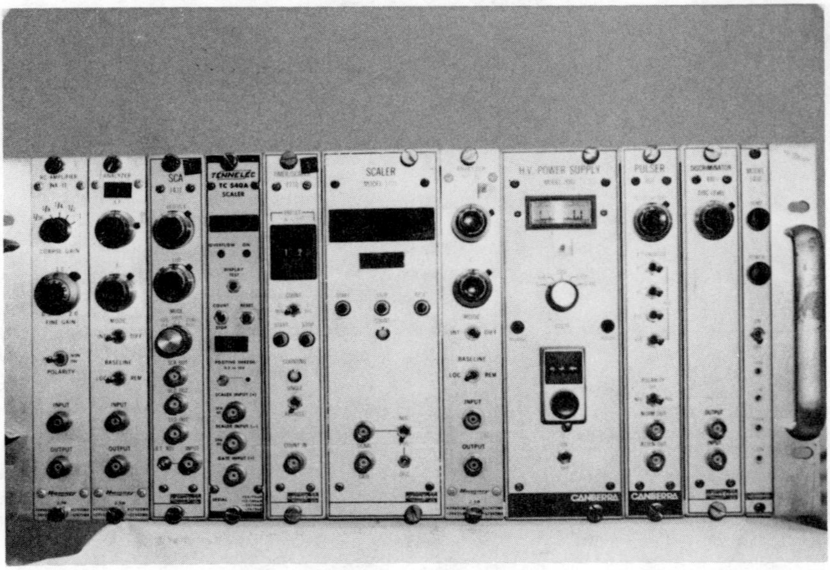

FIG. 1.5 A typical bin filled with a combination of NIMs made by different manufacturers.[1]

1.5.5 The Preamplifier

The primary purpose of the preamplifier is to provide an optimized coupling between the output of the detector and the rest of the counting system. The preamplifier is also necessary to minimize any sources of noise that may change the signal.

The signal that comes out of the detector is very weak, in the millivolt (mV) range (Fig. 1.2). Before it can be recorded, it will have to be amplified by a factor of a thousand or more. To achieve this, the signal will have to be transmitted through a cable to the next instrument of the counting system, which is the amplifier. Transmission of any signal through a cable attenuates it to a certain extent. If it is weak at the output of the detector, it might be lost in the electronic noise that accompanies the transmission. This is avoided by placing the preamplifier as close to the detector as possible. The preamplifier shapes the signal and reduces its attenuation by matching the impedance of the detector with that of the amplifier. After going through the

TABLE 1.2 High Voltage Needed for Certain Common Detectors

Detector	High voltage (V)
Ionization counters	HV < 1000
Proportional counters	500 < HV < 1500
GM counters	800 < HV < 2400
Semiconductor detectors	
Surface-barrier	HV < 100
Li-drifted	100 < HV < 3000

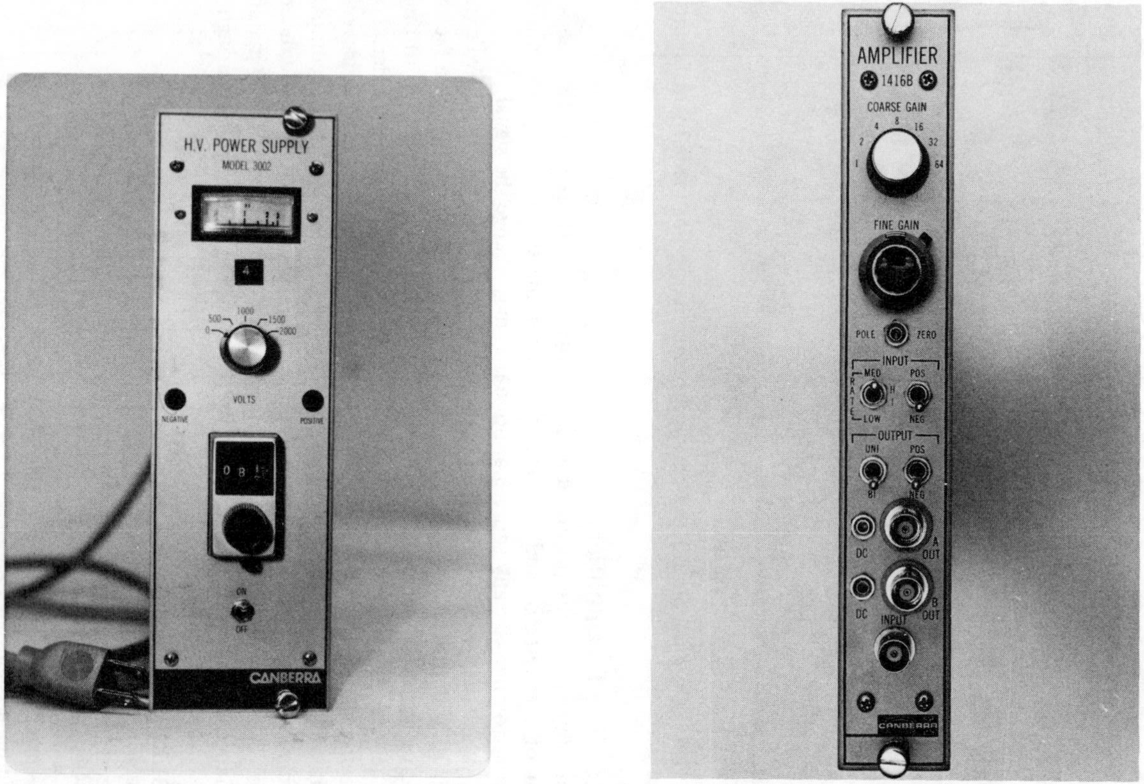

(a) (b)

FIG. 1.6 Typical commercial NIMs: (*a*) HV power supply, (*b*) amplifier (made by Canberra and Tennelec).

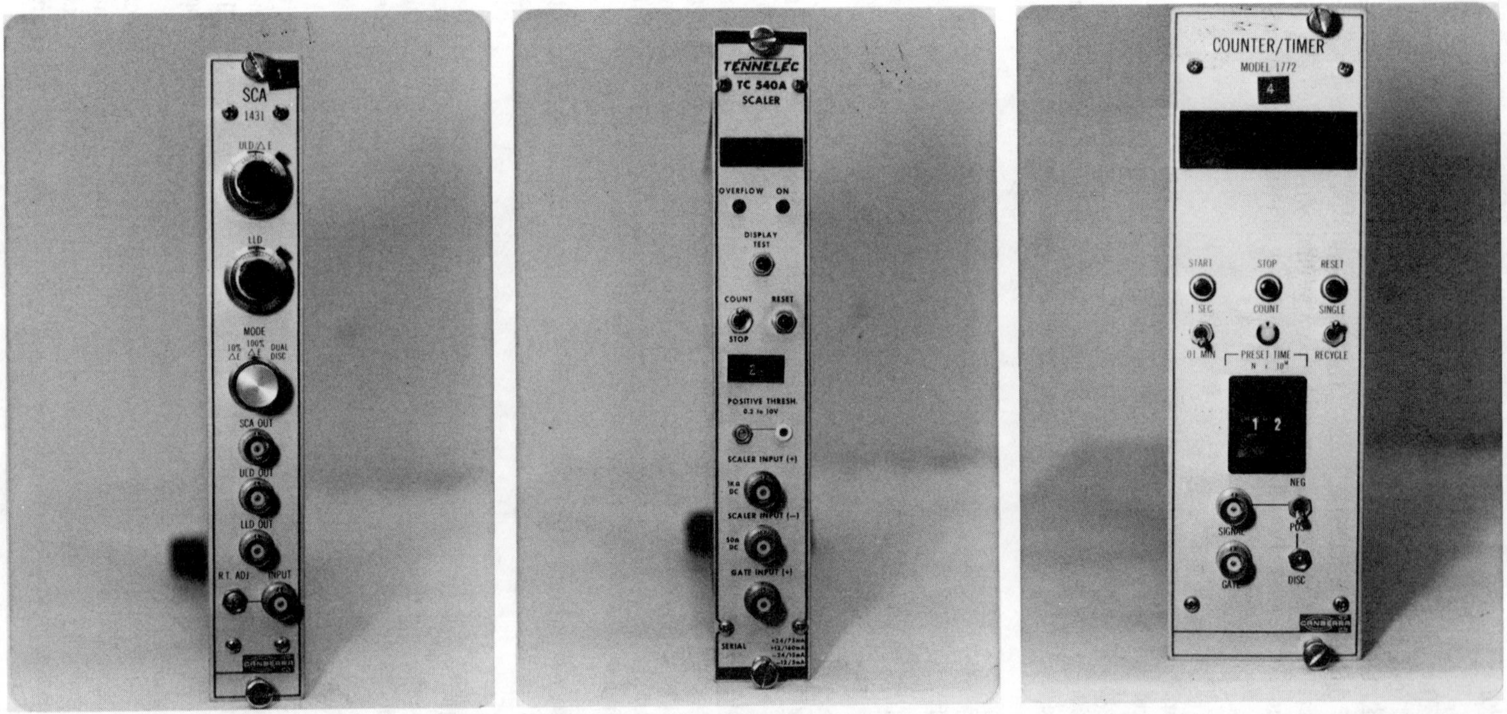

(c) (d) (e)

FIG. 1.6 Typical commercial NIMs (*Continued*): (*c*) single-channel analyzer, (*d*) scaler, (*e*) timer (made by Canberra and Tennelec).

preamplifier, the signal may be safely transmitted to the amplifier, which may be located at a considerable distance away. Although some preamplifiers amplify the signal slightly, their primary function is that of providing electronic matching between the output of the detector and the input of the amplifier.

There are many types of commercial preamplifiers, two of which are shown in Fig. 1.7. In most cases, the HV is fed to the detector through the preamplifier.

1.5.6 The Amplifier

The main amplification unit is the amplifier. It increases the signal by as many as 1000 times or more. Modern commercial amplifiers produce a maximum signal of 10 V, regardless of the input and the amplification. For example, consider a preamplifier that gives at its output three pulses with heights 50 mV, 100 mV, and 150 mV. Assume that the amplifier is set to 100. At the output of the unit the three pulses will be

$$50 \times 10^{-3} \times 100 = 5 \text{ V}$$

$$100 \times 10^{-3} \times 100 = 10 \text{ V}$$

$$150 \times 10^{-3} \times 100 = 10 \text{ V}$$

Note that the third value should be 15 V, but since the amplifier produces a maximum signal of 10 V, the three different input pulses will show, erroneously, as two different pulses at the output. If only the number of particles is measured, there is no error introduced–but if the energy of the particles is measured, then the error is very serious. In the example given above, if gammas of three different energies produce the pulses at the output of the preamplifier, the pulses at the output of the amplifier will be attributed erroneously to gammas of two different energies. To avoid such an error, an observed should follow this rule:

> Before any measurement of particle energy, make certain that the highest pulse of the spectrum to be measured is less than *10 V* at the output of the amplifier.

In addition to signal amplification, an equally important function of the amplifier is to convert the signal at the output of the preamplifier into a form suitable for the measurement desired. More details on this subject are given in Chap. 10. The front panel of a typical commercial amplifier is shown in Fig. 1.6.

Commercial amplifiers have two dials for adjusting the amplification.

1. *Coarse gain:* This dial adjusts the amplification in *steps.* Each step is a fraction of the maximum amplification. For example, the dial may show the numbers 1, 2, 4, 8, 16. If the maximum amplification is 100, then the coarse gain on 16 will give a maximum of 100, the coarse gain on 8 will give 50, etc. Some amplifiers have the numbers $\frac{1}{16}, \frac{1}{8}, \frac{1}{4}, \frac{1}{2}$, 1, and some newer ones have 1, 10, 100, 1000, etc.

2. *Fine gain:* This dial adjusts the amplification *continuously* within each step of the coarse gain. The numbers, in most units, go from 0 to 10. The highest number provides the maximum amplification indicated by the coarse gain. As an example, consider the maximum amplification to be 100. If the coarse gain is 8 (highest number

(a)

Front

Rear

(b)

FIG. 1.7 Two typical commercial preamplifiers: (*a*) type used with a photomultiplier tube (made by Harshaw), and (*b*) type used with semiconductor detectors (made by Canberra).

16) and the fine gain 5 (highest number 10), the amplification will be $100 \times \frac{1}{2}$ (coarse gain) $\times \frac{1}{2}$ (fine gain) = 25.

Most commercial amplifiers provide at the output two types of pulses called unipolar and bipolar (Fig. 1.8).

1.5.7 The Oscilloscope

The oscilloscope is an instrument that permits the study of rapidly changing phenomena, such as a sinusoidal voltage or the pulse of a counter. The phenomenon is observed on a fluorescent screen as shown in Fig. 1.9. The horizontal axis of the screen measures time. The vertical axis gives volts.

In radiation measurements the oscilloscope is used to check the quality of the signal as well as the level and type of the electronic noise. It is always a good practice before any measurement is attempted to examine the signal at the output of the amplifier. A few examples of good and bad pulses are shown in Fig. 1.10. In Figure 1.10, *a* and *b* represent good pulses, and Fig. 1.10*c* is probably an electrical discharge, not good for counting. Figure 1.10*d* is no good either, because a high-frequency signal is "riding" on the output of the preamplifier. If the pulse is not good, the observer should not proceed with the measurement unless the source of noise is identified and eliminated.

1.5.8 The Discriminator or Single-Channel Analyzer (SCA)

The SCA is used to eliminate the electronic noise and, in general, to reject unwanted pulses. When a pulse is amplified, the electronic noise that is always present in a circuit is also amplified. If one attempts to count all the pulses present, the counting rate may be exceedingly high. But electronic noise is a nuisance and it should not be counted.

In some cases, one may want to count only pulses above a certain height, i.e., particles with energy above certain threshold energy. Pulses lower than that height should be rejected. The discriminator or single-channel analyzer is the unit that can make the selection of the desired pulses. Figure 1.6 shows the front panel of a typical

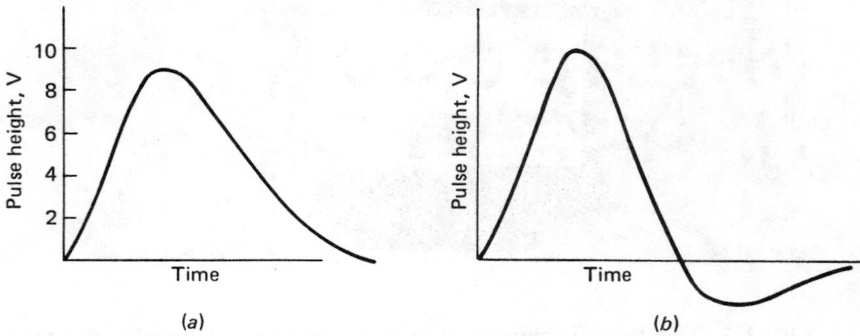

FIG. 1.8 The pulse at the output of the amplifier: (*a*) unipolar pulse and (*b*) bipolar pulse.

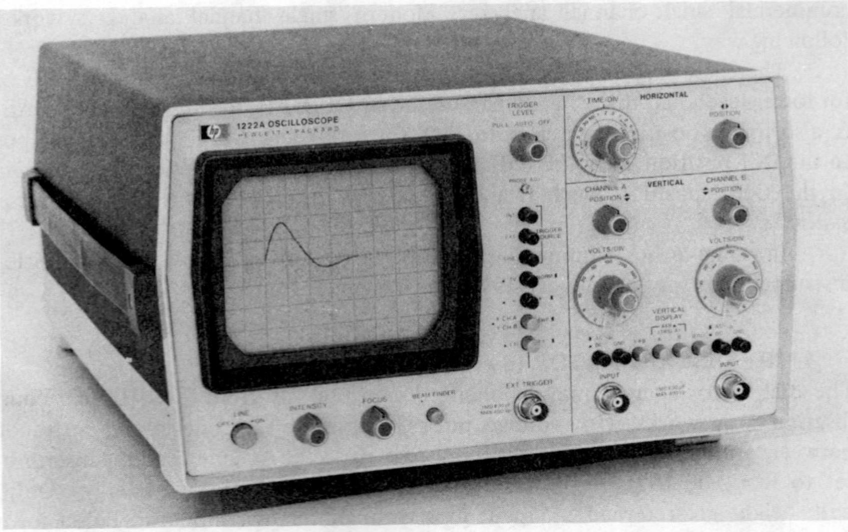

FIG. 1.9 A typical commercial oscilloscope (made by Hewlett-Packard).

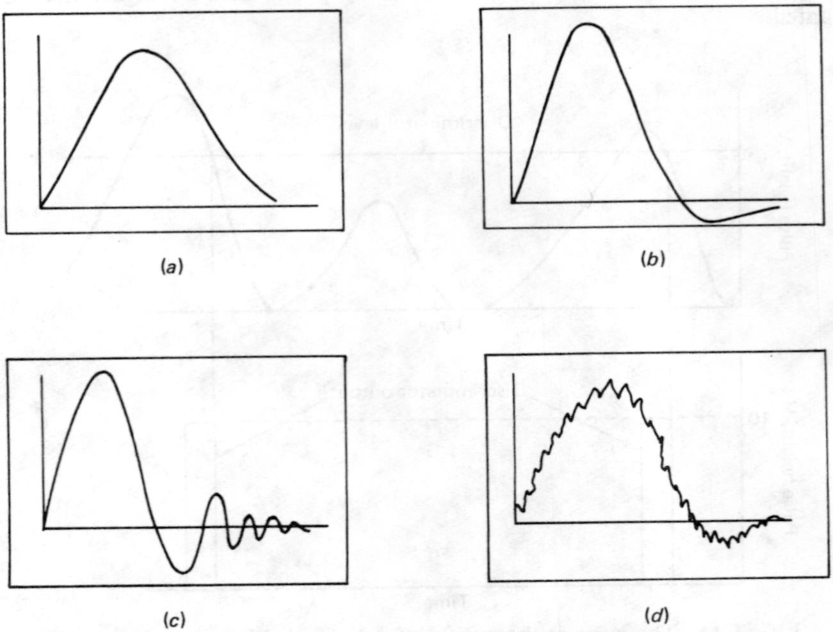

FIG. 1.10 Samples of good (*a* and *b*) and bad (*c* and *d*) pulses as seen on the screen of the oscilloscope.

commercial single-channel analyzer. Modern single-channel analyzers work in the following way.

There are two dials on the front panel of the unit. One is marked E or LLD, for lower-level dial; the other is marked ΔE or ULD/ΔE, for upper-level dial/ΔE. There is also a two-position switch with INT (integral) and DIFF (differential) positions. In the INT position, only the E dial operates, and the unit functions as a *discriminator*. In the DIFF position, both E and ΔE operate, and the unit is then a *single-channel analyzer*.

In some other commercial models, instead of INT and DIFF positions, the instrument has special connectors for the desired output.

THE DISCRIMINATOR (Switch position: INT)

The dial E (for energy) may be changed continuously from 0 to 100. Of course, the discriminator works with voltage pulses, but there is a one-to-one correspondence between a pulse height and the energy of a particle. Assume that the discriminator is set to E = 2.00 V (the 2 V may also correspond to 2 MeV of energy). Only pulses with height greater than 2 V will pass through the discriminator. Pulses lower than 2 V will be rejected. For every pulse that is larger than 2 V, the discriminator will provide at the output a rectangular pulse with height equal to 10 V (Fig. 1.11) regardless of the actual height of the input pulse. The output pulse of the discriminator is a pulse that triggers the unit (scaler), which counts individual pulses and tells it, "a pulse with height bigger than 2 V has arrived; count 1." Thus, the discriminator eliminates all pulses below E and allows only pulses that are higher than E to be counted.

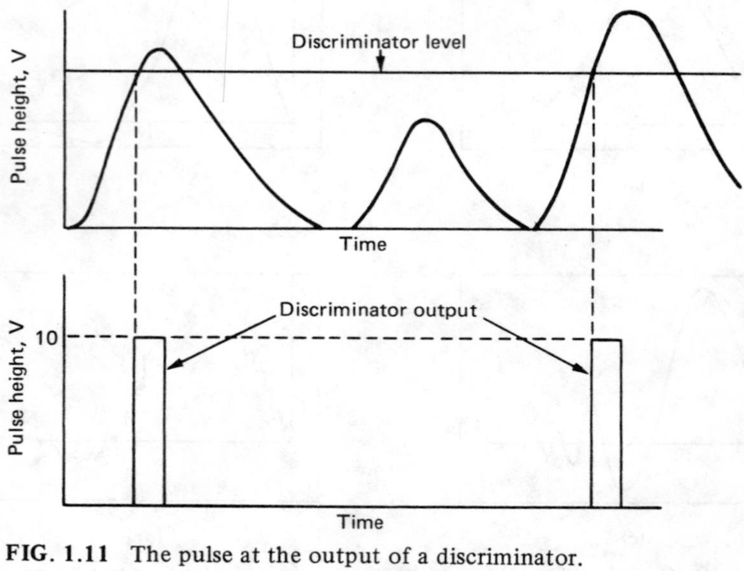

FIG. 1.11 The pulse at the output of a discriminator.

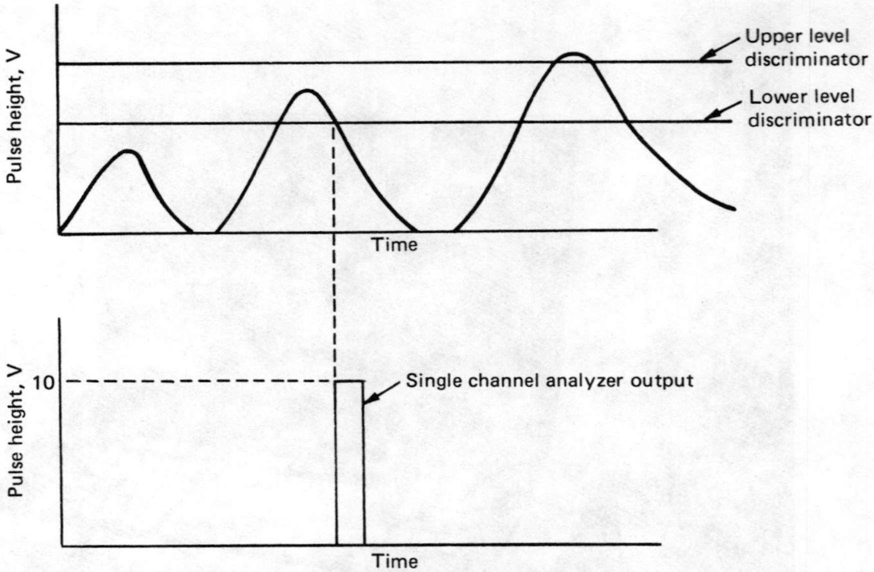

FIG. 1.12 The operation of a single-channel analyzer.

THE SINGLE-CHANNEL ANALYZER (SCA)
(Switch position: DIFF)

Both E and ΔE dials operate. Only pulses with heights between E and E + ΔE are counted (Fig. 1.12). The two dials form a "channel"; hence the name single-channel analyzer. If the E dial is changed to E_1, then pulses with heights between E_1 and $E_1 + \Delta E$ will be counted. In other words, the width ΔE, or window, of the channel is always added to E.

1.5.9 The Scaler

The scaler is a recorder of pulses. For every pulse entering the scaler, a count of 1 is added to the previous total. At the end of the counting period, the total number of pulses recorded is displayed. Figure 1.6 shows the front panel of a typical commercial scaler.

1.5.10 The Timer

The timer is connected to the scaler, and its purpose is to start and stop the scaler at desired counting time intervals. The front panel of a typical timer is shown in Fig. 1.6. Some models combine the timer with the scaler in one module.

1.5.11 The Multichannel Analyzer

The multichannel analyzer (MCA) records and stores pulses according to their height. Each storage unit is called a channel.

The height of the pulse has some known relationship–usually proportional–to the energy of the particle that enters into the detector. Each pulse is in turn stored in

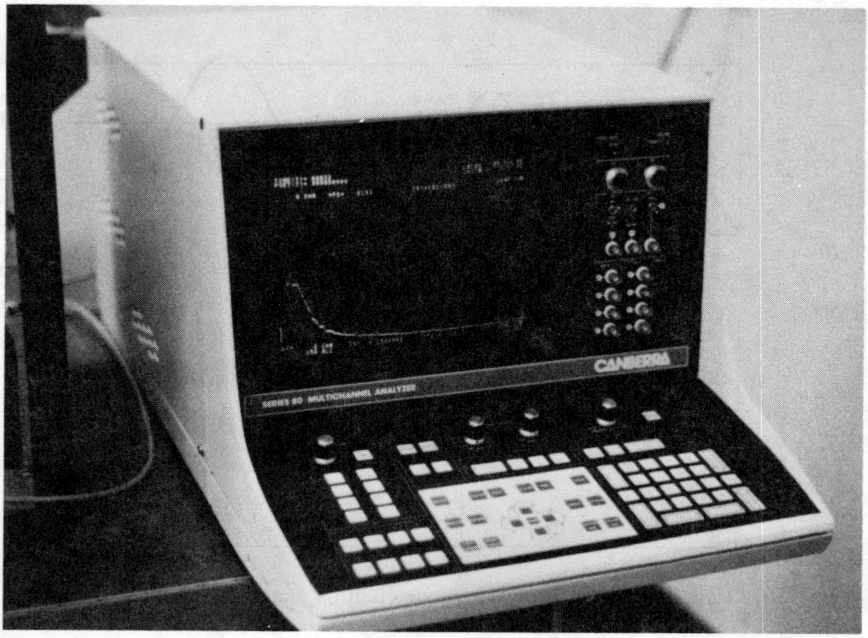

FIG. 1.13 An energy spectrum shown on the screen of a commercial multichannel analyzer (made by Canberra).

a particular channel correspond to a certain energy. The distribution of pulses in the channels is an image of the distribution of the energies of the particles. At the end of a counting period, the spectrum that was recorded may be displayed on the screen of a commercial MCA (Fig. 1.13). The horizontal axis is a channel number, or particle energy. The vertical axis is a number of particles recorded per channel. More details about the MCA and its use are given in Chaps. 9 and 10.

BIBLIOGRAPHY

Beers, Y., *Introduction to the Theory of Error*, Addison-Wesley, Reading, Mass., 1957.

Bevington, P. R., *Data Reduction and Error Analysis for the Physical Sciences*, McGraw-Hill, New York, 1969.

Jaech, J. L., "Statistical Methods in Nuclear Material Control," TID-26298, U.S. Atomic Energy Commission, 1973.

REFERENCE

1. "Standard Nuclear Instrument Modules," TID-20893 (rev. 2), U.S. Atomic Energy Commission, 1968.

2

statistics and errors

2.1 INTRODUCTION

This chapter discusses statistics at the level needed for radiation measurements and analysis of their results. People who perform experiments need statistics for analysis of experiments that are statistical in nature, treatment of errors, and fitting a function to the experimental data. The first two parts are presented in this chapter. Data fitting is discussed in Chap. 11.

2.2 DEFINITION OF PROBABILITY

Assume that one repeats an experiment many times and observes whether or not a certain event x is the outcome. The event is a certain observable result defined by the experimenter. If the experiment was performed N times, and n results were of type x, the probability $P(\mathrm{x})$ that any single event will be of type x is equal to

$$P(\mathrm{x}) = \lim_{N \to \infty} \frac{n}{N} \tag{2.1}$$

The ratio n/N is sometimes called the relative frequency of occurrence of x in the first N trials.

There is an obvious difficulty with the definition given by Eq. 2.1—the requirement of an infinite number of trials. Clearly, it is impossible to perform an infinite number of experiments. Instead, the experiment is repeated N times, and if the event x occurs n times out of N, the probability $P(\mathrm{x})$ is

$$P(\mathrm{x}) = \frac{n}{N} \tag{2.2}$$

Equation 2.2 will not make a mathematician happy, but it is extensively used in practice because it is in accord with the idea behind Eq. 2.1 and gives useful results.

As an illustration of the use of Eq. 2.2, consider the experiment of tossing a coin 100 times and recording how many times the result is "heads" and how many it is "tails." Assume that the result is

Heads: 48 times

Tails: 52 times

On the basis of Eq. 2.2, the probability to have heads or tails if the coin is tossed once more is

$$P(\text{heads}) = \tfrac{48}{100} = 0.48$$

$$P(\text{tails}) = \tfrac{52}{100} = 0.52$$

For this simple experiment, the correct result is known to be

$$P(\text{tails}) = P(\text{heads}) = 0.5$$

and one expects to approach the correct result as the number of trials increases. That is, Eq. 2.2 does not give the correct probability, but as $N \to \infty$, Eq. 2.2 approaches Eq. 2.1.

Since both n and N are positive numbers, $0 \leqslant n/N \leqslant 1$, so

$$0 \leqslant P(\text{x}) \leqslant 1 \tag{2.3}$$

that is, the probability is measured on a scale from 0 to 1.

If the event x occurs every time the experiment is performed, then $n = N$ and $P(\text{x}) = 1$. Thus the probability of a certain (sure) event is equal to 1.

If the event x never occurs, then $n = 0$ and $P(\text{x}) = 0$. In this case the probability of an impossible event is 0.

If the result of a measurement has N possible outcomes, each having equal probability, then the probability for the individual event x_i to occur is

$$P(\text{x}_i) = \frac{1}{N} \qquad i = 1, \ldots, N$$

For example, in the case of coin tossing there are two events of equal probability; therefore

$$P(\text{heads}) = P(\text{tails}) = \tfrac{1}{2}$$

2.3 BASIC PROBABILITY THEOREMS

In the language of probability, an "event" is an outcome of one or more experiments or trials and is defined by the experimenter. Some examples of events are:

1 Tossing a coin once
2 Tossing a coin twice and getting heads both times
3 Tossing a coin 10 times and getting heads for the first five times and tails for the other five
4 Picking up one card from a deck of cards and that card being red
5 Picking up 10 cards from a deck and all of them being hearts
6 Watching the street for 10 min and observing two cyclists pass by
7 Counting a radioactive sample for 10 s and recording 100 counts
8 Inspecting all the fuel rods in a nuclear reactor and finding faults in two of them

Given enough information, one can calculate the probability that any one of these events will occur. In some cases, an event may consist of simpler components, and one would like to know how to calculate the probability of the complex event from the probabilities of its components.

Consider two events x and y and a series of N trials. The result of each trial will be only one of the following four possibilities:

1 x occurred but not y
2 y occurred but not x
3 Both x and y occurred
4 Neither x nor y occurred

Let n_1, n_2, n_3, n_4 be the number of times in the N observations that the respective possibilities occurred. Then,

$$n_1 + n_2 + n_3 + n_4 = N \tag{2.4}$$

The following probabilities are defined with respect to the events x and y:

$P(\mathrm{x})$ = probability that x occurred

$P(\mathrm{y})$ = probability that y occurred

$P(\mathrm{x+y})$ = probability that either x or y occurred

$P(\mathrm{xy})$ = probability that both x and y occurred

$P(\mathrm{x|y})$ = conditional probability of x given y = probability of x occurring given that y has occurred

$P(\mathrm{y|x})$ = conditional probability of y given x = probability of y occurring given that x has occurred

Using Eq. 2.2, these probabilities are

$$P(\mathrm{x}) = \frac{n_1 + n_3}{N} \tag{2.4}$$

$$P(\mathrm{y}) = \frac{n_2 + n_3}{N} \tag{2.5}$$

$$P(\mathrm{x+y}) = \frac{n_1 + n_2 + n_3}{N} \tag{2.6}$$

$$P(\mathrm{xy}) = \frac{n_3}{N} \tag{2.7}$$

$$P(\mathrm{x|y}) = \frac{n_3}{n_2 + n_3} \tag{2.8}$$

$$P(\mathrm{y|x}) = \frac{n_3}{n_1 + n_3} \tag{2.9}$$

For the six probabilities given by Eqs. 2.4 to 2.9, the following two relations hold:

$$P(x + y) = P(x) + P(y) - P(xy) \tag{2.10}$$

$$P(xy) = P(x)P(y|x) = P(y)P(x|y) \tag{2.11}$$

Equation 2.10 is called the *addition law of probability.* Equation 2.11 is called the *multiplication law of probability.*

■ *EXAMPLE 2.1*

Consider two well-shuffled decks of cards. What is the probability of drawing one card from each deck with both of them being the ace of spades?

ANSWER

The events of interest are,

Event x = event y = (drawing one card and that card being ace of spades)

Since each deck has only one ace of spades,

$$P(x) = P(y) = P(\text{ace of spades}) = \tfrac{1}{52}$$

The conditional probability is

$$P(x|y) = P(\text{1st card ace of spades when 2nd card is ace of spades}) = \tfrac{1}{52}$$

In this case, $P(x|y) = P(x)$ because the two events are independent. The fact that the first card from the first deck is the ace of spades has no influence on what the first card from the second deck is going to be. Similarly, $P(y|x) = P(y)$.

Therefore, using Eq. 2.11 one has

$$P(xy) = P(x)P(y) = (\tfrac{1}{52})(\tfrac{1}{52}) = 0.00037$$

■ *EXAMPLE 2.2*

Consider two well-shuffled decks of cards and assume one card is drawn from each of them. What is the probability of one of the two cards being the ace of spades?

ANSWER

Using Eq. 2.10,

$$P(x + y) = \tfrac{1}{52} + \tfrac{1}{52} - (\tfrac{1}{52})(\tfrac{1}{52}) = 0.038$$

Under certain conditions, the addition and multiplication laws expressed by Eqs. 2.10 and 2.11 are simplified.

If the events x and y are mutually exclusive–i.e., they cannot occur simultaneously–then $P(xy) = 0$ and the addition law becomes

$$P(x + y) = P(x) + P(y) \tag{2.12}$$

If the probability that x occurs is independent of whether or not y occurs, and vice versa, then as shown in Ex. 2.1,

$$P(y|x) = P(y)$$
$$P(x|y) = P(x)$$

In that case, the events x and y are called stochastically independent and the multiplication law takes the form

$$P(xy) = P(x)P(y) \tag{2.13}$$

Equations 2.12 and 2.13 are also known as the addition and multiplication laws of probability, but the reader should keep in mind that 2.12 and 2.13 are special cases of 2.10 and 2.11.

EXAMPLE 2.3

What is the probability that a single throw of a die will result in either 2 or 5?

ANSWER

$$P(2) = \tfrac{1}{6} \qquad P(5) = \tfrac{1}{6}$$
$$P(2 + 5) = P(2) + P(5) = \tfrac{1}{6} + \tfrac{1}{6} = \tfrac{1}{3}$$

EXAMPLE 2.4

Consider two well-shuffled decks of cards and assume one card is drawn from each deck. What is the probability of both cards being spades?

ANSWER

$$P(\text{one spade}) = \tfrac{13}{52}$$
$$P[(\text{spade})(\text{spade})] = (\tfrac{13}{52})(\tfrac{13}{52}) = \tfrac{1}{16}$$

Equations 2.12 and 2.13 hold for any number of events, provided the events are mutually exclusive or stochastically independent. Thus, if we have N such events $x_n|_{n=1,\ldots,N}$

$$P(x_1 + x_2 + \cdots + x_N) = P(x_1) + P(x_2) + \cdots + P(x_N) \tag{2.14}$$
$$P(x_1 x_2 \cdots x_N) = P(x_1)P(x_2)\cdots P(x_N) \tag{2.15}$$

2.4 PROBABILITY DISTRIBUTIONS AND RANDOM VARIABLES

When an experiment is repeated many times under identical conditions, the results of the measurement will not necessarily be identical. In fact, as a rule rather than as an exception, the results will be different. Therefore, it is very desirable to know if there is a law that governs the individual outcomes of the experiment. Such a law, if it exists and is known, would be helpful in two ways. First, from a small number of measurements, the experimenter may obtain information about expected results of subsequent measurements. Second, a series of measurements may be checked for faults. If it is known that the results of an experiment obey a certain law and a given series of outcomes of such an experiment does not follow that law, then that series of outcomes is suspect and should be thoroughly investigated before it becomes acceptable.

There are many such laws governing different types of measurements. The three most frequently used will be discussed in later sections of this chapter, but first some general definitions and the concept of the random variable are introduced.

A quantity x that can be determined quantitatively and that in successive but similar experiments can assume different values is called a random variable. Examples of random variables are the result of drawing one card from a deck of cards, the result of the throw of a die, the result of measuring the length of a nuclear fuel rod, and the result of counting the radioactivity of a sample. There are two types of random variables, discrete and continuous.

A discrete random variable takes one of a set of discrete values. Discrete random variables are especially useful in representing results that take integer values—for example, number of persons, number of defective batteries, or number of counts recorded in a scaler.

A continuous random variable can take any value within a certain interval—for example, weight or height of people, the length of a rod, or the temperature of the water coming out of a reactor.

For every random variable x, one may define a function $f(x)$ as follows:

DISCRETE RANDOM VARIABLES

$f(x_i)$ = probability that the value of the random variable is x_i $\quad i = 1, 2, \ldots, N$

where N = number of possible (discrete) values of x. Since x takes only one value at a time, the events represented by the probabilities $f(x_i)$ are mutually exclusive; therefore, using Eq. 2.14,

$$\sum_{i=1}^{N} f(x_i) = 1 \tag{2.16}$$

CONTINUOUS RANDOM VARIABLES

Assume that a random variable may take any value between a and b ($a \leqslant x \leqslant b$). Then

$f(x)\,dx$ = probability that the value of x lies between x and $x + dx$

One should notice that for a continuous variable what is important is not the probability that x will take a specific value, but only the probability that x falls within an interval defined by two values of x. The equation corresponding to 2.16 is now

$$\int_a^b f(x)\,dx = 1 \tag{2.17}$$

Equations 2.16 and 2.17 give the probability of a sure event, because x will certainly have one of the values $x_1, x_2, \ldots, x_N$ and will certainly have a value between a and b.

The function $f(x)$ is called the probability density function† (pdf).

Consider now the following function:

$$F(x_j) = \int_a^{x_j} f(x')\,dx' \tag{2.18}$$

For a discrete variable,

$$F(x_j) = \sum_{x=1}^{j} f(x_i) \tag{2.19}$$

Thus,

$$F(x_j) = \text{probability that the value of } x \text{ is less than or equal to } x_j$$

The function $F(x)$ is called the cumulative distribution function‡ (cdf). The cdf has the following properties:

$$F(x_N) = 1 \tag{2.20}$$

$$F(a) = 0 \tag{2.21}$$

$$F(b) = 1 \tag{2.22}$$

The cdf is a positive monotonously increasing function. There is a relationship between the cdf and the pdf obtained from Eq. 2.18, namely,

$$f(x) = \frac{dF(x)}{dx} \tag{2.23}$$

2.5 LOCATION INDEXES (MODE, MEDIAN, MEAN)

If the distribution function $F(x)$ or $f(x)$ is known, a great deal of information can be obtained about the values of the random variable x. Conversely, if $F(x)$ or $f(x)$ is not

†It has also been called the frequency function.

‡It has also been called the integral or total distribution function.

completely known, certain values of x provide valuable information about the distribution functions. In most practical applications the important values of x are clustered within a relatively narrow interval. To obtain a rough idea about the whole distribution, it is often adequate to indicate the position of this interval by "location indexes" providing typical values of x.

In theory, an infinite number of location indexes† may be constructed, but in practice the following three are most frequently used: the mode, the median, and the mean of a distribution. Their definitions and physical meanings will be presented with the help of an example.

Consider the continuous probability density function shown in Fig. 2.1. The function $f(x)$ satisfies Eq. 2.17, i.e., the total area under the curve of Fig. 2.1 is equal to 1.

The *mode* is defined as the most probable value of x. Therefore, the mode x_1 is that x for which $f(x)$ is maximum and is obtained from

$$\frac{df(x)}{dx} = 0 \tag{2.24}$$

The *median* is the value of x for which

$$F(x_2) = \int_{-\infty}^{x_2} f(x)\,dx = \frac{1}{2} \tag{2.25}$$

i.e., the probability of x taking a value less than x_2 is equal to the probability of x taking a value greater than x_2.

The *mean*, also known as the "average" or the "expectation value" of x, is defined by the equation

$$\bar{x} = m = \int_{-\infty}^{\infty} xf(x)\,dx \tag{2.26}$$

†Measure of location is another name for location indexes.

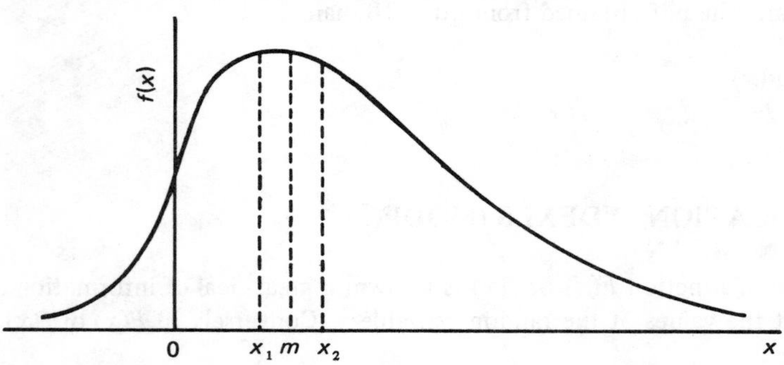

FIG. 2.1 The mode (x_1), the median (x_2), and the mean (m) for a continuous probability distribution function.

The limits of integration cover the range over which the function $f(x)$ is defined.

Which of these or other location indexes one uses is a matter of personal choice and convenience, depending on the problem under study. The mean is by far the most frequently used index and for this reason only the mean will be discussed further.

An expression more general than Eq. 2.26 that gives the mean or average of any function $g(x)$ is

$$\overline{g(x)} = \frac{\int_{-\infty}^{\infty} g(x)f(x)\,dx}{\int_{-\infty}^{\infty} f(x)\,dx} \tag{2.27}$$

Equation 2.27 is true even if $f(x)$ does not satisfy Eq. 2.17. Some elementary but useful properties of the mean that can be easily proven using Eq. 2.27 are the following:

$$\begin{aligned} &\overline{ax} = a\bar{x} = am \qquad a = \text{constant} \\ &\overline{a + x} = a + \bar{x} = a + m \\ &\overline{g_1(x) + g_2(x) + \cdots + g_i(x)} = \overline{g_1(x)} + \overline{g_2(x)} + \cdots + \overline{g_i(x)} \end{aligned} \tag{2.28}$$

For a discontinuous pdf, the location indexes are defined in a similar way. If the pdf satisfies Eq. 2.16, the mean is given by

$$m = \bar{x} = \sum_{i=1}^{N} x_i f(x_i) \tag{2.29}$$

Equation 2.29 is an approximation because the true mean can only be determined with an infinite number of measurements. But, in practice, it is always a finite number of outcomes that is available, and the average $\bar{x}$ instead of the true mean m is determined. Equation 2.29 is analogous to Eq. 2.2, which defines the probability based on a finite number of events. The general expression for the average of a discontinuous pdf, equivalent to Eq. 2.27, is

$$\overline{g(x)} = \frac{\sum_{i=1}^{N} g(x_i)f(x_i)}{\sum_{i=1}^{N} f(x_i)} \tag{2.30}$$

EXAMPLE 2.5

Calculation of the mean. The probability that a radioactive nucleus will not decay for time t is equal to

$$f(t) = \lambda e^{-\lambda t}$$

where λ is a constant. What is the mean life of such a nucleus?

ANSWER

Using Eq. 2.26, the mean life $\bar{t}$ is

$$\bar{t} = \int_0^\infty t\lambda e^{-\lambda t}\,dt = \frac{1}{\lambda}$$

EXAMPLE 2.6

Consider the throw of a die. The probability of getting any number between 1 and 6 is $\frac{1}{6}$. What is the average number?

ANSWER

Using Eq. 2.29,

$$m = \bar{x} = \sum_{i=1}^{6} (i)\frac{1}{6} = \frac{1}{6}\sum_{i=1}^{6} i = \left(\frac{1}{6}\right)\frac{6(6+1)}{2} = 3.5$$

EXAMPLE 2.7

Consider an experiment repeated N times giving the results $x_i|_{i=1,\ldots,N}$. What is the average of the results?

ANSWER

Since the experiments were identical, all the results have the same probability of occurring, a probability that is equal to $1/N$. Therefore, the mean is

$$\bar{x} = \bar{m} = \sum_{i=1}^{N} x_i\left(\frac{1}{N}\right) \tag{2.31}$$

Equation 2.31 defines the so-called arithmetic mean of a series of N random variables. It is used extensively when the results of several measurements of the same variable are combined.

An extension of Eq. 2.31 is the calculation of the "means of means." Assume that one has obtained the averages $\bar{x}_1, \bar{x}_2, \ldots, \bar{x}_M$ by performing a series of M measurements, each involving $N_1, N_2, \ldots, N_M$ events, respectively. The arithmetic mean of *all* the measurements, $\bar{X}$, is

$$\bar{X} = \frac{\bar{x}_1 + \bar{x}_2 + \cdots + \bar{x}_M}{M} \tag{2.32}$$

where

$$\bar{x}_j = \sum_{i=1}^{N_j} \frac{x_{ji}}{N_j} \qquad j = 1, \ldots, M$$

2.6 DISPERSION INDEXES, VARIANCE, AND STANDARD DEVIATION

A pdf or cdf is determined only approximately by any location index. For practical purposes it is sufficient to know the value of one location index–e.g. the mean–together with a measure indicating how the probability density is distributed around the chosen location index. There are several such measures called *dispersion indexes.* The dispersion index most commonly used and the only one to be discussed here is the variance $V(x)$ and its square root, which is called the standard deviation σ.

The variance of a pdf is defined as shown by Eqs. 2.33 and 2.34. For continuous distributions,

$$V(x) = \sigma^2 = \int_{-\infty}^{\infty} (x - m)^2 f(x)\, dx \tag{2.33}$$

For discontinuous distributions,

$$V(x) = \sigma^2 = \sum_{i=1}^{N} (x_i - m)^2 f(x_i) \tag{2.34}$$

It is assumed that $f(x)$ satisfies Eq. 2.16 or 2.17 and N is a large number. It is worth noting that the variance is nothing more than the average of $(x - m)^2$. The variance of a linear function of x, $a + bx$, is

$$V(a + bx) = b^2\, V(x) \tag{2.35}$$

where a and b are constants.

2.7 COVARIANCE AND CORRELATION

Consider the random variables $X_1, X_2, \ldots, X_M$ with means $m_1, m_2, \ldots, m_M$ and variances $\sigma_1^2, \sigma_2^2, \ldots, \sigma_M^2$. A question that arises frequently is, what is the average and the variance of the linear function

$$Q = a_1 X_1 + a_2 X_2 + \cdots + a_M X_M \tag{2.36}$$

where the values of $a_i|_{i=1,\ldots,M}$ are constants?

The average is simply (using Eq. (2.28)

$$\bar{Q} = a_1 m_1 + a_2 m_2 + \cdots + a_M m_M = \sum_{i=1}^{M} a_i m_i \tag{2.36a}$$

The variance is

$$V(Q) = \sigma^2 = \overline{(Q - \bar{Q})^2} = \overline{\left[\sum_{i=1}^{M} a_i(X_i - m_i)\right]^2}$$

$$= \sum_{i=1}^{M} a_i^2 \sigma_i^2 + 2\sum_{j>i}^{M} \sum_{j>i}^{M} a_i a_j \overline{(X_i - m_i)(X_j - m_j)} \tag{2.37}$$

The quantity $\overline{(X_i - m_i)(X_j - m_j)}$ is called the "covariance" between X_i and X_j:

$$\text{cov}(X_i, X_j) = \overline{(X_i - m_i)(X_j - m_j)} \tag{2.38}$$

The covariance, as defined by Eq. 2.38, suffers from the serious drawback that its value changes with the units used for the measurement of X_i, X_j. To eliminate this effect, the covariance is divided by the product of the standard deviations σ_i, σ_j and the resulting ratio is called the correlation coefficient $\rho(X_i, X_j)$. Thus,

$$\rho_{ij} = \rho(X_i, X_j) = \frac{\text{cov}(X_i, X_j)}{\sigma_i \sigma_j} \tag{2.39}$$

Using Eq. 2.39, the variance of Q becomes

$$\sigma^2 = V(Q) = V\left(\sum_{i=1}^{M} a_i X_i\right) = \sum_{i=1}^{M} a_i^2 \sigma_i^2 + 2\sum_{j>i}^{M}\sum^{M} a_i a_j \rho_{ij} \sigma_i \sigma_j \tag{2.40}$$

Random variables for which $\rho_{ij} = 0$ are said to be uncorrelated.

If the X_i's are mutually uncorrelated, Eq. 2.40 takes the simpler form

$$\sigma^2 = V\left(\sum_{i=1}^{M} a_i X_i\right) = \sum_{i=1}^{M} a_i^2 \sigma_i^2 \tag{2.41}$$

Consider now a second linear function of the variables $X_1, X_2, X_3, \ldots, X_M$, namely $R = b_1 X_2 + \cdots + b_M X_M$. The average of R is

$$\bar{R} = b_1 m_1 + b_2 m_2 + \cdots + b_M m_M = \sum_{i=1}^{M} b_i m_i$$

The covariance of Q, R is

$$\text{cov}(Q, R) = \overline{\left[\sum_{i=1}^{M} a_i(X_i - m_i)\right]\left[\sum_{j=1}^{M} b_j(X_j - m_j)\right]}$$

$$= \sum_{i=1}^{M} a_i b_i \sigma_i^2 + \sum_{i \neq j}^{M}\sum^{M} a_i b_j \rho_{ij} \sigma_i \sigma_j \tag{2.42}$$

If all the X's are mutually uncorrelated, then $\rho_{ij} = 0$ and

$$\operatorname{cov}(Q, R) = \sum_{i=1}^{M} a_i b_i \sigma_i^2 \tag{2.43}$$

If all the X's have the same variance σ^2,

$$\operatorname{cov}(Q, R) = \sigma^2 \sum_{i=1}^{M} a_i b_i \tag{2.44}$$

Equations 2.40 to 2.44 will be applied in Sec. 2.16 for the calculation of the *propagation of errors.*

2.8 THE BINOMIAL DISTRIBUTION

The *binomial distribution* is a pdf that applies under the following conditions:

1 The experiment has two possible outcomes, A and B.
2 The probability that any given observation results in outcome of type A is constant, independent of the number of observations.
3 The occurrence of a type A event in any given observation does not affect the probability that the event A or B will occur again in subsequent observations.

Examples of such experiments are tossing a coin (heads or tails is the outcome), inspecting a number of similar items for defects (items are defective or not), and picking up objects from a box containing two types of objects.

The binomial distribution will be introduced with the help of the following experiment.

Suppose that a box contains a large number of two types of objects, type A and type B. Let

p = probability that an object selected at random from this box is type A

$1 - p$ = probability that the randomly selected object is type B

An experimenter selects N objects at random.† The binomial distribution, giving the probability P_n that n out of the N objects are of type A, is

$$P_n = \frac{N!}{(N-n)!n!} p^n (1-p)^{N-n} \tag{2.45}$$

†It is assumed that the box has an extremely large number of objects so that the removal of N of them does not change their number appreciably, or, after an object is selected and its type recorded, it is thrown back into the box. If the total number of objects is small, instead of Eq. 2.45, the hypergeometric density function should be used (see Johnson-Leone and Jaech).

EXAMPLE 2.8

A box contains a total of 10,000 small metallic spheres, of which 2000 are painted white and the rest are painted black. A person removes 100 spheres from the box one at a time at random. What is the probability that 10 of these spheres are white?

ANSWER

The probability of picking one white sphere is

$$p = P(\text{white}) = \frac{2000}{10{,}000} = 0.2$$

The probability that 10 out of 100 selected spheres will be white is, according to Eq. 2.45,

$$P_{10} = \frac{100!}{(100-10)!\,10!}(0.2)^{10}(0.8)^{90} = 0.0034$$

EXAMPLE 2.9

A coin is tossed three times. What is the probability that the result will be heads in all three tosses?

ANSWER

The probability of getting heads in one throw is 0.5. The probability of tossing the coin three times ($N = 3$) and getting heads in all three tosses ($n = 3$) is

$$P_3 = \frac{3!}{(3-3)!3!}(0.5)^3(1-0.5)^{3-3} = 0.125$$

Of course, the same result could have been obtained in this simple case by using the multiplication law, Eq. 2.13:

$$P(\text{heads three times}) = (0.5)(0.5)(0.5) = 0.125$$

It is easy to show that the binomial distribution satisfies

$$\sum_{n=0}^{N} P_n = \sum_{n=0}^{N} \frac{N!}{(N-n)!\,n!} p^n(1-p)^{N-n} = 1 \tag{2.46}$$

The mean m is equal to

$$m = \bar{n} = \sum_{n=0}^{N} nP_n = pN \tag{2.47}$$

The variance $V(n)$ is

$$V(n) = \overline{(m-n)^2} = \sum_{n=0}^{N} (m-n)^2 P_n = m(1-p) \tag{2.48}$$

The standard deviation σ is

$$\sigma = \sqrt{V(n)} = \sqrt{m(1-p)} \tag{2.49}$$

Figure 2.2 shows three binomial distributions for $N = 10$ and $p = 0.1$, 0.4, and 0.8. Notice that as $p \to 0.5$ the distribution tends to be symmetric around the mean.

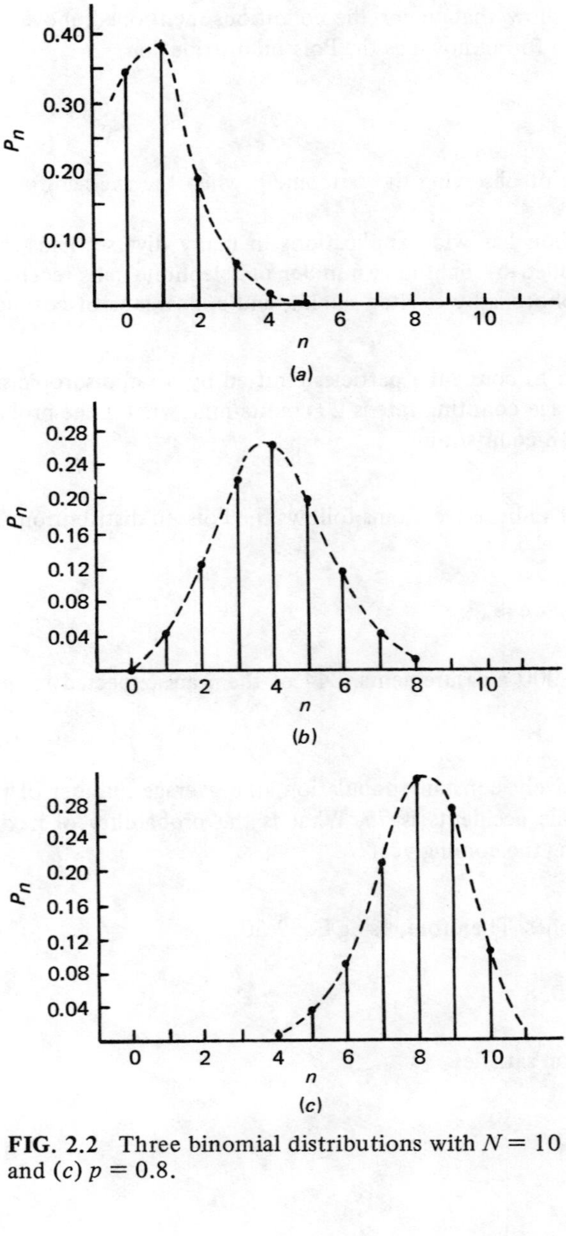

FIG. 2.2 Three binomial distributions with $N = 10$ and (*a*) $p = 0.1$, (*b*) $p = 0.4$, and (*c*) $p = 0.8$.

2.9 THE POISSON DISTRIBUTION

The Poisson distribution applies to events whose probability of occurrence is small and constant. It can be derived from the binomial distribution by letting

$$N \to \infty$$

$$p \to 0$$

in such a way that the value of the average $m = Np$ stays constant. It is left as an exercise for the reader to show that under the conditions mentioned above, the binomial distribution takes the form known as the Poisson distribution,

$$P_n = \frac{m^n}{n!} e^{-m} \tag{2.50}$$

where P_n is the probability of observing the outcome n when the average for a large number of trials is m.

The Poisson distribution has wide applications in many diverse fields, such as decay of nuclei, persons killed by lightning, number of telephone calls received in a switchboard, emission of photons by excited nucldi, and appearance of cosmic rays.

■ *EXAMPLE 2.10*

A radiation detector is used to count the particles emitted by a radioisotopic source. If it is known that the average counting rate is 20 counts/min, what is the probability that the next trial will give 18 counts/min?

ANSWER

The probability of decay of radioactive atoms follows the Poisson distribution. Therefore, using Eq. 2.50,

$$P_{18} = \frac{20^{18}}{18!} e^{-20} = 0.0844 \approx 8\%$$

That is, if one performs 10,000 measurements, 844 of them are expected to give the result 18 counts/min.

■ *EXAMPLE 2.11*

In a certain city with relatively constant population, the average number of people killed per year in automobile accidents is 75. What is the probability of having 80 auto-accident fatalities during the coming year?

ANSWER

The Poisson distribution applies. Therefore, using Eq. 2.50,

$$P_{100} = \frac{75^{80}}{80!} e^{-75} = 0.038 \approx 4\%$$

The Poisson distribution satisfies

$$\sum_{n=0}^{\infty} P_n = 1 \tag{2.51}$$

The mean m is equal to

$$m = \bar{n} = \sum_{n=0}^{\infty} nP_n = m \tag{2.52}$$

The variance is

$$V(n) = \overline{(m-n)^2} = \sum_{n=0}^{\infty} (m-n)^2 P_n = m \tag{2.53}$$

The standard deviation σ is

$$\sigma = \sqrt{V(n)} = \sqrt{m} \tag{2.54}$$

Figure 2.3 shows the Poisson distribution for three different means. It should be pointed out that as the mean increases, the Poisson distribution becomes symmetric around the mean. For $m = 20$, the distribution is already for all practical purposes symmetric around the mean, and it resembles the normal distribution, which is discussed next.

2.10 THE NORMAL (GAUSSIAN) DISTRIBUTION

Both the binomial and Poisson distributions apply to discrete variables whereas most of the random variables involved in experiments are continuous. In addition, the use of discrete distributions necessitates the use of long or infinite series for the calculation of such parameters as the mean and the standard deviation (see Eqs. 2.47, 2.48, 2.52, 2.53). It would be desirable, therefore, to have a pdf that applies to continuous variables. Such a distribution is the normal or Gaussian distribution.

The normal distribution $G(x)$ is given by

$$G(x)\,dx = \frac{1}{(\sqrt{2\pi})\sigma} \exp\left[-\frac{(x-m)^2}{2\sigma^2}\right] dx \tag{2.55}$$

where $G(x)\,dx$ = probability that the value of x lies between x and $x + dx$
m = average of the distribution
σ^2 = variance of the distribution

Notice that this distribution, shown in Fig. 2.4, has a maximum at $x = m$, is symmetric around m, is defined uniquely by the two parameters σ and m, and extends from $x = -\infty$ to $x = +\infty$. Equation 2.55 represents the shaded area under the curve of Fig. 2.4. In general, the probability of finding the value of x between any two limits x_1 and x_2 is given by

$$G(x_1 \leqslant x \leqslant x_2) = \int_{x_1}^{x_2} G(x)\,dx \tag{2.56}$$

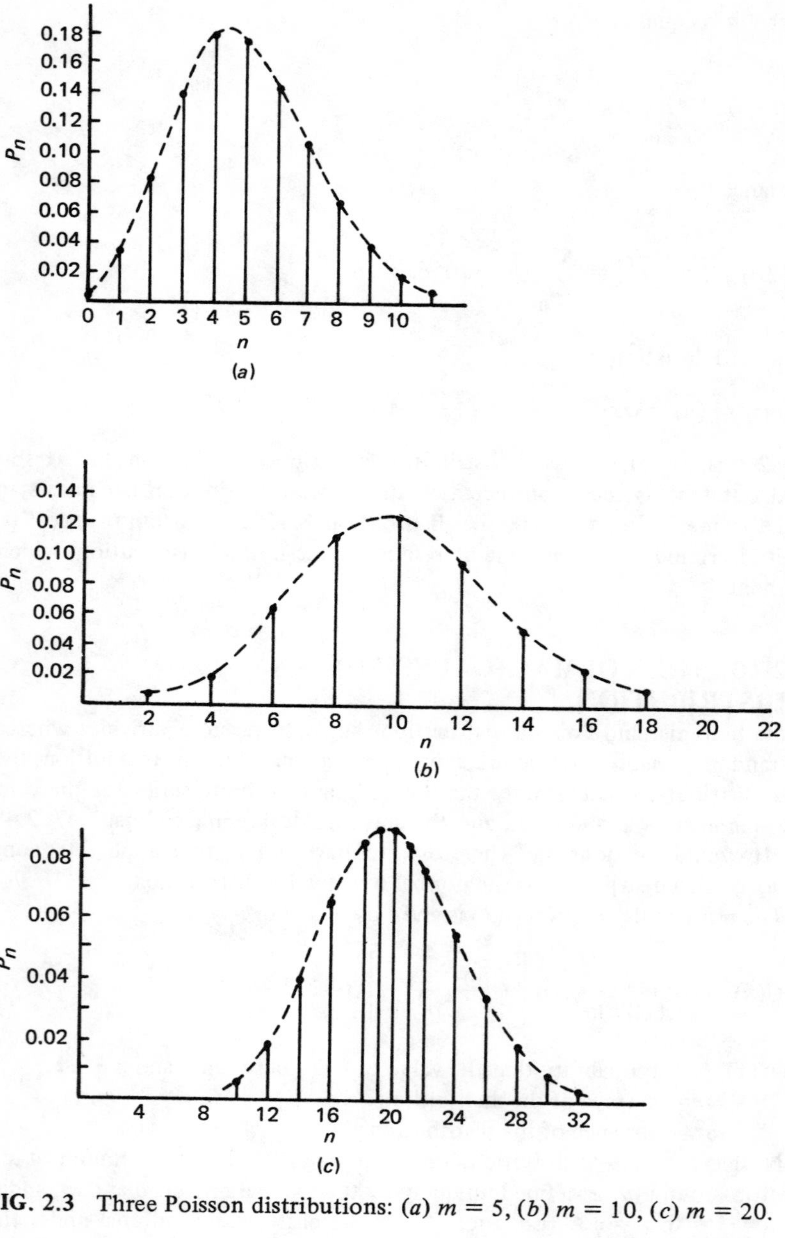

FIG. 2.3 Three Poisson distributions: (*a*) $m = 5$, (*b*) $m = 10$, (*c*) $m = 20$.

The Gaussian given by Eq. 2.55 satisfies

$$\int_{-\infty}^{\infty} G(x)\, dx = 1 \tag{2.57}$$

The average of the distribution is

$$\bar{x} = m = \int_{-\infty}^{\infty} xG(x)\,dx = m \tag{2.58}$$

The variance is

$$V(x) = \int_{-\infty}^{\infty} (x-m)^2 G(x)\,dx = \sigma^2 \tag{2.59}$$

The standard deviation is

$$\sigma = \sqrt{V(x)} \tag{2.60}$$

Three very important items associated with the Gaussian distribution are the following.

1. The *cumulative normal distribution function*, defined by

$$E(x) = \int_{-\infty}^{x} G(x')\,dx' = \int_{-\infty}^{x} \frac{1}{(\sqrt{2\pi})\sigma} \exp\left[-\frac{(x'-m)^2}{2\sigma^2}\right] dx' \tag{2.61}$$

The function $E(x)$ is very useful and is generally known as the error function (see also Sec. 2.10.1). Graphically, the function $E(x)$ (Eq. 2.61) is equal to the shaded area of Fig. 2.5. The function is sketched in Fig. 2.6.

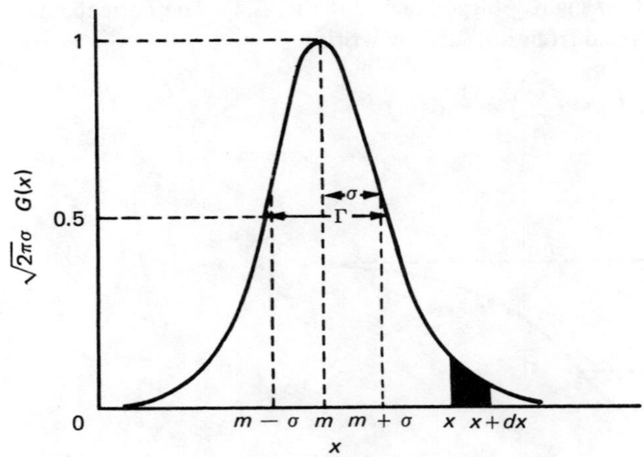

FIG. 2.4 A normal (Gaussian) distribution.

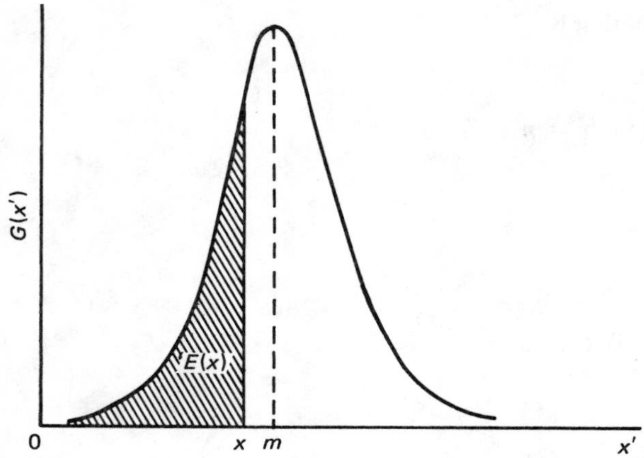

FIG. 2.5 The cumulative normal distribution is equal to the shaded area under the Gaussian curve.

2. The *area under the curve of Fig. 2.4 from* $x = m - \sigma$ *to* $x = m + \sigma$, given by

$$A_\sigma = \int_{m-\sigma}^{m+\sigma} G(x)\,dx = 0.683 \tag{2.62}$$

Equation 2.62 indicates that 68.3 percent of the total area under the Gaussian is included between $m - \sigma$ and $m + \sigma$. Another way of expressing this statement is to say that if a series of events follows the normal distribution, then it should be expected that 68.3 percent of the events will be located between $m - \sigma$ and $m + \sigma$. As discussed later in Sec. 2.13, Eq. 2.62 is the basis for the definition of the "standard" error.

3. The *full width at half maximum (FWHM)*. The FWHM, usually denoted by the symbol Γ, is the width of the Gaussian distribution at the position of half of its maximum. The width Γ is slightly bigger than 2σ (Fig. 2.4). The correct relationship between the two is obtained from Eq. 2.55 by writing

$$G\left(m - \frac{\Gamma}{2}\right) = G\left(m + \frac{\Gamma}{2}\right) = \frac{1}{2}\,G(m)$$

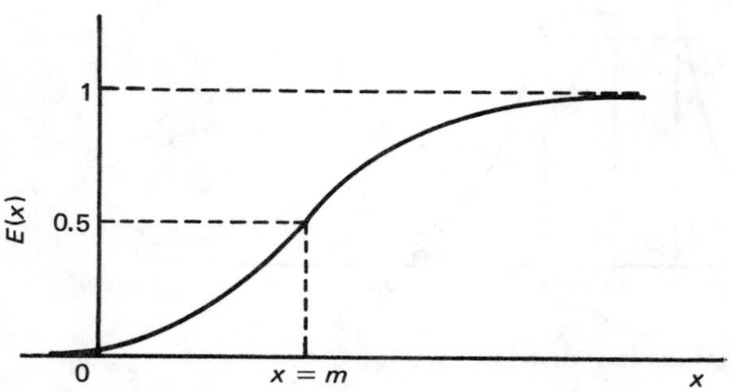

FIG. 2.6 The error function.

Solving this equation for Γ gives

$$\Gamma = (2\sqrt{2\ln 2})\sigma \approx 2.35\sigma \tag{2.63}$$

The width Γ is an extremely important parameter in measurements of the energy distribution of particles.

2.10.1 The Standard Normal Distribution

The evaluation of integrals involving the Gaussian distribution such as those of Eqs. 2.56, 2.61, and 2.62, requires tedious numerical integration. The result of such integrations is a function of m and σ. Therefore, the calculation should be repeated every time m or σ changes. To avoid this repetition, the normal distribution is rewritten in such a way that

$$m = 0 \qquad \text{and} \qquad \sigma = 1$$

The resulting function is called the *standard normal distribution.* Integrals involving the Gaussian distribution, such as that of Eq. 2.61, have been tabulated based on the standard normal distribution for a wide range of x values. With the help of a simple transformation, it is very easy to obtain the integrals for any value of m and σ.

The standard normal distribution is obtained by defining the new variable

$$t = \frac{x - m}{\sigma} \tag{2.64}$$

Substituting into Eq. 2.55, one obtains

$$G(t)\,dt = \frac{1}{\sqrt{2\pi}}\,e^{-t^2/2}\,dt \tag{2.65}$$

It is very easy to show that the Gaussian given by Eq. 2.65 has mean

$$\bar{t} = m = \int_{-\infty}^{\infty} tG(t)\,dt = 0$$

and variance

$$V(t) = \sigma^2 = \int_{-\infty}^{\infty} t^2 G(t)\,dt = 1$$

The cumulative standard normal distribution function, Eq. 2.61, is now written as

$$E(x) = \int_{-\infty}^{x} G(t)\,dt = \int_{-\infty}^{x} \frac{1}{\sqrt{2\pi}}\,e^{-t^2/2}\,dt \tag{2.66}$$

or, in terms of the error function that is found tabulated

$$E(x) = \frac{1}{2}\left(1 + \operatorname{erf}\frac{x}{\sqrt{2}}\right)$$

where

$$\operatorname{erf}\frac{x}{\sqrt{2}} = \sqrt{\frac{2}{\pi}} \int_0^x e^{-t^2/2}\, dt$$

EXAMPLE 2.12

The uranium fuel of light-water reactors is enclosed in metallic tubes with an average outside diameter (OD) equal to 20 mm. It is assumed that the OD is normally distributed around this average with a standard deviation $\sigma = 0.5$ mm. For safety reasons, no tube should be used with OD $>$ 21.5 mm or OD $<$ 18.5 mm. If 10,000 tubes are manufactured, how many of them are expected to be discarded because they do not satisfy the requirements given above?

ANSWER

The probability that the OD of a tube is going to be less than 18.5 mm or greater than 21.5 mm is

$$G(x < 18.5) + G(x > 21.5) = \int_{-\infty}^{18.5} \frac{dx}{\sqrt{2k}(0.5)} \exp\left[-\frac{(x-20)^2}{2(0.5)^2}\right]$$

$$+ \int_{21.5}^{\infty} \frac{dx}{\sqrt{2\pi}(0.5)} \exp\left[-\frac{(x-20)^2}{2(0.5)}\right]$$

Graphically, the sum of these two probabilities is equal to the two shaded areas shown in Fig. 2.7.

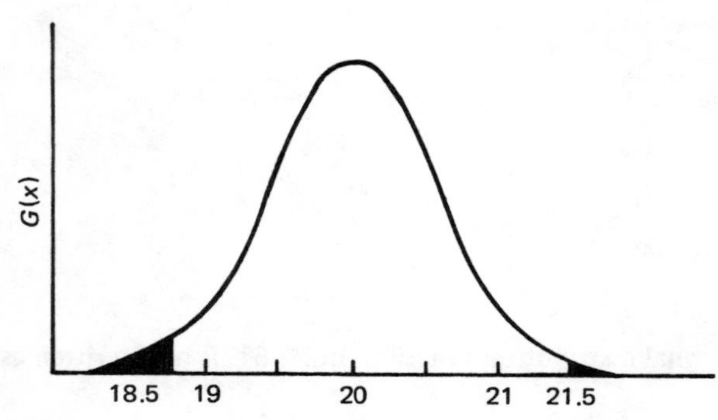

FIG. 2.7 The shaded areas represent the fraction of defective rods, Ex. 2.12.

In terms of the standard normal distribution and also because the two integrals are equal, one obtains

$$G(x < 18.5) + G(x > 21.5) = 2\left[1 - \int_{-\infty}^{3} \frac{1}{\sqrt{2\pi}} e^{-t^2/2}\, dt\right]$$

where

$$t = \frac{x - 20}{0.5}$$

This last integral is tabulated in many books, handbooks, and mathematical tables (see bibliography of this chapter). From such tables one obtains

$$\int_{-\infty}^{3} \frac{1}{\sqrt{2\pi}} e^{-t^2/2}\, dt = 0.99865$$

which gives

$$G(x < 18.5) + G(x > 21.5) = 0.0027$$

Therefore, it should be expected that under the manufacturing conditions of this example, 27 tubes out of 10,000 would be rejected.

2.10.2 Importance of the Gaussian Distribution for Radiation Measurements

The normal distribution is the most important distribution for applications in measurements. It is extremely useful because for almost any type of measurement that has been taken many times, the frequency with which individual results occur forms, to a very good approximation, a Gaussian distribution centered around the average value of the results. The greater the number of trials, the better their representation by a Gaussian. Furthermore, statistical theory shows that even if the original population of the results under study does not follow a normal distribution, their average does. That is, if a series of measurements of the variable $x_i|_{i=1,\ldots,N}$ is repeated M times, the average values $\bar{x}_N|_{N=1,\ldots,M}$ follow a normal distribution even though the x_i's may not. This result is known as the *central limit theorem* and holds for any random sample of variables with finite standard deviation.

In reality, no distribution of experimental data can be exactly Gaussian, since the Gaussian extends from $-\infty$ to $+\infty$. But for all practical purposes, the approximation is good and it is widely used because it leads to excellent results.

It is worth reminding the reader that both the binomial (Fig. 2.2) and the Poisson (Fig. 2.3) distributions resemble a Gaussian under certain conditions. This observation is particularly important in radiation measurements.

The results of radiation measurements are, in most cases, expressed as the number of counts recorded in a scaler. These counts indicate that particles have interacted with a detector and produced a pulse that has been recorded. The particles, in turn, have been produced either by the decay of a radioisotope or as a result of a nuclear reaction. In either case, the emission of the particle is statistical in nature and follows the Poisson distribution. However, as indicated in Sec. 2.9, if the average of the number of counts involved is more than about 20, the Poisson approaches the Gaussian distribution. For this reason, the individual results of such radiation measurements are treated as members of a normal distribution.

Consider now a Poisson and a Gaussian distribution having the same average, $m = 25$. Obviously, there is an infinite number of Gaussians with that average but with different standard deviations. The question one may ask is: "What is the standard deviation of the Gaussian that may represent the Poisson distribution with the same average?" The answer is that the Gaussian with $\sigma = m = 5$ is almost identical with the Poisson. Table 2.1 presents values of the two distributions, and Fig. 2.8 shows them plotted.

The following very important conclusion is drawn from this result:

> The outcomes of a series of radiation measurements are members of a Poisson distribution. They may be treated as members of a Gaussian distribution if the average result is more than $m = 20$. The standard deviation of that Gaussian distribution is $\sigma = \sqrt{m}$.

Use of this conclusion is made in Sec. 2.19, which discusses statistics of radiation counting.

TABLE 2.1 Comparison between a Poisson and a Gaussian Distribution Having the Same Mean ($m = 25$)

n	P_n (Poisson)	$G(n)$ (Gaussian) $\sigma = 5$
10	0.0004	0.0009
12	0.0017	0.0027
14	0.0059	0.0071
16	0.0154	0.0168
18	0.0316	0.0299
20	0.0519	0.0484
22	0.0702	0.0666
24	0.0795	0.0782
25	0.0795	0.0798
26	0.0765	0.0782
28	0.0632	0.0666
30	0.0454	0.0484
32	0.0286	0.0299
34	0.0159	0.0168
36	0.0079	0.0071
38	0.0035	0.0027

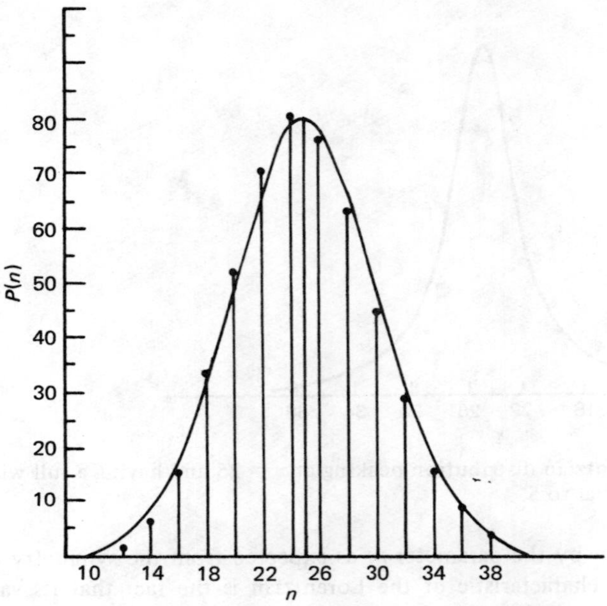

FIG. 2.8 Comparison between a Poisson distribution with $m = 25$ and a Gaussian distribution with the same average and standard deviation $\sigma = \sqrt{m} = 5$.

2.11 THE LORENTZIAN DISTRIBUTION

The Lorentzian distribution, which describes the resonances of nuclear reactions–in particular how the probability of interaction (cross section, see Chap. 3) changes as a function of particle energy–is given by

$$L(x)\,dx = \frac{1}{\pi}\frac{\Gamma/2}{(x-m)^2 + \Gamma^2/4}\,dx \tag{2.67}$$

where $L(x)\,dx$ is the probability that the value of x lies between x and $x + dx$. The Lorentzian is a symmetric function (Fig. 2.9) centered around the value $x = m$. It can be easily shown that

$$\int_{-\infty}^{\infty} L(x)\,dx = 1$$

and that

$$\bar{x} = \int_{-\infty}^{\infty} xL(x)\,dx = m$$

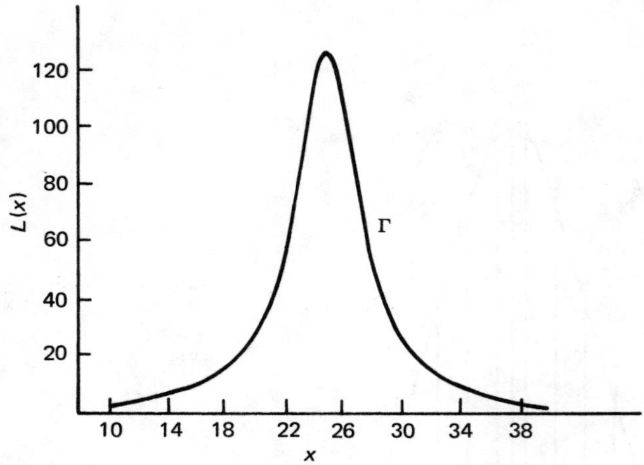

FIG. 2-9 A Lorentzian distribution peaking at $x = 25$ and having a full width half maximum equal to 5.

Thus, the mean is given by the parameter m as expected from the symmetry of the function. One peculiar characteristic of the Lorentzian is the fact that its variance cannot be calculated. Indeed, the integral

$$\sigma^2 = V(x) = \int_{-\infty}^{\infty} (x - m)^2 L(x)\, dx$$

does not converge, which is the result of the slow decrease of the function away from the peak.

In the absence of a standard deviation, the parameter Γ is used for the description of the Lorentzian. The parameter Γ is *equal to* the full width at half maximum of the function.

2.12 THE STANDARD, PROBABLE, AND OTHER ERRORS

Consider a measurement or series of measurements that gave the result R and its estimated error E. The experimenter reports the result as

$$R \pm E \tag{2.68}$$

in which case E is the absolute error (R and E have the same units) or as

$$R \pm \epsilon\% \tag{2.69}$$

where $\epsilon = (E/R)100$ = relative error (dimensionless). In most cases, the relative rather than the absolute error is reported.

When either Eq. 2.68 or 2.69 is used, the important thing to understand is that $R \pm E$ *does not mean* that the correct result has been bracketed between $R - E$ and $R + E$. It means only that *there is a probability* that the correct result has a value between $R - E$ and $R + E$. What is the value of this probability? There is no unanimous agreement on this matter and different people use different values. However, over the years, two probability values have been used more frequently than others and have led to the definition of two corresponding errors, the standard and the probable error.

THE STANDARD ERROR

If the result of a measurement is reported as $R \pm E_s$ and E_s is the standard error, then there is a 68.3 percent chance for the true result to have a value between $R - E_s$ and $R + E_s$.

THE PROBABLE ERROR

By definition, the probable error is equally likely to be exceeded or not. Therefore, if the result of a measurement is $R \pm E_p$ and E_p is the probable error, then there is a 50 percent chance for the true result to have a value between $R - E_p$ and $R + E_p$.

Both standard and probable errors are based on a Gaussian distribution. That is, it is assumed that the result R is the average of individual outcomes that belong to a normal distribution. This does not introduce any limitation in practice because, as stated in Sec. 2.10.2, the individual outcomes of a long series of any type of measurement are members of a Gaussian distribution.† With the Gaussian distribution in mind, it is obvious that the definition of the standard error is based on Eq. 2.62. If a result is R and the standard error is E_s, then $E_s = \sigma$.

$$\int_{R-E_s}^{R+E_s} \frac{1}{(\sqrt{2\pi})\sigma} \exp\left[-\frac{(x-R)^2}{2\sigma^2}\right] dx = 0.683$$

Correspondingly, the probable error E_p satisfies

$$\int_{R-E_p}^{R+E_p} \frac{1}{(\sqrt{2\pi})\sigma} \exp\left[-\frac{(x-R)^2}{2\sigma^2}\right] dx = 0.5$$

It can be shown that

$$E_p = 0.6745 E_s \tag{2.70}$$

The standard and probable errors are the most commonly used in reporting experimental results. Individual researchers may define other errors that represent a different percentage of the Gaussian. For example, the 95 percent error, E_{95}, is that which gives a 95 percent chance to have the true result bracketed between $R - E_{95}$ and $R + E_{95}$.

†Exception: Radiation counting measurements with $m < 20$ obey the Poisson distribution.

2.13 THE ARITHMETIC MEAN AND ITS STANDARD ERROR

Although the true value of a quantity can never be determined, the error of the measurement can be reduced if the experiment is repeated many times.

Consider an experiment that has been repeated N times, where N is a large number, and produced the individual outcomes $n_i|_{i=1,\ldots,N}$. Let the frequency of occurrence of n_i be P_{n_i}.† If one plots P_{n_i} versus n_i, the resulting curve resembles a Gaussian distribution as shown in Fig. 2.10. The larger the value of N, the more the histogram of Fig. 2.10 coincides with a normal distribution. Assume that the dashed line of Fig. 2.10 is an acceptable representation of the experimental results. Under these circumstances, how should the result of the measurement be reported and what is its standard error?

The result of the measurement is reported as the arithmetic average defined by

$$\bar{n} = \frac{n_1 + n_2 + \cdots + n_N}{N} = \sum_{i=1}^{N} \frac{n_i}{N} \tag{2.71}$$

This equation is the same as Eq. 2.31. As N increases, a better estimate of the true value of n is obtained–i.e., the error of the measurement becomes smaller. The true value of n, which is also called the true mean, can only be obtained with an infinite number of measurements. Since it is impossible to perform an infinite number of trials, n is always calculated from Eq. 2.71.

The error of $\bar{n}$ depends on the way the individual measurements are distributed around $\bar{n}$–i.e., it depends on the width of the Gaussian of Fig. 2.10. As the width becomes smaller, the error gets smaller, and therefore the measurement is better. The

†If $N = 1000$ and n_i has occurred 15 times, $P_{n_i} = 15/1000$.

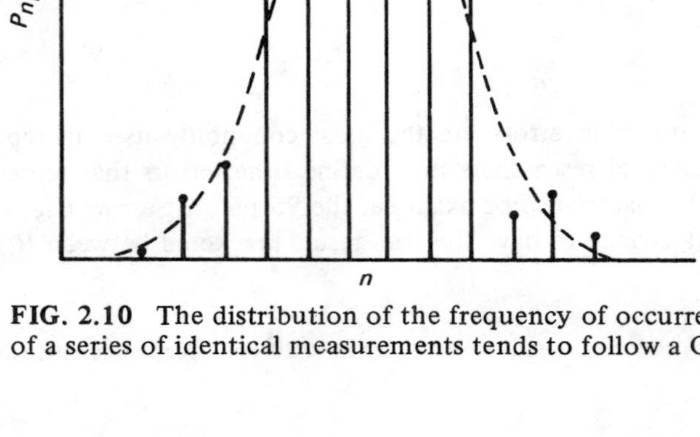

FIG. 2.10 The distribution of the frequency of occurrence of individual results of a series of identical measurements tends to follow a Gaussian distribution.

standard error of $\bar{n}$ is defined in terms of the standard deviation of the distribution. Using Eq. 2.34 and setting $f(x_i) = 1/N$, the standard deviation of the distribution becomes

$$\sigma^2 = \sum_{i=1}^{N} \frac{(n_i - m)^2}{N} \tag{2.72}$$

With a finite number of measurements at our disposal, this equation for σ has to be modified in two ways. First, because the true mean m is never known, it is replaced by its best estimate, which is $\bar{n}$ (Eq. 2.71). Second, it can be generally shown that the best estimate of the standard deviation of N measurements is given by the following equation:

$$\sigma^2 = \frac{1}{N-1} \sum_{i=1}^{N} (n_i - \bar{n})^2 \tag{2.73}$$

The differences between Eq. 2.72 and Eq. 2.73 are the use of $\bar{n}$ instead of m and the use of $N-1$ in the denominator instead of N.† For a large number of measurements, it does not make any practical difference if one divides by N or $N-1$. But it makes a difference for small values of N. Using the extreme value of $N = 1$, one can show that division by N gives the wrong result. Indeed, dividing by N, one obtains

$$\sigma^2 = \frac{1}{N} \sum (n_1 - \bar{n})^2 = \frac{1}{1} \sum (n_1 - n_1)^2 = \frac{0}{1} = 0$$

Zero σ means zero error, which is obviously wrong. The error is never zero, certainly not in the case of one measurement. Division by $N-1$, on the other hand, gives

$$\sigma^2 = \frac{1}{N-1} \sum (n_1 - n_1)^2 = \frac{0}{0}$$

which, being indeterminate, is a more realistic value of the error based on a single measurement.

Since the N results are distributed as shown in Fig. 2.10, 68.3 percent of the outcomes fall between $\bar{n} - \sigma$ and $\bar{n} + \sigma$ (see Eq. 2.62). Therefore, one additional measurement has a 68.3 percent chance of providing a result within $\bar{n} \pm \sigma$. For this reason, x is called the standard deviation or *the standard error of a single measurement.* Is this equal to the standard error of $\bar{n}$? No, and here is why.

According to the definition of the standard error, if $\sigma_{\bar{n}}$ is the standard error of $\bar{n}$, it ought to have such a value that a new average $\bar{n}$ would have a 68.3 percent

†The factor $N-1$ is equal to the "degrees of freedom" or the number of independent data or equations provided by the results. The N independent outcomes constitute, originally, N independent data. However, after $\bar{n}$ is calculated, only $N-1$ independent data are left for the calculation of σ.

chance of falling between $\bar{n} - \sigma_{\bar{n}}$ and $\bar{n} + \sigma_{\bar{n}}$. To obtain the standard error of $\bar{n}$, consider Eq. 2.71 as a special case of Eq. 2.36. The quantity $\bar{n}$ is a linear function of the uncorrelated random variables $n_1, n_2, \ldots, n_N$, each with standard deviation σ. Therefore

$$\bar{n} = \sum_{i=1}^{N} a_i n_i$$

where $a_i = 1/N$. Using Eq. 2.41, the standard deviation of $\bar{n}$ is†

$$\sigma_{\bar{n}} = \sqrt{\sum_{i=1}^{N} a_i^2 \sigma_i^2} = \sqrt{\sum_{i=1}^{N} \frac{1}{N^2} \sigma^2} = \frac{\sigma}{\sqrt{N}} \tag{2.74}$$

If the series of N measurements is repeated, the new average will probably be different from $\bar{n}$, but it has a 68.3 percent chance of having a value between $\bar{n} - \sigma_{\bar{n}}$ and $\bar{n} + \sigma_{\bar{n}}$. The result of the N measurements is

$$\bar{n} \pm \sigma_{\bar{n}} = \bar{n} \pm \frac{\sigma}{\sqrt{N}} \tag{2.75}$$

When a series of measurements is performed, it would be desirable to calculate the result in such a way that the error is a minimum. It can be shown that the average $\bar{n}$ as defined by Eq. 2.71 minimizes the quantity

$$\sum_{i=1}^{N} (\bar{n} - n_i)^2$$

which is proportional to the standard error. Finally, Eq. 2.75 shows that the error is reduced if the number of trials increases. However, that reduction is proportional to $1/\sqrt{N}$, which means that the number of measurements should be increased by a factor of 100 to be able to reduce the error by a factor of 10.

2.14 CONFIDENCE LIMITS

Consider a variable x_i that represents the value of the ith sample of a large population of specimens. The variable x_i may be the diameter of a sphere or the thickness of the cladding of a fuel rod or the length of the fuel rod. A designer may desire a certain diameter of the sphere or a certain thickness of the fuel cladding or a certain length of the fuel length. What happens during actual fabrication is that the individual units are not exactly the same. The person who examines individual units as they are con-

†If the population of the events n_i is finite in size, then it can be shown that $\sigma_{\bar{n}}^2 = [(M - N)/(M - 1)]\,\sigma^2/N$, where M = total number of n_i's (see Jaech).

structed, machined, or fabricated will find that there is a distribution of values for the quantity being examined. The average value is equal to that specified in the blueprints and is called the nominal value. Individual specimens, however, have values of x distributed around the nominal value x_n according to a Gaussian distribution:

$$G(x) = \frac{1}{(\sqrt{2\pi})\sigma} \exp\left[-\frac{(x-x_n)^2}{2\sigma^2}\right]$$

where x_n = nominal value of x = average value of x
σ = standard deviation of the distribution

The manufacturer of any product would like to know what the probability is that any one item will deviate from the nominal value by a certain amount. Or, setting some acceptable value of x, call it x_a, the manufacturer would like to know what is the probability that x will be bigger than x_a. Questions of this type come under the subject of "quality control."

The probability that x will exceed x_a is given by

$$P(x > x_a) = \int_{x_a}^{\infty} \frac{dx}{(\sqrt{2\pi})\sigma} \exp\left[-\frac{(x-x_n)^2}{2\sigma^2}\right] \tag{2.76}$$

The acceptable value of x is usually expressed as

$$x_a = x_n + k\sigma \tag{2.77}$$

i.e., the extreme acceptable value of x, x_a, is allowed to be k standard deviations different from x_n.

In terms of the standard normal distribution, Eq. 2.76 takes the form

$$P(t > k) = \int_{k}^{\infty} \frac{1}{\sqrt{2\pi}} e^{-t^2/2}\, dt \tag{2.78}$$

where

$$t = \frac{x - x_n}{\sigma} \tag{2.79}$$

and

$P(t > k)$ = probability that x will exceed x_a by k standard deviations

Table 2.2 gives values of $P(t > k)$ for several values of k. The values in Table 2.2 are interpreted as follows:

Consider $k = 1$. The probability that x will exceed x_a where $x_a = x_n + \sigma$ is 15.9 percent. If x is some property of a manufactured product, it is said that the confidence

TABLE 2.2 Probability Values and Confidence Limits

Number of standard deviations (k)	$P(x > x_a)$	Confidence limit
0	0.500	50.0
1	0.159	84.1
1.5	0.067	93.3
2	0.023	97.7
2.5	0.006	99.4
3	0.0013	99.87
4	0.00003	99.997

limit is, in this case, $1 - 0.159 = 0.841$ or 84.1 percent, i.e., 84.1 percent of the specimens will have $x < x_a$ (Fig. 2.11). If $k = 2$, the probability that x will exceed x_a is equal to 2.3 percent; therefore, the confidence limit is 97.7 percent.

In actual construction or fabrication of an item, the Gaussian distribution is determined by checking the variable x for a large number of specimens. An average value of x is calculated,

$$\bar{x} = \frac{1}{N} \sum_{i=1}^{N} x_i$$

and a standard deviation

$$\sigma = \frac{1}{N-1} \sum_{i}^{N} (x_i - \bar{x})^2$$

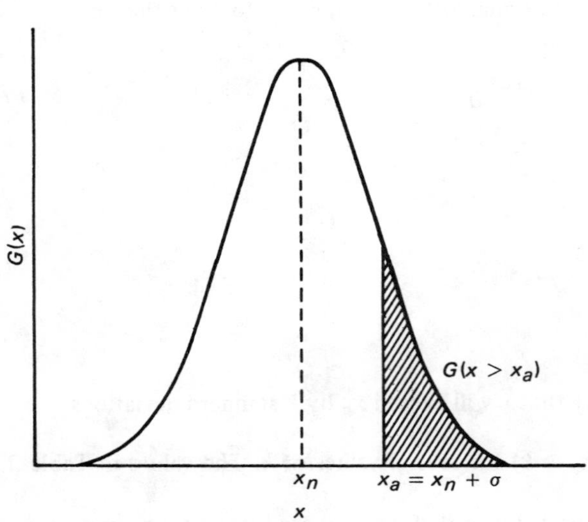

FIG. 2.11 The probability that x will exceed x_a, where $x_a = x_n + \sigma$, is 15.9 percent (shaded area). The confidence limit is $1 - 0.159$, or 84.1 percent.

is obtained. The average $\bar{x}$ should be almost equal to the nominal value of x. A Gaussian distribution for this sample peaks at $\bar{x}$ and has a standard deviation σ. Knowing σ, the value of x_a is calculated from Eq. 2.77 after the confidence limit–the value of k–has been decided upon.

The use of the concept of confidence limits is widespread in industry. As a specific example, let us assume that x is the thickness of the cladding of a reactor fuel rod. The average (nominal) thickness is x_n. The reactor designer would like to be certain that a certain fraction of fuel rods will always have thickness within prescribed limits. Let us say that the designer desires a confidence limit of 99.87 percent. This means that no more than 13 rods out of 10,000 will be expected to have cladding thickness exceeding the nominal one by more than three standard deviations (Table 2.2).

2.15 PROPAGATION OF ERRORS

Sometimes an investigator has to determine a quantity that is a function of more than one random variable. In such cases, it is very important to know how to calculate the error of the complex quantity in terms of the errors of the individual random variables. This procedure is generally known as propagation of errors and is described in this section.

Consider the function $f(x_1, x_2, \ldots, x_M)$, which depends on the random variables $x_1, x_2, \ldots, x_M$. Generally, the values of $x_1, x_2, \ldots, x_M$ are determined experimentally and then the value of $f(x_1, x_2, \ldots, x_M)$ is calculated. For example,

1 $f(x_1, x_2) = x_1 \pm x_2$
2 $f(x_1, x_2) = x_1 x_2$
3 $f(x_1, x_2) = x_1 / x_2$
4 $f(x_1, x_2) = \ln(x_1 + x_2)$
5 $f(x) = x^2$
6 $f(x_1, x_2, x_3) = x_1 + x_2 / x_3$

It has already been mentioned that the x_i's are determined experimentally, which means that average values $\bar{x}_1, \bar{x}_2, \bar{x}_3, \ldots, \bar{x}_M$ are determined along with their standard errors $\sigma_1, \sigma_2, \ldots, \sigma_M$.

1 What is the value of $f(x_1, \ldots, x_M)$ that should be reported?
2 What is the standard error of $f(x_1, \ldots, x_M)$?

It is assumed that the function $f(x_1, \ldots, x_M)$ can be expanded in a Taylor series around the averages $\bar{x}_i|_{i=1,\ldots,M}$.

$$f(x_1, x_2, \ldots, x_M) = f(\bar{x}_1, \bar{x}_2, \ldots, \bar{x}_M) + \sum_{i=1}^{M} (x_i - \bar{x}_i) \frac{\partial f}{\partial \bar{x}_i} + O(x_i - \bar{x}_i)^2$$

The notation used is that

$$\frac{\partial f}{\partial \bar{x}_i} = \left. \frac{\partial f}{\partial x_i} \right|_{x_i = \bar{x}_i}$$

The term $O(x_i - \bar{x}_i)^2$ includes all the terms of order higher than first, and it will be ignored. Thus, the function is written,

$$f(x_1, x_2, \ldots, x_M) = f(\bar{x}_1, \bar{x}_2, \ldots, \bar{x}_m) + \sum_{i=1}^{M} (x_i - \bar{x}_i) \frac{\partial f}{\partial \bar{x}_i} \tag{2.80}$$

Equation 2.80 is a special case of Eq. 2.36. The average value of $f(x_1, \ldots, x_M)$, which is the value to be reported, is

$$\bar{f} = f(\bar{x}_1, \bar{x}_2, \ldots, \bar{x}_M) \tag{2.81}$$

The variance of $f(x_1, \ldots, x_M)$ is given by Eq. 2.40:

$$\sigma_{\bar{f}}^2 = V(f) = \sum_{i=1}^{M} \left(\frac{\partial f}{\partial \bar{x}_i}\right)^2 \sigma_i^2 + 2 \sum_{j>i}^{M} \sum^{M} \left(\frac{\partial f}{\partial \bar{x}_i}\right)\left(\frac{\partial f}{\partial \bar{x}_j}\right) \rho_{ij}\sigma_i\sigma_j \tag{2.82}$$

where ρ_{ij} is the correlation coefficient given by Eq. 2.39.

The standard error of $f(x_1, \ldots, x_M)$ is equal to the standard deviation

$$\sigma_{\bar{f}} = \sqrt{\sum_{i=1}^{M} \left(\frac{\partial f}{\partial x_i}\right)^2 \sigma_i^2 + 2 \sum_{j>i}^{M} \sum^{M} \left(\frac{\partial f}{\partial \bar{x}_i}\right)\left(\frac{\partial f}{\partial \bar{x}_j}\right) \rho_{ij}\sigma_i\sigma_j} \tag{2.83}$$

Equations 2.81 and 2.83 are the answers to questions 1 and 2 stated earlier. They indicate first that the average of the function is calculated using the average values of the random variables and second that its standard error is given by Eq. 2.83. Equation 2.83 looks complicated, but fortunately, in most practical cases, the random variables are uncorrelated–i.e., $\rho_{ij} = 0$ and Eq. 2.83 reduces to

$$\sigma_{\bar{f}} = \sqrt{\sum_{i=1}^{M} \left(\frac{\partial f}{\partial \bar{x}_i}\right)^2 \sigma_i^2} \tag{2.84}$$

Unless otherwise specified, the discussion in the rest of this chapter will concern *only uncorrelated variables.* Therefore, Eqs. 2.81 and 2.84 will be used. The reader, however, should always keep in mind the assumption under which Eq. 2.84 is valid.

2.15.1 Examples of Error Propagation in Common Functions (with Uncorrelated Variables)

EXAMPLE 2.13

$f(x_1, x_2) = a_1 x_1 \pm a_2 x_2$, where a_1 and a_2 are constants

$$\bar{f} = a_1 \bar{x}_1 \pm a_2 \bar{x}_2$$

$$\sigma_{\bar{f}} = \sqrt{\left(\frac{\partial f}{\partial \bar{x}_i}\right)^2 \sigma_1^2 + \left(\frac{\partial f}{\partial \bar{x}_2}\right)^2 \sigma_2^2} = \sqrt{a_1^2 \sigma_1^2 + a_2^2 \sigma_2^2} \tag{2.85}$$

If $a_1 = a_2 = 1$, this example applies to the very common case of summation or difference of two variables.

EXAMPLE 2.14

$f(x_1, x_2) = ax_1x_2$, where a is a constant

$$\bar{f} = a\bar{x}_1\bar{x}_2$$

$$\sigma_{\bar{f}} = a\sqrt{\bar{x}_2^2 \sigma_1^2 + \bar{x}_1^2 \sigma_2^2}$$

EXAMPLE 2.15

$f(x_1, x_2) = ax_1/x_2$

$$\bar{f} = a\frac{\bar{x}_1}{\bar{x}_2}$$

$$\sigma_{\bar{f}} = a\sqrt{\frac{1}{\bar{x}_2^2}\sigma_1^2 + \frac{\bar{x}_1^2}{\bar{x}_2^4}\sigma_2^2}$$

The standard error for Examples 2.14 and 2.15 takes a simpler and easy-to-remember form for both the product and the quotient if it is expressed as the relative error. It is trivial to show that

$$\frac{\sigma_{\bar{f}}}{\bar{f}} = \sqrt{\left(\frac{\sigma_1}{\bar{x}_1}\right)^2 + \left(\frac{\sigma_2}{\bar{x}_2}\right)^2} \tag{2.86}$$

Thus, the relative error of the product ax_1x_2 or the quotient ax_1/x_2 is equal to the square root of the sum of the squares of the relative errors of the variables x_1 and x_2.

EXAMPLE 2.16

$f(x) = x^m$, where m is some real number

$$\bar{f} = (\bar{x})^m$$

$$\sigma_{\bar{f}} = \left(\frac{\partial f}{\partial \bar{x}}\right)\sigma_{\bar{x}} = m(\bar{x})^{m-1}\sigma_{\bar{x}}$$

or $$\frac{\sigma_{\bar{f}}}{\bar{f}} = m\frac{\sigma_{\bar{x}}}{\bar{x}}$$

EXAMPLE 2.17

$f(x) = e^{ax}$

$$\bar{f} = e^{a\bar{x}}$$

$$\sigma_{\bar{f}} = \left(\frac{\partial f}{\partial \bar{x}}\right)\sigma_{\bar{x}} = ae^{a\bar{x}}\sigma_{\bar{x}}$$

or $$\frac{\sigma_{\bar{f}}}{\bar{f}} = a\sigma_{\bar{x}}$$

There is another very important use of Eq. 2.84, which has to do with the calculation of the variation of a function in terms of changes of the independent variables. Consider again the function $f(x_1, x_2, \ldots, x_M)$ and assume that the variables $x_1, x_2, \ldots, x_M$ have changed by the amounts $\Delta x_1, \Delta x_2, \ldots, \Delta x_M$. The variation or change of $f(x_1, \ldots, x_M)$, Δf, is given by

$$\Delta f = \sqrt{\sum_{i=1}^{M} \left(\frac{\partial f}{\partial x_i}\right)^2 \Delta x_i^2} \tag{2.87}$$

Equation 2.87 should not be used if it is specified what the change of variable is, i.e., if the change is a decrease or an increase. If the change is known, one should calculate the function $f(x_1, x_2, \ldots, x_M)$ using the new values of the x's and obtain Δf by subtracting the new from the old value.

EXAMPLE 2.18

The speed of sound is obtained by measuring the time it takes for a certain sound signal to travel a certain distance. What is the speed of sound and its standard error if it takes the sound 2.5 ± 0.125 s to travel 850 ± 5 m?

ANSWER

$$f(x_1, x_2) = v = \frac{x}{t} = \frac{850}{2.5} = 340 \text{ m/s}$$

To calculate the error, use Eq. 2.86:

$$\frac{\sigma_v}{v} = \sqrt{\left(\frac{\sigma_x}{x}\right)^2 + \left(\frac{\sigma_t}{t}\right)^2} = \sqrt{\left(\frac{5}{850}\right)^2 + \left(\frac{125}{2.5}\right)^2} = 0.05 = 5\%$$

The result is 340 ± 17 m/s.

EXAMPLE 2.19

A beam of photons going through a material of thickness t is attenuated in such a way that the fraction of photons traversing the material is $e^{-\mu x}$ where the constant μ is called the attenuation coefficient. If the thickness of the material changes by 10 percent, by how much will the emerging fraction of photons change? Take $x = 0.01$ m and $\mu = 15 \text{ m}^{-1}$.

ANSWER

This is a case requiring the use of Eq. 2.87.

$$f(x) = e^{-\mu x}$$

$$\Delta f = \left(\frac{\partial f}{\partial x}\right) \Delta x = -\mu e^{-\mu x} \Delta x$$

$$\frac{\Delta f}{f} = -\mu \Delta x = -\mu x \left(\frac{\Delta x}{x}\right) = -(15)(0.01)(0.10) = -0.015$$

Therefore, if the thickness increases by 10 percent, the fraction of emerging photons decreases by 1.5 percent.

2.16 GOODNESS OF DATA–χ^2 CRITERION–REJECTION OF DATA

It is desirable when data are obtained during an experiment to be able to determine if the recording system works well or not. The experimenter should ask the question: Are all the obtained data true (due to the phenomenon studied), or are some or all due to extraneous disturbances that have nothing to do with the measurement? A number of tests have been devised for the purpose of checking how reliable the results are, i.e., checking the "goodness of data."

Before any tests are applied, an investigator should use common sense and try to avoid erroneous data. First of all, a good observer will never rely on a single measurement. He or she should repeat the experiment as many times as is feasible (but at least twice) and observe whether the results are reproducible or not. Second, the observer should check the results to see how they deviate from their average value. Too large or too small deviations are suspicious. The good investigator should be alert and should check such data very carefully. For example, if for identical, consecutive measurements one gets the following counts in a scaler:

10,000 10,000 10,000 10,002 9999 9998

the apparatus is not necessarily very accurate and should be checked.

The test that is used more frequently than any other to check the goodness of data is the χ^2 criterion (chi square), or Pearson's χ^2 test.

The χ^2 test is based on the quantity

$$\chi^2 = \frac{\sum_{i=1}^{N} (\bar{n} - n_i)^2}{\bar{n}} \tag{2.88}$$

where $n_i|_{i=1,\ldots,N}$ represents the results of N measurements with $\bar{n}$ being the average.

The test compares the value of χ^2 with the distribution† of χ^2 values that would be expected if the set of N measurements is repeated many times. Using the probability distribution of the χ^2's one can calculate the probability that, upon repeating the series of N measurements, larger deviations from the expected values would be observed. Table 2.3 gives probability values used with the χ^2 test. In general, it is desirable to have χ^2 such that the probability given by Table 2.3 is close to 0.5. Then there is an equal chance for a new set of measurements to lead to χ^2, which is smaller or larger than the first one. Notice in Table 2.3 that when the probability is close to 0.5, $\chi^2 \approx N-1$. Usually the test is applied by selecting a range of acceptable χ^2 values. The data fail the test if the calculated χ^2 falls outside the preselected range.

†For more details about the χ^2 distribution, see Johnson-Leone and Jaech.

TABLE 2.3 Probability Table for χ^2 Criterion†

Degrees of freedom‡ ($N-1$)	Probability						
	0.99	0.95	0.90	0.50	0.10	0.05	0.01
2	0.020	0.103	0.211	1.386	4.605	5.991	9.210
3	0.115	0.352	0.584	2.366	6.251	7.815	11.345
4	0.297	0.711	1.064	3.357	7.779	9.488	13.277
5	0.554	1.145	1.610	4.351	9.236	11.070	15.086
6	0.872	1.635	2.204	5.348	10.645	12.592	16.812
7	1.239	2.167	2.833	6.346	12.017	14.067	18.475
8	1.646	2.733	3.490	7.344	13.362	15.507	20.090
9	2.088	3.325	4.168	8.343	14.684	16.919	21.666
10	2.558	3.940	4.865	9.342	15.987	18.307	23.209
11	3.053	4.575	5.578	10.341	17.275	19.675	24.725
12	2.571	5.226	6.304	11.340	18.549	21.026	26.217
13	4.107	5.892	7.042	12.340	19.812	22.363	27.688
14	4.660	6.571	7.790	13.339	21.064	23.685	29.141
15	5.229	7.261	8.547	14.339	22.307	24.996	30.578
16	5.812	7.962	9.312	15.338	23.542	26.296	32.000
17	6.408	8.672	10.085	16.338	24.769	27.587	33.409
18	7.015	9.390	10.865	17.338	25.989	28.869	34.805
19	7.633	10.117	11.651	18.338	27.204	30.144	36.191
20	8.260	10.851	12.443	19.337	28.412	31.410	37.566
21	8.897	11.591	13.240	20.337	29.615	32.671	38.932
22	9.542	12.338	14.041	21.337	30.813	33.924	40.289
23	10.196	13.091	14.848	22.337	32.007	35.172	41.638
24	10.856	13.848	15.659	23.337	33.196	36.415	42.980
25	11.534	14.611	16.473	24.337	34.382	37.382	44.314
26	12.198	15.379	17.292	25.336	35.563	38.885	45.642
27	12.879	16.151	18.114	26.336	36.741	40.113	46.963
28	13.565	16.928	18.939	27.336	37.916	41.337	48.278
29	14.256	17.708	19.768	28.336	39.087	42.557	49.588

†Calculated values of χ^2 will be equal to or greater than the values given in the table.
‡See footnote on p. 47.

What should one do if the data fail the test? Should all, some, or none of the data be rejected? The answer to these questions is not unique, but rather depends on the criteria set by the observer and the type of measurement. If the data fail the test, the experimenter should be on the lookout for trouble. Some possible reasons for trouble are the following:

1 Unstable equipment may give inconsistent results, e.g., spurious counts generated by a faulty component of an instrument.
2 External signals may be picked up by the apparatus and be "recorded." Sparks, radio signals, welding machines, etc., produce signals that may be recorded by a pulse-type counting system.

3 If a number of samples are involved, widely scattered results may be caused by lack of sample uniformity.
4 A large χ^2 may result from one or two measurements that fall far away from the average. Such results are called the "outliers." Since the results are governed by the normal distribution, which extends from $-\infty$ to $+\infty$, in theory, at least, all results are possible. In practice, it is somewhat disturbing to have a few results that seem to be way out of line.

Should the outliers be rejected? And by what criterion? One should be conservative when rejecting data for three reasons:

1 The results are random variables following the Gaussian distribution. Therefore, outliers are possible.
2 As the number of measurements increases, the probability of an outlier increases.
3 In a large number of measurements, the rejection of an outlier has small effect on the average although it makes the data look better by decreasing the dispersion.

One of the criteria used for data rejection is Chauvenet's criterion, stated as follows:

> A reading may be rejected if it has a deviation from the mean greater than that corresponding to the $1 - 1/2N$ error, where N is the number of measurements.

Data used with Chauvenet's criterion are given in Table 2.4. For example, in a series of 10 measurements, $1 - 1/2N = 1 - 1/20 = 0.95$. If $n_i - \bar{n}$ exceeds the 0.95 percent error (1.96σ), then that reading could be rejected. In that case, a new mean should be calculated without this measurement and also a new standard deviation.

The use of Chauvenet's, or any other, criterion is not mandatory. It is up to the observer to decide if a result should be rejected or not.

2.17 THE STATISTICAL ERROR OF RADIATION MEASUREMENTS

Radioactive decay is a truly random process that obeys the Poisson distribution, according to which the standard deviation of the true mean m is $\sqrt{m}$. However, the true mean is never known and can never be found from a finite number of measurements. But is there a need for a large number of measurements?

Suppose one performs only one measurement and the result is n counts. The

TABLE 2.4 Data for Chauvenet's Criterion

Number of measurements	Number of standard deviations away from average
2	1.15
3	1.38
4	1.54
5	1.65
10	1.96
15	2.13
25	2.33

best estimate of the true mean, as a result of this single measurement, is this number n. If one takes this to be the mean, its standard deviation will be $\sqrt{n}$.

Indeed, this is what is done in practice. The result of a single count n is reported as $n \pm \sqrt{n}$, which implies that

1 The outcome n is considered the true mean.
2 The standard deviation is reported as the standard error of n.

The relative standard error of the count n is

$$\frac{\sigma_n}{n} = \frac{\sqrt{n}}{n} = \frac{1}{\sqrt{n}} \tag{2.89}$$

which shows that the relative error decreases if the number of counts obtained in the scaler increases. Table 2.5 gives several values of n and the corresponding percent standard error. To increase the number n, one either counts for a long time or repeats the measurement many times and combines the results. Repetition of the measurement is preferable to one single long count because by performing the experiment many times the reproducibility of the results is checked.

Consider now a series of N counting measurements with the individual results $n_i|_{i=1,\ldots,N}$. It is assumed that the counts n_i were obtained under identical conditions and for the same counting time; thus their differences are solely due to the statistical nature of radiation measurements. Each number n_i has a standard deviation $\sigma_i = \sqrt{n_i}$. The average of this series of measurements is, using Eq. 2.31,

$$\bar{n} = \frac{1}{N} \sum_{i=1}^{N} n_i \tag{2.31}$$

The standard error of $\bar{n}$ can be calculated in two ways:

1 The average $\bar{n}$ is the best estimate of a Poisson distribution of which the outcomes $n_i|_{i=1,\ldots,N}$ are members. The standard deviation of the Poisson distribution is (see Sec. 2.9) $\sigma = \sqrt{m} = \sqrt{\bar{n}}$. The standard error of the average is (see Eq. 2.75)

$$\sigma_{\bar{n}} = \frac{\sigma}{\sqrt{N}} = \sqrt{\frac{\bar{n}}{N}} \tag{2.90}$$

TABLE 2.5 Percent Standard Error of *n* Counts

n	% Standard error of n
100	10
1,000	3.16
10,000	1
100,000	0.316
1,000,000	0.1

2 The average $\bar{n}$ may be considered a linear function of the independent variables n_i, each with standard error $\sqrt{n_i}$. Then, using Eq. 2.84, one obtains

$$\sigma_{\bar{n}} = \sqrt{\sum_{i=1}^{N} \left(\frac{\partial \bar{n}}{\partial n_i}\right)^2 \sigma_{n_i}^2} = \sqrt{\sum_{i=1}^{N} \frac{1}{N^2} (\sqrt{n_i})^2} = \sqrt{\sum_{i=1}^{N} \frac{1}{N^2} n_i}$$

$$= \frac{1}{N} \sqrt{n_{tot}} \tag{2.91}$$

where

$$n_{tot} = n_1 + n_2 + \cdots + n_N = \text{total number of counts obtained from } N \text{ measurements}$$

It is not difficult to show that Eqs. 2.90 and 2.91 are identical.

In certain cases, the observer needs to combine results of counting experiments with quite different statistical uncertainties. For example, one may have to combine the results of a long and short counting measurement. Then the average should be calculated by weighting the individual results according to their standard deviations (see Bevington and Eadie et al.). The equation for the average is

$$\bar{n} = \frac{\sum_{i=1}^{N} n_i/\sigma_i^2}{\sum_{i-1}^{N} 1/\sigma_i^2} \tag{2.92}$$

EXAMPLE 2.20

Table 2.6 presents typical results of 10 counting measurements. Using these data, the average count and its standard error will be calculated using Eqs. 2.31, 2.90, and 2.91. The average is

$$\bar{n} = \frac{1}{N} \sum_{i-1}^{N} n_i = \frac{2060}{10} = 206$$

Using Eq. 2.90 or Eq. 2.91, the standard error of $\bar{n}$ is

$$\sigma_{\bar{n}} = \sqrt{\frac{206}{10}} = \frac{\sqrt{2060}}{10} = 4.5$$

2.18 THE STANDARD ERROR OF COUNTING RATES

In practice, the number of counts is usually recorded in a scaler, but what is reported is the counting rate, i.e., counts recorded per unit time. The following symbols and definitions will be used for counting rates.

G = number of counts recorded by the scaler in time t_G *with* the sample present
= gross count

B = number of counts recorded by the scaler in time t_B *without* the sample
= background count

$$g = \frac{G}{t_G} = \text{gross counting rate}$$

$$b = \frac{B}{t_B} = \text{background counting rate}$$

$$r = \text{net counting rate}^{\dagger} = \frac{G}{t_G} - \frac{B}{t_B} = g - b \tag{2.93}$$

The standard error of the net counting rate can be calculated based on Eq. 2.84 and by realizing that r is a function of four independent variables G, t_G, B, and t_B:

$$\sigma_r = \sqrt{\left(\frac{\partial r}{\partial G}\right)^2 \sigma_G^2 + \left(\frac{\partial r}{\partial t_G}\right)^2 \sigma_{t_G}^2 + \left(\frac{\partial r}{\partial B}\right)^2 \sigma_B^2 + \left(\frac{\partial r}{\partial t_B}\right)^2 \sigma_{t_B}^2} \tag{2.94}$$

†When the counting rate is extremely high, the counter may be missing some counts. Then a "dead time" correction is necessary, in addition to background subtraction; see Sec. 2.21.

TABLE 2.6 Typical Results of a Counting Experiment†

Observation, i	Number of counts obtained in the scaler, n_i	Square of deviation, $(n_i - \bar{n})^2$
1	197	81
2	210	16
3	200	36
4	198	64
5	205	1
6	195	121
7	190	256
8	220	196
9	215	81
10	230	576
Totals	2060	1428

†One could use Eqs. 2.73 and 2.74 for the calculation of σ and $\sigma_{\bar{n}}$. The result is

$$\sigma_{\bar{n}} = \sqrt{\frac{\sum_{i=1}^{N} (n_i - \bar{n})^2}{N(N-1)}} \equiv \sqrt{\frac{1428}{10(9)}} = 3.98 \approx 4$$

For radiation measurements, use of Eqs. 2.90 and 2.91 is preferred.

The electronic equipment available today is such that the error in the measurement of time is, in almost all practical cases, much smaller than the error in the measurement of G and B.† Unless otherwise specified, σ_{tG} and σ_{tA} will be taken as zero. Then Eq. 2.94 takes the form

$$\sigma_r = \sqrt{\left(\frac{\partial r}{\partial G}\right)^2 \sigma_G^2 + \left(\frac{\partial r}{\partial B}\right)^2 \sigma_B^2} \tag{2.95}$$

The standard errors of G and B are

$$\sigma_G = \sqrt{G} \qquad \sigma_B = \sqrt{B}$$

Using Eqs. 2.93 and 2.95, one obtains for the standard error of the net counting rate:

$$\sigma_r = \sqrt{\frac{G}{t_G^2} + \frac{B}{t_B^2}} \tag{2.96}$$

It is important to notice that in the equation for the net counting rate, the quantities G, B, t_G, and t_B are the independent variables, not g and b. The error of r will be calculated from the error in G, B, t_G, and t_B. It is very helpful to remember the following rule: *The statistical error of a certain count is determined from the number recorded by the scaler. That number is G and B, not the rates g and b.*

EXAMPLE 2.21

A radioactive sample gave the following counts:

$$G = 1000 \qquad t_G = 2 \text{ min} \qquad B = 500 \qquad t_B = 10 \text{ min}$$

What is the net counting rate and its standard error?

ANSWER

$$r = \frac{G}{t_G} - \frac{B}{t_B} = \frac{1000}{2} - \frac{500}{10} = 500 - 50 = 450 \text{ counts/min}$$

$$\sigma_r = \sqrt{\frac{G}{t_G^2} + \frac{B}{t_B^2}} = \sqrt{\frac{1000}{2^2} + \frac{500}{10^2}} = \sqrt{250 + 5} = 16 \text{ counts/min}$$

$$r = 450 \pm 16 = 450 \pm 3.5\%$$

A common error is that, since $r = g - b$, one is tempted to write

$$\sigma_r = \sqrt{\sigma_g^2 + \sigma_b^2} = \sqrt{(\sqrt{g})^2 + (\sqrt{b})^2} = \sqrt{g + b} = \sqrt{500 + 50} = 23 \text{ counts/min}$$

†The errors σ_{t_G} and σ_{t_B} may become important in experiments where very accurate counting time is paramount for the measurement.

This result, $\sigma_r = 23$, is wrong because

$$\sigma_g \neq \sqrt{g} \quad \text{and} \quad \sigma_b \neq \sqrt{b}$$

The correct way to calculate the standard error based on g and b is to use

$$\sigma_g = \frac{\sqrt{G}}{t_a} = \frac{\sqrt{1000}}{2} \qquad \sigma_b = \frac{\sqrt{B}}{t_B} = \frac{\sqrt{500}}{10}$$

Then

$$\sigma_r = \sqrt{\sigma_g^2 + \sigma_b^2} = \sqrt{\frac{G}{t_G^2} + \frac{B}{t_B^2}} = 16 \text{ counts/min}$$

Usually one determines G and B, in which case σ_r is calculated from Eq. 2.96. However, sometimes the background counting rate and its error have been determined earlier. In such a case σ_r is calculated as shown in Example 2.22.

EXAMPLE 2.22

A radioactive sample gave $G = 1000$ counts in 2 min. The background rate of the counting system is known to be $b = 100 \pm 6$ counts/min. What is the net counting rate and its standard error?

ANSWER

$$r = \frac{G}{t_G} - b = \frac{1000}{2} - 100 = 400 \text{ counts/min}$$

$$\sigma_r = \sqrt{\left(\frac{\partial r}{\partial G}\right)^2 \sigma_G^2 + \left(\frac{\partial r}{\partial b}\right)^2 \sigma_b^2} = \sqrt{\frac{G}{t_G^2} + \sigma_b^2} = \sqrt{\frac{1000}{4} + \sigma_b^2} = \sqrt{286}$$

$$= 17 \text{ counts/min}$$

In this problem, b and σ_b are given, not B and t_B. The standard error of the background rate has been determined by an earlier measurement. Obviously, b was not determined by counting for 1 min, because in that case one would have:

$$B = 100 \qquad t_B = 1 \text{ min} \qquad b = 100 \text{ counts/min}$$

$$\sigma_b = \frac{\sqrt{B}}{t_B} = \frac{\sqrt{100}}{1} = 10$$

2.18.1 Combining Counting Rates

If the experiment is performed N times with results

$$G_1, G_2, G_3, \ldots, G_N \qquad B_1, B_2, \ldots, B_N$$

for gross and background counts, the average net counting rate is:

$$\bar{r} = \frac{1}{N} \sum_{i=1}^{N} r_i = \frac{1}{N} \sum_{i=1}^{N} \left(\frac{G_i}{t_{G_i}} - \frac{B_i}{t_{B_i}} \right)$$

In most cases, t_{G_i} and t_{B_i} are kept constant for all N measurements. That is, $t_{G_i} = t_G$ and $t_{B_i} = t_B$. Then

$$\bar{r} = \frac{1}{N} \sum_{i=1}^{N} \left(\frac{G_i}{t_{B_i}} - \frac{B_i}{t_{B_i}} \right) = \frac{1}{N} \left(\frac{G}{t_G} - \frac{B}{t_B} \right) \tag{2.97}$$

where

$$G = \sum_{i-1}^{N} G_i \quad \text{and} \quad B = \sum_{i-1}^{N} B_i$$

The standard error of the average counting rate is, using Eqs. 2.84 and 2.96,

$$\sigma_{\bar{r}} = \frac{1}{N} \sqrt{\sum_{i=1}^{N} \sigma_{r_i}^2} = \frac{1}{N} \sqrt{\sum_{i=1}^{N} \left(\frac{G_i}{t_G^2} + \frac{B_i}{t_B^2} \right)} = \frac{1}{N} \sqrt{\frac{G}{t_G^2} + \frac{B}{t_B^2}} \tag{2.98}$$

A SPECIAL CASE

Sometimes the background rate is negligible compared to the gross counting rate. Then, Eq. 2.97 becomes

$$\sigma_{\bar{r}} = \frac{1}{N} \frac{\sqrt{G}}{t_G}$$

The percent standard error is

$$\frac{\sigma_r}{\bar{r}} = \frac{(1/N)\sqrt{G}/t_G}{G/Nt_G} = \frac{1}{\sqrt{G}}$$

This is the same as Eq. 2.89. Therefore, if the background is negligible, the percent standard error is the same for either the total count or the counting rate.

2.19 METHODS OF ERROR REDUCTION

In every radiation measurement it is extremely important to perform it in such a way that the result is determined with the minimum possible error. In general, the first task of the investigator is to improve the counting apparatus by reducing the background as much as possible. Actually, the important quantity is the ratio b/g or b/r

and not the absolute value of the background. Assuming that all possible improvements of background have been achieved, there is a procedure that if followed will result in a smaller error. Two such procedures will be discussed below. In addition, a method will be presented for the calculation of the counting time necessary to measure a counting rate with a desired degree of accuracy.

2.19.1 The Background Is Constant and there Is No Time for the Performance of the Measurement

In this case, the background is measured for a long period of time to minimize the error introduced by it, i.e., t_B is so long that:

$$\sigma_r = \sqrt{\frac{G}{t_G^2} + \frac{B}{t_B^2}} \approx \sqrt{\frac{G}{t_G^2}}$$

EXAMPLE 2.23

Suppose one obtains the following data:

$$G = 400 \qquad t_G = 5 \text{ min}$$

$$B = 100 \qquad t_B = 2.5 \text{ min}$$

Then:

$$r = \frac{400}{5} - \frac{100}{2.5} = 40 \text{ counts/min}$$

$$\sigma_r = \sqrt{\frac{400}{5^2} + \frac{100}{2.5^2}} = 5.65 \text{ counts/min}$$

$$\frac{\sigma_r}{r} = \frac{5.65}{40} = 0.14 = 14\%$$

If the background is constant, this result can be improved by counting background for a long period of time, e.g., 250 min. In that case, the result is

$$B = \frac{100}{2.5}\ 250 = 10{,}000 \text{ counts} \qquad t_B = 250 \text{ min}$$

$$r = \frac{400}{5} - \frac{10{,}000}{250} = 40 \text{ counts/min}$$

$$\sigma_r = \sqrt{\frac{400}{5^2} + \frac{10{,}000}{250^2}} = \sqrt{16 + 0.16} \approx 4$$

$$\frac{\sigma_r}{r} = \frac{4}{40} = 10\%$$

2.19.1 There Is a Fixed Time T That Should be Devoted in Counting Both Background and Gross Count

In this case, the question is, "What is the optimum time to be used for gross and background counting?" Optimum time results in minimum statistical error for the net counting rate. The optimum time is determined as follows.

An estimate of the counting rates at the time of the measurement is obtained with a short count (not the final one). Assume that one obtained the approximate counting rates

$$g = \frac{G}{t_G} \qquad b = \frac{B}{t_B}$$

Then, from Eq. 2.96 and also using $G = gt_G, B = bt_B$,

$$\sigma_r^2 = \frac{g}{t_G} + \frac{b}{t_B}$$

The best times t_G and t_B are those that minimize σ_r or $(\sigma_r)^2$ subject to the constraint

$$t_B + t_G = T = \text{constant} \tag{2.99}$$

Considering σ_r^2 as a function of t_B and t_G, the minimum will be found by differentiating $(\sigma_r)^2$ and setting the differential equal to zero:

$$d(\sigma_r)^2 = -\frac{g}{t_G^2}\,dt_G - \frac{b}{t_B^2}\,dt_B = 0$$

Differentiating the constraint, Eq. 2.99, one finds

$$dt_G = -dt_B$$

Substituting this value of dt_G into $d(\sigma_r)^2$ gives

$$\frac{t_B}{t_G} = \sqrt{\frac{b}{g}}$$

Therefore, if there is a fixed time T for the measurement the optimum counting times are determined from the two equations

$$t_G + t_B = T \tag{2.99}$$

$$\frac{t_B}{t_G} = \sqrt{\frac{b}{g}} \tag{2.100}$$

2.19.3 Calculation of the Counting Time Necessary to Measure a Counting Rate with a Desired Degree of Accuracy

Assume that the net counting rate of a radioactive sample should be measured with an accuracy of a percent, i.e., $\sigma_r/r = a$ percent. Also assume that a counting system is provided with a background counting rate b and standard error σ_b. Both b and σ_b have been reduced as much as possible for this system and have been determined earlier. The task is to determine the counting time t_G necessary to result in a percent standard error for the net counting rate. This time is calculated as follows.

The net counting rate and its standard error are

$$r = \frac{G}{t_G} - b \qquad \sigma_r = \sqrt{\frac{G}{t_G^2} + \sigma_b^2}$$

Therefore

$$\frac{\sigma_r}{r} = \frac{a}{100} = \frac{\sqrt{G/t_G^2 + \sigma_b^2}}{G/t_G - b} = \frac{\sqrt{g/t_G + \sigma_b^2}}{g - b} \tag{2.101}$$

Equation 2.101 solved for t_G gives

$$t_G = \frac{g}{(g-b)^2(a/100)^2 - \sigma_b^2} \tag{2.102}$$

It is assumed that an approximate gross counting rate is known.

EXAMPLE 2.24

How long should a sample be counted to obtain the net counting rate with an accuracy of 1 percent? It is given that the background for the counting system is 100 ± 2 counts/min.

ANSWER

The first step is to obtain an approximate gross counting rate. Assume that the sample gave 800 counts in 2 min. Then $g = 800/2 = 400$ counts/min and, using Eq. 2.102,

$$t_G = \frac{400}{(400-100)^2(0.01)^2 - 2^2} = 80 \text{ min}$$

Indeed, if one counts for 80 min, the error of r is going to be

$$\sigma_r = \sqrt{\frac{g}{t_G} + \sigma_b^2} = \sqrt{\frac{400}{80} + 2^2} = 3$$

$$\frac{\sigma_r}{r} = \frac{3}{300} = 0.01 = 1\%$$

2.19.4 Relative Importance of Error Components

In every measurement, the observer tries to reduce the experimental error as much as possible. If the quantity of interest depends on many variables, each with its own error, the effort to reduce the error should be directed toward the variable with the largest contribution to the final error.

Consider the quantity $Q = x + y - z$ and assume $x = 3$, $y = 2$, and $z = 1$. Also assume that the corresponding standard errors are

$$\sigma_x = 0.1 \qquad \sigma_y = 0.23 \qquad \sigma_z = 0.05$$

The standard error of Q is

$$\sigma_Q = \sqrt{\sigma_x^2 + \sigma_y^2 + \sigma_z^2} = \sqrt{0.1^2 + 0.23^2 + 0.05^2}$$
$$= \sqrt{0.01 + 0.529 + 0.0025} = 0.25$$

From the relative magnitude of the errors, one can see that if it is necessary to reduce the error further, the effort should be directed toward reduction of σ_y first, σ_x second, and σ_z third. In fact, there is no need to reduce σ_z further before σ_x and σ_y reach the same magnitude as σ_z.

2.20 MINIMUM DETECTABLE ACTIVITY

The minimum detectable level of activity depends on many factors, the primary one being the acceptable maximum error of the measured result. Although there is no general rule for the maximum acceptable error σ_r, an upper limit is set from the requirement that $\sigma_r \leqslant r$. Usually, only measurements that satisfy the constraint

$$\sigma_r \leqslant fr \qquad f < 1 \tag{2.103}$$

will be accepted. Then the minimum acceptable (i.e., minimum detectable) counting rate is calculated as follows.

Assume that the counting system has a constant and known background equal to $b \pm \sigma_b$ counts/min. Also assume that the counting time with the sample present is T min. Using the appropriate forms of Eqs. 2.93 and 2.96 for r and σ_r, respectively, Eq. 2.103 takes the form

$$\sqrt{\frac{G}{T^2} + \sigma_b^2} \leqslant fr \tag{2.104}$$

The gross count G can be written (Eq. 2.93) as

$$G = (r + b)T$$

If this value of G is substituted into Eq. 2.104 and the resulting equation is solved for the net counting rate r, one obtains

$$r \geqslant \frac{1 + \sqrt{1 + 4f^2 bT + 4f^2 T^2 \sigma_b^2}}{2f^2 T} \tag{2.105}$$

Equation 2.105 shows that the minimum detectable counting rate depends on three factors.

1 The factor f, determined by the investigator
2 The background rate b (and σ_b), which depends on the counting system
3 The counting time T, which is limited by the half-life of the isotope being counted and the stability of the electronics (modern electronic equipment stays stable for several hours)

EXAMPLE 2.25

A sample is counted by a detector with a known and constant background rate of 20 ± 1 counts/min. The maximum counting time is 1 h and the maximum acceptable error is 80 percent. What is the minimum detectable counting rate?

ANSWER

Using Eq. 2.105,

$$r \geqslant \frac{1 + \sqrt{1 + 4(0.8)^2(20)(60) + 4(0.8)^2(60)^2(1)^2}}{2(0.8)^2(60)}$$

$$r \geqslant \frac{1 + \sqrt{1 + 3072 + 9216}}{77} = \frac{1 + 110.8}{77} = 1.45 \text{ counts/min}$$

Therefore, under the circumstances of this example, the observer should not accept any gross counting rate that is less than 20 + 1.45 = 21.45 counts/min, or any sample that will give less than 1287 gross counts per hour (21.45 × 60 = 1287).

In most cases, as in Example 2.25, Eq. 2.105 can be simplified because the last two factors under the square root are much greater than 1. Then

$$r \geqslant \frac{\sqrt{4f^2 bT + 4f^2 T^2 \sigma_b^2}}{2f^2 T} = \frac{1}{f}\sqrt{\frac{b}{T} + \sigma_b^2} \tag{2.106}$$

If $b/T \ll \sigma_b^2$, Eq. 2.106 is further simplified and takes the form

$$r \geqslant \frac{\sigma_b}{f} \tag{2.107}$$

2.21 COUNTER DEAD-TIME CORRECTION AND MEASUREMENT OF DEAD TIME

Dead time, or resolving time of a counter, is defined as the minimum time that can elapse between the arrival of two successive particles at the detector so that two distinct pulses are produced.

That is, dead time is the time needed from the moment the particle strikes the detector to the time the voltage pulse goes into the scaler. What is important is the dead time of the system (counter–preamplifier–amplifier–discriminator–scaler) and not of the counter alone. But the dead time of the counter is so much longer than the dead time of the electronics that the latter can be ignored (except when a MCA is used; see Chap. 10).

Because of counter dead time, the possibility exists that some particles will not be recorded since the counter will not produce pulses for them. Pulses will not be produced because the counter will be "occupied" with the formation of the signal generated by particles arriving earlier. The loss of particles is particulary important in the case of high counting rates. Obviously, the observed counting rate should be corrected for the loss of counts due to counter dead time. The rest of this section presents the method for correction as well as a method for the measurement of the dead time.

Suppose τ is the dead time of the system and g the observed counting rate. The fraction of time during which the system is insensitive is $g\tau$. If n is the true counting rate, the number of counts lost is $n(g\tau)$. Therefore

$$n = g + ng\tau$$

and

$$n = \frac{g}{1 - g\tau} \tag{2.108}$$

Equation 2.108 corrects the observed gross counting rate g for the loss of counts due to the dead time of the counter.

EXAMPLE 2.26

Suppose $\tau = 200$ μs and $g = 30{,}000$ counts/min. What fraction of counts is lost because of dead time? What is the true counting rate?

ANSWER

The true counting rate is

$$n = \frac{g}{1 - g\tau} = \frac{30{,}000/60}{1 - (30{,}000/60)(200 \times 10^{-6})}$$

or $\quad n = 555.5$ counts/s

Therefore, dead time is responsible for loss of

$$\frac{555 - 500}{555} = \frac{55}{555} = 10\% \text{ of the counts}$$

Notice that the product $g\tau = 0.10$, i.e., the product of the dead time and the gross counting rate, is a good indicator of the fraction of counts lost because of dead time.

The dead time is measured with the "two-source" method as follows. Let n_1, n_2, n_{12} be the true gross counting rates from the first source only, from the second source only, and from both sources, respectively, and let n_b be the true background rate. Let the corresponding observed counting rates be g_1, g_2, g_{12}, b.

The following equation holds:

$$\left(\begin{array}{c}\text{True net}\\ \text{counting rate}\end{array}\right)_1 + \left(\begin{array}{c}\text{True net}\\ \text{counting rate}\end{array}\right)_2 = \left(\begin{array}{c}\text{True net}\\ \text{counting rate}\end{array}\right)_{1+2}$$

$$(n_1 - n_b) \quad + \quad (n_2 - n_b) \quad = \quad (n_{12} - n_b)$$

or $\quad n_1 + n_2 = n_{12} + n_b$

Using Eq. 2.108,

$$\frac{g_1}{1 - g_1\tau} + \frac{g_2}{1 - g_2\tau} = \frac{g_{12}}{1 - g_{12}\tau} + \frac{b}{1 - b\tau} \tag{2.109}$$

It will be assumed now that $b\tau \ll 1$, in which case,

$$\frac{b}{1 - b\tau} = b$$

(If $b\tau$ is not much less than 1, the instruments should be thoroughly checked for possible malfunction before proceeding with the measurement.)

The dead time τ can be determined from Eq. 2.109 after g_1, g_2, g_{12}, and b are measured. This is achieved by counting radioactive source 1, then sources 1 and 2 together, then only source 2, and finally the background after removing both sources. Equation 2.109 can be rearranged to give:

$$(g_1 g_2 g_{12} + g_1 g_2 b - g_1 g_{12} b - g_2 g_{12} b)\tau^2 - 2(g_1 g_2 - g_{12} b)\tau$$
$$+ g_1 + g_2 - g_{12} - b = 0 \tag{2.110}$$

Equation 2.110 is a second-degree algebraic equation that can be solved for τ. It was derived without any approximations. One can show that it becomes simpler for the following two cases:

1 If the background is negligible, Eq. 2.110 takes the form

$$g_1 g_2 g_{12} \tau^2 - 2 g_1 g_2 \tau + g_1 + g_2 - g_{12} = 0 \tag{2.111}$$

2 If the counting rates are such that

$$\frac{1}{1 - g_i\tau} \approx 1 + g_i\tau \qquad i = 1, 2, 12$$

then the value of τ is (from Eq. 2.109)

$$\tau = \frac{g_1 + g_2 - g_{12} - b}{g_{12}^2 - g_1^2 - g_2^2} \tag{2.112}$$

When dead-time correction is necessary, the net counting rate, called "true net counting rate," is given by

$$r = n - b = \frac{G/t_G}{1 - (G\tau/t_G)} - b \tag{2.113}$$

It is assumed that the true background rate has been determined earlier with the standard error σ_b. The standard error of r, σ_r, is calculated from Eq. 2.113 using Eq. 2.84. If the only sources of error are the gross count G and the background, the standard error of r is

$$\sigma_r = \sqrt{\left(\frac{1}{1 - g\tau}\right)^4 \frac{G}{t_G^2} + \sigma_b^2} \tag{2.114}$$

If there is an error due to dead-time determination, a third term consisting of that error will appear under the radical of Eq. 2.114.

PROBLEMS

2.1 What is the probability when throwing a die three times of getting a four in any of the throws?

2.2 What is the probability when drawing one card from each of three decks of cards that all three cards will be diamonds?

2.3 A box contains 2000 computer cards. If five faulty cards are expected to be found in the box, what is the probability of finding two faulty cards in a sample of 250?

2.4 Calculate the average and the standard deviation of the probability density function $f(x) = 1/(b - a)$ when $a \leqslant x \leqslant b$. (This pdf is used for the calculation to round off errors.)

2.5 The energy distribution of thermal (slow) neutrons in a light-water reactor follows very closely the Maxwell-Boltzmann distribution:

$$N(E)\,dE = A\sqrt{E}\,e^{-E/kT}\,dE$$

where $N(E)\,dE$ = number of neutrons with kinetic energy between E and $E + dE$
k = Boltzmann constant = 1.38054×10^{-16} erg·k^{-1}
T = temperature, K
A = constant

Show that:

(a) The mode of this distribution is $E = \frac{1}{2}kT$.
(b) The mean is $\bar{E} = \frac{3}{2}kT$.

2.6 If the average for a large number of counting measurements is 15, what is the probability that a single measurement will produce the result 20?

2.7 For the binomial distribution prove:

$$\text{(a)} \quad \sum_{n=0}^{N} P_{(N)}^{(n)} = 1 \qquad \text{(b)} \quad \bar{n} = pN \qquad \text{(c)} \quad \sigma^2 = m(1-p)$$

2.8 For the Poisson distribution, prove:

$$\text{(a)} \quad \sum_{n=0}^{\infty} P_n = 1 \qquad \text{(b)} \quad \bar{n} = m \qquad \text{(c)} \quad \sigma^2 = m$$

2.9 For the normal distribution, show:

$$\text{(a)} \quad \int_{-\infty}^{\infty} P(x)\,dx = 1 \qquad \text{(b)} \quad \bar{x} = m \qquad \text{(c) the variance is } \sigma^2$$

2.10 If $n_1, n_2, \ldots, n_N$ are mutually uncorrelated random variables with a common variance σ^2, show that

$$\overline{(n_i - \bar{n})^2} = \frac{N-1}{N}\,\sigma^2$$

2.11 Show that in a series of N measurements, the result R that minimizes the quantity

$$Q = \sum_{i=1}^{N} (R - n_i)^2$$

is $R = \bar{n}$, where $\bar{n}$ is given by Eq. 2.31.

2.12 Prove Eq. 2.62 using tables of the error function.

2.13 As part of a quality control experiment, the lengths of 10 nuclear fuel rods have been measured with the following results in meters:

2.60	2.62	2.65	2.58	2.61
2.62	2.59	2.59	2.60	2.63

What is the average length? What is the standard deviation of this series of measurements?

2.14 At a uranium pellet fabrication plant the average pellet density is 17×10^3 kg/m^3 with a standard deviation equal to 10^3 kg/m^3. What is the probability that a given pellet has a density less than 14×10^3 kg/m^3?

2.15 A radioactive sample was counted once and gave 500 counts in 1 min. The corresponding number for the background is 480 counts. Is the sample radioactive or not? What should one report based on this measurement alone?

2.16 A radioactive sample gave 750 counts in 5 min. When the sample was removed, the scaler recorded 1000 counts in 10 min. What is the net counting rate and its standard percent error?

2.17 Calculate the average net counting rate and its standard error from the data given below:

G	t_G(min)	B	t_B(min)
355	5	120	10
385	5	130	10
365	5	132	10

2.18 A counting experiment has to be performed in 5 min. The approximate gross and background counting rates are 200 counts/min and 50 counts/min, respectively.

(a) Determine the optimum gross and background counting times.

(b) Based on the times obtained in (a), what is the standard percent error of the net counting rate?

2.19 The strength of a radioactive source was measured with a 2 percent standard error by taking a gross count for time t min and a background for time $2t$ min. Calculate the time t if it is given that the background is 300 counts/min and the gross count 45,000 counts/min.

2.20 The strength of radioactive source is to be measured with a counter that has a background of 120 ± 8 counts/min. The approximate gross counting rate is 360 counts/min. How long should one count if the net counting rate is to be measured with an error of 2 percent?

2.21 The buckling B^2 of a cylindrical reactor is given by

$$B^2 = \left(\frac{2.405}{R}\right)^2 + \left(\frac{\pi}{H}\right)^2$$

where R = reactor radius

H = reactor height

If the radius increases by 2 percent and height by 8 percent–due to thermal expansion–by what percent will B^2 change? Take $R = 1$ m, $H = 2$ m.

2.22 Using Chauvenet's criterion, should any of the scaler readings listed below be rejected?

115	121	103	151
121	105	75	103
105	107	100	108
113	110	101	97
110	109	103	101

2.23 Using the data of Prob. 2.13, what is the value of accepted length x_a if the confidence limit is 99.4 percent?

2.24 Determine the dead time of a counter based on the following data obtained with the two-source method:

$$g_1 = 14{,}000 \text{ counts/min} \qquad g_{12} = 26{,}000 \text{ counts/min}$$

$$g_2 = 15{,}000 \text{ counts/min} \qquad b = 50 \text{ counts/min}$$

2.25 If the dead time of a counter is 100 μs, what is the observed counting rate if the loss of counts due to dead time is equal to 5 percent?

2.26 Calculate the true net activity and its standard percent error for a sample which gave 70,000 counts in 2 min. The dead time of the counter is 200 μs. The background is known to be 100 ± 1 counts/min.

2.27 Calculate the true net activity and its standard error based on the following data:

$$G = 100{,}000 \text{ counts} \qquad \text{obtained in 10 min}$$

$$B = 10{,}000 \text{ counts} \qquad \text{obtained in 100 min}$$

The dead time of the counter is 150 μs.

BIBLIOGRAPHY

Arley, N., and Buck, K. R., *Introduction to the Theory of Probability and Statistics*, Wiley, New York, 1950.

Beers, Y., *Introduction to the Theory of Error*, Addison-Wesley, Reading, Mass., 1957.

Bevington, P. R., *Data Reduction and Error Analysis for the Physical Sciences*, McGraw-Hill, New York, 1969.

Eadie, W. T., Dryard, D., James, F. E., Roos, M., and Sadoulet, B., *Statistical Methods in Experimental Physics*, North-Holland, Amsterdam, 1971.

Handbook of Chemistry and Physics, Chemical Rubber Company, Cleveland, Ohio.

Jaech, J. L., "Statistical Methods in Nuclear Material Control," TID-26298, U.S. Atomic Energy Commission, 1973.

Johnson, N. L., and Leone, F. C., *Statistics and Experimental Design*, Wiley, New York, 1964, Chaps. 1, 3, 4, and 5.

Standard Mathematical Tables, Chemical Rubber Company, Cleveland, Ohio.

3

■ review of atomic and nuclear physics

3.1 INTRODUCTION

This chapter reviews the concepts of atomic and nuclear physics relevant to radiation measurements. It should not be considered a comprehensive discussion of any of the subjects presented. For in-depth study, the reader should consult the references listed at the end of the chapter. If a person has studied and understood this material, this chapter could be skipped without loss of continuity.

This review is not presented from the historical point of view. Atomic and nuclear behavior and the theory and experiments backing it are discussed as we understand them today. Emphasis is given to the fact that the current "picture" of atoms, nuclei, and subatomic particles is only a model that represents our best current theoretical and experimental evidence. This model may change in the future if new evidence is obtained pointing to discrepancies between theory and experiment.

3.2 ELEMENTS OF RELATIVISTIC KINEMATICS

The special theory of relativity developed by Einstein in 1905 is based on two simple postulates.

FIRST POSTULATE **The laws of nature and the results of all experiments performed in a given frame of reference (system of coordinates) are independent of the translational motion of the system as a whole.**

SECOND POSTULATE **The speed of light in vacuum is independent of the motion of its source.**

These two postulates, simple as they are, predict consequences that were unthinkable at that time. The most famous predictions of the special theory of relativity are:

1 The mass of a body changes when its speed changes.
2 Mass and energy are equivalent ($E = mc^2$).

Einstein's predictions were verified by experiment a few years later, and they are still believed to be correct today (1981).

The main results of the special theory of relativity will be presented here without proof, using the following notations:

M = rest mass of a particle (or body)

M^* = mass of a particle in motion

v = speed of the particle

c = speed of light in vacuum = 3×10^8 m/s

T = kinetic energy of the particle

E = total energy of the particle

According to the theory of relativity, the mass of a moving particle (or body) changes with its speed according to the equation

$$M^* = \frac{M}{\sqrt{1-(v/c)^2}} = \frac{M}{\sqrt{1-\beta^2}} \tag{3.1}$$

or $$M^* = \gamma M \tag{3.2}$$

where

$$\beta = \frac{v}{c} \tag{3.3}$$

and

$$\gamma = \frac{1}{\sqrt{1-\beta^2}} \tag{3.4}$$

Equation 3.1 shows that

1 As the speed of a moving particle increases, its mass also increases, thus making additional increase of its speed more and more difficult.
2 It is impossible for any mass to reach a speed equal to or greater than the speed of light in vacuum.†

The total energy of a particle of mass M^* is

$$E = M^*c^2 \tag{3.5}$$

Equation 3.5 expresses the very important concept of equivalence of mass and energy. Since the total energy E consists of the rest mass energy plus the kinetic energy, Eq. 3.5 may be rewritten as

$$E = M^*c^2 = T + Mc^2 \tag{3.6}$$

Combining Eqs. 3.2 and 3.6, one obtains the relativistic equation for the kinetic energy

†The speed of light in a medium with index of reflection n is c/n; thus, it is possible for particles to move faster than with c/n in certain media (see Cerencov radiation, Evans).

$$T = (\gamma - 1)Mc^2 \tag{3.7}$$

The quantity γ, which is defined by Eq. 3.4 ($\gamma = M^*c^2/Mc^2$), indicates how many times the mass of the particle has increased, relative to its rest mass, because of its motion. For large moving masses, the relativistic mass increase is too small to measure. Thus, without the availability of subatomic particles such as electrons and protons, it would be extremely difficult to verify this part of Einstein's theory.

The equation that relates the linear momentum and the total energy of a particle is

$$E^2 = (Mc^2) + (pc)^2 \tag{3.8}$$

where

$$\mathbf{p} = M^*\mathbf{v} = \gamma M\mathbf{v} \tag{3.9}$$

is the linear momentum. Combining Eqs. 3.6 and 3.8, one obtains

$$T = \sqrt{(Mc^2)^2 + (pc)^2} - Mc^2 \tag{3.10}$$

or $$p = \frac{1}{c}\sqrt{T^2 + 2TMc^2} \tag{3.11}$$

Equation 3.10 is used for the determination of the kinetic energy if the momentum is known, while Eq. 3.11 gives the momentum if the kinetic energy is known.

For small values of β (Eq. 3.3)–that is, for small speeds–the equations of relativity reduce to the equations of Newtonian (classical) mechanics. In classical mechanics, the mass is constant, and T and p are given by

$$T = \tfrac{1}{2}Mv^2 \tag{3.12}$$

$$\mathbf{p} = M\mathbf{v} \tag{3.13}$$

If the kinetic energy of a particle is a considerable fraction of its rest mass energy, Eqs. 3.7 and 3.9 should be used for the determination of T and $\mathbf{p}$. Then the particle is *relativistic.* If, on the other hand, $\beta \ll 1$, the particle is *nonrelativistic*, and Eqs. 3.12 and 3.13 may be used.

EXAMPLE 3.1

What is the mass increase of a bullet weighing 0.010 kg and traveling at twice the speed of sound?

ANSWER

The speed of the bullet is $v \approx 700$ m/s. Using Eq. 3.2 and 3.4,

$$\frac{M^*}{M} = \gamma = \frac{1}{\sqrt{1-\beta^2}} \approx 1 + \frac{1}{2}\beta^2 = 1 + \frac{1}{2}\left(\frac{700}{3 \times 10^8}\right)^2 = 1 + 2.72 \times 10^{-12}$$

The mass increase is

$$M^* - M = 2.72 \times 10^{-12} M = 2.72 \times 10^{-14} \text{ kg}$$

which is almost impossible to detect.

■ *EXAMPLE 3.2*

An electron has a kinetic energy of 200 keV. (a) What is its speed? (b) What is its new mass relative to its rest mass?

ANSWER

The rest mass energy of the electron is 511 keV. Since $T/mc^2 = 200/511 = 0.391$, relativistic equations should be used. (a) The speed of the electron is obtained with the help of Eqs. 3.7 and 3.4. Equation 3.7 gives

$$\gamma = 1 + \frac{T}{mc^2} = 1 + 0.391$$

and from Eq. 3.4 one obtains

$$\beta = \sqrt{1 - \frac{1}{\gamma^2}} = 0.695$$

Therefore

$$v = \beta c = (0.695)(3 \times 10^8 \text{ m/s}) = 2.085 \times 10^8 \text{ m/s}$$

(b) The new mass relative to the rest mass has already been determined, because

$$\gamma = \frac{m^* c^2}{mc^2} = 1.391$$

i.e., the mass of this electron increased 39.1 percent.

It is instructive to calculate the speed of this electron using the classical method of Eq. 3.12 to see the difference:

$$v = \sqrt{\frac{2T}{m}} = \sqrt{\frac{2Tc^2}{mc^2}} = \left(\sqrt{\frac{2T}{mc^2}}\right) c = \left(\sqrt{\frac{2(200)}{511}}\right) c = 0.885c$$

Thus, the classical equation determines the speed with an error

$$\frac{v_{\text{cl}} - v_{\text{rel}}}{v_{\text{rel}}} = \frac{0.885c - 0.695c}{0.695c} = 27\%$$

■ *EXAMPLE 3.3*

What is the kinetic energy of a neutron with speed 6×10^7 m/s? What is its mass increase?

ANSWER

For this particle,

$$\beta = \frac{v}{c} = \frac{6 \times 10^7}{3 \times 10^8} = 0.2$$

Using Eqs. 3.4 and 3.7,

$$T = (\gamma - 1)Mc^2 = \left(\frac{1}{\sqrt{1-\beta^2}}\right) Mc^2 = \left(\frac{1}{\sqrt{1-0.2^2}} - 1\right) Mc^2$$
$$= (1.021 - 1)Mc^2 = 0.021Mc^2 = (0.021)939.55 \text{ MeV} = 19.73 \text{ MeV}$$
$$= 3.16 \times 10^{-12} \text{ J}$$

The mass increase is $M^*/M = \gamma = 1.021$, i.e., a 2.1 percent mass increase.

3.3 ATOMS

To the best of our knowledge today, every atom consists of a central positively charged nucleus around which negative electrons revolve moving in stable orbits. Considered as a sphere, the atom has a radius of the order of 10^{-10} m and the nucleus has a radius of the order of 10^{-14} m. The number of electrons is equal to the number of positive charges of the nucleus; thus the atom is electrically neutral (in its normal state).

The number of positive elementary charges in the nucleus is called the *atomic number* and is indicated by Z. The atomic number identifies the chemical element. All atoms of an element have the same chemical properties.

The atomic electrons move around the nucleus as a result of the attractive electrostatic Coulomb force between the positive nucleus and the negative charge of the electron. According to classical electrodynamics, the revolving electrons ought to continuously radiate part of their energy, follow a spiral orbit, and eventually be captured by the nucleus. Obviously, this does not happen: atoms exist and are stable.

The available experimental evidence points toward the following facts regarding the motion of atomic electrons:

1 Bound atomic electrons revolve around the nucleus in stable orbits without radiating energy. Every orbit corresponds to a certain electron energy and is called an *energy state.*

2 Only certain orbits (only certain energies) are allowed. That is, the energy states of the bound electrons form a discrete spectrum, as shown in Fig. 3.1. This phenomenon is called *quantization.*

3 If an electron moves from an orbit (state) of energy E_i to another of energy E_f, then (and only then) electromagnetic radiation, an x-ray, is emitted with frequency ν such that

$$\nu = \frac{E_i - E_f}{h} \tag{3.14}$$

where h is Planck's constant.

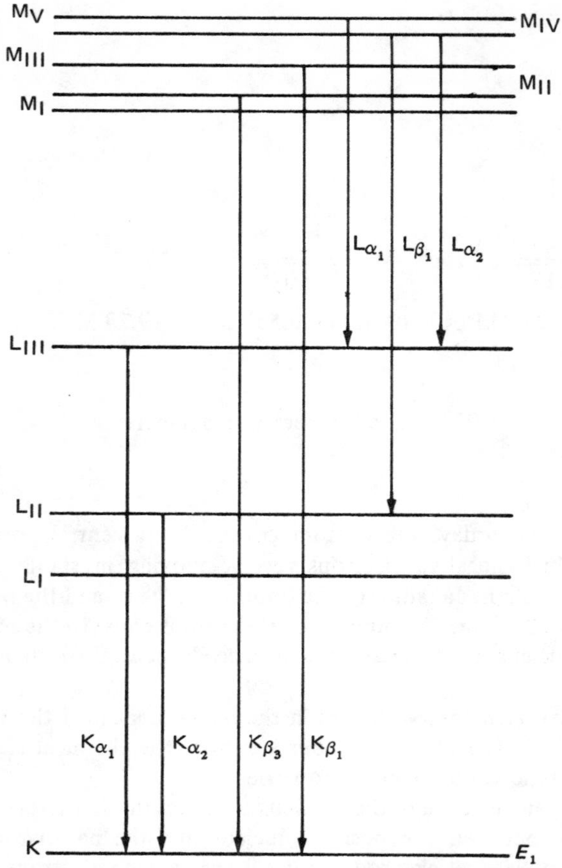

FIG. 3.1 An atomic energy level diagram showing x-ray nomenclature (not drawn to scale). E_1 = lowest energy state = ground state.

The energy of the x-ray depends on the atomic number:

$$E_x = h\nu = E_i - E_f = k(Z - a)^2 \tag{3.15}$$

where k and a are constants. X-ray energies range from a few eV for the light elements to a few hundreds of keV for the heaviest elements.

Every atom emits characteristic x-rays with discrete energies that identify the atom like fingerprints. For every atom, the x-rays are identified according to the final state of the electron transition that produced them. Historically, the energy states of atomic electrons are characterized by the letters K, L, M, N, etc. The K state or K orbit or K shell is the lowest energy state, also called the ground state. The x-rays that are emitted as a result of electronic transitions to the K state, from any other initial state, are called K x-rays (Fig. 3.1). Transitions to the L state give rise to L x-rays and so on. K_α and K_β x-rays indicate transitions from L to K and M to K states, respectively.

A bound atomic electron may receive energy and move from a state of energy E_1 to another of higher energy E_2. This phenomenon is called excitation of the atom (Fig. 3.2). An excited atom moves preferentially to the lowest possible energy state. In times of the order of 10^{-8} s, the electron that jumped to E_2 or another from another state will fall to E_1 and an x-ray will be emitted.

An atomic electron may receive enough energy to leave the atom and become a free particle. This phenomenon is called *ionization*, and the positive entity left behind is called an ion. The energy necessary to cause ionization is the *ionization potential.* The ionization potential is not the same for all the electrons of the same atom because the electrons move at different distances from the nucleus. The closer the electron is to the nucleus, the more tightly bound it is and the greater its ionization potential becomes. Table 3.1 lists ionization potentials of the least bound electron for certain elements.

When two or more atoms join together and form a molecule, their common electrons are bound to the molecule. The energy spectrum of the molecule is also discrete, but more complicated than that shown in Fig. 3.1.

3.4 NUCLEI

At the present time, all experimental evidence indicates that nuclei consist of neutrons and protons, which are particles known as *nucleons.* Nuclei then consist of nucleons. Some of the properties of a neutron, a proton, and an electron, for comparison, are listed in Table 3.2. A free proton—outside the nucleus—will eventually pick up an electron and become a hydrogen atom, or it may be absorbed by a nucleus. A free neutron either will be absorbed by a nucleus or will decay according to the equation

$$n \rightarrow p^{+} + e^{-} + \bar{\nu}$$

i.e., it will be transformed into a proton by emitting an electron and another particle called an *antineutrino.*

A nucleus consists of A particles,

$$A = N + Z$$

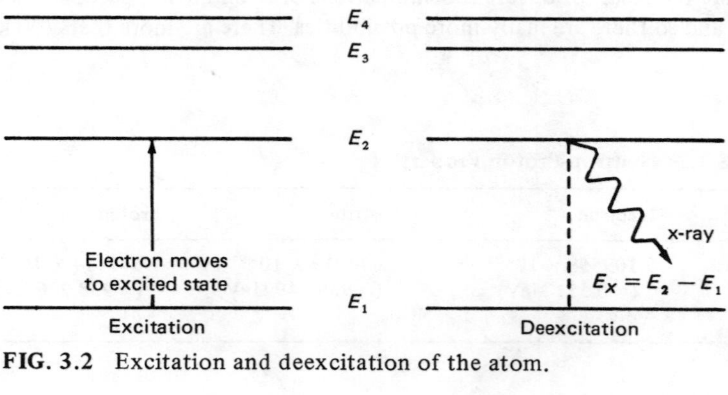

FIG. 3.2 Excitation and deexcitation of the atom.

TABLE 3.1 Ionization Potential for the Least Bound Electron of Certain Elements

Element	Ionization potential (eV)	Element	Ionization potential (eV)
H	13.6	Ne	21.56
He	24.56	Na	5.14
Li	5.4	A	15.76
Be	9.32	Fe	7.63
B	8.28	Pb	7.42
C	11.27	U	4.0

where A = mass number
N = number of neutrons
Z = number of protons = atomic number of the element
A nuclear species X is indicated as

$$^{A}_{Z}\mathrm{X}$$

where X = chemical symbol of the element. For example,

$$^{16}_{8}\mathrm{O}$$

Isobars are nuclides that have the same A.

Isotopes are nuclides that have the same Z. They are nuclei of the same chemical element. They have the same chemical but slightly different physical properties, due to their difference in mass. The nuclear properties change drastically from isotope to isotope.

Isotones are nuclides that have the same N, i.e., same number of neutrons.

Isomers are two different energy states of the same nucleus.

The different atomic species are the result of different combinations of one type of particle–the electron. There are 92 natural elements. Since 1940, 15 more have been artificially produced for a total of 107 elements. The different nuclides, on the other hand, are the result of different combinations of two kinds of particles, neutrons and protons, and so there are many more possibilities. There are more than 700 known nuclides.

TABLE 3.2 Neutron-Proton Properties

	Electron	Neutron	Proton
Rest mass	9.109558×10^{-31} kg (= 0.511 MeV)	1.674919×10^{-27} kg (= 939.549 MeV)	1.672614×10^{-27} kg (= 938.256 MeV)
Charge	$-e$	0	$+e$

Experiments have determined that nuclei are almost spherical, with a volume proportional to the mass number A and a radius approximately equal to†

$$R = 1.3 \times 10^{-15} A^{1/3} \text{ in meters} \tag{3.16}$$

The mass of the nucleus with mass number A and atomic number Z, indicated as $M_N(A, Z)$, is equal to

$$M_N(A, Z) = ZM_p + NM_n - B(A, Z)c^2 \tag{3.17}$$

where M_p = mass of the proton
M_n = mass of the neutron
$B(A, Z)$ = binding energy

The binding energy is equal to the energy that was released when the N neutrons and Z protons formed the nucleus. More details about the binding energy are given in the next section.

The unit used for the measurement of nuclear mass is equal to $\frac{1}{12}$ of the mass of the isotope ${}^{13}_{6}C$. Its symbol is u (formerly amu for atomic mass unit):

$$1 \text{ u} = \tfrac{1}{12} (\text{mass of } {}^{12}_{6}C) = 1.660531 \times 10^{-27} \text{ kg} = 931.478 \text{ MeV}$$

In many experiments, what is normally measured is the atomic, not the nuclear, mass. To obtain the atomic mass, one adds the mass of all the atomic electrons (see next section). A table of atomic masses of many isotopes is given in App. B. The mass may be given in any of the following three ways:

1 Units of u
2 Kilograms
3 Energy units (MeV or J, in view of the equivalence of mass and energy)

3.5 NUCLEAR BINDING ENERGY

The mass of a nucleus is given by Eq. 3.18 in terms of the masses of its constituents. That same equation also defines the binding energy of the nucleus

$$B(A, Z) = [ZM_p + NM_n - M_n(A, Z)]c^2 \tag{3.18}$$

The factor c^2, which multiplies the mass to transform it into energy, will be omitted from now on. It will always be implied that multiplication or division by c^2 is necessary to obtain energy from mass or vice versa. Thus, Eq. 3.18 is rewritten as

$$B(A, Z) = ZM_p + NM_n - M_N(A, Z) \tag{3.19}$$

†For nonspherical nuclei, the radius given by Eq. 3.16 is an average.

The meaning of $B(A, Z)$ may be expressed in two equivalent ways:

1 The binding energy $B(A, Z)$ of a nucleus is equal to the mass transformed into energy when the Z protons and the $N = A - Z$ neutrons got together and formed the nucleus. An amount of energy equal to the binding energy was released when the nucleus was formed. Or,
2 The binding energy $B(A, Z)$ is equal to the energy necessary to break the nucleus apart into its constituents, Z free protons and N free neutrons.

As mentioned in Sec. 3.4, atomic masses rather than nuclear masses are measured in most cases. For this reason, Eq. 3.19 will be expressed in terms of atomic masses by adding the appropriate masses of atomic electrons. If one adds and subtracts Zm in Eq. 3.19,

$$\begin{aligned} B(A, Z) &= ZM_p + Zm + NM_n - M_N(A, Z) - Zm \\ &= Z(M_p + m) + NM_n - [M_N(A, Z) + Zm] \end{aligned} \tag{3.20}$$

Let M_H = mass of the hydrogen atom

B_e = binding energy of the electron in the hydrogen atom

$B_e(A, Z)$ = binding energy of all the electrons of the atom whose nucleus has mass $M_N(A, Z)$

$M(A, Z)$ = mass of the atom with nuclear mass equal to $M_N(A, Z)$

Then

$$M_H = M_p + m - B_e \tag{3.21}$$

$$M(A, Z) = M_N(A, Z) + Zm - B_e(A, Z) \tag{3.22}$$

Combining Eqs. 3.20, 3.21, and 3.22, one obtains

$$B(A, Z) = ZM_H + NM_n - M(A, Z) + B_e(A, Z) - ZB_e \tag{3.23}$$

Unless extremely accurate calculations are involved, the last two terms of Eq. 3.23 are neglected. The error introduced by doing so is insignificant because ZB_e and $B_e(A, Z)$ are less than a few keV and they tend to cancel each other, while $B(A, Z)$ is of the order of MeV. Equation 3.23 is, therefore, usually written as

$$B(A, Z) = ZM_H + NM_n - M(A, Z) \tag{3.24}$$

EXAMPLE 3.4

What is the total binding energy of ${}^4_2\text{He}$?

ANSWER

Using Eq. 3.24 and data from App. B,

$$\begin{aligned} B(4, 2) &= 2M_H - 2M_n - M(4, 2) \\ &= [2(1.00782522) + 2(1.00866544) - 4.00260361]\ \text{u} \\ &= 0.03037771\ \text{u} = (0.0303771\ \text{u})931.478\ \text{MeV/u} \\ &= 28.296\ \text{MeV} = 4.53 \times 10^{-12}\ \text{J} \end{aligned}$$

EXAMPLE 3.5
What is the binding energy of the isotope $^{238}_{92}U$?

ANSWER

$$B(238, 92) = [92(1.00782522) + 146(1.00866544) - 238.05076] \text{ u}$$
$$= 1.93431448 \text{ u} = 1.93431448 \text{ u})931.478 \text{ MeV/u}$$
$$= 1801.771 \text{ MeV} = 2.886 \times 10^{-10} \text{ J}$$

The energy necessary to remove one particle from the nucleus is the *separation* or *binding energy* of that particle for that particular nuclide. A "particle" may be a neutron, a proton, an alpha particle, a deuteron, etc. The separation or binding energy of a nuclear particle is analogous to the ionization potential of an electron. If a particle enters the nucleus, an amount of energy equal to its separation energy is released.

The separation or binding energy of a neutron (B_n) is defined by the equation

$$B_n = M[(A-1), Z] + M_n - M(A, Z) \tag{3.25}$$

Using Eq. 3.24, Eq. 3.25 is written

$$B_n = B(A, Z) - B[(A-1), Z] \tag{3.26}$$

which shows that the binding energy of the last neutron is equal to the difference between the binding energies of the two nuclei involved. Typical values of B_n are a few MeV (less than 10 MeV).

The separation or binding energy of a proton is

$$B_p = M(A-1, Z-1) + m_p - M(A, Z) \tag{3.27}$$

or, using Eq. 3.24,

$$B_p = B(A, Z) - B(A-1, Z-1) \tag{3.28}$$

The separation energy for an alpha particle is

$$B_\alpha = M(A-4, Z-2) + M_{\text{He}} - M(A, Z) \tag{3.29}$$

or, using Eq. 3.24,

$$B_\alpha = B(A, Z) - B(A-4, Z-2) - B(4, 2) \tag{3.30}$$

EXAMPLE 3.6
What is the separation energy of the last neutron of ^4_2He?

ANSWER

Using data from App. B and Eq. 3.25, one obtains

$$B_n = M(3,2) + M_n - M(4,2)$$
$$= [(3.016030 + 1.008665 - 4.002604)\text{ u}]931.478\text{ MeV/u}$$
$$= 0.022091(931.478\text{ MeV}) = 20.58\text{ MeV} = 3.3 \times 10^{-12}\text{ J}$$

If the average binding energy per nucleon,

$$b(A,Z) = \frac{B(A,Z)}{A} \tag{3.31}$$

is plotted as a function of A, one obtains the result shown in Fig. 3.3. The average binding energy changes relatively little, especially for $A > 30$. Notice that Fig. 3.3 has a different scale for $A < 30$.

Figure 3.3 is very important because it reveals the processes by which energy may be released in nuclear reactions. If one starts with a very heavy nucleus ($A = 240$) and breaks it into two medium-size nuclei (fission), energy will be released because the average binding energy per nucleon is larger for nuclides in the middle of the periodic table than it is for heavy nuclides. On the other hand, if one takes two very small nuclei ($A = 2, 3$) and fuses them into a bigger one, energy is again released due to similar increase in the average binding energy per nucleon.

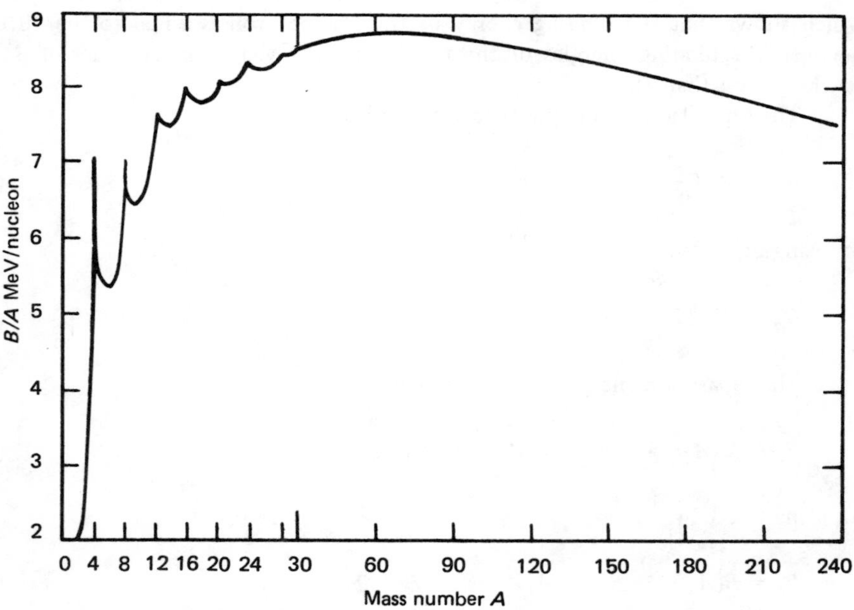

FIG. 3.3 The change of the average binding energy per nucleon with mass number A. Notice the change in scale after $A = 30$. (From *The Atomic Nucleus* by Evans.)

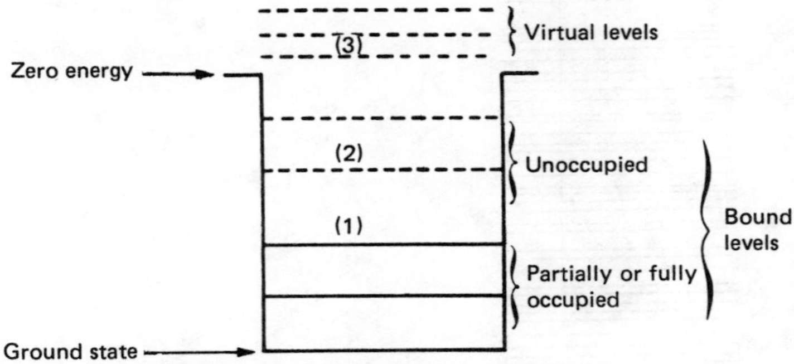

FIG. 3.4 Bound and virtual nuclear energy levels.

3.6 NUCLEAR ENERGY LEVELS

Neutrons and protons are held together in the nucleus by nuclear forces. Although the exact nature of nuclear forces is not known, scientists have successfully predicted many characteristics of nuclear behavior by assuming a certain form for the force and constructing nuclear models based on that form. The success of these models is measured by how well their predicted results agree with the experiment. Many nuclear models have been proposed, each of them explaining certain features of the nucleus; but as of today, no model exists that explains all the facts about all the known nuclides.

All the nuclear models assume that the nucleus, like the atom, can exist only in certain discrete energy states. Depending on the model, the energy states may be assigned to the nucleons–neutrons and protons–or the nucleus as a whole. The present discussion of nuclear energy levels will be based on the second approach.

The lowest possible energy state of a nucleus is called the ground state (Fig. 3.4). In Fig. 3.4, the ground state is shown as having negative energy to indicate a bound state. The ground state and all the excited states below the zero energy level are called bound states. If the nucleus finds itself in any of the bound states, it deexcites after a time of the order of 10^{-12} to 10^{-10} s by dropping to a lower state. Deexcitation is accompanied by the emission of a photon with energy equal to the difference between the energies of the initial and final states. Energy states located above the zero energy level are called virtual energy levels. If the nucleus obtains enough energy to be raised to a virtual level, it may deexcite either by falling to one of the bound levels or by emitting a nucleon.

Studies of the energy levels of all the known nuclides reveal the following:

1 The distance between nuclear energy levels is of the order of keV to MeV. By contrast, the distance between atomic levels is of the order of eV.
2 The distance between levels descreases as the excitation energy decreases (Fig. 3.5). For very high excitation energies, the density of levels becomes so high that it is difficult to distinguish individual energy levels.
3 As the mass number A increases, the number of levels increases; i.e., heavier nuclei have more energy levels than lighter nuclei (in general–there may be exceptions).

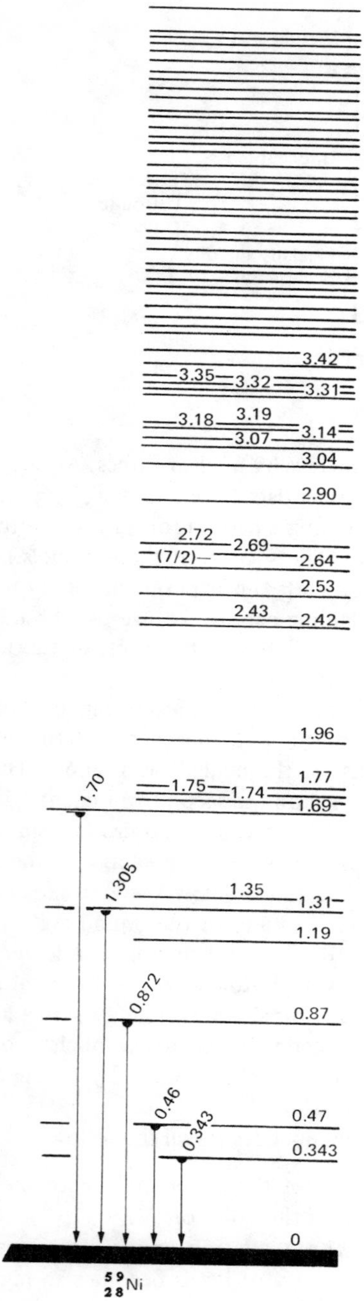

FIG. 3.5 The energy levels of $^{59}_{28}$Ni. In this diagram, the zero of energy has been switched to the ground state of the nucleus. Numbers are energy level energies in MeV. (From *Table of Isotopes.*)

4 As A increases, the energy of the first excited state decreases (again, in general—exceptions exist). For example,

^{9}Be: first excited state is at 1.67 MeV

^{55}Fe: first excited state is at 0.847 MeV

^{238}U: first excited state is at 0.045 MeV

3.7 ENERGETICS OF NUCLEAR DECAYS

This section discusses the energetics of α, β, and γ decay, demonstrating how the kinetic energies of the products of the decay can be calculated from the masses of the particles involved. In all cases, it will be assumed that the original unstable nucleus is at rest—i.e., it has zero kinetic energy and linear momentum. This assumption is very realistic because the actual kinetic energies of nuclei due to thermal motion are of the order of kT (on the order of eV), where k is the Boltzmann constant and T the temperature (Kelvin), while the energy released in most decays is of the order of MeV.

In writing the equation representing the decay, the following notation will be used:

M = atomic mass (or Mc^2 = rest mass energy)

E_γ = energy of a photon

T_i = kinetic energy of a particle type i

P_i = linear momentum of a particle type i

3.7.1 Gamma Decay

In γ decay, a nucleus goes from an excited state to a state of lower energy and the energy difference between the two states is released in the form of a photon. Gamma decay is represented by

$${}^{A}_{Z}\mathrm{X}^* \rightarrow {}^{A}_{Z}\mathrm{X} + \gamma$$

where ${}^{A}_{Z}\mathrm{X}^*$ indicates the excited nucleus.

Applying conservation of energy and momentum for the states before and after the decay, we have†

Conservation of energy:

$$M^*(A,Z) = M(A,Z) + T_M + E_\gamma \tag{3.32}$$

†Equations in this chapter are written in terms of atomic, not nuclear, masses. This notation introduces a slight error because the binding energy of the atomic electrons is not taken into account (see Sec. 3.5).

Conservation of momentum:

$$0 = \mathbf{P}_M + \mathbf{P}_\gamma \tag{3.33}$$

Using these two equations and the nonrelativistic form of the kinetic energy of the nucleus,

$$T_M = \frac{1}{2} MV^2 = \frac{P_M^2}{2M} = \frac{P_\gamma^2}{2M} = \frac{E_\gamma^2}{2Mc^2} \tag{3.34}$$

Use has been made of the relationship $E_\gamma = P_\gamma c$ (the photon rest mass is zero). Equation 3.34 gives the kinetic energy of the nucleus after the emission of a photon of energy E_γ. This energy is called *recoil energy.*

The recoil energy is small. Consider a typical photon of 1 MeV emitted by a nucleus with $A = 50$. Then, from Eq. 3.34,

$$T_M \approx \frac{1^2 \text{ (MeV)}^2}{2(50)(940)(\text{MeV})} \approx 10 \text{ eV}$$

Most of the time, this energy is neglected and the gamma energy is written as

$$E_\gamma = M^*(A, Z) - M(A, Z)$$

However, there are cases where the recoil energy may be important, e.g., in radiation damage studies.

Sometimes the excitation energy of the nucleus is given to an atomic electron instead of being released in the form of a photon. This type of nuclear transition is called internal conversion (IC), and the ejected atomic electron is called an internal conversion electron.

Let T_i be the kinetic energy of an electron ejected from shell i and B_i be the binding energy of an electron in shell i. Equation 3.32 now takes the form

$$M^*(A, Z) = M(A, Z) + T_i + B_i + T_M \tag{3.35}$$

Even though the electron has some nonzero rest mass energy, it is so much lighter than the nucleus that $T_M \ll T_i$. Consequently, T_M is neglected and Eq. 3.35 is written as

$$T_i = M^*(A, Z) - M(A, Z) - B_i \tag{3.36}$$

If $\quad Q = M^*(A, Z) - M(A, Z) =$ energy released during the transition

then

$$T_i = Q - B_i$$

When internal conversion occurs, there is a probability that an electron from the K shell, L shell, or another shell, may be emitted. The corresponding equations for the electron kinetic energies are

$$
\begin{aligned}
T_K &= Q - B_K \\
T_L &= Q - B_L \\
T_M &= Q - B_M \quad \text{etc.}
\end{aligned}
\tag{3.37}
$$

Therefore, a nucleus that undergoes internal conversion is a source of groups of monoenergetic electrons with energies given by Eqs. 3.37. A typical internal conversion electron spectrum is shown in Fig. 3.6. The two peaks correspond to K and L electrons. The diagram on the right (Fig. 3.6) shows the transition energy to be 392 keV. The K-shell energy is then $B_K = 392 - 363 = 29$ keV and the L-shell binding energy is $B_L = 392 - 387 = 5$ keV. Let:

λ_e = probability that internal conversion will occur

λ_γ = probability that a photon will be emitted

λ_i = probability that an electron from shell i will be emitted

λ = total probability for γ decay

Then†

$$\lambda_e = \lambda_K + \lambda_L + \lambda_M = \cdots \tag{3.38}$$

†Tables of isotopes usually give not the values of the different λ's but the so-called IC coefficients, which are the ratios λ_K/λ_γ, λ_L/λ_γ, etc. (see Table of Isotopes).

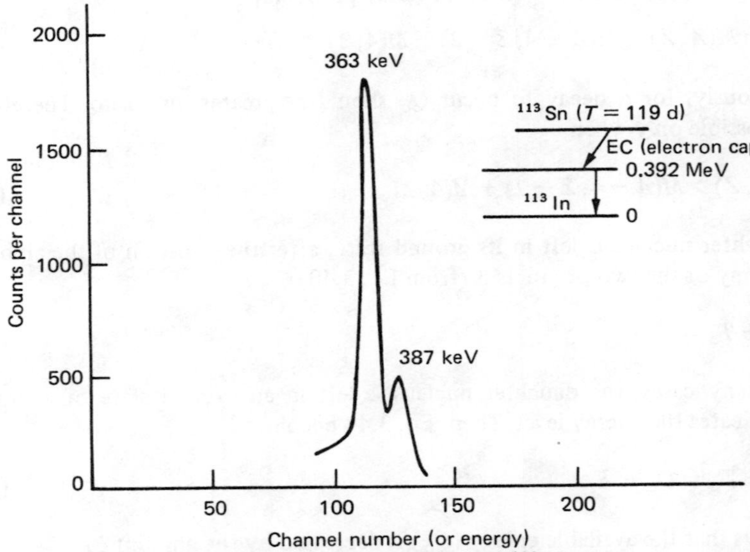

FIG. 3.6 The internal conversion spectrum of ^{113}Sn. The two peaks correspond to K electrons (363 keV) and L electrons (387 keV).

and $\lambda = \lambda_\gamma + \lambda_e$ (3.39)

For most nuclei $\lambda_e = 0$, but there is no γ-decaying nucleus for which $\lambda_\gamma = 0$. This means radioisotopes that internally convert, emit gammas as well as electrons. After an atomic electron is emitted, the empty state that was created will quickly be filled by another electron that "falls in" from the outer shells. As a result of such a transition, an x-ray is emitted. Therefore, internally converting nuclei *emit γ-rays, electrons* and *x-rays.*

Radioisotopes that undergo internal conversion are the only sources of monoenergetic electrons, except for accelerators. They are very useful as instrument calibration sources. Three isotopes frequently used are ^{113}Sn, ^{137}Cs, ^{207}Bi.

3.7.2 Alpha Decay

Alpha decay is represented by the equation

$$^{A}_{Z}\mathrm{X} \rightarrow {}^{A-4}_{Z-2}\mathrm{X} + {}^{4}_{2}\mathrm{He}$$

Applying conservation of energy and momentum,

$$M(A, Z) = M(A-4, Z-2) + M(4, 2) + T_M + T_\alpha \tag{3.40}$$

and

$$0 = \mathbf{P}_\alpha + \mathbf{P}_M \tag{3.41}$$

The energy that becomes available as a result of the emission of the alpha particle is called the *decay energy* Q_α, defined by

$$Q_\alpha = [\text{mass of parent}] - [\text{mass of decay products}]$$

or

$$Q_\alpha = M(A, Z) - M(A-4, Z-2) - M(4, 2) \tag{3.42}$$

Obviously, for α decay to occur Q_α should be greater than zero. Therefore, α decay is possible only when

$$M(A, Z) > M(A-4, Z-2) + M(4, 2) \tag{3.43}$$

If the daughter nucleus is left in its ground state, after the emission of the alpha, the kinetic energy of the two products is (from Eq. 3.40),

$$T_M + T_\alpha = Q_\alpha \tag{3.44}$$

In many cases, the daughter nucleus is left in an excited state of energy E_i, where i indicates the energy level. Then, Eq. 3.44 becomes

$$T_M + T_\alpha = Q_\alpha - E_i \tag{3.45}$$

which shows that the available energy (Q_α) is decreased by the amount E_i.

The kinetic energies T_α and T_M can be calculated from Eqs. 3.41 and 3.44. The result is

$$T_\alpha = \frac{M(A-4,Z-2)}{M(A-4,Z-2)+M(4,2)} Q_\alpha \approx \frac{A-4}{A}(Q_\alpha - E_i) \tag{3.46}$$

$$T_M = \frac{M_\alpha}{M(A-4,Z-2)+M(4,2)}(Q_\alpha - E_i) \approx \frac{4}{A}(Q_\alpha - E_i) \tag{3.47}$$

EXAMPLE 3.7

What are the kinetic energies of the alphas emitted by ^{238}U?

ANSWER

The decay scheme of $^{238}_{92}U$ is shown in Fig. 3.7. After the alpha is emitted, the daughter nucleus, $^{234}_{90}Th$, may be left in one of the two excited states at 0.16 MeV and 0.048 MeV or go to the ground state.

The decay energy Q_α is (Eq. 3.42)

$$\begin{aligned} Q_\alpha &= M(238,92) - M(234,90) - M_\alpha = 238.05076 - 234.04361 - 4.0026036 \\ &= 0.004514 \text{ u} = 0.004514(931.478 \text{ MeV}) = 4.2 \text{ MeV} \end{aligned}$$

Depending on the final state of $^{234}_{90}Th$, the energy of the alpha particle is

$$T_\alpha = \tfrac{234}{238} Q_\alpha = 4.13 \text{ MeV}$$

$$T_\alpha = \tfrac{234}{238}(Q_\alpha - 0.048) = 4.08 \text{ MeV}$$

$$T_\alpha = \tfrac{234}{238}(Q_\alpha - 0.16) = 3.97 \text{ MeV}$$

3.7.3 Beta Decay

In β decay, a nucleus emits an electron or a positron and is transformed into a new element. In addition to the electron or the positron, a neutral particle with rest mass zero is also emitted. There are two types of β decay, β^- and β^+.

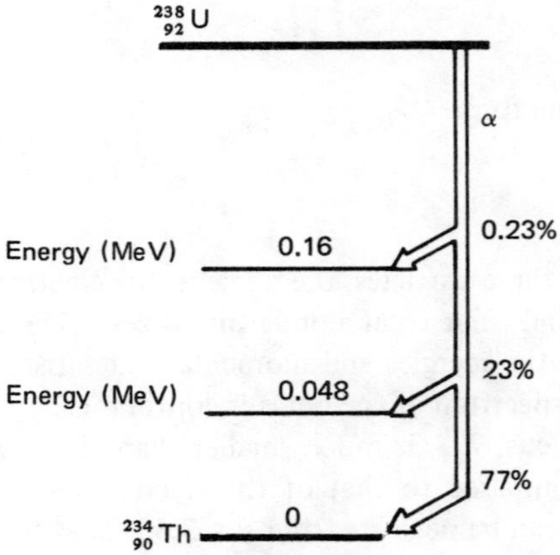

FIG. 3.7 The alpha decay scheme of ^{238}U. The numbers in percent give the probability of decay through the corresponding level. (From *Table of Isotopes.*)

β^- DECAY

This type of decay is represented by

$$^{A}_{Z}X \rightarrow {}^{A}_{Z+1}X + \beta^- + \bar{\nu}$$

where β^- = negative beta particle = electron
$\bar{\nu}$ = antineutrino

Historically, the name β particle has been given to electrons that are emitted by nuclei undergoing beta decay. The antineutrino ($\bar{\nu}$) is a neutral particle with rest mass equal to zero.

The energy equation of β^- decay is

$$M_N(A, Z) = M_N(A, Z + 1) + m + T_{\beta^-} + T_{\bar{\nu}} + T_M \tag{3.48}$$

where $M_N(A, Z)$ is the *nuclear* mass and m is the electron rest mass. Using atomic masses, Eq. 3.48 becomes (see Sec. 3.5)

$$M(A, Z) = M(A, Z + 1) + T_{\beta^-} + T_{\bar{\nu}} + T_M \tag{3.49}$$

The momentum equation is

$$0 = \mathbf{P}_M + \mathbf{P}_{\beta^-} + \mathbf{P}_{\bar{\nu}} \tag{3.50}$$

The β^- decay energy, Q_{β^-}, is defined as

$$Q_{\beta^-} = M(A, Z) - M(A, Z + 1) \tag{3.51}$$

The condition for β^- decay to be possible is

$$M(A, Z) - M(A, Z + 1) > 0 \tag{3.52}$$

In terms of Q_{β^-}, Eq. 3.49 is rewritten in the form

$$T_{\beta^-} + T_{\bar{\nu}} + T_M = Q_\beta \tag{3.53}$$

Equations 3.50 and 3.53 show that three particles, the nucleus, the electron, and the antineutrino, share the energy Q_{β^-}, and their total momentum is zero. There is an infinite number of combinations of kinetic energies and momenta that satisfy these two equations and as a result, the energy spectrum of the betas is continuous.

In Eq. 3.53, the energy of the nucleus, T_M, is much smaller than either T_{β^-} or $T_{\bar{\nu}}$, because the nuclear mass is huge compared to that of the electron or the antineutrino. For all practical purposes, T_M can be neglected and Eq. 3.53 takes the form

$$T_{\beta^-} + T_{\bar{\nu}} = Q_{\beta^-} \tag{3.54}$$

As in the case of α decay, the daughter nucleus may be left in an excited state after the emission of the β^- particle. Then, the energy available to become kinetic energy of the emitted particles is less. If the nucleus is left in the ith excited state E_i, Eq. 3.54 takes the form

$$T_{\beta^-} + T_{\bar{\nu}} = Q_{\beta^-} - E_i = E_{\max} \tag{3.55}$$

According to Eq. 3.54, the electron and the antineutrino share the energy Q_{β^-} (or $E_{\max}$) and there is a certain probability that either particle may have an energy within the limits

$$0 \leqslant T_{\beta^-} \leqslant E_{\max} \tag{3.56}$$

$$E_{\max} \geqslant T_{\bar{\nu}} \geqslant 0 \tag{3.57}$$

Therefore, according to Eqs. 3.56, the beta particles have a continuous energy spectrum. Let $\beta(T)\,dT$ be the number of beta particles with kinetic energy between T and $T + dT$. The function $\beta(T)$ has the general shape shown in Fig. 3.8. The energy spectrum of the antineutrinos is the complement of that shown in Fig. 3.8, consistent with Eq. 3.57. The continuous energy spectrum of β^- particles should be contrasted with the energy spectrum of internal conversion electrons shown by Fig. 3.6.

As stated earlier, beta particles are electrons. The practical difference between the terms *electrons* and *betas* is this: A beam of electrons of energy T consists of electrons each of which has the kinetic energy T. A beam of beta particles with energy $E_{\max}$ consists of electrons that have a continuous energy spectrum (Fig. 3.8) ranging from zero up to a maximum kinetic energy $E_{\max}$.

Figure 3.9 shows the β^- decay scheme of the isotope $^{137}_{55}$Cs. An example of a Q_{β^-} calculation, consider the decay of $^{137}_{55}$Cs.

$$Q_{\beta^-} = M(55, 137) - M(56, 137) = (136.90682 - 136.90556 \text{ u} = 0.00126 \text{ u}$$

$$= 0.0012625(931.478 \text{ MeV}) = 1.1760 \text{ MeV} = 1.36 \times 10^{-13} \text{ J}$$

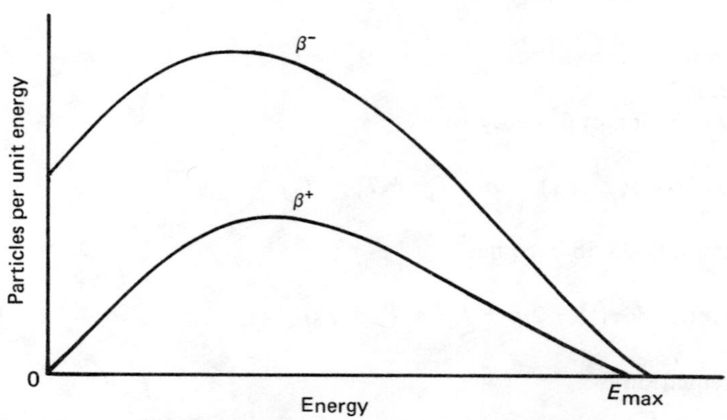

FIG. 3.8 A typical beta energy spectrum (shows shape only; does not mean that β^- is more intense than β^+).

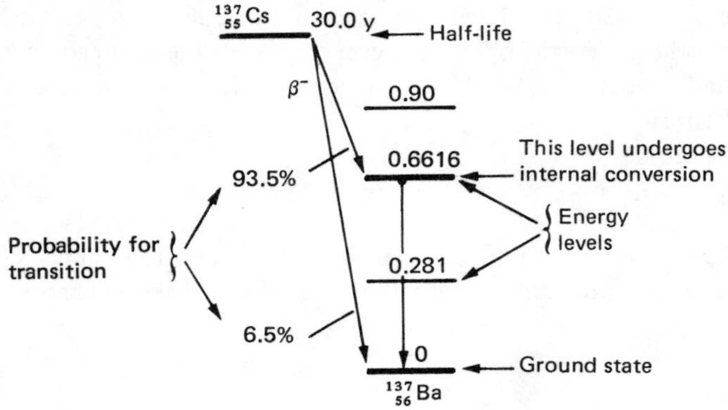

FIG. 3.9 The decay scheme of ^{137}Cs. The Q value of the β^- decay is -1.176 MeV (from *Table of Isotopes*). Probability for each transition is given in percent.

If the $^{137}_{56}$Ba is left in the 0.6616-MeV state (which happens 93.5 percent of the time), the available energy is

$$E_{\max} = 1.1760 - 0.6616 = 0.5144 \text{ MeV}$$

For many calculations it is necessary to use the average energy of the beta particles, $\bar{E}_{\beta^-}$. An accurate equation for $\bar{E}_{\beta^-}$ has been developed,[1] but in practice the average energy is taken to be

$$\bar{E}_{\beta^-} = \frac{E_{\max}}{3}$$

β^+ DECAY

The expression representing β^+ decay is

$$^{A}_{Z}\text{X} \rightarrow {}_{Z-1}^{\;\;A}\text{X} + \beta^+ + \nu$$

where β^+ = positron
ν = neutrino

The energy equation of β^+ decay is

$$M_N(A, Z) = M_N(A, Z-1) + m + T_{\beta^+} + T_\nu + T_M \tag{3.58}$$

Using atomic masses, Eq. 3.58 becomes

$$M(A, Z) = M(A, Z-1) + 2m + T_{\beta^+} + T_\nu + T_M \tag{3.59}$$

The momentum equation is

$$0 = \mathbf{P}_m + \mathbf{P}_\beta + \mathbf{P}_\nu \tag{3.60}$$

The β^+ decay energy is

$$Q_{\beta^+} = M(A,Z) - M(A,Z-1) - 2m \tag{3.61}$$

The condition for β^+ decay to be possible is

$$M(A,Z) - M(A,Z-1) - 2m > 0 \tag{3.62}$$

A comparison of Eqs. 3.53 and 3.62 shows that β^- decay is possible if the mass of the parent is just bigger than the mass of the daughter nucleus, while β^+ decay is possible only if the parent and daughter nuclear masses differ by at least $2mc^2 = 1.022$ MeV.

The energy spectrum of β^+ particles is continuous, for the reasons presented earlier, and similar to that of β^- decay (Fig. 3.8). The average energy of the positrons from β^+ decay $\bar{E}_{\beta^+}$ is also taken to be equal to $E_{max}/3$ unless extremely accurate values are needed, in which case the equation given in reference 1 should be used.

A typical β^+ decay scheme is shown in Fig. 3.10.

ELECTRON CAPTURE

In some cases, an atomic electron is captured by the nucleus and a neutrino is emitted according to the equation

$$M_N(A,Z) + m = M_N(A,Z-1) + T_\nu + B_e \tag{3.63}$$

In Eq. 3.63 all the symbols have been defined before except B_e, binding energy of the electron captured by the nucleus. This transformation is called *electron capture* (EC). In terms of atomic masses, Eq. 3.63 takes the form

$$M(A,Z) = M(A,Z-1) + T_\nu \tag{3.64}$$

The energy Q_{EC} released during EC is

$$Q_{EC} = M(A,Z) - M(A,Z-1) \tag{3.65}$$

The condition for EC to be possible is

$$M(A,Z) - M(A,Z-1) > 0 \tag{3.66}$$

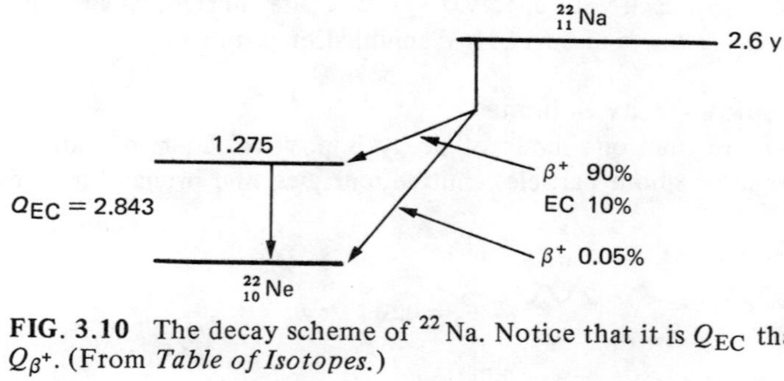

FIG. 3.10 The decay scheme of ^{22}Na. Notice that it is Q_{EC} that is listed, not Q_{β^+}. (From *Table of Isotopes.*)

Electron capture is an alternative to β^+ decay. Comparison of Eqs. 3.61 and 3.66 shows that nuclei that cannot experience β^+ decay can undergo EC, since a smaller mass difference is required for the latter process. Of course, EC is always possible if β^+ decay is. For example, ^{22}Na (Fig. 3.10) decays both by β^+ and EC.

After EC, there is a vacancy left behind that is filled by an electron falling in from a higher orbit. Assuming that a K electron was captured, an L electron may fill the empty state left behind. When this happens an energy approximately equal to $B_K - B_L$ becomes available (where B_K and B_L are the binding energy of K or L electron, respectively). The energy $B_K - B_L$ may be emitted as a K x-ray called *fluorescent radiation*, or it may be given to another atomic electron. If this energy is given to an L electron, that particle will be emitted with kinetic energy equal to $(B_K - B_L) - B_L = B_K - 2B_L$. Atomic electrons emitted in this way are called *Auger electrons.*

Whenever an atomic electron is removed and the vacancy left behind is filled by an electron from a higher orbit, there is a competition between the emission of Auger electrons and fluorescent radiation. The number of x-rays emitted per vacancy in a given shell is the *fluorescent yield.* The fluorescent yield increases with atomic number.

3.7.4 Particles, Antiparticles, and Electron-Positron Annihilation

Every known subatomic particle has a counterpart called the *antiparticle.* A charged particle and an antiparticle have the same mass, and opposite charge. If a particle is neutral–for example, the neutron–its antiparticle is still neutral. Then their difference is due to some other property, such as magnetic moment. Some particles, like the photon, are identical with their own antiparticles. An antiparticle cannot exist together with the corresponding particle: when an antiparticle meets a particle, the two react and new particles appear.

Consider the example of the electron and the "antielectron," which is the positron. The electron and the positron are identical particles except for their charge, which is equal to e but negative and positive respectively. The rest mass of either particle is equal to 0.511 MeV. A positron moving in a medium loses energy continuously, as a result of collisions with atomic electrons (see Chap. 4). Close to the end of its track, the positron combines with an atomic electron, the two annihilate, and photons appear with a total energy equal to $2mc^2$. At least two photons should be emitted for conservation of energy and momentum to be satisfied (Fig. 3.11). Most of the time two photons, each with energy 0.511 MeV, are emitted. As a result, every positron emitter is also a source of 0.511-MeV annihilation gammas.

3.7.5 Complex Decay Schemes

For many nuclei, more than one mode of decay is positive. Users of radioisotopic sources need information about particles emitted, energies, and probabilities of emis-

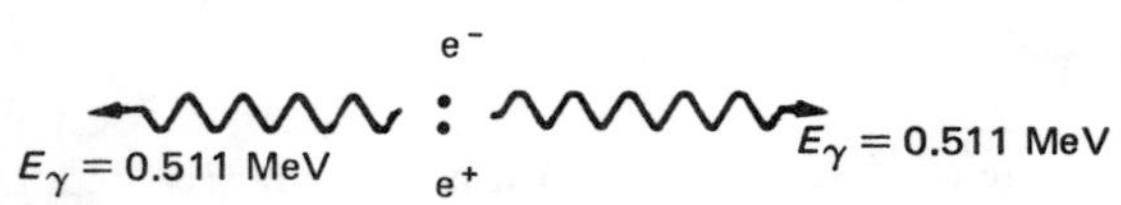

FIG. 3.11 Electron-positron annihilation.

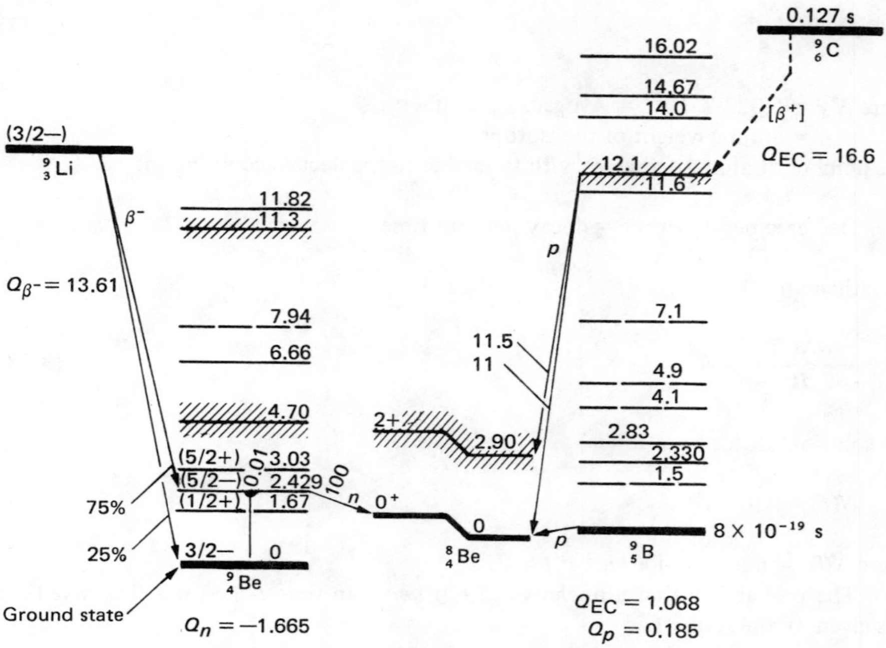

FIG. 3.12 A complex decay scheme. For complete explanation of all the symbols and numbers see *Table of Isotopes.* Half-life is given for each element's ground state, and energy of each level is given at intermediate states. Q_n is the neutron separation energy. Transition probabilities are indicated as percentages. (From *Table of Isotopes.*)

sion. Many books on atomic and nuclear physics contain such information, and the most comprehensive collection of data on this subject can be found in the Table of Isotopes by Lederer, Hollander, and Perlman.[4] Figure 3.12 shows an example of a complex decay scheme taken from that book. The arrows explain the most important quantities shown in the diagram.

3.8 THE RADIOACTIVE DECAY LAW

Radioactive decay is spontaneous change of a nucleus. The change may result in a new nuclide or simply change the energy of the nucleus. If there is a certain amount of a radioisotope at hand, there is no certainty that in the next second "so many nuclei will decay" or "none will decay." One can talk of the probability that a nucleus will decay in a certain period of time.

The probability that a given nucleus will decay per unit time is called the *decay constant* and is indicated by the letter λ. For a certain species, λ is

1 The same for all the nuclei
2 Constant, independent of the number of nuclei present
3 Independent of the age of the nucleus

Consider a certain mass m of a certain radioisotope with decay constant λ. The number of atoms (or nuclei) in the mass m is equal to

$$N = m\frac{N_A}{A} \tag{3.67}$$

where $N_A = 6.023 \times 10^{23}$ = Avogadro's number
A = atomic weight of the isotope
This number of atoms decreases with time, due to the decay according to:

Decrease per unit time = decay per unit time

or mathematically,

$$-\frac{dN(t)}{dt} = \lambda N(t) \tag{3.68}$$

The solution of this equation is

$$N(t) = N(0)e^{-\lambda t} \tag{3.69}$$

where $N(0)$ = number of atoms at $t = 0$.

The probability that a nucleus *will not decay* in time t–i.e., it will survive time t–is given by the ratio of

$$\frac{\text{Atoms not decaying in time } t}{\text{atoms at } t = 0} = \frac{N(0)e^{-\lambda t}}{N(0)} = e^{-\lambda t} \tag{3.70}$$

The probability that the nucleus will decay between t and $t + dt$ is

$$p(t)\,dt = (\text{probability to survive to time } t)(\text{probability to decay in } dt) = e^{-\lambda t}\,\lambda\,dt$$

The average lifetime $\bar{t}$ of the nucleus is given by

$$\bar{t} = \frac{\int_0^\infty tp(t)\,dt}{\int_0^\infty p(t)\,dt} = \frac{\int_0^\infty te^{-\lambda t}\,dt}{\int_0^\infty e^{-\lambda t}\,dt} = \frac{1}{\lambda} \tag{3.71}$$

One concept used extensively with radioisotopes is the *half-life* T, defined as the time it takes for half of a certain number of nuclei to decay. Thus, using Eq. 3.69,

$$\frac{N(T)}{N(0)} = \frac{1}{2} = e^{-\lambda T}$$

which then gives the relationship between λ and T:

$$T = \frac{\ln 2}{\lambda} \tag{3.72}$$

For a sample of $N(t)$ nuclei at time t, each having decay constant λ, the expected number of nuclei decaying per unit time is

$$A(t) = \lambda N(t) \tag{3.73}$$

where $A(t)$ = activity of the sample at time t.

The units of activity are the Becquerel (Bq), equal to 1 decay/s, or the Curie (Ci) equal to 3.7×10^{10} Bq. The Becquerel is the SI unit defined in 1977.

The term *specific activity* (SA) is used frequently. It may have one of the two following meanings.

1 For solids:

$$SA = \frac{\text{activity}}{\text{mass}} \ (\text{Bq/kg or Ci/g})$$

2 For gases or liquids:

$$SA = \frac{\text{activity}}{\text{volume}} \ (\text{Bq/m}^3 \text{ or Ci/cm}^3)$$

EXAMPLE 3.8

What is the SA of ^{60}Co?

ANSWER

The SA is

$$SA = \frac{A}{m} = \frac{\lambda N}{m} = \frac{\ln 2}{Tm} m \frac{N_A}{A} = \frac{\ln 2(6.022 \times 10^{23})}{(5.2 \text{ y})(3.16 \times 10^7 \text{ s/y})(0.060 \text{ kg})}$$

$$= 4.23 \times 10^{16} \text{ Bq/kg} = 1.14 \times 10^3 \text{ Ci/g}$$

EXAMPLE 3.9

What is the SA of a liquid sample of 10^{-3} m^3 containing 10^{-6} kg of ^{32}P?

ANSWER

The SA is

$$SA = \frac{A}{V} = \frac{\lambda N}{V} = \frac{\ln 2}{VT} m \frac{N_A}{A} = \frac{(\ln 2)(10^{-6} \text{ kg})(6.022 \times 10^{23})}{(10^{-3} \text{ m}^3)(14.3 \text{ d})(86400 \text{ s/d})(0.032 \text{ kg})}$$

$$= 1.05 \times 10^{16} \text{ Bq/m}^3 = 0.285 \text{ Ci/cm}^3$$

There are isotopes that decay by more than one mode. Consider such an isotope decaying by the modes 1, 2, 3, . . . , i, (e.g., alpha, beta, gamma, etc., decay) and let

λ_i = probability per unit time that the nucleus will decay by the ith mode

The total probability of decay (total decay constant) is

$$\lambda = \lambda_1 + \lambda_2 + \cdots + \lambda_i + \cdots \tag{3.74}$$

If the sample contains $N(t)$ atoms at time t, the number of decays per unit time by the ith mode is

$$A_i(t) = \lambda_i N(t) = \lambda_i N(0) e^{-\lambda t} \tag{3.73a}$$

The term ***partial half-life*** is sometimes used to indicate a different decay mode. If T_i is the partial half-life for the ith decay mode, using Eqs. 3.72 and 3.74 one obtains

$$\frac{1}{T} = \frac{1}{T_1} + \frac{1}{T_2} + \frac{1}{T_3} + \cdots + \frac{1}{T_i} \tag{3.75}$$

It should be pointed out that it is the total decay constant that is used by Eqs. 3.69 and 3.73a, and not the partial decay constants.

EXAMPLE 3.10

The isotope ^{252}Cf decays by alpha decay and by spontaneous fission. The total half-life is 2.57 years and the half-life for alpha decay is 2.65 years. What is the number of spontaneous fissions per second per 10^{-3} kg (1 g) of ^{252}Cf?

ANSWER

The spontaneous fission activity is

$$A_{sf} = \lambda_{sf} N = \frac{\ln 2}{T_{sf}} m \frac{N_A}{A}$$

The spontaneous fission half-life is, using Eq. 3.75,

$$T_{sf} = \frac{T_t T_\alpha}{T_\alpha - T_t} = \frac{(2.57)(2.65)}{2.65 - 2.57} = 85.1 \text{ y} = 2.69 \times 10^9 \text{ s}$$

Therefore,

$$A_{sf} = \frac{\ln 2}{2.69 \times 10^9 \text{ s}} (10^{-3} \text{ kg}) \frac{6.022 \times 10^{22}}{252 \times 10^3} \text{ atoms/kg} = 6.16 \times 10^5 \text{ sf/s}$$

Sometimes the daughter of a radioactive nucleus may also be radioactive and decay to a third radioactive nucleus. Thus, a radioactive chain

$$N_1 \rightarrow N_2 \rightarrow N_3 \rightarrow \text{etc.}$$

is generated. An example of a well-known series is that of ^{235}U, which through combined α and β^- decays ends up as an isotope of lead. The general equation giving the number of atoms of the ith isotope at time t in terms of the decay constants of all the other isotopes in the chain was developed by Bateman.[2] If $N_i(0)$ is the number of atoms of the first isotope of the series at time $t = 0$ and

$$N_i(0) = 0 \qquad i > 1$$

then the Bateman equation takes the form

$$N_i(t) = \lambda_1 \lambda_2 \cdots \lambda_{i-1} N_1(0) \sum_{j=1}^{i} \frac{e^{-\lambda_j t}}{\Pi_{k \neq j} (\lambda_k - \lambda_j)} \tag{3.76}$$

EXAMPLE 3.11

Apply the Bateman equation for the second and third isotope in a series.

ANSWER

$$\text{(a)} \quad N_2(t) = \lambda_1 N_1(0) \left(\frac{e^{-\lambda_1 t}}{\lambda_1 - \lambda_1} + \frac{e^{-\lambda_2 t}}{\lambda_1 - \lambda_2} \right) = \frac{\lambda_1 N_1(0)}{(\lambda_2 - \lambda_1)} (e^{-\lambda_1 t} - e^{-\lambda_2 t})$$

$$\text{(b)} \quad N_3(t) = \lambda_1 \lambda_2 N_1(0) \left[\frac{e^{-\lambda_1 t}}{(\lambda_2 - \lambda_1)(\lambda_3 - \lambda_1)} + \frac{e^{-\lambda_2 t}}{(\lambda_1 - \lambda_2)(\lambda_3 - \lambda_2)} + \frac{e^{-\lambda_3 t}}{(\lambda_1 - \lambda_3)(\lambda_2 - \lambda_3)} \right]$$

3.9 NUCLEAR REACTIONS

3.9.1 General Remarks

A *nuclear reaction* is an interaction between two particles, a fast bombarding particle, called the *projectile*, and a slower or stationary *target*. The products of the reaction may be two or more particles. For the energies considered here (< 20 MeV), the products are also two particles (with the exception of fission, which is discussed in the next section).

If x_1, X_2 are the colliding particles and x_3, X_4 are the products, the reaction is indicated as

$$^{A_1}_{Z_1}x_1 + ^{A_2}_{Z_2}X_2 \rightarrow ^{A_3}_{Z_3}x_3 + ^{A_4}_{Z_4}X_4$$

or $\quad X_2(x_1, x_3)X_4$

The particles in parentheses are the light particles, x_1 being the projectile. Another representation for the reaction is based on the light particles only, in which case the reaction shown above is indicated as an (x_1, x_3) reaction. For example, the reaction

$$^1_0n + ^{10}_5B \rightarrow ^4_2He + ^7_3Li$$

may also be indicated as $^{10}_5B(n, \alpha)^7_3Li$ or simply as an (n, α) reaction.

Certain quantities are conserved when a nuclear reaction takes place. Four are considered here. For the reaction shown above, the following quantities are conserved:

Charge:

$$Z_1 + Z_2 = Z_3 + Z_4$$

Mass number:

$$A_1 + A_2 = A_3 + A_4$$

Total energy:

$$E_1 + E_2 = E_3 + E_4 \qquad \text{(rest mass plus kinetic energy)}$$

Linear momentum:

$$\mathbf{P}_1 + \mathbf{P}_2 = \mathbf{P}_3 + \mathbf{P}_4$$

Many nuclear reactions proceed through the formation of a *compound nucleus.* The compound nucleus, formed after particle x_1 collides with X_2, is highly excited and lives for a time of the order of 10^{-12} to 10^{-14} s before it decays to x_3 and X_4. A compound nucleus may be formed in more than one way and may decay by more than one mode that does not depend on the mode of formation. Consider the example of the compound nucleus $^{14}_{7}N$:

Formation	Compound nucleus	Decay
$^{4}_{2}He + ^{10}_{5}B$		$^{1}_{0}n + ^{13}_{7}N$
$^{4}_{2}He + ^{12}_{6}C$	$^{14}_{7}N$	$^{1}_{1}p + ^{13}_{6}C$
$^{13}_{6}C + ^{1}_{1}H$		$^{4}_{2}He + ^{10}_{5}B$
$^{1}_{0}n + ^{13}_{7}N$		$^{2}_{1}H + ^{12}_{6}C$

The modes of formation and decay of ^{14}N are shown in the form of an energy-level diagram in Fig. 3.13. No matter how the compound nucleus is formed, it has an excitation energy equal to the separation energy of the projectile (α, n, p, etc.) plus a fraction of the kinetic energy of the two particles. Since the separation energy is of the order of MeV, it is obvious that the compound nucleus has considerable excitation energy even if the projectile and the target have zero kinetic energy.

Exactly what happens inside the compound nucleus is not known. It is believed—and experiment does not contradict this idea—that the excitation energy of the compound nucleus is shared quickly by all the nucleons ($A_1 + A_2$). There is continuous exchange of energy among all the nucleons until one of them (or a cluster of them) obtains energy greater than its separation energy and is able to leave the compound nucleus, becoming a free particle.

3.9.2 Kinematics of Nuclear Reactions

In this section, two questions will be answered:

1 Given the masses m_1, M_2, m_3, M_4, and the kinetic energies of the projectile (m_1) and the target (M_2), how can one calculate the kinetic energies of the products with masses m_3 and M_4?

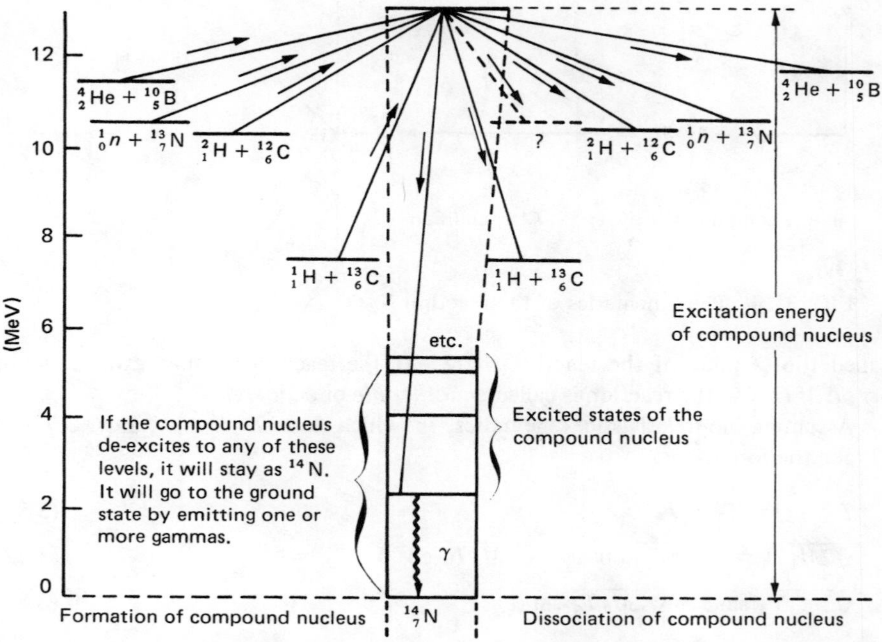

FIG. 3.13 Different modes of formation and decay of the compound nucleus. For clarity, the diagram shows that the compound nucleus has the same excitation energy regardless of the way it is formed. This is not necessarily the case.

2 What is the minimum kinetic energy the particles with masses m_1, M_2 ought to have to be able to initiate the reaction?

The discussion will be limited to the case of a stationary target, the most commonly encountered in practice.

Consider a particle of mass m_1 having speed v_1 (kinetic energy T_1) hitting a stationary particle of mass M_2. The particles m_3, M_4 are produced as a result of this reaction with speeds v_3, v_4 (kinetic energies T_3, T_4), as shown in Fig. 3.14. Applying conservation of energy and linear momentum, one has

Energy:

$$m_1 + T_1 + M_2 = m_3 + T_3 + M_4 + T_4 \tag{3.77}$$

Momentum, x axis:

$$m_1 v_1 = m_3 v_3 \cos\theta + M_4 V_4 \cos\phi \tag{3.78}$$

Momentum, y axis:

$$m_3 v_3 \sin\theta = M_4 V_4 \sin\phi \tag{3.79}$$

The quantity

$$Q = m_1 + M_2 - m_3 - M_4 \tag{3.80}$$

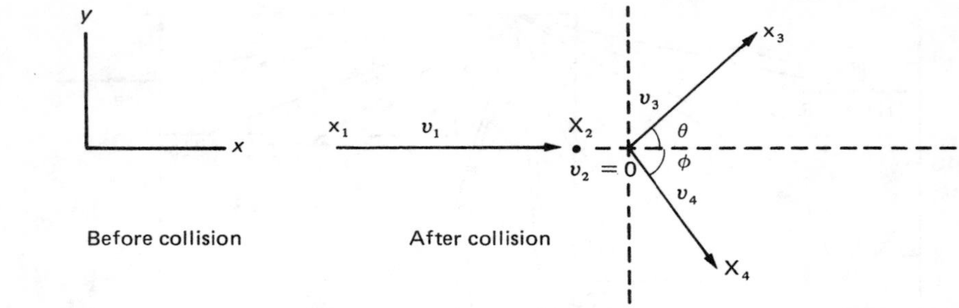

FIG. 3.14 The kinematics of the reaction $X_2(x_1, x_3)X_4$.

is called the Q value of the reaction. If $Q > 0$, the reaction is called *exothermic* or *exoergic.* If $Q < 0$, the reaction is called *endothermic* or *endoergic.*

Assuming nonrelativistic kinematics, in which case $T = \frac{1}{2}mv^2$, Eqs. 3.77 to 3.79 take the form

$$T_1 + Q = T_3 + T_4 \tag{3.81}$$

$$\sqrt{2M_1 T_1} = \sqrt{2m_3 T_3} \cos\theta + \sqrt{2M_4 T_4} \cos\phi \tag{3.82}$$

$$\sqrt{2m_3 T_3} \sin\theta = \sqrt{2M_4 T_4} \sin\phi \tag{3.83}$$

Equations 3.81 to 3.83 have four unknowns T_3, T_4, ϕ, and θ, so they cannot be solved to give a unique answer for any single unknown. In practice, one expresses a single unknown in terms of a second one–e.g., T_3 as a function of θ, after eliminating T_4 and ϕ. Such an expression, although straightforward, is complicated. Two cases of special interest are the following.

1. $\theta = 0$, $\phi = 180°$. In this case, the particles m_3 and M_4 are emitted along the direction of motion of the bombarding particle (Fig. 3.15). Equations 3.81 and 3.82 take the form

$$T_1 + Q = T_3 + T_4 \tag{3.84}$$

$$\sqrt{m_1 T_1} = \sqrt{m_3 T_3} = \sqrt{M_4 T_4} \tag{3.85}$$

and they can be solved for T_3 and T_4. These values of T_3 and T_4 give the maximum and minimum kinetic energies of particles m_3 and M_4.

EXAMPLE 3.12

Consider the reaction

$${}^1_0n + {}^{14}_7\text{N} \rightarrow {}^4_2\text{He} + {}^{11}_5\text{B}$$

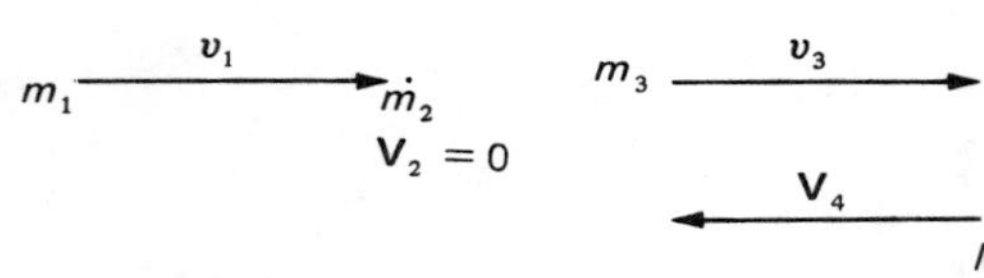

FIG. 3.15 A case where the reaction products are emitted 180° apart.

with the nitrogen being at rest and the neutron having energy 2 MeV. What is the maximum kinetic energy of the alpha particle?

ANSWER
The Q value of the reaction is

$$Q = 14.003074 + 1.008665 - 4.002604 - 11.001305 = -0.163 \text{ MeV}$$

Solving Eqs. 3.84 and 3.85 for T_3, one obtains a quadratic equation for T_3 (T_3 in MeV),

$$T_3^2 - 2.574T_3 + 1.4786 = 0$$

which gives two values of T_3:

$$T_{3,1} = 1.709 \text{ MeV} \qquad T_{3,2} = 0.865 \text{ MeV}$$

The corresponding values of T_4 are:

$$T_{4,1} = 0.131 \text{ MeV} \qquad T_{4,2} = 0.975 \text{ MeV}$$

The two pairs of values correspond to the alpha being emitted at $\theta = 0$ ($T_3 = 1.709$ MeV = max. kin. energy) or $\theta = 180°$ ($T_3 = 0.865$ MeV = min. kin. energy). Correspondingly, the boron nucleus is emitted at $\phi = 180°$ or $\phi = 0°$. One can use the momentum balance equation (Eq. 3.85) to verify this conclusion.

2. $\theta = 90°$. In this case, the reaction looks as shown in Fig. 3.16. The momentum vectors form a right triangle as shown on the right of Fig. 3.16. Therefore, Eqs. 3.81 and 3.82 take the form

$$T_1 + Q = T_3 + T_4 \tag{3.86}$$

$$m_1 T_1 + m_3 T_3 = M_4 T_4 \tag{3.87}$$

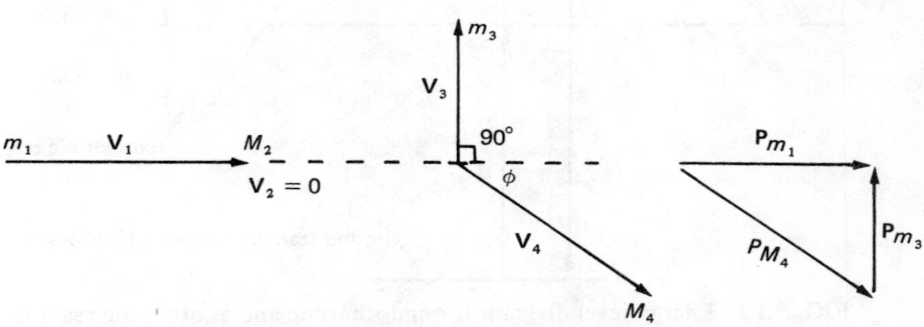

FIG. 3.16 A case where the reaction products are emitted 90° apart.

Again, one can solve for T_3 and T_4. The value of T_3 is

$$T_3 = \frac{(M_4 - m_1)T_1 + M_4 Q}{m_3 + M_4} \tag{3.88}$$

EXAMPLE 3.13

What is the energy of the alpha particle in the reaction

$$^1_0n + {}^{14}_7\text{N} \rightarrow {}^4_2\text{He} + {}^{11}_5\text{B}$$

if it is emitted at 90°? Use $T_1 = 2$ MeV, the same as in Example 3.12.

ANSWER

Using Eq. 3.88,

$$T_3 = \frac{(11-1)2 + 11(-0.163)}{15} \text{ MeV} = 1.214 \text{ MeV}$$

The value of the minimum (threshold) energy necessary to initiate a reaction can be understood with the help of Fig. 3.17. When the particle m_1 enters the target nucleus M_2, a compound nucleus if formed with excitation energy equal to

$$B_{m_1} + \frac{M_2}{m_1 + M_2} T_1 \tag{3.89}$$

where B_{m_1} = binding energy of particle m_1
$(M_2/M_1 + M_2)T_1$ = fraction of the incident particle kinetic energy available as excitation energy of the compound nucleus

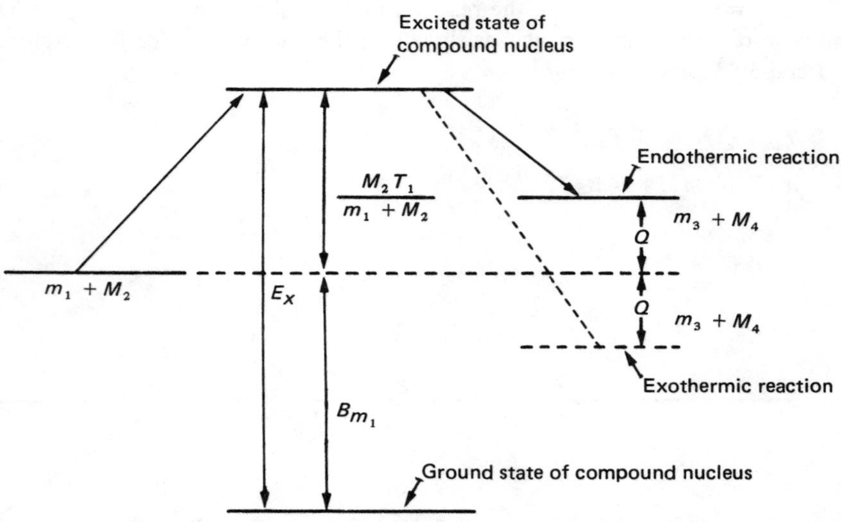

FIG. 3.17 Energy-level diagram for endothermic and exothermic reactions. For endothermic reactions, the threshold energy is equal to $[(m_1 + M_2)/m_2]|Q|$.

Only a fraction of the kinetic energy T_1 is available as excitation energy, because the part

$$\left(\frac{m_1}{m_1 + M_2}\right) T_1$$

becomes kinetic energy of the compound nucleus (see Evans or any other book on nuclear physics), and as such is not available for excitation.

If the reaction is exothermic ($Q > 0$), it is energetically possible for the compound nucleus to deexcite by going to the state $(m_3 + M_4)$ (Fig. 3.17), even if $T_1 \approx 0$. For an endothermic reaction, however, energy at least equal to Q should become available (from the kinetic energy of the projectile). Therefore, the kinetic energy T_1 should be such that

$$\frac{M_2}{m_1 + M_2} T_1 \geqslant Q \tag{3.90}$$

or the threshold kinetic energy for the reaction is

$$T_{1,\mathrm{th}} \geqslant \frac{m_1 + M_2}{M_2} |Q| \tag{3.91}$$

3.10 FISSION

Fission is the reaction in which a heavy nucleus splits into two heavy fragments. In the fission process, net energy is released, because the heavy nucleus has less binding energy per nucleon than the fission fragments, which belong to the middle of the periodic table. In fact, for $A > 85$ the binding energy per nucleon decreases (Fig. 3.3); therefore any nucleus with $A > 85$ would go to a more stable configuration by fissioning. Such "spontaneous" fission is possible but very improbable. Only very heavy nuclei ($Z > 92$) undergo spontaneous fission at a considerable rate.

For many heavy nuclei ($Z \geqslant 90$), fission takes place if an amount of energy at least equal to a critical energy E_c is provided in some way, as by neutron or gamma absorption. Consider, as an example, the nucleus ^{235}U (Fig. 3.18). If a neutron with kinetic energy T_n is absorbed, the compound-nucleus ^{236}U has excitation energy equal to (Eq. 3.89)

$$B_n + \frac{A}{A+1} T_n$$

If $B_n + (A/A+1)T_n \geqslant E_c$, fission may occur and the final state is the one shown as fission products in Fig. 3.18. For ^{236}U, $E_c = 5.3$ MeV and $B_n = 6.4$ MeV. Therefore, even a neutron with zero kinetic energy may induce fission, if it is absorbed. For ^{239}U, which is formed when a neutron is absorbed by ^{238}U, $B_n = 4.9$ MeV and $E_c =$

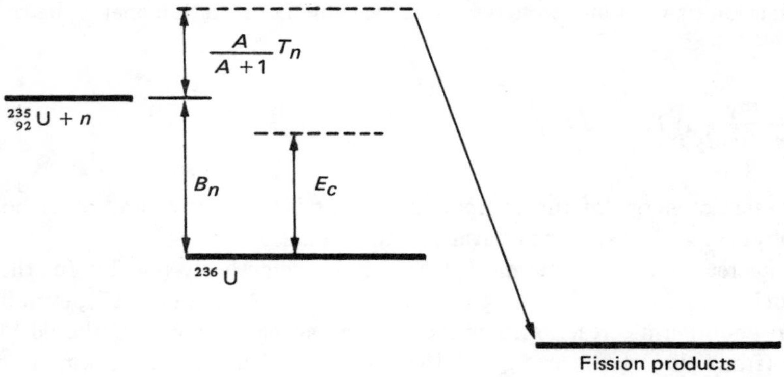

FIG. 3.18 The fission of ^{235}U induced by neutron absorption.

5.5 MeV. Therefore, fission cannot take place unless the neutron kinetic energy satisfies

$$T_n > \frac{A+1}{A}(E_c - B_n) = \frac{239}{238}(5.5 - 4.9) \approx 1 \text{ MeV}$$

The fission fragments are nuclei in extremely excited states with mass numbers in the middle of the periodic system. They have a positive charge of about $20e$ and they are neutron-rich. This happens because the heavy nuclei have a much higher neutron-proton ratio than nuclei in the middle of the periodic system.

Consider as an example ^{236}U (Fig. 3.19). Assume that it splits into two fragments as follows:

$$Z_1 = 48 \quad N_1 = 80 \quad A_1 = 128$$
$$Z_2 = 44 \quad N_2 = 64 \quad A_2 = 108$$

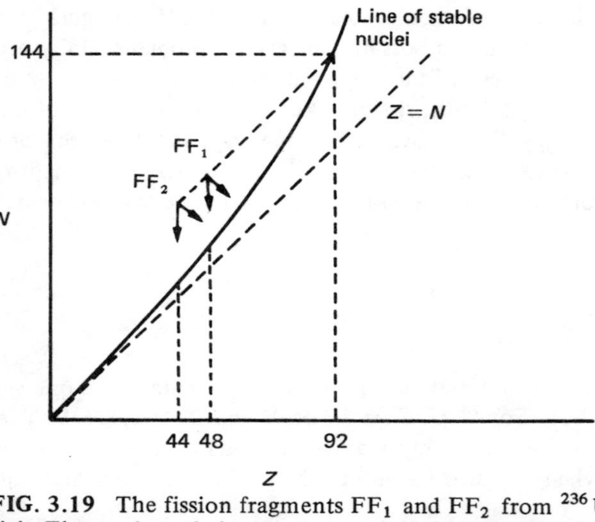

FIG. 3.19 The fission fragments FF_1 and FF_2 from ^{236}U fission are neutron-rich. They reduce their neutron number either by beta decay or by neutron emission.

The two fission fragments have a neutron-proton ratio higher than what stability requires for their atomic mass. They get rid of the extra neutron either by directly emitting neutrons or by β^- decay.

A nucleus does not always split in the same fashion. There is a probability that each fission fragment (A, Z) will be emitted, called the *fission yield.* Figure 3.20 shows the fission yield for ^{235}U fission. For thermal neutrons, the "asymmetric" fission is favored. It can be shown that asymmetric fission yields more energy. As the neutron energy increases, the excitation energy of the compound nucleus increases. The possibilities for fission are such that it does not make much difference, from an energy point of view, whether the fission is symmetric or asymmetric. Therefore, the probability of symmetric fission increases.

The fission fragments deexcite by emitting neutrons, betas, and gammas, and most of the fragments stay radioactive long after the fission took place. The important characteristics of the particles emitted by fission fragments are:

1 Betas. About six β^- particles are emitted per fission, carrying a total average energy of 7 MeV.
2 Gammas. About seven gammas are emitted at the time of fission. These are called *prompt gammas.* At later times, about seven to eight more gammas are released, called *delayed gammas.* Photons carry a total of about 15 MeV per fission.
3 Neutrons. The number of neutrons per fission caused by thermal neutrons is between two and three. This number increases linearly with the kinetic energy of the neutron inducing the fission. The average energy of a neutron emitted in fission is about 2 MeV. More than 99 percent of the neutrons are emitted at the time of fission and are called *prompt neutrons.* A very small fraction is emitted as *delayed neutrons.* Delayed neutrons are very important for the control of nuclear reactors.
4 Neutrinos. About 11 MeV are taken away by neutrinos, which are also emitted

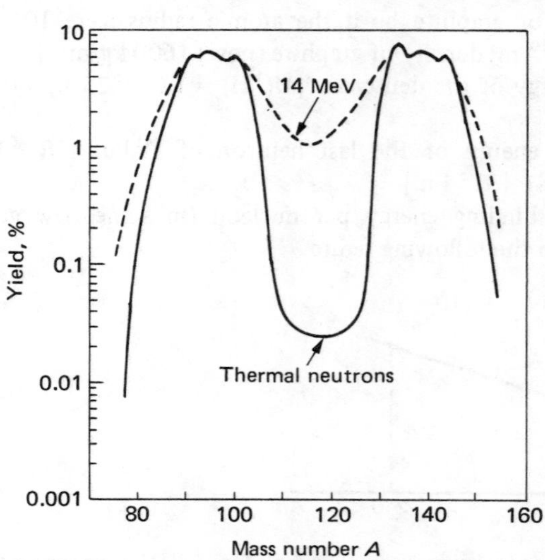

FIG. 3.20 ^{235}U fission yield for thermal neutron induced fission. Data from Ref. 3. The dashed line indicates the yield when the fission is induced by 14-MeV neutrons.

TABLE 3.3 Fission Products

Particle	Number/fission	MeV/fission
Fission fragments	2	160 to 170
Neutrons	2 to 3	5
Gammas (prompt)	7	8
Gammas (delayed)	7	7
Betas	6	7
Neutrinos	6	11
Total		198 to 208

during fission. This energy is the only part of the fission energy yield which completely escapes. It represents about 5 percent of the total fission energy.

Table 3.3 summarizes the particles and energies involved in fission.†

PROBLEMS

3.1 What is the speed of a 10-MeV electron? What is its total mass, relative to its rest mass?

3.2 What is the speed of a proton with a total mass equal to $2Mc^2$? (M is the proton rest mass.)

3.3 What is the kinetic energy of a neutron that will result in 1 percent error difference between relativistic and classical calculation of its speed?

3.4 What is the mass of an astronaut traveling with speed $v = 0.8c$? Mass at rest is 70 kg.

3.5 What is the kinetic energy of an alpha particle with a total mass 10 percent greater than its rest mass?

3.6 What would the density of graphite be if the atomic radius were 10^{-13} m? [Atomic radius (now) 10^{-10} m; density of graphite (now) 1600 kg/m^3.]

3.7 Calculate the binding energy of the deuteron. [$M(^1\text{H}) = 1.007825$ u; $M(^2\text{H}) = 2.01410$ u.]

3.8 Calculate the separation energy of the last neutron of ^{241}Pu. [$M(^{240}\text{Pu}) = 240.0540$ u; $M(^{241}\text{Pu}) = 241.13154$ u.]

3.9 Assume that the average binding energy per nucleon (in some new galaxy) changes with A as shown in the following figure.

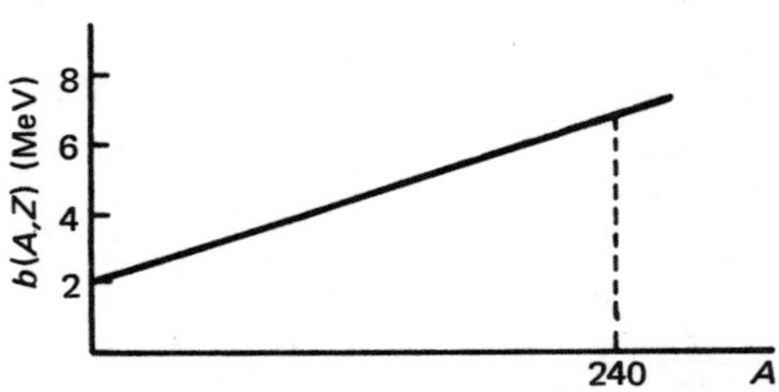

†Tritium is sometimes produced in fission. In reactors fueled with ^{235}U it is produced at the rate of 8.7×10^{-5} tritons per fission. The most probable kinetic energy of the tritons is about 7.5 MeV.

(a) Would fission or fusion or both release energy in such a world?
(b) How much energy would be released if a tritium (^3H) nucleus and a helium (^4He) nucleus combined to form a lithium nucleus? [$M(^3H) = 3.01605$ u; $M(^4He) = 4.00260$u; $M(^7Li) = 7.01600$ u.]

3.10 A simplified diagram of the ^{137}Cs decay is shown in the figure below. What is the recoil energy of the nucleus when the 0.6616-MeV gamma is emitted?

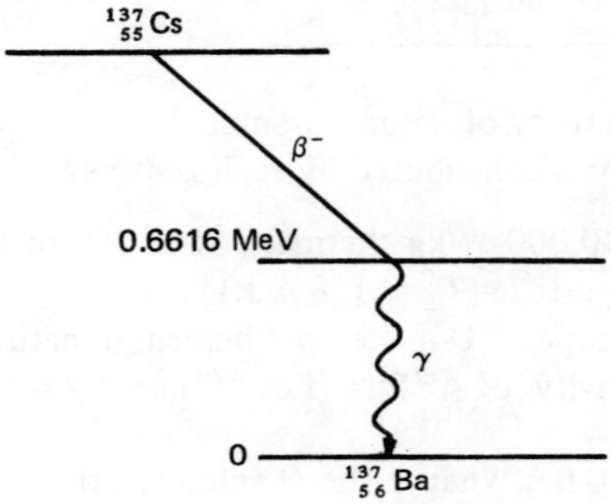

3.11 The isotope ^{239}Pu decays by alpha emission to ^{235}U as shown in the figure below.

(a) What is Q_α?
(b) What is the kinetic energy of the alphas if the ^{235}U nucleus is left in the third excited state? What is the kinetic energy of the alphas if the ^{235}U nucleus is left in the ground state? [$M(^{235}U) = 235.0439$ u; $M(^{239}Pu) = 239.0522$ u.]

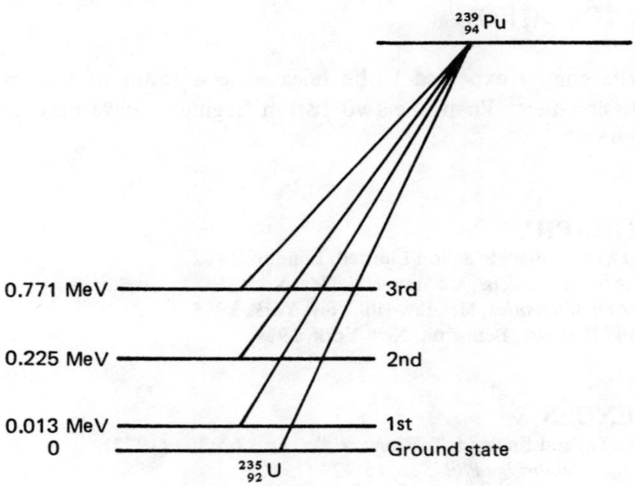

3.12 Consider the isotopes ^{63}Zn and ^{63}Cu. Is β^+ decay possible? Is EC possible? What is Q_{β^+}? What is Q_{EC}? [$M(^{63}Cu) = 62.9298$ u? $M(^{63}Zn) = 62.9330$ u.]

3.13 The isotope $^{11}_4Be$ decays to $^{11}_5B$. What are the maximum and average kinetic energy of the betas? $M(^{11}Be) = 11.0216$ u; $M(^{11}B) = 11.00931$ u.]

3.14 Natural uranium contains the isotopes ^{234}U, ^{235}U, and ^{238}U, with abundances and half lives as shown below:

	Half-life (years)	Abundance (%)
^{234}U	2.47×10^5	0.0057
^{235}U	7.10×10^8	0.71
^{238}U	4.51×10^9	99.284

(a) What is the alpha specific activity of natural uranium?
(b) What fraction of the activity is contributed by each isotope?

3.15 The isotope ^{210}Po generates 140,000 W/kg thermal power due to alpha decay. What is the energy of the alpha particle ($T_\alpha = 138.4$ d.)

3.16 How many years ago did the isotope ^{235}U make up 3 percent of natural uranium?

3.17 What is the specific alpha activity of ^{239}Pu? (For ^{239}Pu: $T_{sf} = 5.5 \times 10^5$ y, $T_{tot} = 2.44 \times 10^4$ y.)

3.18 Consider the reaction $^7_3\text{Li}(p, n)^7_4\text{Be}$. What is the Q value for this reaction? If a neutron is emitted at 90° (in LS) with kinetic energy 2 MeV, what is the energy of the incident photon? [$M(^7\text{Li}) = 7.01600$ u; $M(^7\text{Be}) = 7.0169$ u.]

3.19 What is the necessary minimum kinetic energy of a proton to make the reaction $^4_2\text{He}(p, d)^3_2\text{H}$ possible (^{4}He at rest.)

3.20 A 1-MeV neutron collides with a stationary $^{13}_7$N nucleus. What is the maximum kinetic energy of the emerging proton?

3.21 What is the threshold gamma energy for the reaction

$$\gamma + {}^{12}_6\text{C} \rightarrow 3({}^4_2\text{He})$$

3.22 What is the energy expected to be released as a result of a thermal-neutron-induced fission in ^{239}Pu if the two fission fragments have masses $M_1 = 142$ u and $M_2 = 95$ u?

BIBLIOGRAPHY

Born, M., *Atomic Physics*, Blackie & Son Limited, London, 1962.
Evans, R. D., *The Atomic Nucleus*, McGraw-Hill, New York, 1972.
Schiff, L. I., *Quantum Mechanics*, McGraw-Hill, New York, 1955.
Segre, E., *Nuclei and Particles*, Benjamin, New York, 1964.

REFERENCES

1 Stamatelatos, M. G., and England, T. R., *Nucl. Sci. Eng.* **63**:304 (1977).
2 Bateman, H., *Proc. Cambridge Phil. Soc.* **15**:423 (1910).
3 DiIorio, G. J., "Direct Physical Measurement of Mass Yields in Thermal Fission of Uranium 235," Ph.D. thesis, University of Illinois at Urbana-Champaign, 1976.
4 Lederer, C. M., and Shirley, V. S. (eds.), *Table of Isotopes*, 7th ed., Wiley-Interscience Publication, New York, 1978.

4

■ energy loss and penetration of radiation through matter

4.1 INTRODUCTION

This chapter discusses the mechanisms by which ionizing radiation interacts and loses energy as it moves through matter. The study of this subject is extremely important for radiation measurements because the detection of radiation is based on its interactions and the energy deposited in the material of which the detector is made. Therefore, to be able to build detectors and interpret the results of the measurement we need to know how radiation interacts and what the consequences are of the various interactions.

The topics presented here should be considered only an introduction to this extensive subject. Emphasis is given to that material considered important for radiation measurements. The range of energies considered is shown in Table 1.1.

For the discussion that follows, ionizing radiation is divided into three groups:

1 Charged particles: electrons (e^-), positrons (e^+), protons (p), deuterons (d), alphas (α), heavy ions ($A > 4$)
2 Photons: Gammas (γ) or x-rays
3 Neutrons (n)

The division into three groups is convenient because each group has its own characteristic properties and can be studied separately.

A charged particle moving through a material interacts, through Coulomb forces, with the negative electrons and the positive nuclei that constitute the atoms of that material. As a result of these interactions, the charged particle loses energy continuously and finally stops after traveling a finite distance, called the *range. The r*ange depends on the type and energy of the particle and on the material through which the particle moves. The probability of a charged particle going through a piece of material without an interaction is practically zero. This fact is very important for the operation of charged-particle detectors.

Neutrons and gammas have no charge. They interact with matter in ways that will be discussed later, but there is a finite nonzero probability that a neutron or a γ ray may go through any thickness of any material without having an interaction. As a result, no finite range can be defined for neutrons or gammas.

4.2 MECHANISMS OF CHARGED-PARTICLE ENERGY LOSS

Charged particles traveling through matter lose energy

1 In Coulomb interactions with electrons and nuclei
2 By emission of electromagnetic radiation (bremsstrahlung)

3 In nuclear interactions
4 By emission of Cerenkov radiation

For charged particles with kinetic energies considered here, nuclear interactions may be neglected.

Cerenkov radiation constitutes a very small fraction of the energy loss. It is important only because it has a practical application in the operation of Cerenkov counters (see Evans). Cerenkov radiation is visible electromagnetic radiation emitted by particles traveling in a medium, with speed larger than the speed of light in that medium.

4.2.1 Coulomb Interactions

Consider a charged particle traveling through a certain material, and consider an atom of that material. As shown in Fig. 4.1, the fast charged particle may interact with the atomic electrons or the nucleus of the atom. Since the radius of the nucleus is approximately 10^{-14} m and the radius of the atom is 10^{-10} m, one might expect that

$$\frac{\text{Number of interactions with electrons}}{\text{number of interactions with nuclei}} = \frac{(R^2)\text{ atom}}{(R^2)\text{ nucleus}} = \frac{(10^{-10})^2}{(10^{-14})^2} \approx 10^8$$

This simplified argument indicates that collisions with atomic electrons are more important than with nuclei. Nuclear collisions will not be considered here.

Looking at Fig. 4.1, at a certain point in time the particle is at point P and the electron at E. If the distance between them is r, the coulomb force is $F = k(ze^2/r^2)$, where ze is the charge of the particle and k is a constant that depends on the units. The action of this force on the electron, over a period of time, may result in the transfer of energy from the moving charged particle to the bound electron. Since a bound atomic electron is in a quantized state, the result of the passage of the charged particle may be ionization or excitation.

Ionization occurs when the electron obtains enough energy to leave the atom and become a free particle with kinetic energy equal to:

$$(\text{KE})_e = (\text{energy given by particle}) - (\text{ionization potential})$$

The electron freed from the atom acts like any other moving charged particle. It may cause ionization of another atom if its energy is high enough. It will interact with matter, lose its kinetic energy, and finally stop. Fast electrons produced by ionizing collisions are called δ rays.

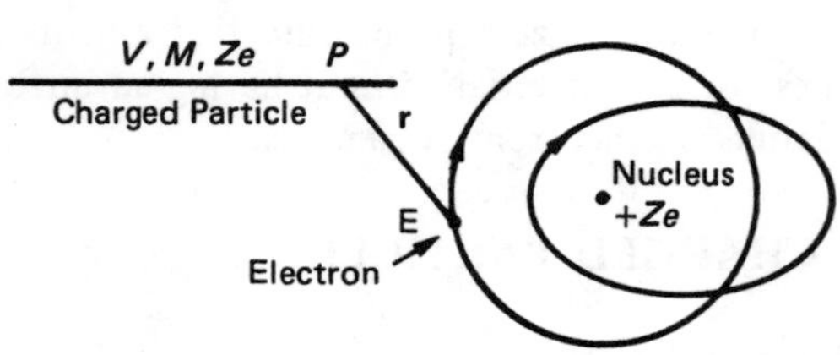

FIG. 4.1 A fast charged particle of mass M and charge Ze interacts with the electrons of an atom.

The ionization leaves behind a positive ion, which is a massive particle compared to an electron. If an ion and an electron move in a gas, the ion will move much slower than the electron. Eventually, the ion will pick up an electron from somewhere and will become a neutral atom again.

Excitation takes place when the electron acquires enough energy to move to an empty state in another orbit of higher energy. It is still bound, but it has moved from a state with energy E_1 to one with E_2, thus producing an excited atom. In a short period of time, of the order of 10^{-8} to 10^{-10} s, the electron will move to a lower energy state, provided there is one empty. If the electron falls from E_2 to E_1, the energy $E_2 - E_1$ is emitted in the form of an x-ray.

Collisions that result in ionization or excitation are called *inelastic collisions.* A charged particle moving through matter may also have elastic collisions with nuclei or atomic electrons. In such a case, the incident particle loses the energy required for conservation of kinetic energy and linear momentum. Elastic collisions are not important for charged particle energy loss and detection.

4.2.2 Emission of Electromagnetic Radiation (Bremsstrahlung)

Every free charged particle that accelerates or decelerates loses part of its kinetic energy by emitting electromagnetic radiation. This radiation is called *bremsstrahlung*, which in German means braking radiation. Bremsstrahlung is not a monoenergetic radiation. It consists of photons with energies from zero up to a maximum equal to the kinetic energy of the particle.

Emission of bremsstrahlung is predicted not only by quantum mechanics but also by classical physics. Theory predicts that a charge that is accelerated radiates energy with intensity proportional to the square of its acceleration. Consider a charged particle with charge ze and mass M moving in a certain material of atomic number Z. The Coulomb force between the particle and a nucleus of the material is $F \sim zeZe/r^2$, where r = distance between the two charges. The acceleration of the incident charged particle is $a = F/M \sim zZe^2/M$. Therefore the intensity of the emitted radiation I is:

$$I \propto a^2 \sim \left(\frac{zZe^2}{M}\right)^2 \sim \frac{z^2Z^2}{M^2} \tag{4.1}$$

This expression indicates that:

1 For two particles traveling in the same medium, the lighter particle will emit a much greater amount of bremsstrahlung than the heavier particle (other things being equal).
2 More bremsstrahlung is emitted if a particle travels in a medium with high atomic number Z than in one with low atomic number.

For charged particles with energies considered here, the kinetic energy lost as bremsstrahlung might be important for electrons only. Even for electrons, it is important for high-Z materials like lead ($Z = 82$). For more detailed treatment of the emission of bremsstrahlung, the reader should look up the references listed at the end of the chapter.

4.3 STOPPING POWER DUE TO IONIZATION AND EXCITATION

A charged particle moving through a material exerts Coulomb forces on many atoms simultaneously. Every atom has many electrons with different ionization and excitation potentials. As a result of this, the moving charged particle interacts with a tremendous number of electrons–millions. Each interaction has its own probability for occurrence and for a certain energy loss. It is impossible to calculate the energy loss by studying individual collisions. Instead, an average energy loss is calculated per unit distance traveled. The calculation is slightly different for electrons or positrons than for heavier charged particles like p, d, and α, for the following reason.

It was mentioned earlier that most of the interactions of a charged particle involve the particle and atomic electrons. If the mass of the electron is taken as 1, then the masses of the other common heavy† charged particles are the following:

Electron mass = 1

Proton mass ≈ 1840

Deuteron mass ≈ 2(1840)

Alpha mass ≈ 4(1840)

If the incoming charged particle is an electron or a positron, it may collide with an atomic electron and lose all its energy in a single collision because the collision involves two particles of the same mass. Hence, incident electrons or positrons may lose a large fraction of their kinetic energy in one collision. They may also be easily scattered to large angles, as a result of which their trajectory is zig-zag (Fig. 4.2). Heavy charged particles, on the other hand, behave differently. On the average, they lose smaller amounts of energy per collision. They are hardly deflected by atomic electrons, and their trajectory is almost a straight line.

Assuming that all the atoms and their atomic electrons act independently, and considering only energy lost to excitation and ionization, the average energy loss‡ per unit distance traveled by the particle is given by Eqs. 4.2, 4.3, and 4.4. (For their derivation see the chapter bibliography: Evans, Segré, and Roy and Reed.)

STOPPING POWER DUE TO IONIZATION-EXCITATION FOR p, d, t, α

$$\frac{dE}{dx}(\text{MeV/m})^{\S} = 4\pi r_0^2 z^2 \frac{mc^2}{\beta^2} NZ \left[\ln\left(\frac{2mc^2}{I}\beta^2\right) - \ln(1-\beta^2) - \beta^2 \right] \quad (4.2)$$

†In this discussion, "heavy" particles are all charged particles except electrons and positrons.

‡Since $E = T + Mc^2$ and Mc^2 = constant, $dE/dx = dT/dx$; thus Eqs. 4.2 to 4.4 express the kinetic as well as the total energy loss per unit distance.

§In SI units, the result would be J/m; 1 MeV = 1.602×10^{-13} J.

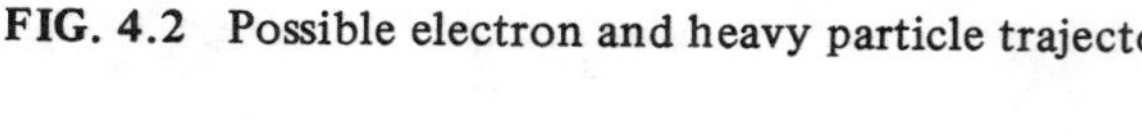

FIG. 4.2 Possible electron and heavy particle trajectories.

STOPPING POWER DUE TO IONIZATION-EXCITATION FOR ELECTRONS

$$\frac{dE}{dx}\,(\text{MeV/m}) = 4\pi r_0^2 \frac{mc^2}{\beta^2} NZ \left\{ \ln\left(\frac{\beta\gamma\sqrt{\gamma-1}\,mc^2}{I}\right) + \frac{1}{2\gamma^2}\left[\frac{(\gamma-1)^2}{8} + 1 - (2\gamma^2 + 2\gamma - 1)\ln 2\right]\right\} \tag{4.3}$$

STOPPING POWER DUE TO IONIZATION-EXCITATION FOR POSITRONS

$$\frac{dE}{dx}\,(\text{MeV/m}) = 4\pi r_0^2 \frac{mc^2}{\beta^2} NZ \left\{ \ln\left(\frac{\beta\gamma\sqrt{\gamma-1}}{I}\,mc^2\right) - \frac{\beta^2}{24}\left[23 + \frac{14}{\gamma+1} + \frac{10}{(\gamma+1)^2} + \frac{4}{(\gamma+1)^3}\right] + \ln 2\right\} \tag{4.4}$$

where $r_0 = e^2/mc^2 = 2.818 \times 10^{-15}$ m = classical electron radius
mc^2 = rest mass energy of the electron = 0.511 MeV
$\gamma = T + Mc^2/Mc^2 = 1/\sqrt{1-\beta^2}$
T = kinetic energy = $(\gamma - 1)Mc^2$
M = rest mass of the particle
$\beta = v/c \quad c = 3 \times 10^8$ m/s
N = number of atoms/m^3 for the material through which the particle moves
Z = atomic number of the material
z = charge of the incident particle ($z = 1$ for e^-, e^+, p, d; $z = 2$ for α)
I = mean excitation potential of the material

An approximate equation for I, which gives good results for $Z > 12$,[1] is

$$I\,(\text{eV}) = (9.76 + 58.8Z^{-1.19})Z \tag{4.5}$$

Table 4.1 gives values of I for many common materials.

Many different names have been used for the quantity dE/dx: names like energy loss, specific energy loss, differential energy loss, or stopping power. In this text, the term *stopping power* will be used for dE/dx given by Eq. 4.2 to 4.4, as well as for a similar equation for heavier charged particles presented in Sec. 4.7.2.

It should be noted that the stopping power

1 Is independent of the mass of the particle
2 Is proportional to z^2 ((charge)^2] of particle
3 Depends on the speed of particle
4 Is proportional to the density of the material (N)

For low kinetic energies, dE/dx is almost proportional to $1/V^2$. For relativistic energies, the term in the bracket predominates and dE/dx increases with kinetic energy. Figure 4.3 shows the general behavior of dE/dx as a function of kinetic energy. For

TABLE 4.1 Values of Mean Excitation Potentials for Common Elements and Compounds†

Material	I (eV)	Material	I (eV)
C	78	Ag	469*
N	85	Au	771*
O	89	Pb	773
Mg	156	U	839*
Al	160*	H_2O	65.1
Cr	258*	CO_2	85.9
Mn	273*	NaI	433
Fe	281*	Lucite	65.6
Ni	303*	Methane	44.6
Cu	321*	Muscle	66.2
Zn	323*	Air	86.8

†Those with * are from experimental results of refs. 2 and 3; others, from ref. 1.

all particles, dE/dx exhibits a minimum that occurs approximately at $\gamma \approx 3$. For electrons, $\gamma = 3$ corresponds to $T = 1$ MeV; for alphas, $\gamma = 3$ corresponds to $T \simeq$ 3750 MeV.

Equations 4.3 and 4.4, giving the stopping power for electrons and positrons, respectively, are essentially the same. Their difference is due to the second term in the bracket, which is always much smaller than the logarithmic term. For an electron and positron with the same kinetic energy, Eqs. 4.3 and 4.4 provide results which are different by about 10 percent or less. For low kinetic energies, dE/dx for positrons is larger than that for electrons; at about 2000 keV, the energy loss is the same; for higher kinetic energies, dE/dx for positrons is less than that for electrons.

As stated earlier, Eqs. 4.2 to 4.4 disregard the effect of forces between atoms and atomic electrons of the attenuating medium. A correction for this *density effect*[4,5] has been made, but it is small and it will be neglected here. The density effect reduces the stopping power slightly.

EXAMPLE 4.1

What is the stopping power of air for a 5-MeV alpha particle?

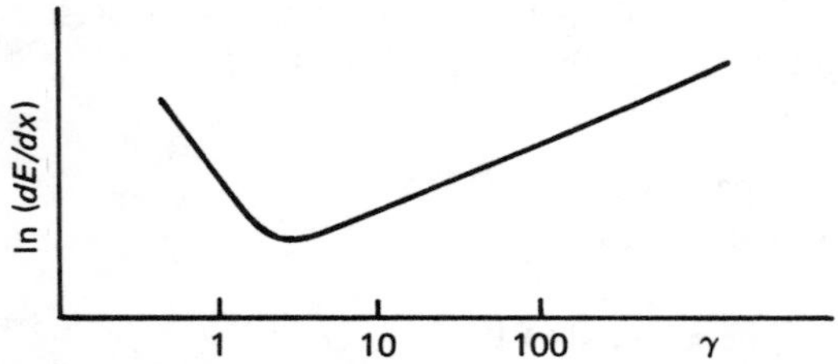

FIG. 4.3 Change of stopping power with the kinetic energy of the particle: $T = (\gamma - 1)mc^2$.

ANSWER

For this example, air will be considered to be 100 percent nitrogen. Therefore, $Z = 7$, $I = 85$ eV, and:

$$N = \rho \frac{N_A}{A} = (1.29 \text{ kg/m}^3) \frac{6.023 \times 10^{23} \text{ at/mol}}{14 \times 10^{-3} \text{ kg/mol}} = 5.55 \times 10^{-3} \times 10^{28} \text{ at/m}^3$$

$$\beta^2 = 1 - \frac{1}{\gamma^2} = 1 - \left[\frac{4(939)}{5 + (4)(939)}\right]^2 = 0.00266$$

$$\frac{dE}{dx} = 10^{-28}(2^2)\frac{0.511}{0.00266}(5.55 \times 10^{-3})(10^{28})7 \left[\ln\left(\frac{2 \times 0.511 \times 0.00266}{85 \times 10^{-6}}\right) - \ln(0.997) - 0.00266\right] = 103.4 \text{ MeV/m} = 1.66 \times 10^{-11} \text{ J/m}$$

4.4 ENERGY LOSS DUE TO BREMSSTRAHLUNG EMISSION

The calculation of energy loss due to emission of bremsstrahlung is more involved than the calculation of energy loss due to ionization and excitation. Here, an approximate equation will be given for electrons or positrons only, because it is for these particles that energy loss due to radiation emission may be important.

For electrons or positrons with kinetic energy T (MeV) moving in a material with atomic number Z, the energy loss due to bremsstrahlung emission, $(dE/dx)_{rad}$, is given in terms of the ionization and excitation energy loss by Eq. 4.6 (see Evans).

$$\left(\frac{dE}{dx}\right)_{rad} = \frac{ZT\,(\text{MeV})}{750}\left(\frac{dE}{dx}\right)_{ion} \tag{4.6}$$

where $(dE/dx)_{ion}$ is the stopping power due to ionization-excitation (Eq. 4.3 or 4.4).

EXAMPLE 4.2

Consider an electron with $T = 5$ MeV. What fraction of its energy is lost as bremstrahlung as it starts moving (a) in aluminum and (b) in lead?

ANSWER

(a) If it travels in aluminum ($Z = 13$),

$$\left(\frac{dE}{dx}\right)_{rad} = \frac{13(5)}{750}\left(\frac{dE}{dx}\right)_{ion} = 0.09\left(\frac{dE}{dx}\right)_{ion}$$

That is, the rate of energy loss due to radiation is about 9 percent of $(dE/dx)_{ion}$.

(b) For the same electron moving in lead ($Z = 82$),

$$\left(\frac{dE}{dx}\right)_{rad} = \frac{82(5)}{750}\left(\frac{dE}{dx}\right)_{ion} = 0.55\left(\frac{dE}{dx}\right)_{ion}$$

In this case, the rate of radiation energy loss is 55 percent of $(dE/dx)_{ion}$.

Equation 4.6, relating radiation to ionization energy loss, is a function of the kinetic energy of the particle. As the particle slows down, T decreases and $(dE/dx)_{rad}$ also decreases. The total energy radiated as bremsstrahlung is approximately equal in MeV to†

$$T_{rad} = 4.0 \times 10^{-4} ZT^2 \tag{4.7}$$

EXAMPLE 4.3

What is the total energy radiated by the electron of Example 4.2?

ANSWER

Using Eq. 4.7,

(a) In aluminum: $T_{rad} = (4.0 \times 10^{-4})(13)5^2 = 0.130$ MeV

(b) In lead: $T_{rad} = (4.0 \times 10^{-4})(82)5^2 = 0.820$ MeV

The total stopping power for electrons or positrons will be given by the sum of Eqs. 4.3 or 4.4 and 4.6:

$$\left(\frac{dE}{dx}\right)_{tot} = \left(\frac{dE}{dx}\right)_{ion} + \left(\frac{dE}{dx}\right)_{rad} = \left(1 + \frac{ZT}{750}\right)\left(\frac{dE}{dx}\right)_{ion} \tag{4.8}$$

If the particle moves in a compound or a mixture instead of a pure element, an effective atomic number Z_{ef} should be used in Eqs. 4.6 and 4.7. The value of Z_{ef} is given by Eq. 4.9.

$$Z_{ef} = \frac{\sum_{i=1}^{L} (w_i/A_i) Z_i^2}{\sum_{i=1}^{L} (w_i/A_i) Z_i} \tag{4.9}$$

where L = number of elements in the compound or mixture
w_i = weight fraction of ith element
A_i = atomic weight of ith element
Z_i = atomic number of ith element

For a compound with molecular weight M, the weight fraction is given by

$$w_i = \frac{N_i A_i}{M} \tag{4.10}$$

where N_i is the number of atoms in the compound of ith element.

†The coefficient 4.0×10^{-4} used in Eq. 4.7 is not universally accepted (see Evans).

4.5 CALCULATION OF dE/dx FOR A COMPOUND OR MIXTURE

Equations 4.2 to 4.4 give the result of the stopping power calculation if the particle moves in a pure element. If the particle travels in a compound or a mixture of several elements, the stopping power is given by

$$\frac{1}{\rho}\left(\frac{dE}{dx}\right)_{\text{compound}} = \sum_i w_i \frac{1}{\rho_i}\left(\frac{dE}{dx}\right)_i \tag{4.11}$$

where ρ = density of compound or mixture

ρ_i = density of ith element

$1/\rho_i(dE/dx)_i$ = stopping power in MeV/kg/m^2 for ith element, as calculated using Eqs. 4.2 to 4.4

EXAMPLE 4.4

What is the energy loss of a 10-MeV electron moving in air? Assume that air consists of 21 percent oxygen and 79 percent nitrogen.

ANSWER

Equation 4.11 will be used, but first dE/dx will have to be calculated for the two pure gases. Using Eq. 4.3,

$$\gamma = \frac{T + mc^2}{mc^2} = \frac{10 + 0.511}{0.511} = 20.569$$

$$\beta = \sqrt{\frac{\gamma^2 - 1}{\gamma^2}} = 0.9988 \qquad \beta^2 = 0.9976$$

For oxygen:

$$\frac{1}{\rho}\left(\frac{dE}{dx}\right) = 10^{-28}\left(\frac{0.511}{0.9976}\right)\frac{0.6023 \times 10^{24}}{16 \times 10^{-3}}(8)$$

$$\times\left(\ln\frac{0.9988(20.569)\sqrt{19.569}\,(0.511)}{80 \times 10^{-6}} + \frac{1}{2(20.569)^2}\left\{\frac{19.569^2}{8}\right.\right.$$

$$\left.\left. + 1 - [2(20.569)^2 + 2(20.569) - 1]\ln 2\right\}\right)$$

$$= 3.09 \times 10^{-14}\ \text{J/kg/m2} = 1.928\ \text{MeV/g/cm}^2$$

For nitrogen:

$$\frac{1}{\rho}\left(\frac{dE}{dx}\right) = 10^{-28}\left(\frac{0.511}{0.9976}\right)\frac{0.6023 \times 10^{24}}{14 \times 10^{-3}}(7)$$

$$\times\left(\ln\frac{0.9988(20.569)\sqrt{19.569}\,(0.511)}{85 \times 10^{-6}} + \frac{1}{2(20.569)^2}\left\{\frac{19.569^2}{8}\right.\right.$$

$$\left.\left. + 1 - [2(20.569)^2 + 2(20.569) - 1]\ln 2\right\}\right)$$

$$= 3.098 \times 10^{-14}\ \text{J/(kg/m}^2) = 1.9\ \text{MeV/(g/cm}^2)$$

For air:

$$\frac{1}{\rho}\left(\frac{dE}{dx}\right)_{\text{air}} = 0.21\left(\frac{1}{\rho}\right)\left(\frac{dE}{dx}\right)_{\text{O}} + 0.79\left(\frac{1}{\rho}\right)\left(\frac{dE}{dx}\right)_{\text{N}}$$

$$= 0.21(3.09 \times 10^{-14}) + 0.79(3.098 \times 10^{-14}) \text{ J/kg/m}^2$$

$$= 3.096 \times 10^{-14} \text{ J/kg/cm}^2 = 1.933 \text{ MeV/g/cm}^2$$

$$\left(\frac{dE}{dx}\right)_{\text{air}} = 3.096 \times 10^{-14} \text{ J/kg/m}^2 (1.29 \text{ kg/m}^3)$$

$$= 3.994 \times 10^{-14} \text{ J/m} = 0.249 \text{ MeV/m}$$

4.6 RANGE OF CHARGED PARTICLES

A charged particle moving through a certain material loses its kinetic energy through interactions with the electrons and nuclei of the material. Eventually the particle will stop, pick up the necessary number of electrons from the surrounding matter, and become neutral. For example,

$$p^+ + e^- \rightarrow \text{hydrogen atom}$$

$$\alpha^{2+} + 2e^- \rightarrow \text{He atom}$$

The total distance traveled by the particle is called the *pathlength.* The pathlength S, shown in Fig. 4.4, is equal to $\Sigma_i S_i$. The thickness of material that just stops a particle of kinetic energy T, mass M, and charge z is called the *range* R of the particle in that material. It is obvious that $R \leqslant S$. For electrons, which have a zig-zag path, $R < S$. For heavy charged particles, which are very slightly deflected, $R \approx S$.

Range is distance, and its basic dimension is length (m). In addition to meters, another unit used for range is kg/m² (or g/cm²). The relationship between the two is:

$$R\ (\text{kg/m}^2) = [R\ (\text{m})]\ [\rho\ (\text{kg/m}^3)]$$

where ρ is the density of the material in which the particle travels. The range measured in kg/m² is independent of the state of matter. That is, a particle will have the same range in kg/m² whether it moves in ice, water, or steam. Of course, the range measured in meters will be different.

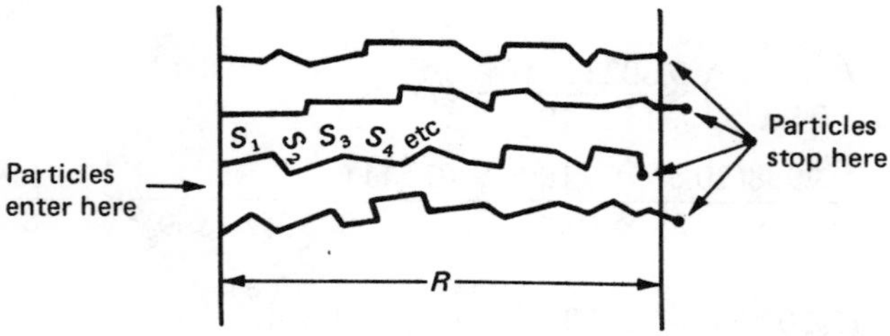

FIG. 4.4 Pathlength (S) and range (R). The end points of the pathlengths are distributed around an average thickness that is the range.

FIG. 4.5 Particle transmission experiment.

The range is an average quantity. Particles of the same type with the same kinetic energy moving in the same medium will not stop after traveling exactly the same thickness R. Their pathlength will not be the same either. What actually happens is that the end points of the pathlengths will be distributed around an average thickness called the range. To make this point more clear, two experiments will be discussed dealing with transmission of charged particles. Heavy particles and electrons-positrons will be treated separately.

4.6.1 Range of Heavy Particles ($Z \leqslant 2, A \leqslant 4$: p, d, t, α)

Consider a parallel beam of heavy charged particles all having the same energy and impinging upon a certain material (Fig. 4.5). The thickness of the material may be changed at will. On the other side of the material, a detector records the particles that traverse it. It is assumed that the particle direction does not change and that the detector will record all particles that go through the material, no matter how low their energy is. The number of particles $N(t)$ traversing the thickness t changes, as shown in Fig. 4.6.

In the beginning, $N(t)$ stays constant, even though t changes. Beyond a certain thickness, $N(t)$ starts decreasing and eventually goes to zero. The thickness for which $N(t)$ drops to half its initial value is called the *mean range* R. The thickness for which $N(t)$ is practically zero is called the *extrapolated range* R_e. The difference between R and R_e is about 5 percent or less. Unless otherwise specified, when range is used it is the mean range R.

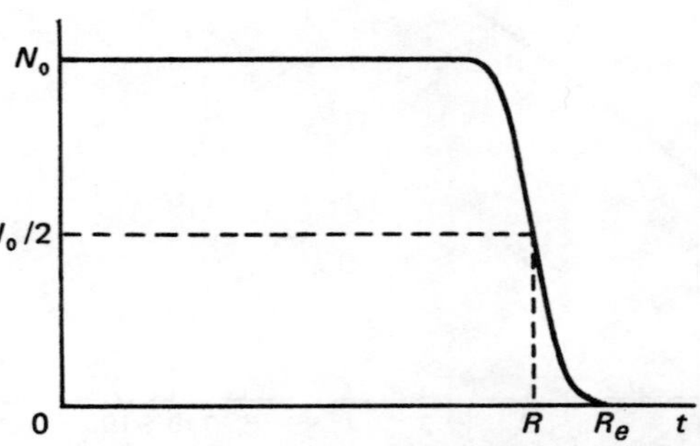

FIG. 4.6 The number of heavy charged particles (α, p, d, t) transmitted through thickness t.

There are semiempirical formulas that give the range as a function of particle kinetic energy. For alpha particles, the range in air at normal temperature and pressure is given by

$$R\ (\mathrm{mm}) = \exp\left[1.61\sqrt{T\ (\mathrm{MeV})}\right] \qquad 1 < T \leqslant 4\ \mathrm{MeV}$$
$$R\ (\mathrm{mm}) = (0.05T + 2.85)T^{3/2}\ (\mathrm{MeV}) \qquad 4 \leqslant T \leqslant 15\ \mathrm{MeV} \tag{4.12}$$

where T = kinetic energy of the particle in MeV. Figure 7 gives the range of alphas in silicon.

If the range is known for one material, it can be determined for any other by applying the Bragg-Kleeman rule†:

$$\frac{R_1}{R_2} = \frac{\rho_2}{\rho_1}\sqrt{\frac{A_1}{A_2}} \tag{4.13}$$

where ρ_i and A_i are the density and atomic weight, respectively, of material i. For a compound or mixture, an effective molecular weight is used, obtained from the equation

$$\sqrt{A_{ef}} = \left(\sum_{i=1}^{L} \frac{w_i}{\sqrt{A_i}}\right)^{-1} \tag{4.14}$$

where the quantities w_i, A_i, and L have the same meaning as in Eq. 4.9.

†The Bragg-Kleeman rule does not hold for electron or positron ranges.

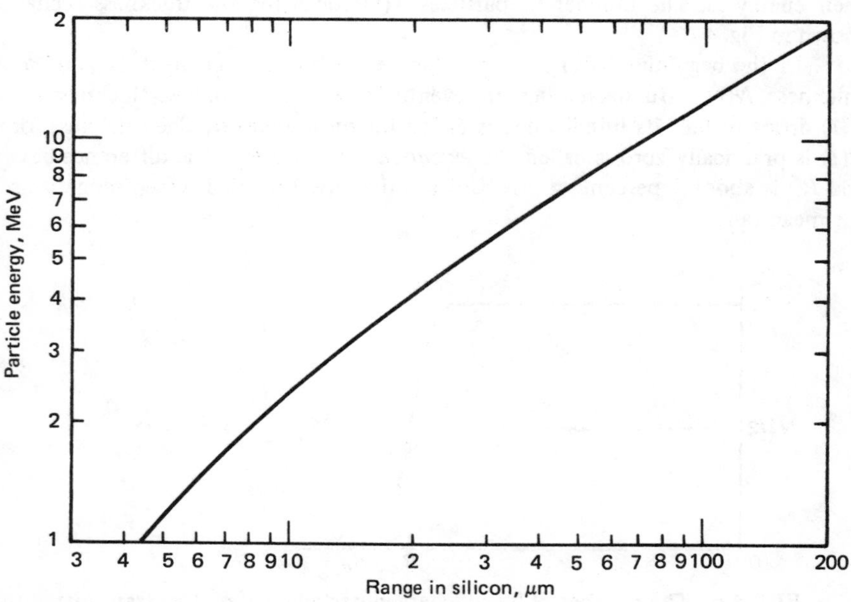

FIG. 4.7 Range-energy curve for alpha particles in silicon (ref. 6).

EXAMPLE 4.5

What is the effective molecular weight for water? What is it for air?

ANSWER

For H_2O (0.11% H, 0.89% O):

$$\sqrt{A_{ef}} = \left(\frac{0.11}{\sqrt{1}} + \frac{0.89}{\sqrt{16}}\right)^{-1} = 3 \quad A_{ef} = 9$$

For air (22.9% O, 74.5% N, 2.6% Ar),

$$\sqrt{A_{ef}} = \left(\frac{0.229}{\sqrt{16}} + \frac{0.745}{\sqrt{14}} + \frac{0.026}{\sqrt{40}}\right)^{-1} = 3.84 \quad A_{ef} = 14.74$$

Using the Bragg-Kleeman rule (Eq. 4.13), with air as one of the materials ($\rho = 1.23\ \text{kg/m}^3$, $\sqrt{A_{ef}} = 3.84$), one obtains

$$R\ (\text{mm}) = (3.2 \times 10^{-1}) \frac{\sqrt{A_{ef}}}{\rho\ (\text{kg/m}^3)} R_{air}\ (\text{mm}) \tag{4.15}$$

There are two ways to calculate the range of alphas in a material other than air and silicon.

1 The range in air should be obtained first, using Eq. 4.12, and then the range in the material of interest should be calculated using Eq. 4.15. Or,
2 The range in silicon should be read off Fig. 4.7, and then the range in the material of interest should be calculated using Eq. 4.13.

EXAMPLE 4.6

What is the range of a 3-MeV alpha particle in gold?

ANSWER

The range of this alpha in silicon is (Fig. 4.7) $R = 12.5\ \mu\text{m} = 12.5 \times 10^{-6}$ m. Using Eq. 4.13, the range in gold is:

$$R_{Au} = (12.5 \times 10^{-6}) \frac{2.33 \times 10^3}{19.32 \times 10^3} \sqrt{\frac{197}{28}} = 4 \times 10^{-6}\ \text{m} = 4\ \mu\text{m}$$

Or, using Eqs. 4.12 and 4.15,

$$R = (3.2 \times 10^{-1}) \frac{\sqrt{197}}{19.32 \times 10^3} \exp(1.61 \sqrt{T}) = 3.8 \times 10^{-3}\ \text{mm} = 3.8\ \mu\text{m}$$

EXAMPLE 4.7

What is the range of a 10-MeV alpha particle in aluminum?

ANSWER

From Fig. 4.7, the range in silicon is $R = 70\ \mu\text{m} = 7.0 \times 10^{-5}$ m. Using Eq. 4.13, the range in aluminum is:

$$R_{\mathrm{Al}} = (70\ \mu\mathrm{m})\,\frac{2.33 \times 10^3}{2.7 \times 10^3}\sqrt{\frac{27}{28}} = 59\ \mu\mathrm{m}$$

Or, using Eqs. 4.12 and 4.15,

$$R_{\mathrm{Al}} = (3.2 \times 10^{-1})\,\frac{\sqrt{27}}{2.7 \times 10^3}\,[0.05(10) + 2.85]\,10^{3/2}\ \mathrm{mm}$$

$$= 6.52 \times 10^{-5}\ \mathrm{m} = 65\ \mu\mathrm{m}$$

The difference of 6 μm is within the range of accuracy of the Bragg–Kleeman rule.

The range of protons in *aluminum* has been measured by Bichsel.[7] His results are represented very well by the following two equations:

$$R\ (\mu\mathrm{m}) = 14.21T^{1.5874} \qquad 1\ \mathrm{MeV} < T \leqslant 2.7\ \mathrm{MeV} \tag{4.16}$$

$$R\ (\mu\mathrm{m}) = 10.5\,\frac{T^2}{0.68 + 0.434 \ln T} \qquad 2.7\ \mathrm{MeV} \leqslant T < 20\ \mathrm{MeV} \tag{4.17}$$

For other materials, Eq. 4.13 should be used after the range in aluminum is determined from Eqs. 4.16 and 4.17.

The range of protons and deuterons can be calculated from the range of an alpha particle *of the same speed* using the formula:

$$R(p, d) = 4\frac{M(p, d)}{M_\alpha} R_\alpha - 2\ (\mathrm{mm, air}) \tag{4.18}$$

where R_α = range in air of an alpha particle having the same speed as the deuteron or the proton

M = mass of the particle (1 for proton, 2 for deuteron)

M_α = mass of alpha particle = 4

For materials other than air, the Bragg–Kleeman rule (Eq. 4.15) should be used.

The fact that the alpha and proton or deuteron ranges are related by the same speed rather than the same kinetic energy is due to the dependence of dE/dx on the speed of the particle.

EXAMPLE 4.8

What is the range of a 5-MeV deuteron in air?

ANSWER

Equation 4.18 will be used, but first the range of an alpha particle will speed equal to that of a 5-MeV deuteron will have to be calculated. The kinetic energy of an alpha particle with the *same speed* as that of the deuteron will be found using the corresponding equations for the kinetic energy. Since $T = \frac{1}{2}MV^2$ for these nonrelativistic particles,

$$T_\alpha = \frac{M_\alpha}{M_d} T_d = 2T_d = 10\ \mathrm{MeV}$$

The range of a 10-MeV alpha particle (in air) is (Eq. 4.12)

$$R = (0.05T + 2.85)T^{3/2} = 106 \text{ mm} = 0.106 \text{ m}$$

The range of the 5-MeV deuteron (in air) is then (Eq. 4.18)

$$R_p = 2R_\alpha - 2 = 210 \text{ mm} = 0.21 \text{ m}$$

EXAMPLE 4.9

What is the range of a 5-MeV deuteron in aluminum?

ANSWER

Using the Bragg-Kleeman rule (Eq. 4.15) and the result of the previous example,

$$R_d\,(\text{Al}) = (3.2 \times 10^{-1})\frac{\sqrt{27}}{2.7 \times 10^3}\,210 \text{ mm} = 0.129 \text{ mm} = 129\ \mu\text{m}$$

4.6.2 Range of Electrons and Positrons

Electrons and positrons behave in essentially the same way with regard to energy loss, slowing down, and penetration through matter. Small differences exist; one was indicated when dE/dx was discussed in Sec. 4.3. Small differences in the values of the range between electrons and positrons should also be expected, and indeed this is the case. Most of the range measurements have been performed with electrons, because electrons are used much more frequently than positrons in radiation measurements. For this reason, from this point on only electrons will be discussed. The reader should be aware that the results are equally applicable for positrons, to a first approximation, but for very accurate results the references listed at the end of this chapter should be consulted.

If the experiment shown in Fig. 4.5 and discussed in Sec. 4.6.1 is repeated with the incident beam consisting of monoenergetic electrons, the result will look as shown in Fig. 4.8. For electrons, the transmission curve does not have a flat part. It decreases gradually to a level which is the background. The range† is equal to the thickness

†In many texts, this is called the "extrapolated" range. Since only one type of range is used, there is no need to carry along the word "extrapolated."

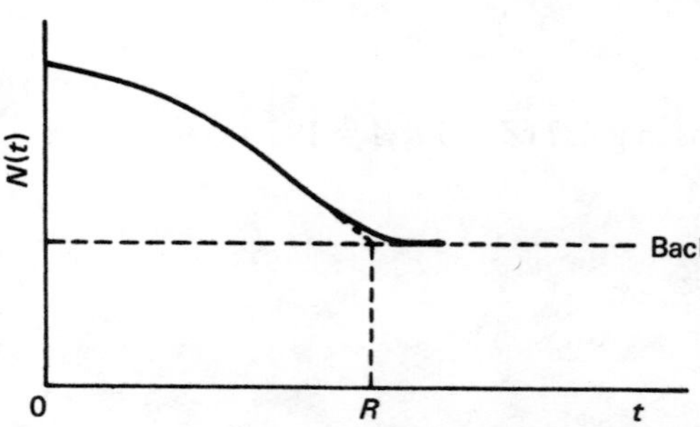

FIG. 4.8 The number of electrons transmitted through thickness t. Experiment set-up shown in Fig. 4.5.

of the material, which is defined by the point where the linear extrapolation of the transmission curve meets the background.

The semiempirical equation giving the range of electrons for the energy range 0.3 keV to 30 MeV has been developed by Tabata, Ito, and Okabe,[8] based on the experimental results available until 1972. This equation, indicated from now on as the TIO equation, has the following form:

$$R\ (\text{kg/m}^2) = a_1 \left\{ \frac{\ln[1 + a_2(\gamma - 1)]}{a_2} - \frac{a_3(\gamma - 1)}{1 + a_4(\gamma - 1)^{a_5}} \right\} \tag{4.19}$$

where

$$\left.\begin{aligned} a_1 &= \frac{2.335A}{Z^{1.209}} & a_3 &= 0.9891 - (3.01 \times 10^{-4} Z) \\ a_2 &= 1.78 \times 10^{-4} Z & a_4 &= 1.468 - (1.180 \times 10^{-2} Z) \\ & & a_5 &= \frac{1.232}{Z^{0.109}} \end{aligned}\right\} \tag{4.20}$$

A, Z, and γ have been defined in Sec. 4.3.

Figures 4.9 and 4.10 show results based on Eq. 4.19, as well as experimental points.

In the case of absorbers that are mixtures or compounds, the atomic number Z and atomic weight A to be used in Eq. 4.20 are given by

$$Z_{ef} = \sum_i^L w_i Z_i \tag{4.21}$$

$$A_{ef} = Z_{ef} \left(\sum_i^L w_i \frac{Z_i}{A_i} \right)^{-1} \tag{4.22}$$

where w_i is the weight fraction of element with atomic number Z_i and atomic weight A_i.

EXAMPLE 4.10

What is the range of 1-MeV electrons in gold? ($Z = 79, A = 197$.)

ANSWER

Using Eq. 4.19 and 4.20,

$$a_1 = \frac{2.335(197)}{79^{1.209}} = 2.336$$

$$a_2 = (1.78 \times 10^{-4})(79) = 0.01406$$

$$a_3 = 0.9891 - (3.01 \times 10^{-4})(79) = 0.965$$

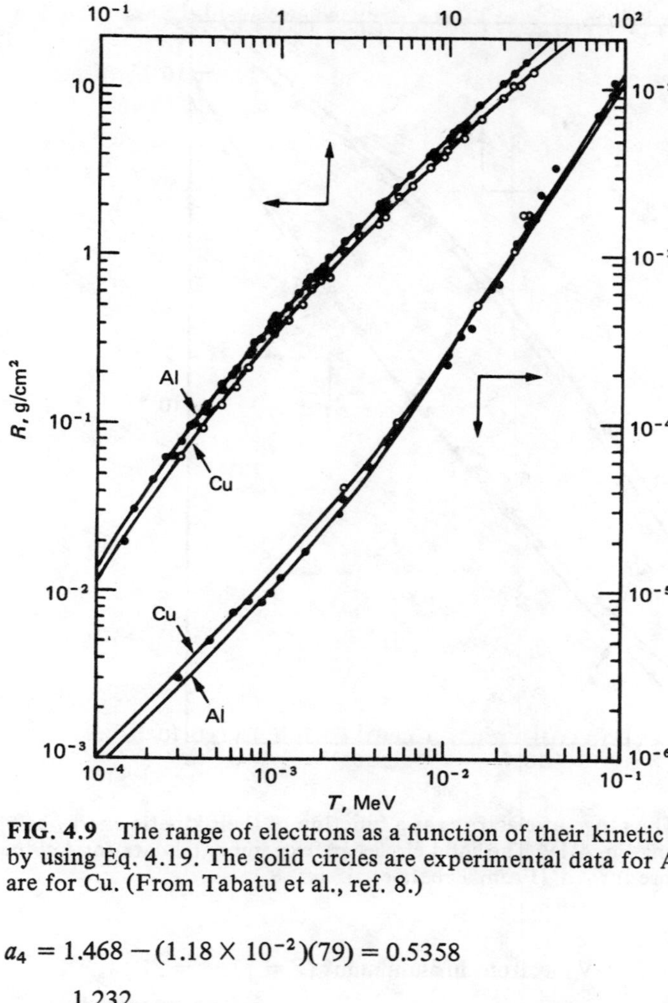

FIG. 4.9 The range of electrons as a function of their kinetic energy as obtained by using Eq. 4.19. The solid circles are experimental data for Al; the open circles are for Cu. (From Tabatu et al., ref. 8.)

$$a_4 = 1.468 - (1.18 \times 10^{-2})(79) = 0.5358$$

$$a_5 = \frac{1.232}{79^{0.109}} = 0.765$$

$$\gamma = \frac{1.511}{0.511} = 2.957$$

$$R = 2.336 \left[\frac{\ln(1 + 0.0275)}{0.01406} - \frac{1.8885}{1.895} \right] = 2.18 \text{ kg/m}^2 = 0.219 \text{ g/cm}^2$$

Since the density of gold is 19.3×10^3 kg/m^3, the range in μm is

$$R = \frac{2.18 \text{ kg/m}^2}{19.3 \times 10^9 \text{ kg/m}^3} = 1.13 \times 10^{-4} \text{ m} = 113\ \mu\text{m}$$

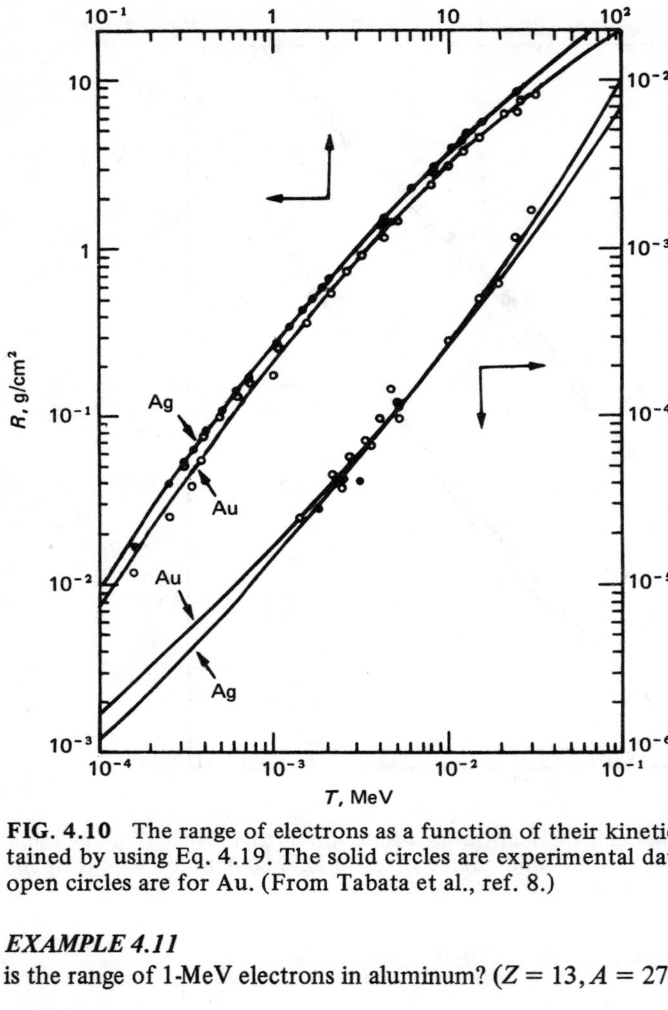

FIG. 4.10 The range of electrons as a function of their kinetic energy as obtained by using Eq. 4.19. The solid circles are experimental data for Ag; the open circles are for Au. (From Tabata et al., ref. 8.)

EXAMPLE 4.11

What is the range of 1-MeV electrons in aluminum? ($Z = 13, A = 27$.)

ANSWER

Again, using Eq. 4.19 and 4.20,

$$a_1 = \frac{2.335(27)}{13^{1.209}} = 2.837$$

$$a_2 = (1.78 \times 10^{-4})(13) = 2.314 \times 10^{-3}$$

$$a_3 = 0.9891 - (3.01 \times 10^{-4})(13) = 0.985$$

$$a_4 = 1.468 - (1.18 \times 10^{-2})(13) = 1.3146$$

$$a_5 = \frac{1.232}{13^{0.109}} = 0.9315$$

$$R = 2.837\left(\frac{\ln 1.0045}{0.00231} - \frac{1.928}{3.457}\right) = 3.93 \text{ kg/m}^2 = 0.393 \text{ g/cm}^2$$

Since the density of aluminum is 2.7×10^3 kg/m³, the range in μm is

$$R = \frac{3.93 \text{ kg/m}^2}{2.7 \times 10^3 \text{ kg/m}^3} = 1.46 \times 10^{-3} \text{ m} = 1460 \ \mu\text{m}$$

4.6.3 Transmission of Beta Particles

Beta particles have a continuous energy spectrum extending from zero energy up to maximum kinetic energy E_{max} (see Sec. 3.7.3). If the transmission experiment shown in Fig. 4.5 is repeated with an incident beam of β particles, the result will look as shown in Fig. 4.11. The number of betas $N(t)$ transmitted through a thickness t is very closely represented by

$$N(t) = N(0)e^{-\mu t} \tag{4.23}$$

where μ is the mass absorption coefficient.

The value of μ has been determined experimentally as a function of the maximum beta energy and is given by

$$\mu \, (\text{m}^2/\text{kg}) = 1.7 E_{max}^{-1.14} \tag{4.24}$$

where E_{max} is in MeV. Notice that μ is given in units of m²/kg; therefore the thickness t in the exponent of Eq. 4.23 should be in kg/m². The exponential transmission law represented by Eq. 4.23 is the result of experimental observation. There is no theory

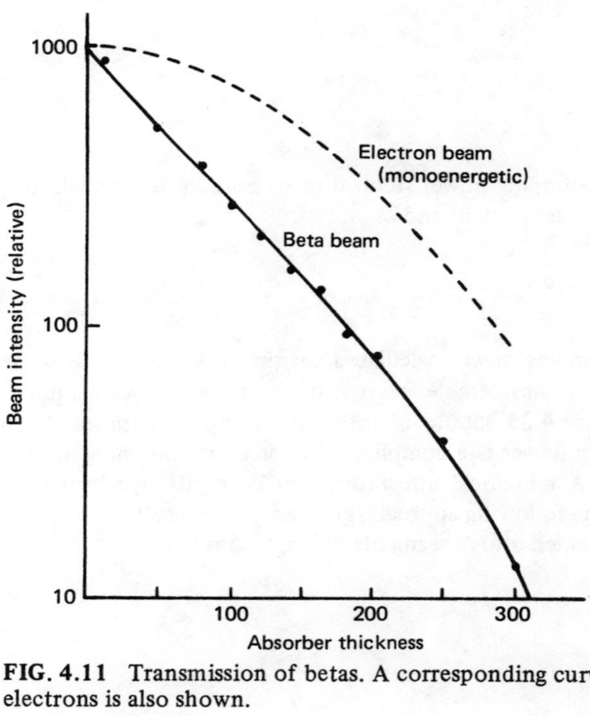

FIG. 4.11 Transmission of betas. A corresponding curve for monoenergetic electrons is also shown.

predicting it. The range of β particles is calculated using Eq. 4.19 for kinetic energy equal to E_{max}.

EXAMPLE 4.12

What fraction of 2-MeV betas will go through a single Al foil of thickness 0.1 mm?

ANSWER

The mass absorption coefficient is, using Eq. 4.24,

$$\mu = 1.7 \times 2^{-1.14} = 0.7714 \text{ m}^2/\text{kg}$$

The fraction transmitted is, using Eq. 4.23,

$$\frac{N(t)}{N(0)} = e^{-\mu t} = \exp\left[-0.7714 \text{ m}^2/\text{kg}(0.1 \times 10^{-3} \text{ m})(2.7 \times 10^3 \text{ kg/m}^3)\right]$$

$$= \exp(-0.208) = 0.81$$

Therefore, 81 percent of the betas will go through this foil.

4.6.4 Energy Loss After Traversing a Material of Thickness t

One is often required to calculate the energy loss of a charged particle after it traverses a material of thickness t. The first step in solving such a problem is to calculate the range of the particle in that medium. If the range $R < t$, the particle stopped in the medium and the total energy loss is equal to the initial energy of the particle. If $R > t$, the energy loss ΔE is given by

$$\Delta E = \int_0^t \frac{dE}{dx}\, dx \tag{4.25}$$

where dE/dx is the total stopping power (ionization-excitation plus radiation loss). If $t \ll R$, one may take dE/dx as constant and obtain

$$\Delta E = \left(\frac{dE}{dx}\right)_0 t \qquad t \ll R \tag{4.26}$$

where $(dE/dx)_0$ is the stopping power calculated for the initial energy of the particle.

If the thickness t is a considerable fraction of the range, dE/dx cannot be considered constant. Then Eq. 4.25 should be integrated using the appropriate form of dE/dx. Since the stopping power is a complicated expression, the integration cannot be carried out by hand. A numerical integration can be performed by a computer. In most cases, however, the following approach gives adequate results.

The thickness t is divided into N segments of length Δx_i where

$$\sum_{i=1}^{N} \Delta x_i = t$$

Equation 4.25 takes the form

$$\Delta E = \sum_{i=1}^{N} \left(\frac{dE}{dx}\right)_i \Delta x_i \tag{4.27}$$

where $(dE/dx)_i$ is the stopping power calculated for the kinetic energy of the particle at the beginning of the segment Δx_i.

There is no general rule as to the best value of the number of segments N. Obviously, N should be such that $(dE/dx)_i$ changes by a small but acceptable amount as the particle travels the segment Δx_i.

EXAMPLE 4.13

What is the energy loss of a 10-MeV electron going through 15 mm of aluminum?

ANSWER

Using Eq. 4.19 or Fig. 4.9, the range of a 10-MeV electron in aluminum is $R = 20.4$ mm. The particle will emerge, but the thickness of the absorber is a considerable fraction of the range. Therefore, one should use Eq. 4.27.

If one chooses $N = 5$ and equal segments, Eq. 4.27 takes the form

$$\Delta E = \sum_{i=1}^{5} \left(\frac{dE}{dx}\right)_i \Delta x_i \qquad \Delta x_i = 3 \text{ mm}$$

The table below shows how the calculation proceeds.

i	T_i (MeV)	dE/dx (MeV/mm)	$(\Delta E)_i$ (MeV)	$T_{i+1} = T_i - (\Delta E)_i$
1	10	0.605	1.815	8.185
2	8.185	0.568	1.704	6.481
3	6.481	0.530	1.590	4.891
4	4.891	0.492	1.476	3.415
5	3.415	0.457	1.373	2.042

Total energy loss is 7.958 MeV.

4.7 STOPPING POWER AND RANGE OF HEAVY IONS ($Z > 2, A > 4$)

4.7.1 Introduction

The equations presented in Secs. 4.3 to 4.6 for energy loss and range of charged particles were derived with the assumption that the charge of the particle does not change as the particle traverses the medium. This assumption is certainly valid for electrons, positrons, protons, and deuterons ($Z = 1$). It holds well for alphas too ($Z = 2$). However, for $Z > 2$, the charge of the particle can not be assumed constant, and for this reason the energy-loss and range calculations require special treatment.

Consider an atom or an ion with speed greater than the orbital velocity of its own electrons. If this particle enters a certain medium, the atomic electrons will be quickly removed from the atom or ion leaving behind a bare nucleus. The nucleus will keep moving through the medium, continuously losing energy in collisions with the electrons of the medium.† It is probable that the ion will capture an electron in one of these collisions. It is also probable that the electron will be lost in the next collision. As the ion slows down and its speed becomes of the same order of magnitude as the orbital speeds of the atomic electrons, the probability for electron capture increases, while the probability for electron loss decreases. When the ion slows down even farther and is slower than the orbiting electrons, the probability of losing an electron becomes essentially zero while the probability of capturing one becomes significant. As the speed of the ion continues to decrease, a third electron is captured, then a fourth, and so on. At the end, the ion is slower than the least bound electron. By that time, it is a neutral atom. What is left of its kinetic energy is exchanged through nuclear and not electronic collisions. The neutral atom is considered as stopped when it either combines chemically with one of the atoms of the material or is in thermal equilibrium with the medium.

4.7.2 The dE/dx Calculation

The qualitative discussion of Sec. 4.7.1 showed how the charge of a heavy ion changes as the ion slows down in the medium. It is this variation of the charge that makes the energy loss calculation very difficult. There is no single equation giving dE/dx for all heavy ions and for all stopping materials. Instead, dE/dx is calculated differently, depending on the speed of the ion relative to the speed of the orbital electrons.

The stopping power is written, in general, as the sum of two terms:

$$\frac{dE}{dx} = \left(\frac{dE}{dx}\right)_e + \left(\frac{dE}{dx}\right)_n \tag{4.28}$$

where $(dE/dx)_e$ = electronic energy loss

$(dE/dx)_n$ = nuclear energy loss

An excellent review of the subject is presented by Northcliffe[9] and Lindhard, Scharff, and Schiott.[10] The results are usually presented as universal curves in terms of two dimensionless quantities, the distance ρ and the energy ϵ, first introduced by Lindhard et al.[10] and defined as follows:

$$\rho = 4\pi a^2 N \frac{M_1 M_2 x}{(M_1 + M_2)^2} \tag{4.29}$$

$$\epsilon = \frac{a}{r_0}\left[\frac{M_2}{Z_1 Z_2 (M_1 + M_2)}\right]\frac{T}{mc^2} \tag{4.30}$$

†Collisions with nuclei are not important if the particle moves much faster than the atomic electrons.

where† $a = 0.8853a_0(Z_1^{2/3} + Z_2^{2/3})^{-1/2}$ (4.31)

x = actual distance traveled

$a_0 = h^2/me^2$ = Bohr radius = 5.29×10^{-11} m

Z_1, M_1 = charge and mass of incident particle

Z_2, M_2 = charge and mass of stopping material

The parameters N, r_0, and mc^2 have been defined in Sec. 4.3.

At high ion velocities, $v \gg v_0 Z_1^{2/3}$, where $v_0 = e^2/\hbar$ = orbital velocity of the electron in the hydrogen atom, the nuclear energy loss is negligible. The particle has an effective charge equal to Z_1, and the energy loss is given by an equation of the form

$$\frac{dE}{dx} \sim \frac{Z_1^2 Z_2}{v^2 A_2} \ln \frac{2mv^2}{I} \tag{4.32}$$

which is similar to Eq. 4.2.

At velocities of the order $v \approx v_0 Z_1^{2/3}$, the ion starts picking up electrons and its charge keeps decreasing. The energy loss through nuclear collisions is still negligible.

In the velocity region $v \lesssim v_0 Z_1^{2/3}$, the electronic energy loss equation takes the form[10]

$$\left(\frac{d\epsilon}{d\rho}\right)_e = k\epsilon^n \tag{4.33}$$

where

$$k = \xi_e \frac{0.0793 Z_1 Z_2 (A_1 + A_2)^{3/2}}{(Z_1^{2/3} + Z_2^{2/3})^{3/4} A_1^{3/2} A_2^{1/2}} \tag{4.34}$$

$$\xi_e \approx Z_1^{1/6}$$

and n has a value very close to $\frac{1}{2}$[10,11]. The constant k depends on Z and A only, not on energy, and its value is less than 1. Some typical values are given in Table 4.2.

Table 4.3 shows the kinetic energy per unit atomic mass, as well as the kinetic energy, of several ions for $v = v_0 Z_1^{2/3}$.

The electronic stopping power for different ions and stopping materials is obtained by using the following semiempirical approach.

†The number $0.8853 = (9\pi^2)^{1/3}/2^{7/3}$ is called the Thomas-Fermi constant.

TABLE 4.2 Values of k (Eq. 4.34)

Z_1	A_1	Z_2	A_2	k
10	50	13	27	0.085
20	60	79	197	0.022
92	238	79	197	0.162

TABLE 4.3 The Kinetic Energy of Heavy Ions for Several Values of $v = v_0 Z_1^{2/3}$

Ion	Z_1	$v_0 Z_1^{2/3}$ ($\times 10^{-7}$) (m/s)	β ($\times 10^2$)	T/A_1	T (MeV)
C	6	0.72	2.4	0.27	3.3
Al	13	1.2	4.04	0.76	20.7
Ni	28	2.0	6.7	2.13	126
Br	35	2.3	7.8	2.87	230
Ag	47	2.8	9.5	4.27	461
I	53	3.1	10.3	5.02	638
Au	79	4.0	13.4	8.6	1694
U	92	4.5	14.9	10.57	2515

The ratio of stopping power for two ions having the same velocity and traveling in the same medium is given by (Eq. 4.32):

$$\frac{(dE/dx)_{T_1 Z_1 A_1}}{(dE/dx)_{T_2 Z_2 A_2}} = \frac{Z_1^2}{Z_2^2} \tag{4.35}$$

The application of Eq. 4.35 to heavy ions should take into account the change of the charge Z_1 as the ion slows down. This is accomplished by replacing Z_1 with an effective charge,

$$Z_{\text{eff}} = \eta Z_1$$

where η is a parameter that depends on energy. The second particle in Eq. 4.35 is taken to be the proton ($Z_2 = A_2 = 1$), thus leading to the form[12,13,14]

$$\frac{(dE/dx)_{Z_1 A_1 T_1}}{(dE/dx)_p T_p} = \frac{\eta^2 Z_1^2}{\eta_p^2} \tag{4.36}$$

where the effective proton charge η_p is given by Eq. 4.37 reported by Booth and Grant[14] and T_p is the proton kinetic energy in MeV.

$$\eta_p^2 = [1 - \exp(-150 T_p)] \exp(-0.835 e^{-14.5} T_p) \tag{4.37}$$

Equations giving the value of η have been reported by many investigators.[12–15] The most recent equation reported by Forster et al.[15] valid for $8 \leqslant Z_1 \leqslant 20$ and for $v/v_0 > 2$ is:

$$\eta = 1 - A(Z_1) \exp\left(-0.879 \frac{v}{v_0} Z_1^{-0.65}\right) \tag{4.38}$$

with

$$A(Z_1) = 1.035 - 0.4 \exp(0.16 Z_1)$$

The proton stopping power is known.[16] Brown[17] has developed an equation of the form:

$$\ln\left(\frac{dE}{dx}\right)_p = a + b \ln T_p + C(\ln T_p)^2 \tag{4.39}$$

by least-squares fitting the data of Northcliffe and Schilling.[16]

The experimental determination of dE/dx is achieved by passing ions of known initial energy through a thin layer of a stopping material and measuring the energy loss. The thickness Δx of the material should be small enough so that $dE/dx \approx \Delta E/\Delta x$. Unfortunately, such a value of Δx is so small, especially for very heavy ions, that the precision of measuring Δx is questionable and the uniformity of the layer has an effect on the measurement. Typical experimental results of stopping power are presented in Fig. 4.12. The data of Fig. 4.12 comes from reference 11. The solid line is based on the following empirical equation proposed by Bridwell and Moak[18,19]:

$$\frac{dE}{dx}\,(\text{MeV/kg}\cdot\text{m}^2) = \frac{2064.5}{A_2}\sqrt{\frac{TA_1 Z_2}{Z_1}} \tag{4.40}$$

where T is the kinetic energy of the ion in MeV.

For a compound or mixture, dE/dx can be obtained by using Eq. 4.11 with $(dE/dx)_i$ obtained from Eq. 4.36 or Eq. 4.40.

At velocities $v < v_0 Z_1^{2/3}$, the energy loss through nuclear elastic collisions becomes important. The so-called *nuclear stopping power* is given by the following approximate expression[10]:

$$\left(\frac{d\epsilon}{d\rho}\right) = \frac{1}{2\epsilon} \ln(1.294\epsilon) \tag{4.41}$$

While the electronic stopping power $(d\epsilon/d\rho)_e$ continuously decreases as the ion speed v decreases, the nuclear stopping power increases as v decreases, goes through a maximum, and then decreases again (Fig. 4.13).

4.7.2 Range of Heavy Ions ($A > 2$)

The range of heavy ions has been measured and calculated for many ions and for different absorbers. But there is no single equation—either theoretical or empirical—giving the range in all cases. Heavy ions are hardly deflected along their path, except very close to the end of their track where nuclear collisions become important. Thus the range R, which is defined as the depth of penetration along the direction of incidence, will be almost equal to the pathlength, the actual distance traveled by the ion. With this observation in mind, the range is given by the equation:

$$R = \int_0^E \frac{dE}{(dE/dx)_e + (dE/dx)_n} \tag{4.42}$$

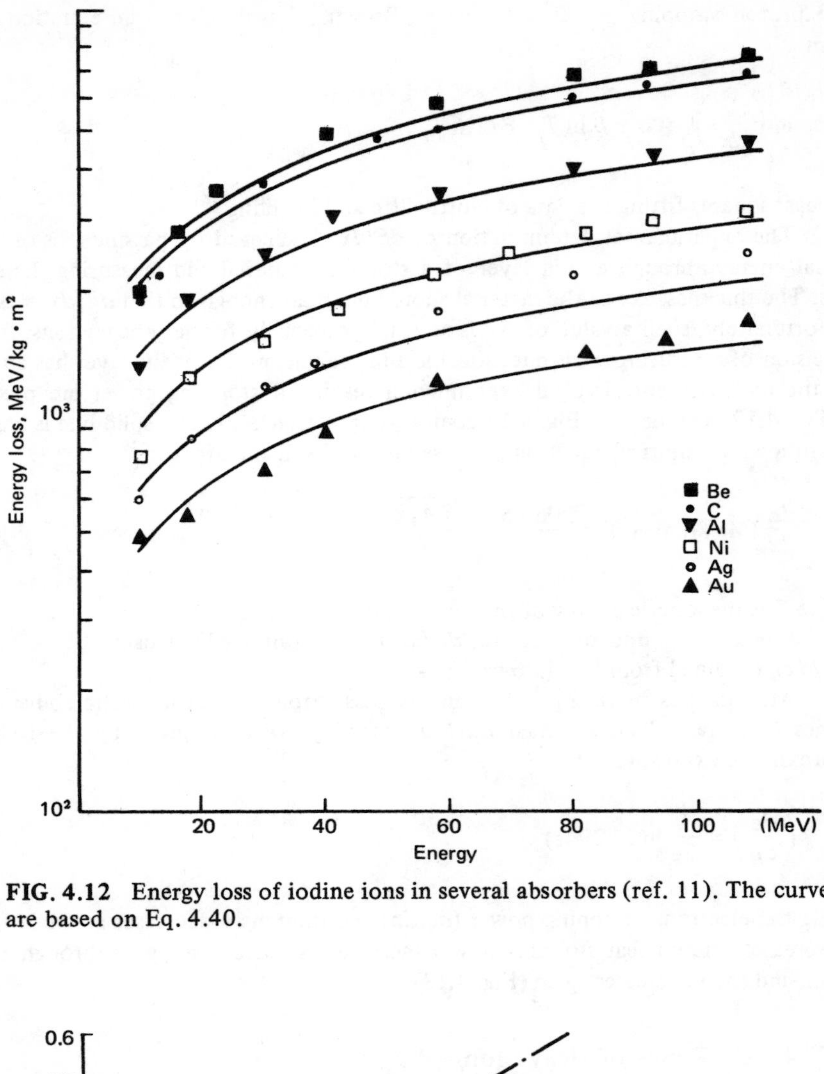

FIG. 4.12 Energy loss of iodine ions in several absorbers (ref. 11). The curves are based on Eq. 4.40.

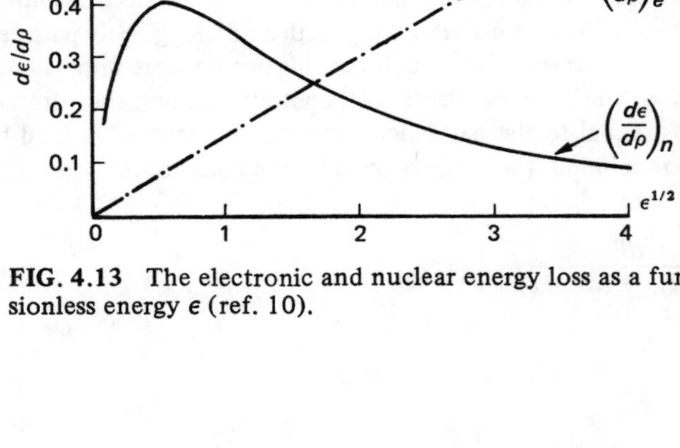

FIG. 4.13 The electronic and nuclear energy loss as a function of the dimensionless energy ϵ (ref. 10).

Results of calculations based on Eq. 4.42 are given by many authors. Based on calculations described in reference 10, Siffert and Coche[20] present universal graphs for several heavy ions in silicon. Figures 4.14 and 4.15 present their results.

The range of a heavy ion in a compound or mixture is calculated from the range in pure elements by using the equation[21,22]:

$$R\ (\mathrm{kg/m^2}) = \left(\sum_i \frac{w_i}{R_i} \right)^{-1} \tag{4.43}$$

where R_i = range, in ig/m², in element i
w_i = weight fraction of ith element

4.8 INTERACTIONS OF GAMMA AND X-RAYS WITH MATTER

X-rays or γ rays are electromagnetic radiation. Considered as particles, when they travel with the velocity of light c, their rest mass is zero and so is their charge. A common name for x-rays and γ rays as particles is photons. The relationship between energy, frequency, and wavelength λ is:

$$E = h\nu = h \frac{c}{\lambda} \tag{4.44}$$

There is no clear distinction between x-rays and γ rays. The term x-rays is applied generally to the photons with $E < 1$ MeV. Gammas are the photons with $E \geqslant 1$ MeV. In what follows, the terms photon, γ, and x-ray will be used interchangeably.

X-rays are generally produced by atomic transitions such as excitation and ionization. Gamma rays are emitted in nuclear transitions. Photons are also produced as bremsstrahlung, by accelerating or decelerating charged particles. X-rays and γ rays emitted by atoms and nuclei are monoenergetic. Bremsstrahlung has a continuous energy spectrum.

There is a long list of possible interactions of photons, but only the three most important ones will be discussed here: the photoelectric effect, Compton scattering, and pair production.

4.8.1 The Photoelectric Effect

The photoelectric effect is an interaction between a photon and a bound atomic electron. As a result of the interaction, the photon disappears and one of the atomic electrons is ejected as a free electron, called the *photoelectron* (Fig. 4.16). The kinetic energy of the electron is

$$T = E_\gamma - BE \tag{4.45}$$

where E_γ = energy of the photon
BE = binding energy of the electron

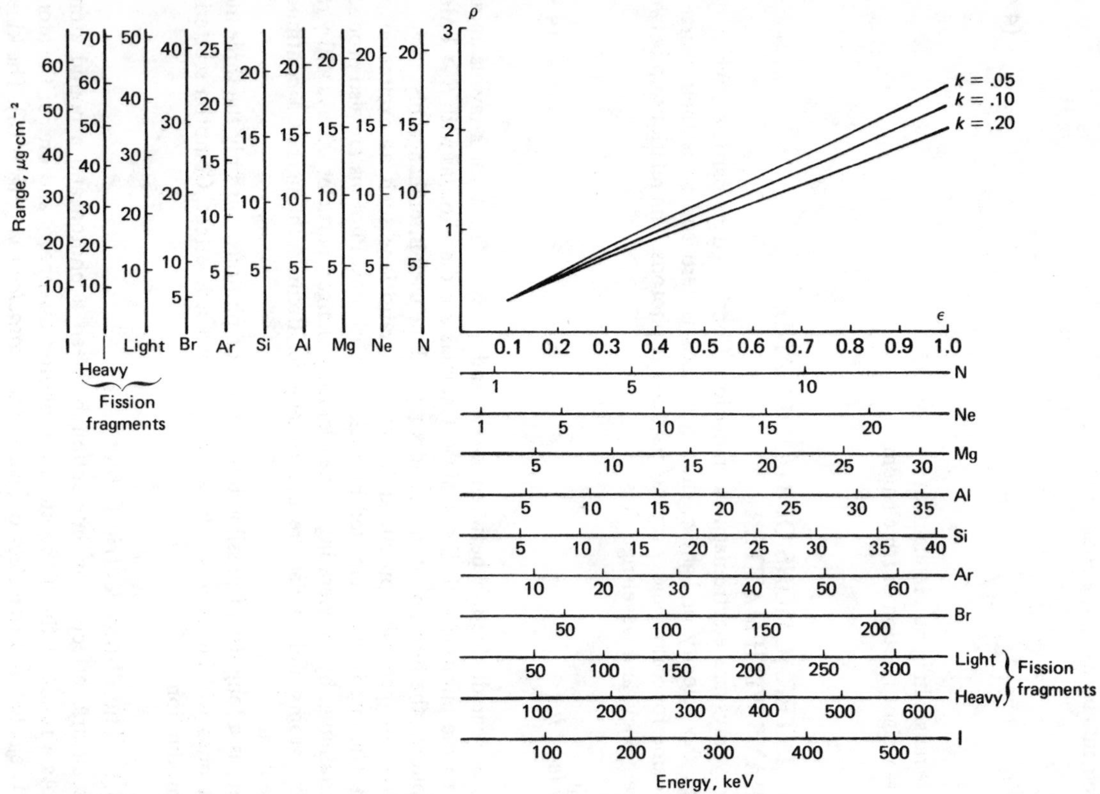

FIG. 4.14 Universal range–energy plot for $\epsilon < 1$. It allows determination of range in silicon for many heavy ions (ref. 20).

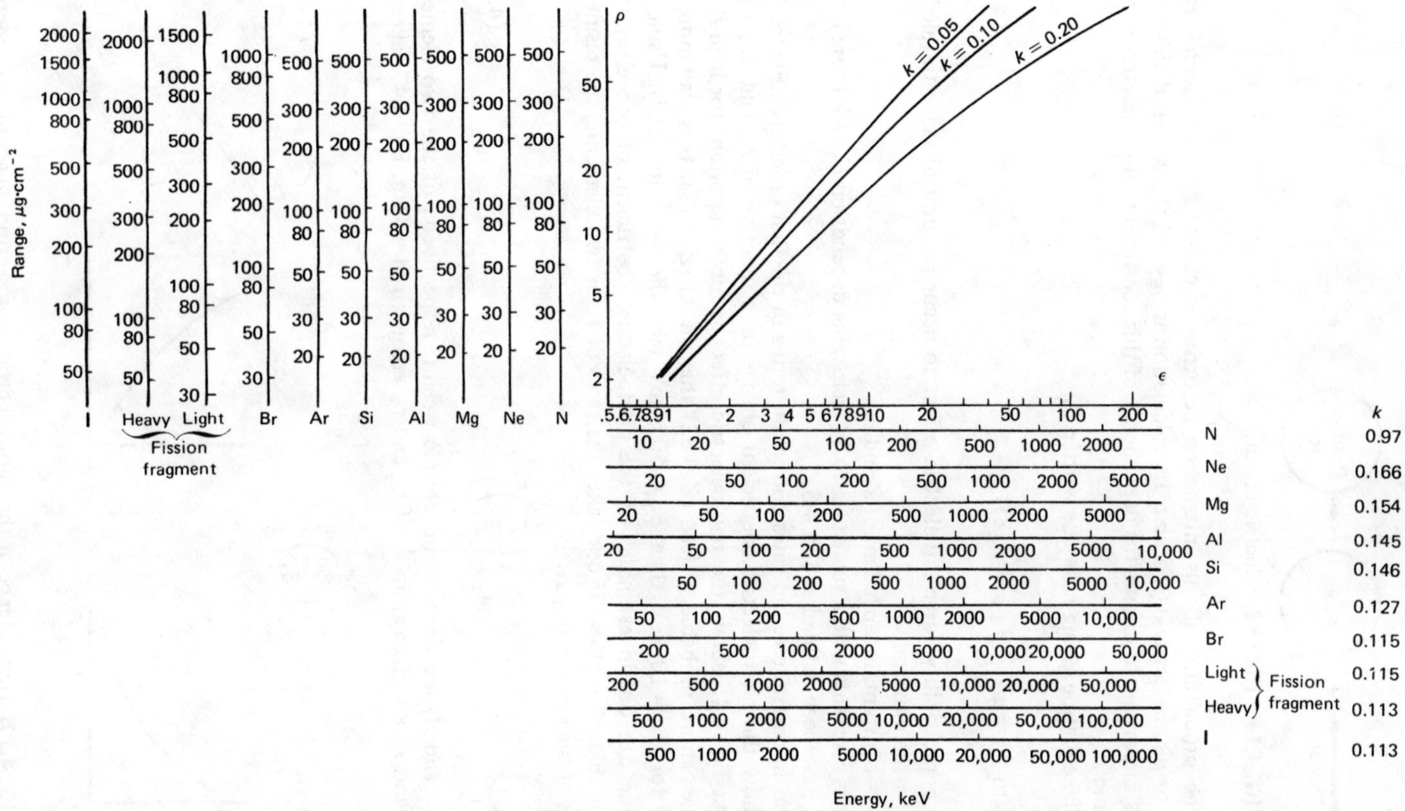

FIG. 4.15 Universal range–energy plot for $\epsilon > 1$. It allows determination of range in silicon for many heavy ions (ref. 20).

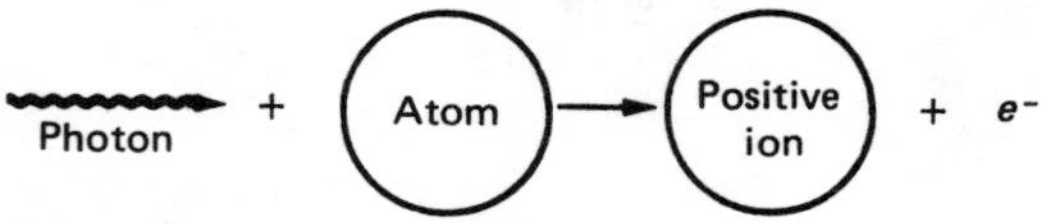

FIG. 4.16 The photoelectric effect.

The probability of this interaction occurring is called the *photoelectric cross section* or *photoelectric coefficient.* Its calculation is beyond the scope of this book, but it is important to discuss the dependence of this coefficient on parameters such as E_γ, Z, and A.

The equation giving τ may be written as:

$$\tau\,(\mathrm{m}^{-1}) = aN\,\frac{Z^n}{E_\gamma^m}\,[1 - \mathcal{O}(Z)] \tag{4.46}$$

where τ = probability for photoelectric effect to occur per unit distance traveled by the photon

a = constant, independent of Z and E_γ

m, n = constants with a value of 3 to 5 (their value depends on E_γ; see Evans)

N, Z have been defined in Sec. 4.3.

The term in parentheses indicates correction terms of the first order in Z. Figure 4.17 shows how the photoelectric coefficient changes as a function of E_γ and Z. As Fig. 4.17 and Eq. 4.46 show, the photoelectric effect is more important for high-Z material, i.e., more probable in lead ($Z = 82$) than in Al ($Z = 13$). It is also more important for $E_\gamma = 10$ keV than $E_\gamma = 500$ keV (for the same material). Using Eq. 4.46, one can obtain estimates of the photoelectric coefficient of one element in terms of that of another. If one takes the ratio of τ for two elements, the result for photons of the same energy is:

$$\tau_2\,(\mathrm{m}^{-1}) = \tau_1\,\frac{\rho_2}{\rho_1}\left(\frac{A_1}{A_2}\right)\left(\frac{Z_2}{Z_1}\right)^n \tag{4.47}$$

where ρ_i and A_i are density and atomic weight, respectively, of the two elements, and τ_1 and τ_2 are given in m^{-1}. If τ_1 and τ_2 are given in m^2/kg, Eq. 4.47 takes the form

$$\tau_2\,(\mathrm{m}^2/\mathrm{kg}) = \tau_1\,\frac{A_1}{A_2}\left(\frac{Z_2}{Z_1}\right)^n \tag{4.47a}$$

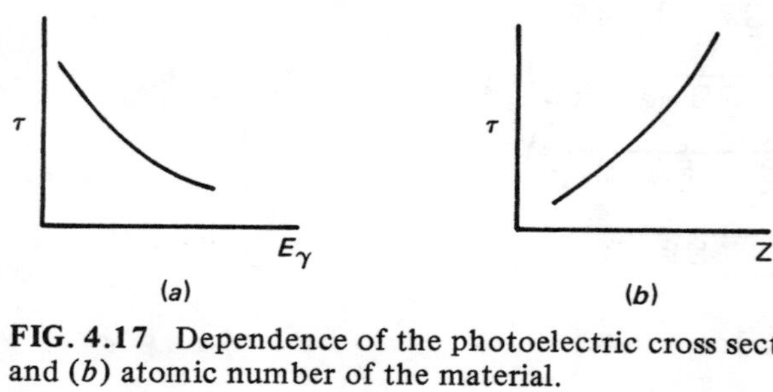

FIG. 4.17 Dependence of the photoelectric cross section on (*a*) photon energy and (*b*) atomic number of the material.

4.8.2 Compton Scattering or Compton Effect

The *Compton effect* is a collision between a photon and a free electron. Of course, under normal circumstances, all the electrons in a medium are not free but bound. If the energy of the photon, however, is of the order of keV or more, while the binding energy of the electron is of the order of eV, the electron may be considered free.

The photon does not disappear after a Compton scattering. Only its direction of motion and energy change (Fig. 4.18). The photon energy is reduced by a certain amount that is given to the electron. Therefore, conservation of energy gives (assuming the electron is stationary before the collision):

$$T = E_\gamma - E_{\gamma'} \tag{4.48}$$

If Eq. 4.48 is used along with the conservation of momentum equations, the energy of the scattered photon as a function of the scattering angle θ can be calculated. The result is:

$$E_{\gamma'} = \frac{E_\gamma}{1 + (1 - \cos\theta)E_\gamma/mc^2} \tag{4.49}$$

Using Eq. 4.48 and 4.49, one obtains the kinetic energy of the electron.

$$T = \frac{(1 - \cos\theta)E_\gamma/mc^2}{1 + (1 - \cos\theta)E_\gamma/mc^2} E_\gamma \tag{4.50}$$

A matter of great importance for radiation measurement is the maximum and minimum energy of the photon and the electron after the collision. The minimum energy of the scattered photon is obtained when $\theta = \pi$. This, of course, corresponds to the maximum energy of the electron. From Eq. 4.49,

$$E_{\gamma',min} = \frac{E_\gamma}{1 + 2E_\gamma/mc^2} \tag{4.51}$$

and $$T_{max} = \frac{2E_\gamma/mc^2}{1 + 2E_\gamma/mc^2} E_\gamma \tag{4.52}$$

The maximum energy of the scattered photon is obtained for $\theta = 0$, which essentially means that the collision did not take place. From Eq. 4.49,

$$E_{\gamma',min} = E_\gamma$$

$$T_{min} = 0$$

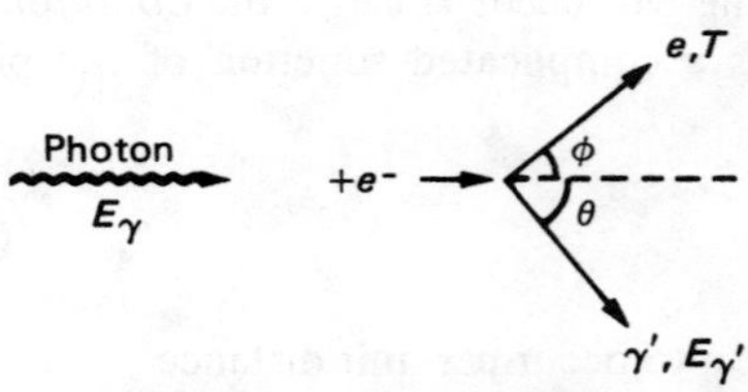

FIG. 4.18 The Compton effect.

The conclusion to be drawn from Eq. 4.51 is that the minimum energy of the scattered photon is greater than zero. Therefore, *in Compton scattering, it is impossible for all the energy of the incident photon to be given to the electron.* The energy given to the electron will be dissipated in the material within a distance equal to the range of the electron. The scattered photon may escape.

EXAMPLE 4.14

A 3-MeV photon interacts by Compton scattering. (a) What is the energy of the photon and the electron if the scattering angle of the photon is 90°? (b) What if the angle of scattering is 180°?

ANSWER

(a) Using Eq. 4.49,

$$E_{\gamma'} = \frac{3}{1+(1-0)3/0.511} = 0.437 \text{ MeV}$$

$$T = 3 - 0.437 = 2.563 \text{ MeV}$$

(b) Using Eq. 4.51,

$$E_{\gamma',\min} = \frac{3}{1+(2)3/0.511} = 0.235 \text{ MeV}$$

$$T = 3 - 0.235 = 2.765 \text{ MeV}$$

EXAMPLE 4.15

What is the minimum energy of the γ ray after Compton scattering if the original photon energy is 0.511 MeV, 5 MeV, 10 MeV, or 100 MeV?

ANSWER

The results are shown in the table below (Eq. 4.51 has been used).

E_γ	$(E_{\gamma'})_{\min}$	T
0.511	0.181	0.330
5	0.243	4.757
10	0.25	9.75
100	0.25	99.75

The probability that Compton scattering will occur is called the *Compton coefficient* or the *Compton cross section.* It is a complicated function of the photon energy, but it may be written in the form:

$$\sigma\,(\text{m}^{-1}) = NZf(E_\gamma) \tag{4.53}$$

where σ = probability for Compton interaction to occur per unit distance
$f(E_\gamma)$ = a function of E_γ

If one writes the atom density N explicitly, Eq. 4.53 takes the form:

$$\sigma \sim \rho \frac{N_A}{A} Zf(E_\gamma) \sim \rho \left(\frac{N_A}{A}\right) \frac{A}{2} f(E_\gamma) \sim \rho \frac{N_A}{2} f(E_\gamma) \tag{4.54}$$

In deriving Eq. 4.54, use has been made of the fact that for most materials, except hydrogen, $A \approx 2Z$ to $A \approx 2.6Z$. According to Eq. 4.54, the probability for Compton scattering to occur is almost independent of the atomic number of the material. Figure 4.19 shows how σ changes as a function of E_γ and Z. If the Compton cross section is known for one element, it can be calculated for any other by using Eq. 4.53 (for photons of the same energy):

$$\sigma_2 \ (\mathrm{m}^{-1}) = \sigma_1 \left(\frac{\rho_2}{\rho_1}\right)\left(\frac{A_1}{A_2}\right)\left(\frac{Z_1}{Z_2}\right) \tag{4.55}$$

where σ_1 and σ_2 are given in m^{-1}. If σ_1 and σ_2 are given in m^2/kg, Eq. 4.55 takes the form

$$\sigma_2 \ (\mathrm{m}^2/\mathrm{kg}) = \sigma_1 \left(\frac{A_1}{A_2}\right)\left(\frac{Z_1}{Z_2}\right) \tag{4.55a}$$

4.8.3 Pair Production

Pair production is an interaction between a photon and a nucleus. As a result of the interaction, the photon disappears and an electron-positron pair appears (Fig. 4.20). Although the nucleus does not undergo any change as a result of this interaction, its presence is necessary for pair production to occur. A γ ray will not disappear in empty space by producing an electron-positron pair.†

Conservation of energy gives the following equation for the kinetic energy of the electron and the positron:

$$T_{e^-} + T_{e^+} = E_\gamma - (mc^2)_{e^-} - (mc^2)_{e^+} = E_\gamma - 1.022 \text{ MeV} \tag{4.56}$$

†Pair production may take place in the field of an electron. The probability for that to happen is much smaller and the threshold for the gamma energy is 4 mc^2 = 2.04 MeV.

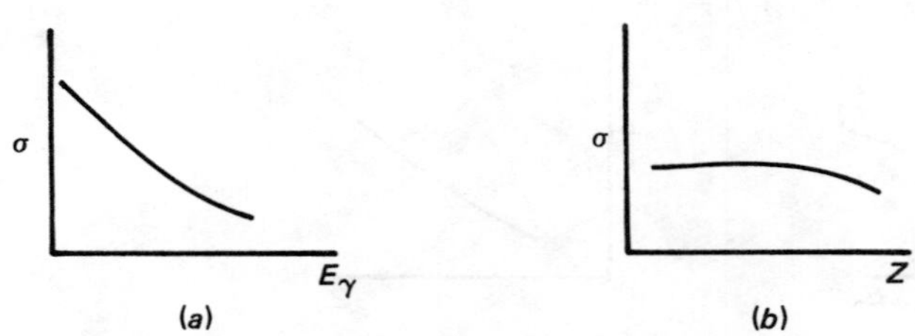

FIG. 4.19 Dependence of the Compton cross section on (*a*) photon energy and (*b*) atomic number of the material.

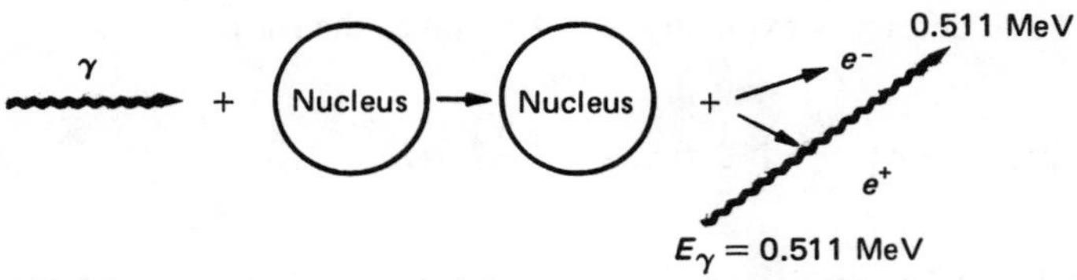

FIG. 4.20 Pair production. The gamma disappears and a positron–electron pair is created. Two 0.511-MeV photons are produced when the positron annihilates.

The available kinetic energy is equal to the energy of the photon minus 1.022 MeV, which is necessary for the production of the two rest masses. Electron and positron share, for all practical purposes, the available kinetic energy, i.e.:

$$T_{e^-} = T_{e^+} = \tfrac{1}{2}(E_\gamma - 1.022 \text{ MeV}) \tag{4.57}$$

Pair production eliminates the original photon, but two photons are created when the positron annihilates (see Sec. 3.7.4). The annihilation gammas are important in constructing shielding against them as well as for the detection of gammas (see Chap. 12).

The probability for pair production to occur, called the *pair production coefficient or cross section*, is a complicated function of E_γ and Z (see Evans and Roy & Reed). It may be written in the form:

$$\kappa\ (\text{m}^{-1}) = NZ^2 f(E_\gamma, Z) \tag{4.58}$$

where κ is the probability for pair production to occur per unit distance traveled and $f(E_\gamma, Z)$ is a function which changes slightly with Z and increases with E_γ.

Figure 4.21 shows how κ changes with E_γ and Z. It is important to note that κ has a threshold at 1.022 MeV and increases with E_γ and Z. Of the three coefficients (τ and σ being the other two), κ is the only one increasing with the energy of the photon.

If the pair production cross section is known for one element, an estimate of its value can be obtained for any other element by using Eq. 4.58 (for photons of the same energy).

$$\kappa_2\ (\text{m}^{-1}) = \kappa_1 \left(\frac{\rho_2}{\rho_1}\right)\left(\frac{A_2}{A_1}\right)\left(\frac{Z_2}{Z_1}\right)^2 \tag{4.59}$$

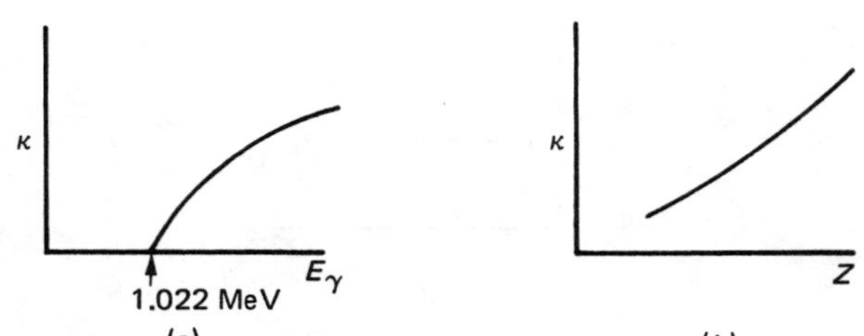

FIG. 4.21 Dependence of the pair-production cross section on (*a*) photon energy and (*b*) atomic number of the material.

where κ_1 and κ_2 are given in m^{-1}. If κ_1 and κ_2 are given in m^2/kg, Eq. 4.59 takes the form

$$\kappa_2\,(m^2/kg) = \kappa_1 \left(\frac{A_2}{A_1}\right)\left(\frac{Z_2}{Z_1}\right)^2 \tag{4.59a}$$

4.8.4 Total Attenuation Coefficient

When a photon travels through matter, it may interact through any of the three major interactions discussed earlier. (For pair production, $E_\gamma > 1.022$ MeV.) There are other interactions, but they are not mentioned here because they are not important in the detection of gammas.

Figure 4.22 shows the relative importance of the three interactions as E_γ and Z change. Consider a photon with $E = 0.1$ MeV. If this particle travels in carbon ($Z = 6$), the Compton effect is the predominant mechanism by which this photon interacts. If the same photon travels in iodine ($Z = 53$), the photoelectric interaction prevails. For a γ of 1 MeV, the Compton effect predominates regardless of Z. If a photon of 10 MeV travels in carbon, it will interact mostly through Compton scattering. The same photon moving in iodine will interact, mainly, through pair production.

The total probability for interaction μ, called the total *linear* attenuation coefficient, is equal to the sum of the three probabilities:

$$\mu\,(m^{-1}) = \tau + \sigma + \kappa \tag{4.60}$$

Physically, μ is the probability of interaction per unit distance.

There are tables that give μ for all the elements, for many photon energies.† Most of the tables provide μ in units of m^2/kg. (or cm^2/g), because in these units the density of the material does not have to be specified. If μ is given in m^2/kg (or cm^2/g),

†Tables of mass attenuation coefficients are given in App. D.

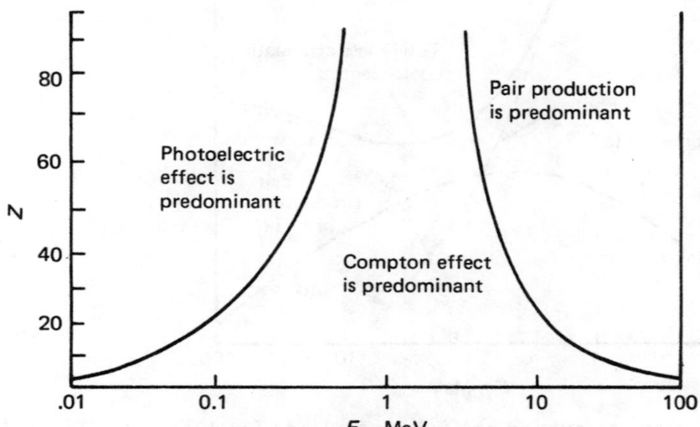

FIG. 4.22 The relative importance of the three major gamma interactions (from Evans).

it is called the *total mass attenuation coefficient.* The relationship between linear and mass coefficients is

$$\mu\,(\mathrm{m^2/kg}) = \frac{\mu\,(\mathrm{m^{-1}})}{\rho\,(\mathrm{kg/m^3})} \tag{4.61}$$

Figure 4.23 shows the individual coefficients as well as the total mass attenuation coefficient for lead, as a function of photon energy. The total mass attenuation coefficient shows a minimum because as E increases, τ decreases, κ increases, and σ does not change appreciably. However, the minimum of μ does not fall at the same energy for all elements. For lead, μ is minimized at $E_\gamma \sim 3.5$ MeV; for aluminum, the minimum is at 20 MeV; and for NaI, the minimum is at 5 MeV.

If a parallel beam of monoenergetic γ rays goes through a material of thickness t, the fraction of the beam that traverses the medium without any interaction is equal to $e^{-\mu t}$ (Fig. 4.24). The probability that a photon will go through thickness t without an interaction is:

$$\begin{array}{l}\text{Probability of}\\ \text{traversing}\\ \text{thickness } t\end{array} = \frac{\text{number transmitted}}{\text{number incident}} = \frac{I_0 e^{-\mu t}}{I_0} = e^{-\mu t} \tag{4.62}$$

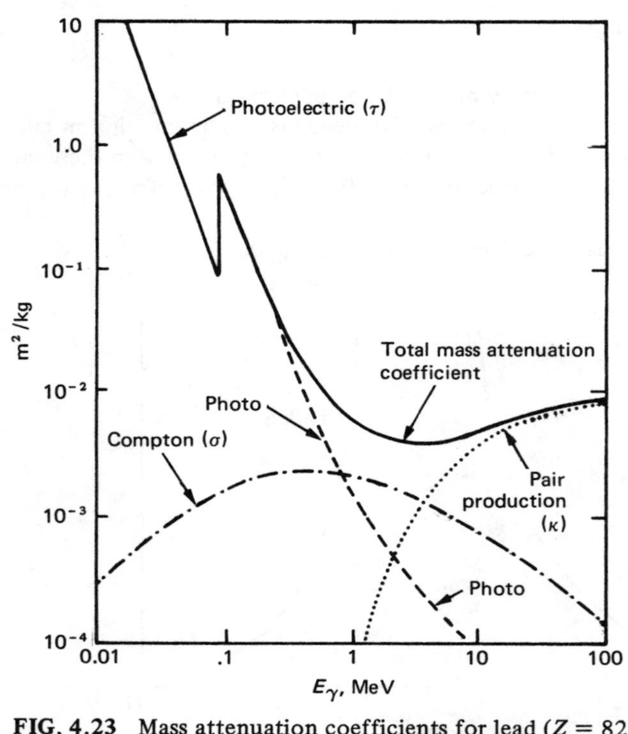

FIG. 4.23 Mass attenuation coefficients for lead ($Z = 82$, $\rho = 11.35 \times 10^3$ kg/m^3).

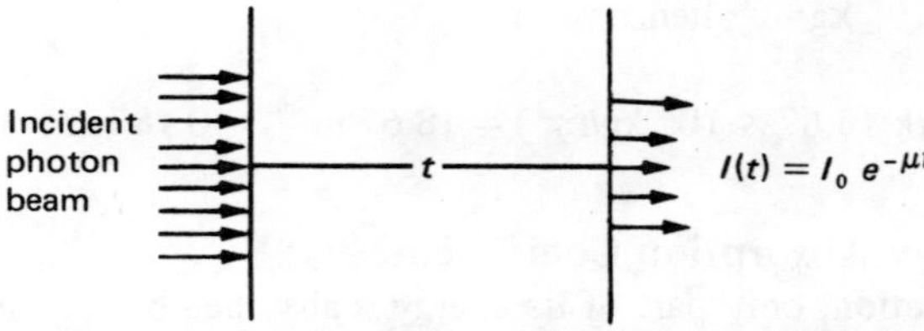

FIG. 4.24 The intensity of the transmitted beam (only particles that did not interact) decreases exponentially with material thickness.

The average distance between two successive interactions, called the *mean free path* (λ), is given by:

$$\lambda\,(\text{m}) = \frac{\int_0^\infty x e^{-\mu x}\,dx}{\int_0^\infty e^{-\mu x}\,dx} = \frac{1}{\mu} \tag{4.63}$$

Thus, the mean free path is simply the inverse of the total linear attenuation coefficient. If $\mu = 10\ \text{m}^{-1}$ for a certain γ ray traveling in a certain medium, then the distance between two successive interactions of this gamma in that medium is $\lambda = 1/\mu = 1/10\ \text{m} = 0.10\ \text{m}$.

The total mass attenuation coefficient for a compound or a mixture is calculated by the same method used for $(dE/dx)_c$ in Sec. 4.5. It is easy to show (see Prob. 4.15) that:

$$\mu_c\,(\text{m}^2/\text{kg}) = \sum_i w_i \mu_i\,(\text{m}^2/\text{kg}) \tag{4.64}$$

where μ_c = total *mass* attenuation coefficient for a compound or a mixture
w_i = weight fraction of ith element in the compound
μ_i = total *mass* attenuation coefficient of ith element

EXAMPLE 4.16
What is the total mass attenuation coefficient for 1.25 gammas in NaI?

ANSWER
For this compound, the following data apply:

Na: $\mu = 0.00546\ \text{m}^2/\text{kg}$ $\quad w = \frac{23}{150} = 0.153$

I: $\mu = 0.00502\ \text{m}^2/\text{kg}$ $\quad w = \frac{127}{150} = 0.847$

Using Eq. 4.64,

$$\mu\,(\text{NaI}) = 0.00546(0.153) + 0.00502(0.847) = 0.00509\ \text{m}^2/\text{kg} = 0.0509\ \text{cm}^2/\text{g}$$

The density of NaI is 3.67×10^3 kg/m^3; hence,

$$\mu\,(\mathrm{m}^{-1}) = 0.00509\ \mathrm{m}^2/\mathrm{kg}(3.67 \times 10^3\ \mathrm{kg/m}^3) = 18.67\ \mathrm{m}^{-1} = 0.187\ \mathrm{cm}^{-1}$$

4.8.5 Photon Energy Absorption Coefficient

When a photon has an interaction, only part of its energy is absorbed by the medium at the point where the interaction took place. Energy given by the photon to electrons and positrons is considered absorbed at the point of interaction because the range of these charged particles is short. However, x-rays, Compton-scattered photons, or annihilation gammas may escape. The fraction of photon energy that escapes is important when one wants to calculate heat generated due to gamma absorption in shielding materials or gamma radiation dose to humans (see Chap. 16). The gamma energy deposited in any material is calculated with the help of an energy absorption coefficient defined in the following way.

The *gamma energy absorption coefficient* is in general that part of the total attenuation coefficient that, when multiplied by the gamma energy, will give the energy deposited at the point of interaction. Equation 4.60 gives the total attenuation coefficient. The *energy absorption coefficient* μ_a is:

$$\mu_a = \tau + \frac{T_{\mathrm{av}}}{E_\gamma}\,\sigma + \kappa \tag{4.65}$$

where T_{av} is the average energy of the Compton electron, and μ_a may be a linear or mass energy absorption coefficient depending on the units (see Sec. 4.8.4).

In writing Eq. 4.65, it is assumed that:

1 If photoelectric effect or pair production takes place, all the energy of the gamma is deposited there.
2 If Compton scattering occurs, only the energy of the electron is absorbed. The Compton-scattered gamma escapes.

In the case of photoelectric effect, assumption (1) is good. For pair production, however, it is questionable because only the energy $E_\gamma - 1.022$ MeV is given to the electron-positron pair. The rest of the energy, equal to 1.022 MeV, is taken by the two annihilation gammas, and it may not be deposited in the medium. There are cases when Eq. 4.65 is modified to account for this effect.[23] Gamma absorption coefficients, as defined by Eq. 4.65, are given in App. E.

EXAMPLE 4.17

A 1Ci ^{137}Cs source is kept in a large water vessel. What is the energy deposited by the gammas in H_2O at a distance 0.05 m from the source?

ANSWER

^{137}Cs emits a 0.662-MeV gamma. The mass absorption coefficient for this photon in water is (App. E) 0.00327 m^2/kg. The total mass attenuation coefficient is 0.00862 m^2/kg. The energy deposited at a distance of 0.05 m from the source is: ($E_d = \phi \mu_a E_\gamma$)

$$E_d\left(\frac{\text{MeV}}{\text{kg}\cdot\text{s}}\right) = \frac{S}{4\pi r^2}\, e^{-\mu r}\left(\frac{\gamma}{\text{m}^2\cdot\text{s}}\right)[\mu_a\,(\text{m}^2/\text{kg})]\,[E\,(\text{MeV}/\gamma)]$$

$$= \frac{3.7\times 10^{10}}{4\pi(0.05)^2}\,(e^{-0.00862(10^3)0.05})(0.00327(0.662)$$

$$= 1.66\times 10^6\ \text{MeV/g}\cdot\text{s} = 2.65\times 10^{-4}\ \text{J/kg}\cdot\text{s}$$

4.8.6 Buildup Factors

Consider a point isotropic gamma source at a certain distance away from a detector, as shown in Fig. 4.25. The total gamma beam hitting the detector consists of two components:

1 The *unscattered beam* (I_u) consists of those γ's that go through the shield of thickness t without any interaction. If the source emits $S(\gamma/s)$, the unscattered beam is

$$I_u\left(\frac{\gamma}{\text{m}^2\cdot\text{s}}\right) = \frac{S}{4\pi r^2}\, e^{-\mu r} \tag{4.66}$$

2 The *scattered beam* (I_s) consists of scattered gammas and gammas generated through interactions of the incident photons (e.g., x-rays and annihilation gammas). The calculation of the scattered beam is not trivial, and there is no simple expression like Eq. 4.62 representing it.

The total beam hitting the detector is:

$$I_{\text{tot}} = I_u + I_s \tag{4.67}$$

Instead of calculating the scattered beam, Eq. 4.67 may be rewritten in terms of a buildup factor B, such that:

$$I_{\text{tot}} = BI_u \tag{4.68}$$

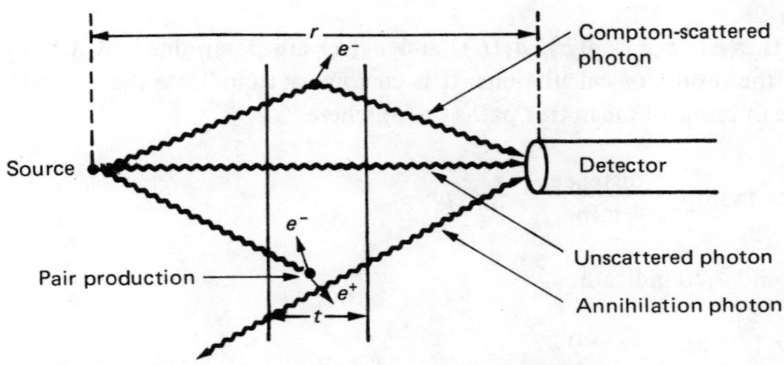

FIG. 4.25 If a point isotropic source is placed behind a shield of thickness t, both scattered and unscattered photons will hit the detector.

In general, the *buildup factor* is defined as:

$$B = \frac{\text{quantity of interest at a point due to the total number of photons}}{\text{quantity of interest at a point due to unscattered photons only}}$$

The "quantity of interest" may be the number of particles, the energy deposited, or the radiation dose† imparted by them, giving rise to number, energy, and dose buildup factors, respectively. The buildup factors depend on the energy of the radiation, the medium being traversed, and the geometry of both source and medium. Their main advantage stems from the fact that they change relatively slowly with changes in geometry and energy of radiation. Thus, the study of a small number of cases can provide information for other problems with different parameters.

More important than the number buildup factor is the *dose buildup factor*, defined as:

$$B_D = \frac{\text{radiation dose at a point due to all photons}}{\text{radiation dose at a point due to unscattered photons only}}$$

Extensive calculations of buildup factors have been performed, and the results have been tabulated for several gamma energies, media, and distances. In addition, attempts have been made to derive empirical equations.[24–29] Two of the more useful formulas are:

Berger formula:

$$B(E, \mu r) = 1 + a(E)\mu r[\exp b(E)\mu r] \tag{4.69}$$

Taylor formula:

$$B(E, \mu r) = A(E) \exp [-a_1(E)\mu r] + (1 - A) \exp [-a_2(E)\mu r] \tag{4.70}$$

The constants $a(E)$, $b(E)$, $A(E)$, $a_1(E)$, and $a_2(E)$ are determined by fitting these equations to the results of calculations. It is customary to indicate the dependence of B on distance in terms of mean free paths (mfp) where

$$\text{Distance in mfp} = \frac{\text{distance}}{\text{mfp}} = \frac{r}{\lambda} = \mu r$$

As Eqs. 4.69 and 4.70 indicate,

$$B(\mu r) \to 1 \quad \text{as} \quad \mu r \to 0$$

†Radiation dose is defined in Sec. 16.2.

For short distances (distance ~ 1 mfp), the value of B is close to unity, but for distances of many mfp's, the value of B becomes quite large.

Tables of buildup factor constants are given in App. E. The use of buildup factors is demonstrated in Chap. 16, in connection with radiation dose rate calculations.

4.9 INTERACTIONS OF NEUTRONS WITH MATTER

Neutrons, with protons, are the constituents of nuclei (see Sec. 3.4). Since a neutron has no charge, it interacts with nuclei only through nuclear forces. When it approaches a nucleus, it does not have to go through a Coulomb barrier, as a charged particle does. As a result, the probability (cross section) for nuclear interactions is higher for neutrons than for charged particles. This section discusses the important characteristics of neutron interactions, with emphasis given to neutron cross sections and calculation of interaction rates.

4.9.1 Types of Neutron Interactions

The interactions of neutrons with nuclei are divided into two categories.

SCATTERING

In this type of interaction, the neutron interacts with a nucleus, but both particles reappear after the reaction. A scattering collision is indicated as a (n, n) reaction or as

$$n + {}^A_Z\text{X} \rightarrow {}^A_Z\text{X} + n$$

Scattering may be elastic or inelastic. In elastic scattering, the total *kinetic* energy of the two colliding particles is conserved. The kinetic energy is simply redistributed between the two particles. In inelastic scattering, part of the kinetic energy is given to the nucleus as an excitation energy. After the collision, the excited nucleus will decay by emitting one or more γ rays.

Scattering reactions are responsible for neutron-slowing in reactors. Neutrons emitted in fission have an average energy of 2 MeV. The probability that neutrons will induce fission is much higher if the neutrons are very slow—"thermal"—with kinetic energies of the order of eV. The fast neutrons lose their kinetic energy as a result of scattering collisions with nuclei of a "moderating" material which is usually water or graphite.

ABSORPTION

If the interaction is an absorption, the neutron disappears, but one or more other particles appear after the reaction takes place. Table 4.4 illustrates some examples of absorptive reactions.

4.9.2 Neutron Cross Sections

Consider a monoenergetic parallel beam of neutrons hitting a thin target† of thickness t (Fig. 4.26). The number of reactions per second, R, taking place in this target may be written as:

†A thin target is one that does not appreciably attenuate the neutron beam (see Eq. 4.74).

$$R \text{ (reactions/s)} = \begin{pmatrix}\text{neutrons per m}^2\cdot\text{s} \\ \text{hitting the target}\end{pmatrix} \begin{pmatrix}\text{targets exposed} \\ \text{to the beam}\end{pmatrix}$$
$$\times \begin{pmatrix}\text{Probability of Interaction} \\ \text{per n/m}^2 \text{ per nucleus}\end{pmatrix}$$

or $$R = I\,(\text{n/m}^2\cdot\text{s})\,[N\,(\text{nuclei/m}^3)]\,[a\,(\text{m}^2)]\,[t\,(\text{m})]\,[\sigma\,(\text{m}^2)] \tag{4.71}$$

where I, a, and t are shown in Fig. 4.26. The parameter σ, called the *cross section*, has the following physical meaning:

$\sigma\,(\text{m}^2)$ = probability that an interaction will occur per target nucleus per neutron per m^2 hitting the target

The unit of σ is the barn (b).

$$1\text{ b} = 10^{-24}\text{ cm}^2 = 10^{-28}\text{ m}^2$$

Since the nuclear radius is approximately 10^{-15} to 10^{-14} m, 1 b is approximately equal to the cross-sectional area of a nucleus.

Neutron cross sections are defined separately for each type of reaction and isotope. For the reactions listed in Sec. 4.9.1, one defines, for example,

σ_s = elastic scattering cross section

σ_i = inelastic scattering cross section

σ_a = absorption cross section

σ_γ = capture cross section

σ_f = fission cross section

The total cross section–i.e., the total probability that a reaction of any type will take place–is equal to the sum of all the σ's.

$$\sigma_{\text{tot}} = \sigma_s + \sigma_i + \sigma_\gamma + \sigma_f + \cdots \tag{4.72}$$

In the notation used here, $\sigma_a = \sigma_\gamma + \sigma_f$.

TABLE 4.4 Absorptive Reactions

Reaction	Name
$n + {}^A_Z\text{X} \rightarrow {}_{Z-1}^{A}\text{Y} + p$	(n, p) reaction
$n + {}^A_Z\text{X} \rightarrow {}^{A-3}_{Z-2}\text{Y} + {}^4_2\alpha$	(n, α) reaction
$n + {}^A_Z\text{X} \rightarrow {}^{A-1}_{Z}\text{X} + 2n$	$(n, 2n)$ reaction
$n + {}^A_Z\text{X} \rightarrow {}^{A+1}_{Z}\text{X} + \gamma$	(n, γ) reaction
$n + {}^A_Z\text{X} \rightarrow {}^{A_1}_{Z_1}\text{Y}_1 + {}^{A_2}_{Z_2}\text{Y}_2 + n + n + \cdots$	fission

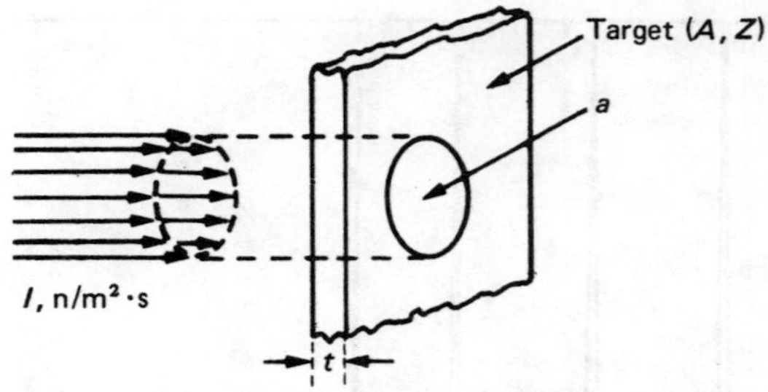

FIG. 4.26 A parallel neutron beam hitting a thin target: a = area of target struck by the beam.

Neutron cross sections depend strongly on the energy of the neutron as well as on the atomic weight and atomic number of the target nucleus.

Figures 4.27 and 4.28 show the total cross section for two isotopes over the same neutron energy range. Notice the vast difference between the two σ's, both in terms of their variation with energy and their value in barns.

The cross section σ (b) is called the *microscopic* cross section. Another form of the cross section also frequently used is the *macroscopic* cross section Σ (m^{-1}), defined by the equation:

$$\Sigma_i\,(m^{-1}) = N\,(\text{atoms/m}^3)\,[\sigma_i\,(m^2)] \tag{4.73}$$

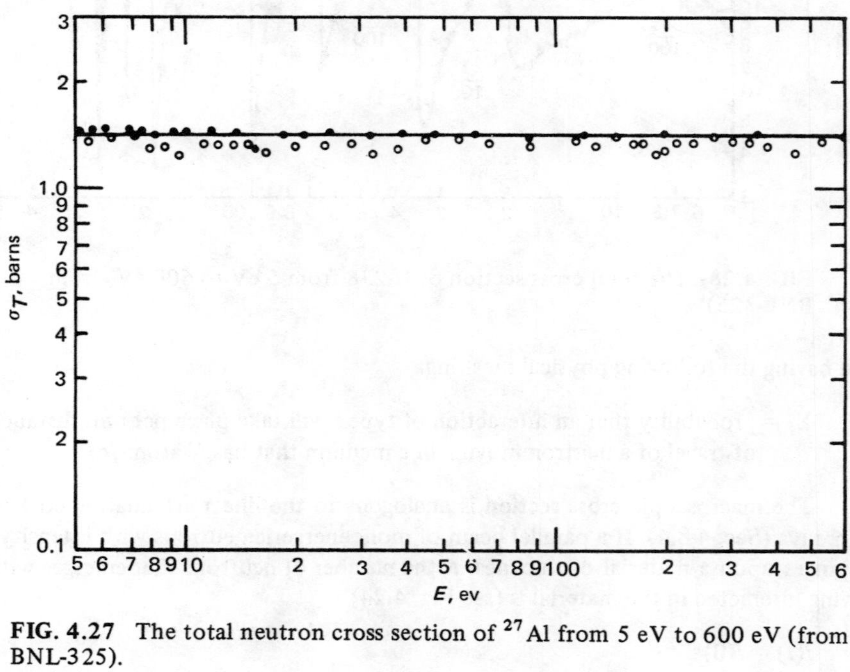

FIG. 4.27 The total neutron cross section of ^{27}Al from 5 eV to 600 eV (from BNL-325).

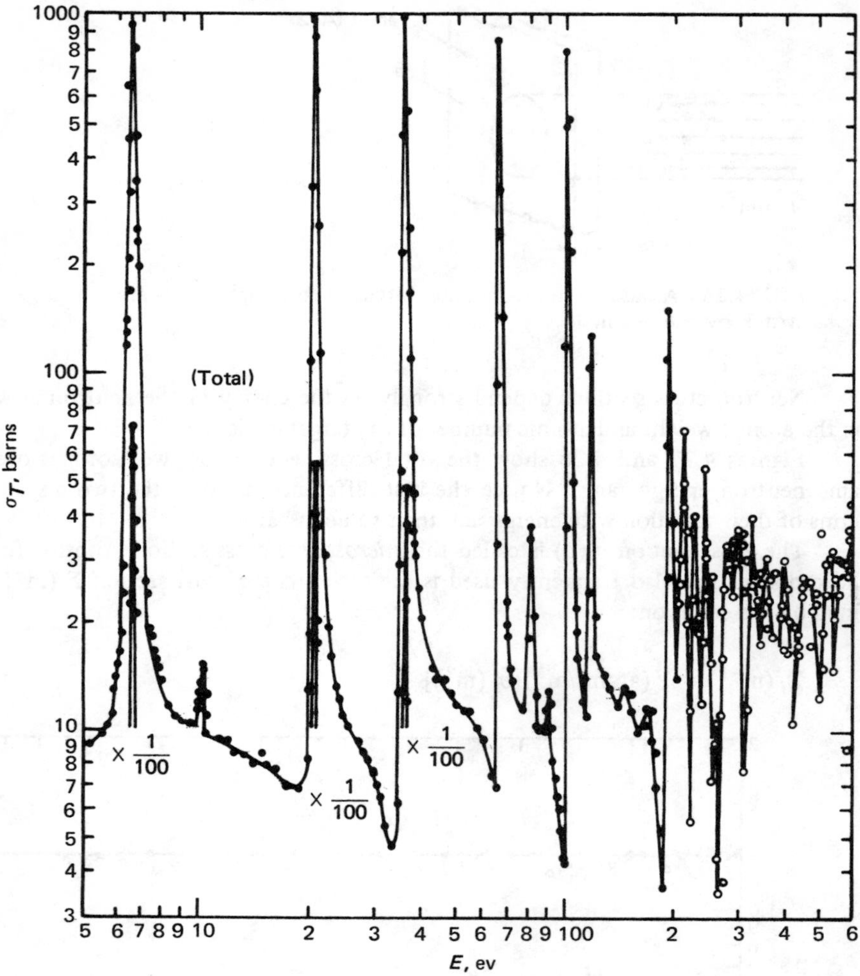

FIG. 4.28 The total cross section of U-238 from 5 eV to 600 eV (from BNL-325).

and having the following physical meaning:

Σ_i = probability that an interaction of type i will take place per unit distance of travel of a neutron moving in a medium that has N atoms/m^3

The macroscopic cross section is analogous to the linear attenuation coefficient of γ rays (Sec. 4.8.4). If a parallel beam of monoenergetic neutrons with intensity $I(0)$ impinges upon a material of thickness t, the number of neutrons that emerges without having interacted in the material is (see Fig. 4.24)

$$I(t) = I(0)e^{-\Sigma_t t} \tag{4.74}$$

where $\Sigma_t = \Sigma_s + \Sigma_i + \Sigma_a + \cdots =$ total macroscopic neutron cross section.

As with γ rays,

$e^{-\Sigma_t t}$ = probability that the neutron will travel distance t without an interaction

The average distance between two successive interactions, the mean free path λ, is

$$\lambda = \frac{\int_0^\infty x e^{-\Sigma x}\, dx}{\int_0^\infty e^{-\Sigma x}\, dx} = \frac{1}{\Sigma} \tag{4.75}$$

EXAMPLE 4.18

What are the macroscopic cross sections Σ_s, Σ_a, and Σ_t for thermal neutrons in graphite? The scattering cross section is $\sigma_s = 4.8$ b and the absorption cross section is $\sigma_a = 0.0034$ b. What is the mean free path?

ANSWER

For graphite, $\rho = 1.6 \times 10^3$ kg/m³ and $A = 12$. Therefore,

$$N = \rho \frac{N_A}{A} = (1.6 \times 10^3) \frac{0.6023 \times 10^{24}}{12 \times 10^{-3}} = 0.0803 \times 10^{30} \text{ atoms/m}^3$$

Using Eq. 4.73:

$$\Sigma_s = (0.0803 \times 10^{30})(4.8 \times 10^{-28}) = 38.5 \text{ m}^{-1} = 0.385 \text{ cm}^{-1}$$

$$\Sigma_a = (0.0803 \times 10^{30})(0.0034 \times 10^{-28}) = 0.027 \text{ m}^{-1} = 0.00027 \text{ cm}^{-1}$$

$$\Sigma_t = \Sigma_s + \Sigma_a = 38.53 \text{ m}^{-1} = 0.3853 \text{ cm}^{-1}$$

The mean free path is:

$$\lambda = \frac{1}{\Sigma} = 0.0259 \text{ m} = 2.59 \text{ cm}$$

For a mixture of several isotopes, the macroscopic cross section Σ_i is calculated by:

$$\Sigma_i = \sum_j N_j \sigma_{ij} \tag{4.76}$$

where σ_{ij} = microscopic cross section of isotope j for reaction type i
$N_j = w_j \rho N_A / A_j$
w_j = weight fraction of jth isotope in the mixture
ρ = density of mixture

Equation 4.76 assumes that all the isotopes act independently, i.e., that the chemical-crystal binding forces are negligible. In certain cases, especially for thermal neutrons, these binding forces play an important role and cannot be neglected.

EXAMPLE 4.19

What is the total macroscopic absorption cross section of natural uranium? Natural uranium consists of 0.711 percent ^{235}U and the rest is ^{238}U. For thermal neutrons, the absorption cross sections are σ_a (^{235}U) = 678 b and σ_a (^{238}U) = 2.73 b.

ANSWER

The density of uranium is 19.1×10^3 kg/m^3. Therefore,

$$\Sigma_a (U) = 0.00711(19.1 \times 10^3) \frac{0.6023 \times 10^{24}}{235 \times 10^{-3}} (678 \times 10^{-28})$$

$$+ 0.99289(19.1 \times 10^3) \frac{0.6023 \times 10^{24}}{238 \times 10^{-3}} (2.73 \times 10^{-28})$$

$$= (23.6 + 13.1) \text{ m}^{-1} = 36.7 \text{ m}^{-1} = 0.367 \text{ cm}^{-1}$$

4.9.3 The Neutron Flux

If the incident neutron beam consists of neutrons traveling in the same direction, as in Fig. 4.26, the reaction rate is calculated by Eq. 4.74. Unfortunately, in most cases the neutron source consists of neutrons that travel in all directions and the meaning of beam intensity shown in Fig. 4.26 will have to be modified.

Consider a medium that contains neutrons of the same speed v, but moving in all directions. Assume that at some point in space the neutron density is n (neutrons/m^3). If a target is placed at that point, the interaction rate R (reactions/m$^3 \cdot$s) will be equal to:

$$R = \begin{pmatrix}\text{distance traveled by all} \\ \text{neutrons in 1 m}^3\end{pmatrix} \begin{pmatrix}\text{probability of interaction per unit} \\ \text{distance traveled by one neutron}\end{pmatrix}$$

or $$R = n \text{ (neutrons/m}^3)[v \text{ (distance/s)}] [\Sigma \text{ (m}^{-1})] = nv\Sigma \text{ (reaction/m}^3 \cdot \text{s)}$$

The product nv, which has the units of neutrons/m$^2 \cdot$s and represents the total pathlength traveled per second by all the neutrons in 1 m^3, is called the *neutron flux* ϕ.

$$\phi = nv \; (n/\text{m}^2 \cdot \text{s}) \tag{4.77}$$

Using Eq. 4.77, the expression for the reaction rate becomes

$$R_i = \phi \Sigma_i \; [(\text{reactions of type } i)/\text{m}^3 \cdot \text{s}] \tag{4.78}$$

EXAMPLE 4.20

What is the fission rate at a certain point inside a nuclear reactor where the neutron flux is known to be $\phi = 2.5 \times 10^{14}$ neutrons/m$^2 \cdot$s, if a thin foil of ^{235}U is placed there? The fission cross section for ^{235}U is $\sigma_f = 577$ b.

ANSWER

The macroscopic fission cross section is:

$$\Sigma_f = N\sigma_f = (19.1 \times 10^3) \frac{0.6023 \times 10^{24}}{235 \times 10^{-3}} (577 \times 10^{-28})$$

$$= 2824 \text{ m}^{-1} = 28.24 \text{ cm}^{-1}$$

and $R_f = \phi\Sigma_f = (2.5 \times 10^{14})2824 = 7.06 \times 10^{17}$ fissions/m³·s

$= 2.06 \times 10^{11}$ fissions/cm³·s

Another quantity related to the flux and used in radiation exposure calculations is the *neutron fluence F*, defined by

$$F\,(\text{n/m}^2) = \phi t$$

where t is the time of exposure to flux ϕ.

4.9.4 Interaction Rates of Polyenergetic Neutrons

Equation 4.78 gives the reaction rate for the case of monoenergetic neutrons. In practice, and especially for neutrons produced in a reactor, the flux consists of neutrons that have an energy spectrum extending from $E = 0$ up to some maximum energy E_{max}. In such a case, the reaction rate is written in terms of an average cross section. Let

$\phi(E)\,dE$ = neutron flux consisting of neutrons with kinetic energy between E and $E + dE$

$\sigma_i(E)$ = cross section for reaction type i for neutrons with kinetic energy E

N = number of targets per m³ (stationary targets)

The reaction rate is:

$$R\,(\text{reactions/m}^3\cdot\text{s}) = \int dE\,\phi(E)N\sigma_i(E) \tag{4.79}$$

where the integration extends over the neutron energies of interest. The total flux is:

$$\phi = \int \phi(E)\,dE$$

An average cross section is defined in such a way that when it is multiplied by the total flux gives the reaction rate of Eq. 4.79, i.e.,

$$R = \int_E dE\,\phi(E)N\sigma_i(E) = \phi\bar{\Sigma}_i = \phi N\bar{\sigma}_i \tag{4.80}$$

from which the definition of the average cross section is

$$\bar{\sigma}_i = \frac{\int dE\,\phi(E)N\sigma_i(E)}{\int \phi(E)\,dE} \tag{4.81}$$

Since the calculation of average cross sections is beyond the scope of this text, the reader should consult the proper books on reactor physics. The main purpose of this short discussion is to alert the reader to the fact that when polyenergetic neutrons are involved, an appropriate *average* cross section should be used for the calculation of reaction rates.

PROBLEMS

4.1 Calculate the stopping power due to ionization and excitation of a 2-MeV electron moving in water. What is the radiation energy loss rate of this particle? What is the total energy radiated?

4.2 Calculate the stopping power in aluminum for a 6-MeV alpha particle.

4.3 The window of a Geiger-Muller counter is made of mica and has a thickness of 0.02 kg/m^2 ($\rho = 2.6 \times 10^3$ kg/m^3).

(a) What is the minimum electron energy that will just penetrate this window?
(b) What is the energy loss, in MeV/mm, of an electron with the kinetic energy determined in (a) moving in mica?
(c) What is the energy loss, in MeV/mm, of a 6-MeV alpha particle moving in mica?
(d) Will a 6-MeV alpha particle penetrate this mica window?

4.4 Beta particles emitted by ^{32}P ($E_{max} = 1.7$ MeV) are counted by a gas counter. Assuming that the window of the counter causes negligible energy loss, what gas pressure is necessary to stop all the betas inside the counter if the length of the detector is 100 mm? Assume that the gas is argon.

4.5 What is the kinetic energy of an alpha particle that will just penetrate the human skin? For the skin, assume $t = 1$ mm; $\rho = 10^3$ kg/m^3; 65 percent O, 18 percent C, 10 percent H, 7 percent N.

4.6 Repeat Prob. 4.5 with an electron.

4.7 Assuming that a charged particle loses energy linearly with distance, derive the function $T = T(x)$ where $T(x)$ = kinetic energy of the particle after going through thickness x. The initial kinetic energy is T_0, and the range is R.

4.8 A beam of 6-MeV alpha particles strikes a gold foil with thickness equal to one-third of the alpha range. What is the total energy loss of the alpha as it goes through this foil?

4.9 What is the energy deposited in a piece of paper by a beam of 1.5-MeV electrons? Assume that the paper has the composition CH_2, thickness 0.1 mm, and density 800 kg/m^3. The incident parallel electron beam consists of 10^8 electrons/$m^2 \cdot s$. Give your result in MeV/$cm^2 \cdot s$ and J/$m^2 \cdot s$.

4.10 What is the range of 10-MeV proton in air at 1 atm? What is the range at 10 atm?

4.11 What is the range of a 4-MeV deuteron in gold?

4.12 A 1.5-MeV gamma undergoes Compton scattering. What is the maximum energy the Compton electron can have? What is the minimum energy of the scattered photon?

4.13 The energy of a Compton photon scattered to an angle of 180° is 0.8 MeV. What is the energy of the incident photon?

4.14 Prove that a gamma scattered by 180°, as a result of a Compton collision, cannot have energy greater than $mc^2/2$, where $mc^2 = 0.511$ MeV is the rest mass energy of the electron.

4.15 Prove that the attenuation coefficient of gammas for a compound or a mixture can be written as:

$$\mu\,(\mathrm{m^2/kg}) = \sum_{i=1}^{H} w_i \mu_i\,(\mathrm{m^2/kg})$$

where w_i = weight fraction of ith element
μ_i = total mass attenuation coefficient of ith element

4.16 A 1.75-MeV γ ray hits a 25-mm-thick NaI crystal. What fraction of the interactions of this photon will be photoelectric? What is the average distance traveled before the first interaction occurs?

4.17 A parallel beam of gammas impinges upon a multiple shield consisting of successive layers of concrete, Fe, and Pb, each layer having thickness 100 mm. Calculate the fraction of gammas traversing this shield. The total attenuation coefficients are $\mu(\text{concrete}) = 0.002\ \mathrm{m^2/kg}$, $\mu(\text{Fe}) = 0.004\ \mathrm{m^2/kg}$, and $\mu(\text{Pb}) = 0.006\ \mathrm{m^2/kg}$; $\rho_{\text{concrete}} = 2.3 \times 10^3\ \mathrm{kg/m^3}$.

4.18 Assume that a parallel beam of 3-MeV gammas and a parallel beam of 2-MeV neutrons impinge upon a piece of lead 50 mm thick. What fraction of γ's and what fraction of neutrons will emerge on the other side of this shield without any interaction? Based on your result, what can you say about the effectiveness of lead as a shield for γ's or neutrons? [$\sigma(2\ \text{MeV}) = 3.5$ b.]

4.19 What are the capture, fission, and total macroscopic cross section of uranium enriched to 90 percent in ^{235}U for thermal neutrons? ($\rho = 19.1 \times 10^3\ \mathrm{kg/m^3}$.)

^{235}U: $\sigma_\gamma = 101$ b $\quad \sigma_f = 577$ b $\quad \sigma_s = 8.3$ b

^{238}U: $\sigma_\gamma = 2.7$ b $\quad \sigma_f = 0$ $\quad \sigma_s = 8$ b

4.20 What is the average distance a thermal neutron will travel in 90 percent enriched uranium (see Prob. 4.19) before it has an interaction?

4.21 The water in a pressurized-water reactor contains dissolved boron. If the boron concentration is 800 parts per million, what is the mean free path of thermal neutrons? The microscopic cross sections are:

H_2O: $\sigma_s = 103$ b $\quad \sigma_a = 0.65$ b

Boron: $\sigma_s = 4$ b $\quad \sigma_a = 759$ b

BIBLIOGRAPHY

Evans, R. D., *The Atomic Nucleus*, McGraw-Hill, New York, 1972.
"Neutron Cross Sections" (and supplements), BNL-325, 1958.
Roy, R. R., and Reed, R. D., *Interactions of Photons and Leptons with Matter*, Academic Press, New York, 1968.
Segré, E., *Nuclei and Particles*, W. A. Benjamin, New York, 1968.
Ziegler, J. F. (organizer), *The Stopping and Ranges of Ions in Matter*, 5 vols., Pergamon Press, New York, 1977.

REFERENCES

1 Berger, M. J., and Seltzer, S. M., NASA SP-3012, 1964.
2 Anderson, H. H., Sorensen, H., and Vadja, P., *Phys. Rev.* **180**:383 (1969).
3 Sorensen, H., and Andersen, H. H., *Phys. Rev.* **8B**:1854 (1973).
4 Sternheimer, R. M., *Phys. Rev.* **88**:851 (1952).
5 Sternheimer, R. M., *Phys. Rev.* **103**:511 (1956).
6 Williamson, C. F., Baujot, J. P., and Picard, J., CEA-R-3042 (1966).
7 Bichsel, H., *Phys. Rev.* **112**:1089 (1958).
8 Tabata, T., Iot, R., and Okabe, S., *Nucl. Instr. Meth.* **103**:85 (1972).
9 Northcliffe, L. C., *Ann. Rev. Nucl. Sci.* **13**:67 (1963).
10 Lindhard, J., Scharff, M., and Schiott, H. E., *Fys. Medd.* **33**:14 (1963).
11 Moak, C. D., and Brown, M. D., *Phys. Rev.* **149**:244 (1966).
12 Brown, M. D., and Moak, C. D., *Phys. Rev.* **6B**:90 (1972).
13 Betz, G., Isele, H. J., Rossle, E., and Hortig, G., *Nucl. Inst. Meth.* **123**:83 (1975).
14 Booth, W., and Grant, I. S., *Nucl. Phys.* **63**:481 (1965).
15 Forster, J. S., Ward, D., Andrews, H. R., Ball, G. G., Costa, G. J., Davies, W. G., and Mitchell, J. V., *Nucl. Instr. Meth.* **136**:349 (1976).
16 Northcliffe, L. C., and Schilling, R. F., *Nucl. Data* **A7**:223 (1970).
17 Brown, M. D., "Interaction of Uranium Ions in Solids," Ph.D. thesis, University of Tennessee, 1972 (unpublished).
18 Bridwell, L., and Bucy, S., *Nucl. Sci. Eng.* **37**:224 (1969).
19 Bridwell, L., and Moak, C. D., *Phys. Rev.* **156**:242 (1967).
20 Siffert, P., and Coche, A., "General Characteristics of the Interactions of Nuclear Radiations with Matter and their Consequences," in G. Bertolini and A. Coche (eds.), *Semiconductor Detectors*, Wiley, New York, 1968, pp. 279–300.
21 Hakim, M., and Shafrir, N. H., *Nucl. Sci. Eng.* **48**:72 (1972).
22 Gesini, G., Lucarini, G., and Rustichelli, F., *Nucl. Inst. Meth.* **127**:579 (1975).
23 Hubbell, J. H., NSRDS-NBS 29, August 1969.
24 Goldstein, H., and Wilkins, J. E., Jr., NYO-3075, Nuclear Development Associates, Inc., 1954.
25 R. G. Jaeger (ed.), *Engineering Compendium on Radiation Shielding* V1, IAEA, New York, 1968.
26 Trubey, D. K., ORNL-RSIC-10, ORNL (1966).
27 Eisenhauer, C. M., and Simmons, G. L., *Nucl. Sci. Eng.* **56**:263 (1975).
28 Kuspa, J. P., and Tsoulfanidis, N., *Nucl. Sci. Eng.* **52**:117 (1973).
29 Chilton, A. B., *Nucl. Sci. Eng.* **59**:436 (1979).

5

■ gas-filled detectors

5.1 INTRODUCTION

Gas-filled detectors operate by utilizing the ionization produced by radiation as it passes through a gas. Typically, such a counter consists of two electrodes to which a certain electrical potential is applied. The space between the electrodes is filled with a gas (Fig. 5.1). Ionizing radiation, passing through the space between the electrodes, dissipates part or all of its energy by generating electron-ion pairs. Both electrons and ions are charge carriers that move under the influence of the electric field. Their motion induces a current on the electrodes, which may be measured (Fig. 5.1*a*). Or, through appropriate electronics, the charge produced by the radiation may be transformed into a pulse, in which case particles are counted individually (Fig. 5.1*b*). The first type of counter (Fig. 5.1*a*) is called *current* or *integrating chamber*; the second type (Fig. 5.1*b*) is called *pulse chamber.* To get an idea of what charges and currents one might expect to measure, consider this representative example.

For most gases, the average energy required to produce an electron-ion pair is about 30 eV. This number takes into account all collisions, including those that lead to excitation. If a 3-MeV alpha or beta particle deposits all its energy in the counter, it will produce, on the average,

$$\frac{3 \times 10^6}{30} = 10^5 \text{ electron-ion pairs}$$

A typical gas counter has a capacitance of about 50 pF, and the charge will be collected in a time of the order of 1 μs. If all the charge created by the 3-MeV particle is collected, the voltage and current expected are of the order

$$V = \frac{Q}{C} \approx \frac{10^5 \times 1.6 \times 10^{-19} \text{ C/el}}{50 \times 10^{-12} \text{ F}} \approx 0.5 \times 10^{-3} \text{ V} \approx 0.5 \text{ mV}$$

$$i = \frac{Q}{t} \approx \frac{10^5 \times 1.6 \times 10^{-19}}{10^{-6}} \text{ A} \sim 1.6 \times 10^{-8} \text{ A}$$

In an ionized gas without an electric field, electrons and positive ions will move at random with an average kinetic energy equal to $\frac{3}{2}kT$ where k = Boltzmann constant† and T = temperature of the gas (Kelvin). When an electric field is present, both electrons and positive ions acquire a net velocity component along the lines of the electric field. Electrons move toward the positive electrode, positive ions toward the negative one. The force on either charge carrier is the same and equal to $F = Ee$, where E = electric field intensity, but the acceleration is quite different. The acceleration a is

†$k = 1.3805 \times 10^{-23}$ J/K.

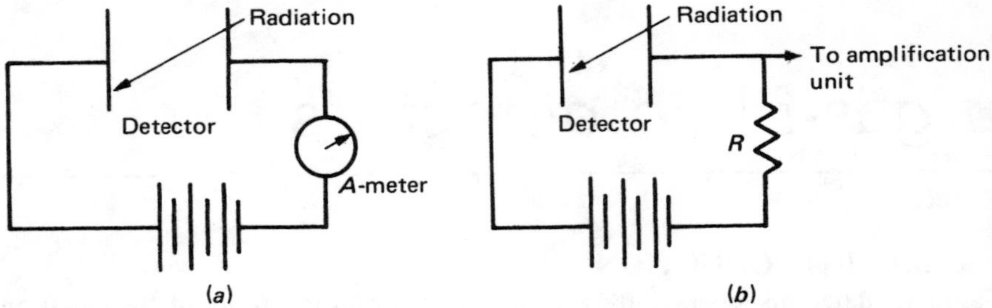

FIG. 5.1 A typical gas-filled detector: (*a*) the direct current produced in the circuit is measured; (*b*) individual pulses are detected.

equal to F/M, where M is the mass of the ion or electron. Therefore, the acceleration of an electron will be thousands of times larger than the acceleration of an ion. The time it takes the electrons to reach the positive electrode of a typical counter is about 1 μs. The corresponding time for the positive ions is about 1 ms, a thousand times longer.

The discussion up to this point has been limited to the effects of the ionization produced directly by the incident particle. This is called *primary* ionization. There are types of gas counters in which the electric field is so strong that the electrons of the primary ionization acquire enough kinetic energy between collisions to produce new electron-ion pairs. These new charges constitute the *secondary* ionization. Primary and secondary ionization are generated within such a short period of time that they contribute to one and the same pulse.

5.2 RELATIONSHIP BETWEEN HIGH VOLTAGE AND CHARGE COLLECTED

Assume that the following experiment is performed (Fig. 5.2). A radioactive source of constant intensity is placed at a fixed distance from a gas counter. The high voltage (HV) applied to the counter may be varied with the help of a potentiometer. An

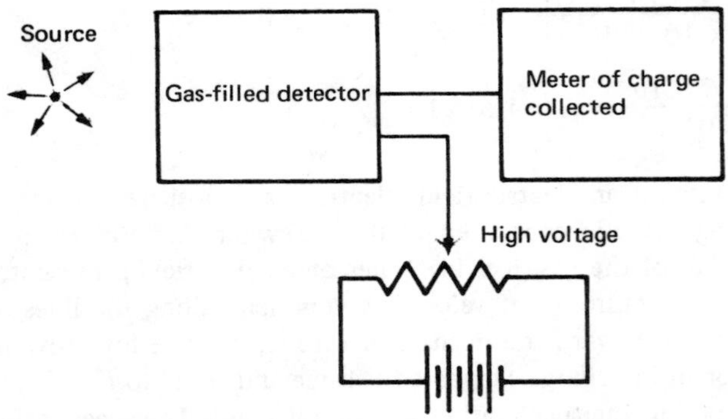

FIG. 5.2 Experimental set-up for the study of the relationship between high voltage applied and charge collected.

appropriate meter measures the charge collected per unit time. If the HV applied to the counter is steadily increased, the charge collected per unit time changes as shown in Fig. 5.3.

The curve of Fig. 5.3 is divided into five regions.

REGION I

When the voltage is very low, the electric field in the counter is not strong, electrons and ions move with relatively slow speeds, and their recombination rate is considerable. As V increases, the field becomes stronger, the carriers move faster, and their recombination rate decreases up to the point where it becomes zero. Then, all the charge created by the ionizing radiation is being collected ($V = V_I$). Region I is called the *recombination* region.

REGION II

In region II, the charge collected stays constant despite a change in the voltage because the recombination rate is zero and no new charge is produced. This is called the *ionization* region.

REGION III

In this region, the collected charge starts increasing because the electrons produce secondary ionization that results in charge multiplication. The electric field is so strong, in a certain fraction of the counter volume, that electrons from the primary ionization acquire enough energy between collisions to produce additional ionization. The *gas multiplication factor*–i.e., the ratio of the total ionization produced divided by the primary ionization–is, for a given voltage, independent of the primary ionization. Thus the output of the counter is proportional to the primary ionization. The pulse height at the output is proportional to the energy dissipated inside the counter; therefore particle identification and energy measurement are possible. This region is, appropriately enough, called the *proportional* region.

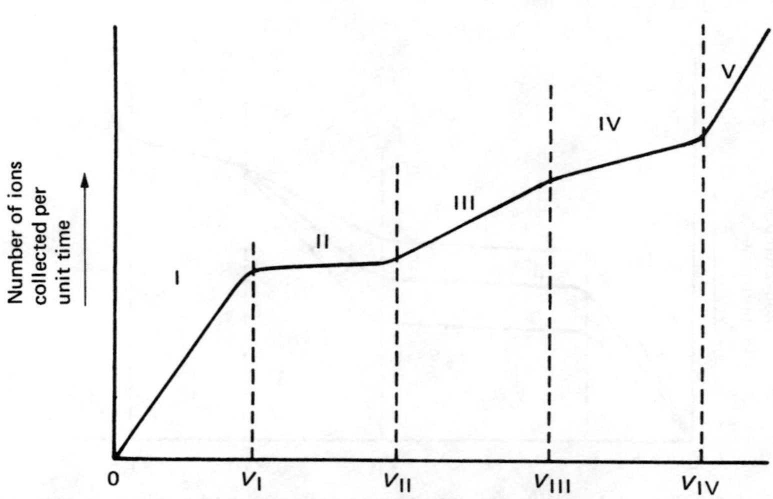

FIG. 5.3 The relationship between voltage applied to the counter and charge collected.

REGION IV

In this region, the electric field inside the counter is so strong that a single electron-ion pair generated in the chamber is enough to initiate an avalanche of electron-ion pairs. This avalanche will produce a strong signal with shape and height independent of the primary ionization and the type of particle, a signal that depends only on the electronics of the counter. Region IV is called the *Geiger-Müller* (GM) region.

REGION V

If the applied voltage is raised beyond the value V_{IV}, a single ionizing event initiates a continuous discharge in the gas, and the device is not a counter anymore. If a commerical gas counter operates with voltage $V > V_{IV}$, it will probably be destroyed.

If the graph discussed above is obtained using an α, β, or γ source, the results will be as shown in Fig. 5.4.

5.3 DIFFERENT TYPES OF GAS-FILLED COUNTERS

Gas counters take their name from the voltage region ion which they operate. No counter operates in region I of Fig. 5.3, because a slight change in voltage will change the signal.

Ionization chambers operate in region II. No charge multiplication takes place. The output signal is proportional to the particle energy dissipated in the detector; therefore measurement of particle energy is possible. Since the signal from an ionization chamber is not large, only strongly ionizing particles such as alphas, protons, fission fragments, and other heavy ions are detected by such counters. The voltage applied is less than 1000 V.

Proportional counters operate in region III. Charge multiplication takes place, but the output signal is still proportional to the energy deposited in the counter. Measurement of particle energy is possible. Proportional counters may be used for the detection of any charged particle.

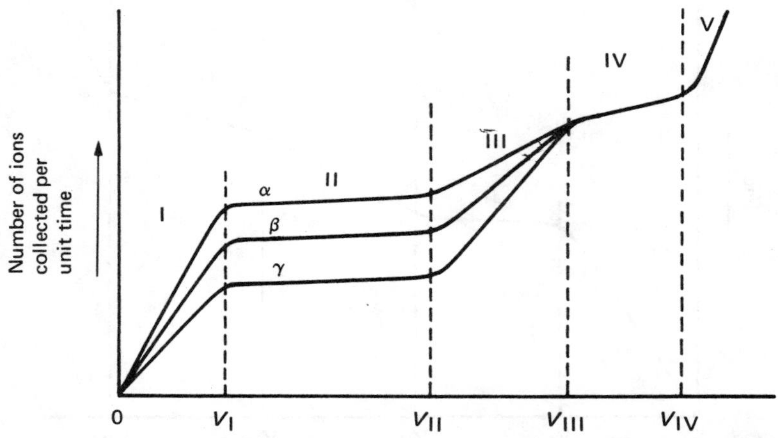

FIG. 5.4 The relationship between charge collected and applied voltage for three different types of particles. In region IV, the curve increases slightly but is the same for all particles.

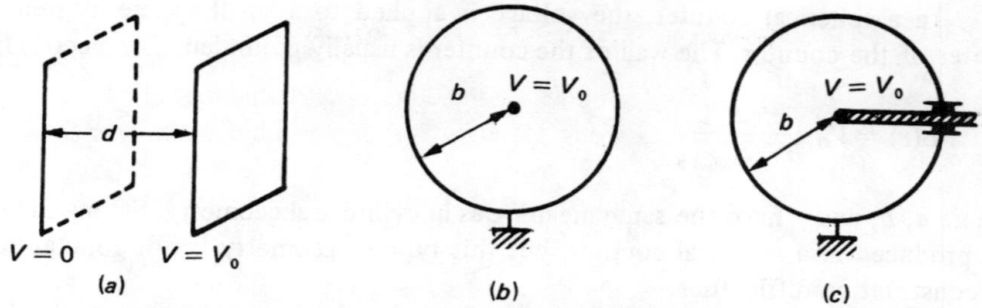

FIG. 5.5 The different geometries of gas-filled detectors: (*a*) parallel plate; (*b*) cylindrical; (*c*) spherical.

Identification of the type of particle is possible with both ionization and proportional counters. An alpha particle and an electron having the same energy and entering either of the counters, will give a different signal. The alpha particle signal will be bigger than the electron signal. The voltage applied to proportional counters ranges between 800 and 2000 V.

Geiger–Müller (GM) counters operate in region IV. GM counters are very useful because their operation is simple and they provide a very strong signal, so strong that a preamplifier is not necessary. They can be used with any kind of ionizing radiation (with different levels of efficiency). The disadvantage of GM counters is that their signal is independent of the particle type and its energy. *Therefore, a GM counter provides information only about the number of particles.* Another minor disadvantage is their relatively long dead time (200 to 300 μs). (For more details about dead time, see Sec. 5.6.2.) The voltage applied to GM counters ranges from 1000 to 3000 V.

Gas counters may be constructed in any of three basic geometries: parallel plate, cylindrical, or spherical (Fig. 5.5). In a parallel-plate chamber, the electric field (neglecting edge effects) is uniform, with strength equal to

$$E = \frac{V_0}{d} \tag{5.1}$$

In the cylindrical chamber, the voltage is applied to a very thin wire, a few mills of an inch in diameter, stretched axially at the center of the cylinder. The cylinder wall is usually grounded. The electric field is, in this case,

$$E(r) = \frac{V_0}{\ln(b/a)} \frac{1}{r} \tag{5.2}$$

where a = radius of the central wire
b = radius of the counter
r = distance from the center of the counter

It is obvious from Eq. 5.2 that very strong electric fields can be maintained inside a cylindrical counter close to the central wire. Charge multiplication is achieved more easily in a cylindrical than in a plate-type gas counter. For this reason, proportional and GM counters are manufactured with cylindrical geometry.

In a spherical counter, the voltage is applied to a small sphere located at the center of the counter. The wall of the counter is usually grounded. The electric field is

$$E(r) = V_0 \frac{ab}{b-a} \frac{1}{r^2} \tag{5.3}$$

where a, b, and r have the same meaning as in cylindrical geometry. Strong fields may be produced in a spherical counter, but this type of geometry is not popular because of construction difficulties.

A counter filled with a gas at a certain pressure may operate in any of the regions I–IV discussed earlier, depending on a combination of the following parameters:

1 Size of the counter
2 Size of wire (in cylindrical counters)
3 Gas type
4 Gas pressure
5 Level of high voltage

Normally, gas counters are manufactured to operate in one region only. The user buys an ionization counter, a proportional counter, or a GM counter. The manufacturer has selected the combination of variables 1–4 listed above that results in the desired type of gas counter. The last variable, the high voltage applied, is not a fixed number but a range of values. The range is specified by the manufacturer, but the user decides on the best possible value of HV.

The rest of this chapter discussed the special characteristics of the three types of gas counters.

5.4 IONIZATION CHAMBERS

5.4.1 Pulse Formation in an Ionization Chamber

The formation and shape of the signal in an ionization chamber will be analyzed for a parallel-plate counter as shown in Fig. 5.1*b*. The analysis is similar for a cylindrical or a spherical chamber.

Consider the ionization chamber shown in Fig. 5.6. The two parallel plates make a capacitor with capacitance C, and with the resistor R an RC circuit is formed. A constant voltage V_0 is applied on the plates. The time dependent voltage $V(t)$ across the resistor R represents the signal. The objective of this section is to obtain the function $V(t)$.

Assume that one electron-ion pair has been formed at a distance x_0 from the collecting plate (collector). The electron and the ion start moving in the electric field, and they acquire kinetic energy at the expense of the electrostatic energy stored in the capacitance of the chamber. If the charge moves a distance dx, conservation of energy requires that

(Work on charges) = (change in electrostatic energy)

$$eE(dx^+ - dx^-) = d\left(\frac{Q^2}{2C}\right) = \frac{Q}{C}\,dQ \approx V_0(dQ^- + dQ^+) \tag{5.4}$$

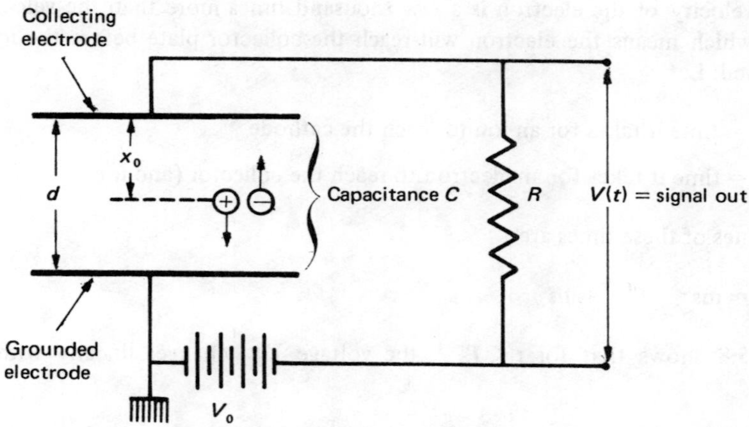

FIG. 5.6 The electronic circuit of a parallel-plate ionization chamber.

where E = electric field intensity
Q = charge on chamber plates
dQ^+, dQ^- = changes in positive, negative charge, respectively

It is assumed that the change in the charge (dQ) is so small that the voltage V_0 stays essentially constant. The voltage $V(t)$ across the resistor R is the result of this change in the charge and is given by

$$V(t) = \frac{1}{C} \int_0^t dQ(t) = \frac{1}{C} \int_0^t (dQ^+ + dQ^-) \tag{5.5}$$

Substituting in Eq. 5.5 the value of dQ from Eq. 5.4, one obtains

$$V(t) = \frac{1}{C} \int_0^t \frac{e}{V_0} E[dx^+(t) - dx^-(t)] \tag{5.6}$$

Let w^+ = drift velocity of positive ions

w^- = drift velocity of electrons

In general, the drift velocity is a function of the *reduced field strength* E/p, where p is the gas pressure in the chamber.

The derivation up to this point is independent of the chamber geometry. To proceed further requires substitution of the value of the electric field from either Eq. 5.1, 5.2, or 5.3. For a plate-type ionization chamber the field is constant (Eq. 5.1), independent of x, and so is the drift velocity. Therefore, Eq. 5.6 becomes

$$V(t) = \frac{e}{Cd} \int_0^t (w^+ - w^-)\, dt = -\frac{e}{Cd}(w^-t - w^+t) \tag{5.7}$$

The drift velocity of the electron is a few thousand times more than the velocity of the ion,† which means the electron will reach the collector plate before the ion has hardly moved. Let

$T^{(+)}$ = time it takes for an ion to reach the cathode

$T^{(-)}$ = time it takes for an electron to reach the collector (anode)

Typical values of these times are

$$T^{(+)} \approx \text{ms} \qquad T^{(-)} \approx \mu\text{s}$$

Equation 5.8 shows that for $t < T^{(-)}$, the voltage $V(t)$ changes linearly with time (Fig. 5.7).

$$V(t) = -\frac{e}{Cd}(w^- - w^+)t \qquad 0 < t \leqslant T^{(-)} \tag{5.8}$$

For $T > T^{(-)}$, the signal is

$$V(t) = -\frac{e}{Cd}(x_0 - w^+ t) \qquad t > T^{(-)} \tag{5.9}$$

Finally, after $t = T^{(+)}$ the ion reaches the grounded cathode and the signal reaches its maximum (negative) value, which is

$$V(T^+) = -\frac{e}{Cd}x_0 \qquad t > T^{(+)} \tag{5.10}$$

If N electron-ion pairs are produced, the final voltage will be

$$V(T^+) = -\frac{Ne}{Cd}x_0 \qquad t = T^{(+)} \tag{5.11}$$

For $t > T^{(+)}$, the pulse decays with decay constant RC (see Sec. 10.3).

†Typical values of drift velocities are $w^+ \approx 10$ m/s, $w^- = 10^4 - 10^5$ m/s.

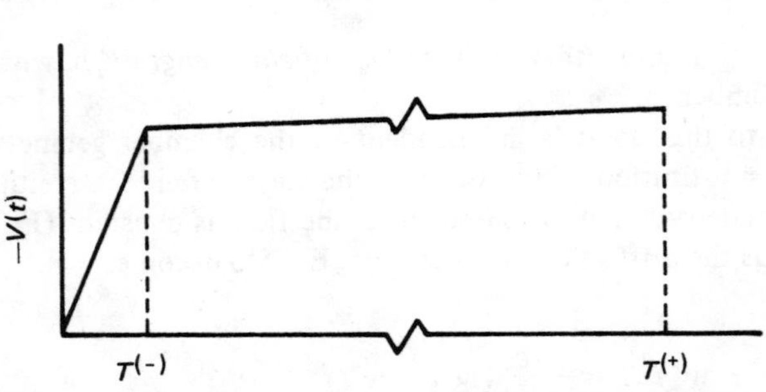

FIG. 5.7 The voltage pulse generated by an ionization chamber.

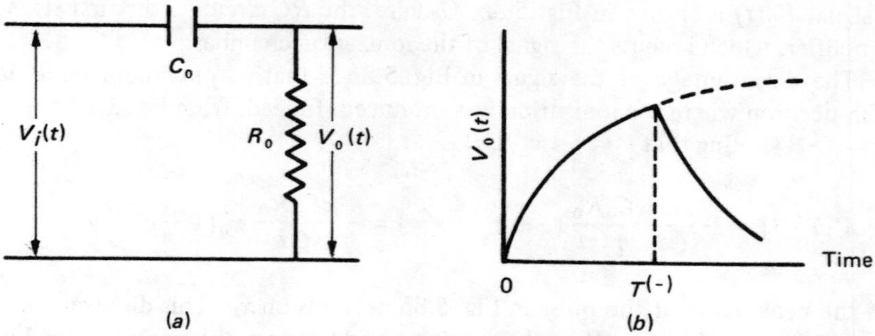

FIG. 5.8 (*a*) The signal $V_i(t)$ is fed into the *RC* circuit. (*b*) The output of the *RC* circuit decays quickly with a decay constant R_0C_0.

The pulse profile of Fig. 5.7 was derived under the assumption that all ion pairs were produced at $x = x_0$. Actually, the ionization is produced along the track traveled by the incident particle. The final pulse will be the result of the superposition of many pulses with different $T^{(-)}$ values. Because of this effect, the sharp change in slope at $t = T^{(-)}$ will disappear and the pulse will be smoother.

The pulse of Fig. 5.7 is not suitable for counting individual particles because it does not decay quickly enough. A pulse-type counter should produce a signal that decays faster than the average time between the arrival of two successive particles. For example, if the counting rate is 1000 counts/min, a particle arrives at the counter on the average of every 1/1000 min (60 ms).

In Fig. 5.7, the pulse could be stopped at time $t = T^{(+)}$ by electronic means. Such a technique would produce pulses with height proportional to the total charge generated in the detector, but with a duration of a few hundreds of microseconds, which is unacceptably long. The method used in practice is to "chop off" the pulse at time $t = T^{(-)}$, which amounts to stopping the pulse after only the electrons are collected. The signal is then fed into an *RC* circuit that, as described in Chap. 10, changes the pulse as shown in Fig. 5.8.

Let $V_i(t)$ be the signal at the output of the detector that is used as an input to an *RC* circuit. It can be shown (see Chap. 10) that the following relationship holds between $V_i(t)$ and the output signal of the *RC* circuit, $V_0(t)$. For $0 \leqslant t \leqslant T^{(-)}$ (Fig. 4.7):

$$V_i(t) = -\frac{e}{Cd}(w^- - w^+)t = kt$$

$$V_0(t) = k\frac{C_0R_0}{T^{(-)}}(1 - e^{-t/C_0R_0}) \tag{5.12}$$

For $t > T^{(-)}$, $V_i(t)$ is essentially constant, and

$$V_0(t) = \frac{kR_0C_0}{T^{(-)}}(e^{T^{(-)}/C_0R_0} - 1)e^{-t/C_0R_0} \tag{5.13}$$

The signal $V_0(t)$ is shown in Fig. 5.8*b*. Usually, the *RC* circuit is the first stage of the preamplifier, which accepts the signal of the ionization chamber.

The disadvantage of the signal in Fig. 5.8*b* is that its maximum value depends on the position where the ionization was produced. Indeed, from Eq. 5.12 one obtains for $t = T^{(-)}$ (noting that $k = -ew^-/Cd$)

$$V[T^{(-)}] = -\frac{ew^-}{Cd}\frac{C_0R_0}{T^{(-)}}(q - e^{-t/R_0C_0}) = -\frac{eC_0R_0}{Cd}x_0(1 - e^{-t/R_0C_0}) \qquad (5.14)$$

Thus the peak value of the pulse in Fig. 5.8*b* depends on x_0. This disadvantage can be corrected in several ways. One is by placing a grid between the two plates and keeping it at an intermediate voltage V_g $(0 < V_g < V_0)$. For more details about the "gridded" ionization chamber, the reader should consult the references.

The analysis of the pulse formation in a cylindrical or a spherical counter follows the same approach. The results are slightly different because the electric field is not constant (see Eqs. 5.2 and 5.3), but the general shape of the signal is that shown in Fig. 5.7. (See Franzen and Cochran, and Kowalski for detailed calculations of the pulse shapes for the three geometries of gas-filled chambers.)

5.4.2 Current Ionization Chambers

An ionization chamber of the *current* type measures the average ionization produced by many incoming particles. This is achieved by measuring directly the electrical current generated in the chamber, using either a sensitive galvanometer for currents of 10^{-8} A or higher (Fig. 5.9), or an electrometer (sometimes with an amplifier) for currents less than 10^{-8} A. In the case of the electrometer, as shown in Fig. 5.10, the current is determined by measuring the voltage drop across the known resistance *R*. The voltage drop may be measured by the electrometer directly or after some amplification.

For current ionization chambers, it is very important to know the relationship between applied voltage and output current (for a constant radiation source). This relationship, which is shown in Fig. 5.11, consists of regions I and II of the graph of Fig. 5.3. The proper operating voltage of the ionization chamber is that for which all the ionization produced by the incident radiation is measured. If this is the case,

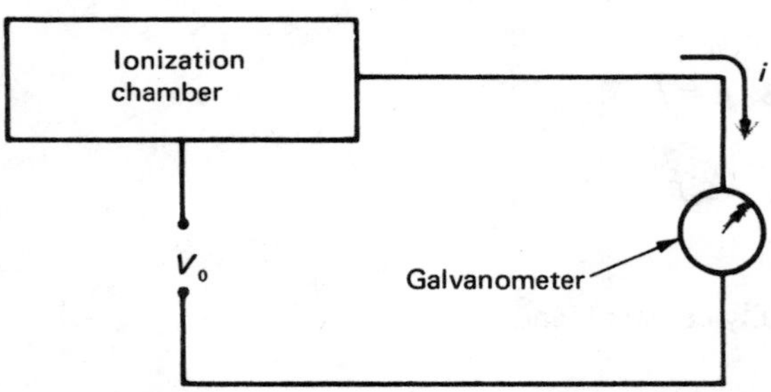

FIG. 5.9 Measurement of the current produced by an ionization chamber by using a galvanometer.

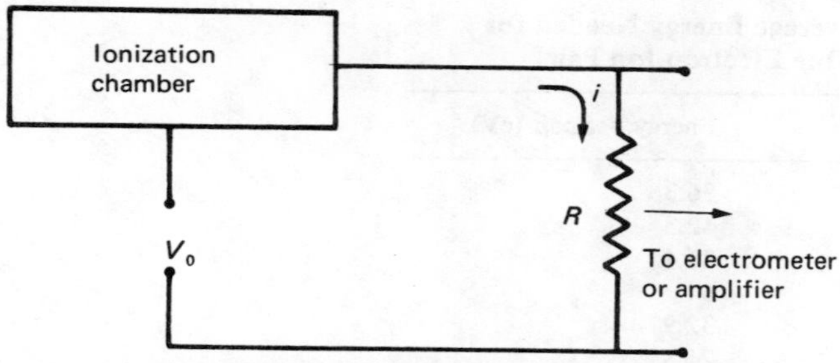

FIG. 5.10 Measurements of the current produced by an ionization chamber by using an electrometer.

a slight increase of the applied voltage will result in negligible change of the measured current. The voltage is then called the saturation voltage (V_s), and the corresponding current is called saturation current. The value of the saturation current depends on the intensity and type of the radiation source (Fig. 5.11). It also depends, for the same radiation source, on the size and geometry of the chamber as well as on the type and pressure of the gas used. If one considers different gases, other things being equal, the highest current will be produced by the gas with the lowest average energy needed for the production of one electron-ion pair. Typical energies for common gases are given in Table 5.1.

During measurements of the ionization current with an electrometer, one would like to know the response of the measuring instrument if the signal from the ionization chamber changes. Assume that the current of the chamber changes suddenly from a value of i_1 to i_2. The response of the electrometer is obtained by considering the equivalent electronic circuit of Fig. 5.10, shown in Fig. 5.12. The capacitor C represents the combined capacitance of the chamber and everything else. The resistor R represents a corresponding total resistance for the circuit. The signal to be measured is the voltage $V(t)$, where *for* $t \leqslant 0$,

$$V_1 = i_R R = i_1 R \tag{5.15}$$

At time $t = 0$, the current changes instantaneously from i_1 to i_2, and the voltage will eventually become

$$V_2 = i_2 R \tag{5.16}$$

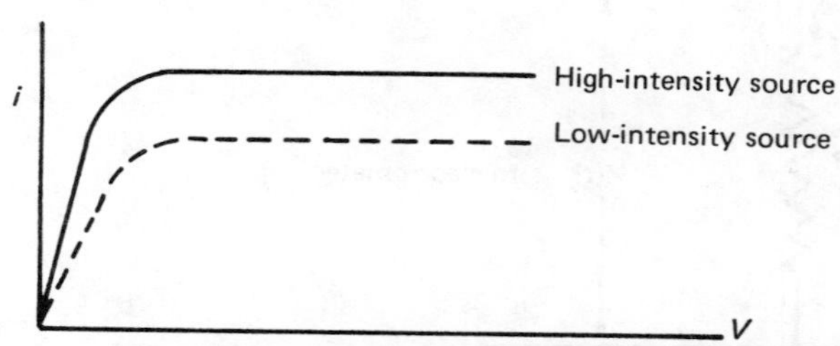

FIG. 5.11 The ionization chamber current as a function of applied voltage.

TABLE 5.1 Average Energy Needed for Production of One Electron-Ion Pair†

Gas	Energy per pair (eV)
H	36.3
He	42.3
A	26.4
Air	34
CO_2	32.9
C_2H_6 (ethane)	24.8
CH_4	27.3

†From Franzen and Cochran.

During the transition period, Kirchoff's first law gives

$$i_2 = i_C + i_R = \frac{dQ(t)}{dt} + \frac{V(t)}{R} = C\frac{dV(t)}{dt} + \frac{V(t)}{R}$$

or
$$\frac{dV(t)}{dt} + \frac{1}{RC} V(t) = i_2 R \tag{5.17}$$

The solution of this differential equation, with the initial condition given by Eq. 5.15, is

$$V(t) = i_2 R + R(i_2 - i_1)e^{-t/RC} \tag{5.18}$$

The function given by Eq. 5.18 is shown in Fig. 5.13. The response of the electrometer is exponential with a rate of change determined by the time constant RC. For fast response, the time constant should be as short as practically possible.

5.5 PROPORTIONAL COUNTERS

5.5.1 Gas Multiplication in Proportional Counters

When the electric field strength inside a gas counter exceeds a certain value, the electrons that move in such a field acquire, between collisions, sufficient energy to pro-

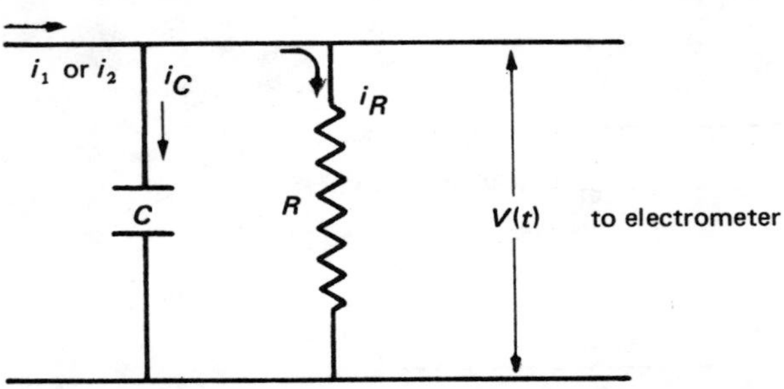

FIG. 5.12 The equivalent electronic circuit of Fig. 5.10.

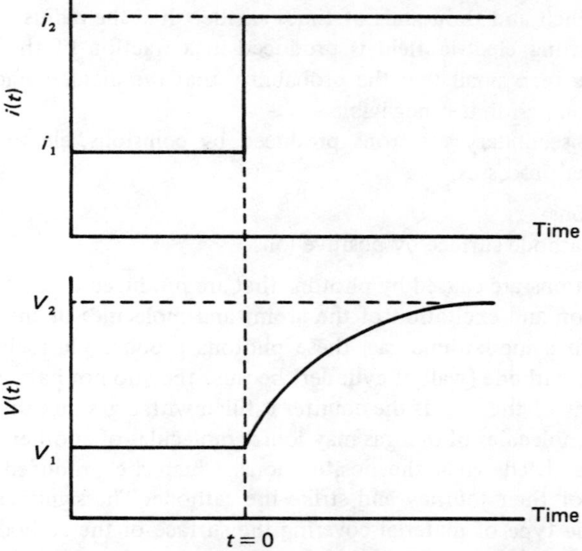

FIG. 5.13 Response of an electrometer to a step change of the ionization current.

duce new ions. Thus, more electrons will be liberated, which in turn will produce more ions. The net effect of this process is multiplication of the primary ionization. The phenomenon is called *gas multiplication.*† To achieve the high field intensity needed for gas multiplication without excessive applied voltage, chambers operating in this mode are usually cylindrical with a very thin wire stretched axially at the center of the counter (Fig. 5.14). The wall of the counter is normally grounded and a positive voltage is applied to the central wire. In such a geometry, the electrostatic field inside the chamber is radial and its intensity is (Eq. 5.2)

$$E(r) = \frac{V_0}{\ln (b/a)} \frac{1}{r}$$

†Also called gas gain or gas amplification.

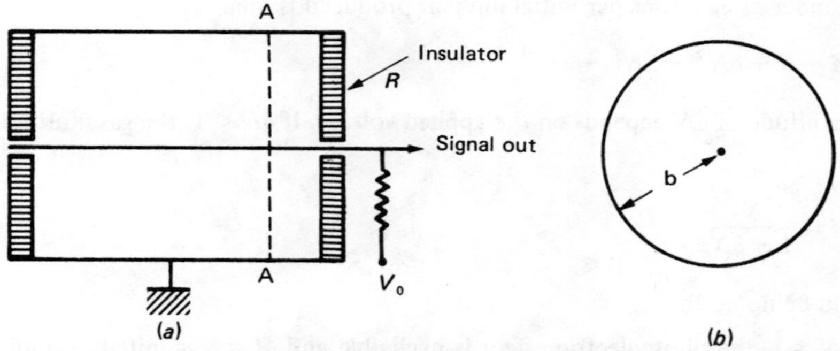

FIG. 5.14 (*a*) A cylindrical gas-filled detector. (*b*) Cross section of the detector at AA.

The field intensity increases rapidly as the wire is approached. Since the radius a of the wire is a few mills of an inch and thousands of times smaller than the radius b of the counter, an extremely strong electric field is produced in a fraction of the chamber's volume. This volume is so small that the probability that the incident radiation will produce an electron ion pair in it is negligible.

In addition to the secondary electrons produced by collisions, electrons are also produced by two other processes:

1 Photoelectric interactions
2 Bombardment of the cathode surface by positive ions

The photoelectric interactions are caused by photons that are produced in the counter as a result of the ionization and excitation of the atoms and molecules of the gas. If the chamber is filled with a monatomic gas, these photons produce photoelectrons only when they strike the cathode (wall of cylinder) because they do not have enough energy to ionize the atoms of the gas. If the counter is filled with a gas mixture, however, photons emitted by molecules of one gas may ionize molecules of another.

Electrons are also emitted when the positive ions, which are produced in the chamber, reach the end of their journey and strike the cathode. The significance of this effect depends on the type of material covering the surface of the cathode and, more important, on the type of the gas filling the chamber.

The production of electrons by these processes results in the generation of successive avalanches of ionization because all the electrons, no matter how they are produced, migrate in the direction of the intense electric field and initiate additional ionization. The gas multiplication factor M, which is equal to the *total* number of free electrons produced in the counter when *one* ion pair is produced by the incident radiation, is calculated as follows. Let

N = total number of electrons set free per primary electron-ion pair

δ = average number of photoelectrons produced per ion pair generated in the counter ($\delta \ll 1$)

The initial avalanche of N electrons will produce δN photoelectrons. Each photoelectron produces a new avalanche of N new electrons; therefore the second avalanche consists of δN^2 electrons. The third avalanche will have δN^3 electrons, and so on. The total number of electrons per initial ion pair produced is then

$$M = N + \delta N^2 + \delta N^3 + \cdots$$

The magnitude of δN depends on the applied voltage. If $\delta N < 1$, the gas multiplication factor is

$$M = \frac{N}{1 - \delta N} \tag{5.19}$$

It should be noted that

1 If $\delta N \ll 1$, the photoelectric effect is negligible and $M = N$ = initial gas multiplication (first avalanche).

2 If $\delta N < 1$, M can become much larger than N.
3 If $\delta N \geqslant 1$, $M \to \infty$, which means that a self-supporting discharge occurs in the counter.

The gas multiplication factor M is a function of the ratio $V_0/\ln(b/a)$ and the product Pa, where P is the pressure of the gas in the counter (Rossi and Staub). Experimental results of M values for two gases are shown in Figs. 5.15 and 5.16. Diethorn[1] has obtained the equation

$$\ln M = \frac{V \ln 2}{\overline{\Delta V} \ln(b/a)} \ln \frac{V}{\bar{K} Pa \ln(b/a)} \tag{5.20}$$

where $\overline{\Delta V}$ and $\bar{K}$ are constants of the gas. Equation 5.20 has been tested and found to be valid.[2-4] As Figs. 5.15 and 5.16 show, M increases almost exponentially with applied voltage.

One method by which the strong dependence of M on applied voltage is reduced is by adding a small amount of a polyatomic organic gas in the gas of the counter. One popular mixture is 10 percent CH_4 and 90 percent argon. The organic gases, called "quenching" gases, stabilize the operation of the counter by reducing the effect of the secondary processes. They achieve this because organic polyatomic molecules

1 dissociate rather than produce electrons when they hit the cathode
2 dissociate when they absorb a photon
3 have lower ionization potential than the molecules of the main gas; as a result, they are ionized in collisions with ions of the main gas and thus prevent the ions from reaching the cathode

The total charge produced in a proportional counter is

$$Q = MNe = M \frac{\Delta E}{w} e \tag{5.21}$$

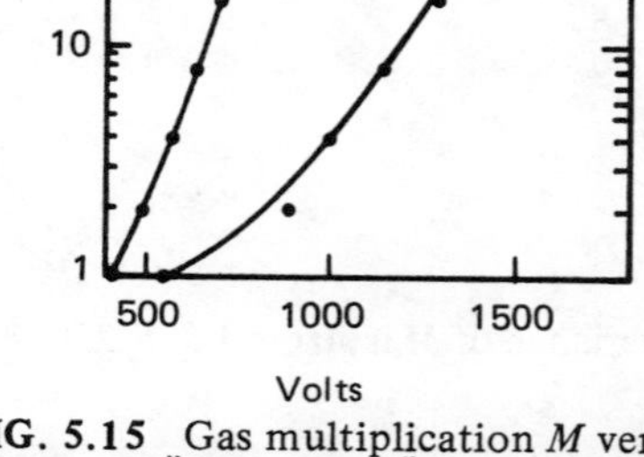

FIG. 5.15 Gas multiplication M versus voltage. Gas is 93.6 percent pure argon ($a = 0.005''$, $b = 0.435''$, at two different pressures). (From Rossi-Staub.)

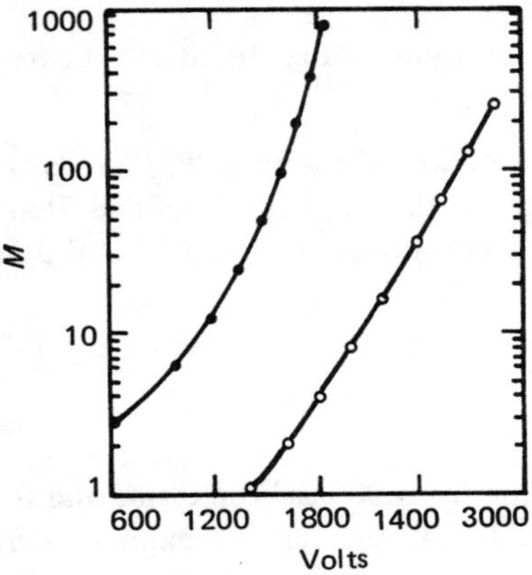

FIG. 5.16 Gas multiplication M versus voltage. Gas is BF_3. (A) $a = 0.005''$, $b = 0.75''$, $P = 10$ cmHg. (B) $a = 0.005''$, $b = 0.78''$, $P = 80.4$ cmHg. (From Rossi-Staub.)

where ΔE = energy of the incident particle dissipated in the counter
w = average energy required for production of one electron-ion pair

Equation 5.21 indicates that Q (output) is proportional to the energy deposited in the counter (ΔE). This is the reason why such counters are called proportional. The proportionality holds, however, only if the gas multiplication factor M is constant, independent of the primary ionization. The question then arises, under what conditions is this true?

A proportional counter is strictly proportional as long as the space charge due to the positive ions does not modify too much the electric field around the wire. The magnitude of the space charge is a function of the primary ionization and the gas multiplication. If the primary ionization is very small, the value of M may be 10^5 to 10^6 before the space charge affects the proportionality. On the other hand, if the primary ionization is too strong, the critical value of M is smaller. It has been reported[5] that there is a critical maximum value of the charge produced by the multiplication process beyond which proportionality does not hold. That number, obviously, depends on the counter (size, types of gas, etc.).

The events that produce the avalanches of electrons in a proportional counter are statistical in nature. The final multiplication factor M will not be constant but will show statistical fluctuations. The probability that the multiplication will have the value M is, according to Snyder,[6] equal to

$$P(M) = \frac{1}{\bar{M}} \exp\left(-\frac{M}{\bar{M}}\right) \tag{5.22}$$

where $\bar{M}$ = mean multiplication factor. The variance of M is, from Eq. 5.22,

$$\sigma_M^2 = \bar{M}^2 \tag{5.23}$$

5.5.2 The Pulse Shape of a Proportional Counter

The shape of the pulse of a proportional counter is understood as one follows the events that lead to the formation of the pulse. A cylindrical counter will be considered, such as that shown in Fig. 5.14*a*.

Assume that the incident particle generated N electron-ion pairs at a certain point inside the counter. The electrons start moving toward the wire (anode). As soon as they reach the region of the strong field close to the wire, they produce secondary ionization. Since all the secondary ionization is produced in the small volume surrounding the wire, the amplitude of the output pulse is independent of the position of the primary ionization. The electrons of the secondary ionization are collected quickly by the wire, before the ions have moved appreciably. Their contribution to the pulse is negligible because they cross only a very small fraction of the potential difference on their way to the anode. The pulse developed in the central wire is almost entirely due to the motion of the ions. As the ions move towards the cathode, the voltage pulse on the wire begins to rise: quickly at first, when the ions are crossing the region of the intense electric field, and slower later when the ions move into the region of low-intensity field. The voltage pulse as a function of time is given by (Kowalski)

$$V(t) = \frac{Q}{2C \ln (b/a)} \ln \left(1 + \frac{b^2}{a^2} \frac{t}{t_{\text{ion}}}\right) \tag{5.24}$$

where Q is given by Eq. 5.21

C = capacitance of the counter

t_{ion} = time it takes the ions to reach the cathode

The equation for t_{ion} is (Kowalski)

$$t_{\text{ion}} = \frac{P \ln (b/a)}{2V_0 \mu_{\text{ion}}} (b^2 - r^2) \tag{5.25}$$

where P = gas pressure

μ_{ion} = ion mobility in the field of the counter†

r = point where the ion was produced

The pulse $V(t)$ is shown by the solid line of Fig. 5.17. The pulse rises quickly and reaches half of its maximum in time of the order of microseconds. Then it bends and rises at a much slower rate, until about a millisecond later it reaches its final value, Q/C.

The pulse of Fig. 5.17 was derived under the assumption that all the ions were produced at the same point. In reality, the ions are produced along the track of the incident particle. This modifies the shape of the pulse during its initial rise but it leaves it virtually unaffected during the later period.

The pulse of Fig. 5.17 is unacceptably long, even for a modest counting rate. As

†The ion mobility is the proportionality constant between the drift velocity and the reduced field; thus $W^+ = \mu^+(E/P)$.

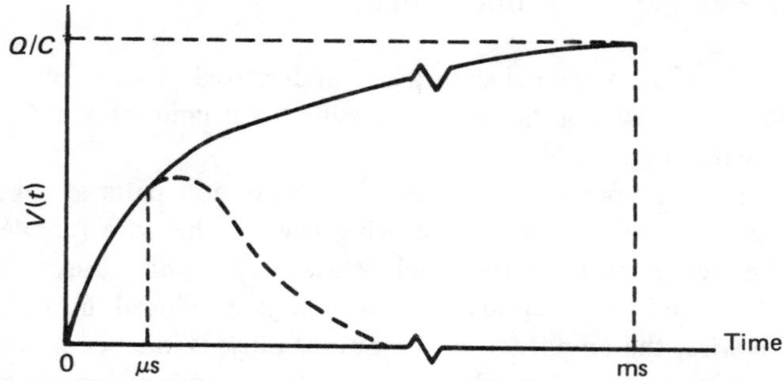

FIG. 5.17 The voltage pulse of a proportional counter.

in the case of the ionization chamber, the pulse is "chopped off" at some convenient time with the help of a differentiating circuit (Chap. 10). The result will be a pulse shown by the dashed line in Fig. 5.17.

5.5.3 The Change of Counting Rate with High Voltage—The High-Voltage Plateau

When a detector is used for the study of a phenomenon involving counting of particles, the investigator would like to be certain that changes in the counting rate are due to changes in the phenomenon under study and not due to changes of the environment such as atmospheric pressure, temperature, humidity, or voltage. For most radiation measurements, all these factors may be neglected except voltage changes.

Consider a gas-filled counter. For its operation it is necessary to apply high voltage (HV), usually positive, which may range from +300 to +3000 V, depending on the counter. For the specific counter used in an experiment, the observer would like to know by what fraction the counting rate will change if the HV changes by a certain amount. It is highly desirable to have a system for which the change in the counting rate is negligible, when the HV changes for a reason beyond the control of the investigator (e.g., change in the 110 V provided by the outlet on the wall, which may, in turn, cause a fluctuation in the output of the HV power supply). For this reason, the response of a counting system to such variations ought to be known. This information is provided by the HV plateau of the counter. The determination of the HV plateau will be discussed below for a proportional counter. However, the experiment and the results are equally applicable for a Geiger-Müller counter.

The HV plateau is obtained by performing the experiment sketched in Fig. 5.18. A radioactive source, emitting a certain type of particles, is placed at a fixed distance from the counter. The signal from the detector is amplified with the help of a pre-

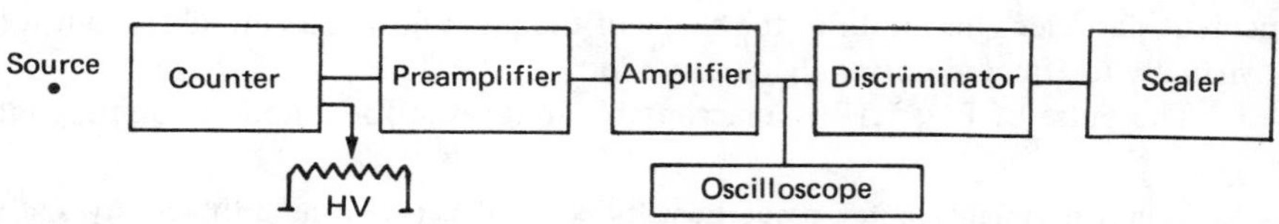

FIG. 5.18 Experimental arrangement for the determination of the HV plateau.

amplifier and an amplifier. It is then fed through a discriminator, and pulses above the discriminator level are counted by the scaler. The counting rate of the scaler is recorded as a function of the HV, the only variable changed. The result of the experiment is shown in Fig. 5.19 (lower curve). Also shown in Fig. 5.19 (upper curve) is a part of the graph of Fig. 5.3 from regions II (ionization) and III (proportional) with the ordinate now shown as pulse height, which is, of course, proportional to the number of ions collected per unit time. The dashed line represents the discriminator level. The shape of the HV plateau is explained as follows.

For very low voltage ($V < V_A$) the counting rate is zero. The source is there, ionization is produced in the counter, pulses are fed into the amplifier and the discriminator, but the scaler does not receive any signal because all the pulses are below the discriminator level. Hence, the counting rate is zero. As the HV increases beyond V_A, more ionization is produced in the counter, some pulse heights generated in it are above the discriminator level, and the counting rate starts increasing. The counting rate keeps increasing with HV, since more and more pulses are produced with a height above the discriminator level. This continues up to the point when $V \simeq V_B$. For $V > V_B$, the ionization is still increasing, the pulse height is also increasing, but all the pulses are now above the discriminator level. Since all the pulses are counted, each *pulse being recorded as one regardless of its height*, the counting rate does not change. This continues up to $V \approx V_C$. Beyond that point, the counting rate will start increasing again because the HV is so high that spurious and double pulses may be generated. The counter should not be operated beyond $V = V_C$.

The region of the graph between V_B and V_C is called the *HV plateau.* It represents the operational range of the counter. Although the manufacturer of the detector provides this information to the investigator, it is standard (and safe) practice to determine the plateau of a newly purchased counter before it is used in an actual measurement for the first time.

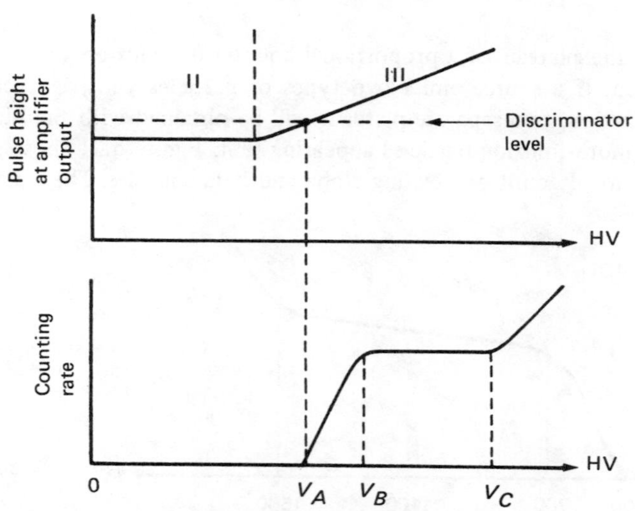

FIG. 5.19 The HV plateau (lower curve).

The plateau of Fig. 5.19 is shown as completely flat. For most counters, the plateau has a positive slope that may be due to spurious counts or to increasing efficiency of the counter, or to both of these effects. Investigation of proportional counters[7] showed that the positive slope is the result of an increase in detector efficiency. For GM counters, on the other hand, the slope of the plateau is due to the production of more spurious counts.

The performance of a counter is expressed in terms of the slope of the plateau given in the form

$$\text{Plateau slope} = \frac{\Delta r/r}{\Delta V} \tag{5.26}$$

where $\Delta r/r$ is the relative change of the counting rate r for the corresponding change in voltage ΔV. Frequently, Eq. 5.26 is expressed in percent change of the counting rate per 100 V change of the high voltage, i.e.,

$$\text{Plateau slope} = \frac{100(\Delta r/r)}{\Delta V}(100) = 10^4 \frac{\Delta r/r}{\Delta V} \tag{5.27}$$

EXAMPLE 5.1

What is the change of counting rate per 100 V of the plateau for a counter having the plateau shown in Fig. 5.20?

ANSWER

The plateau extends from 700 to 1500 V. The slope over the region is (using Eq. 5.27),

$$\frac{10^4(r_2 - r_1)/r_1}{V_2 - V_1} = \frac{10^4(3800 - 3000)/3000}{1500 - 700} = 3.3\% \text{ per } 100 \text{ V}$$

The location of the plateau of a proportional counter depends on the type of particles being detected. If a source emits two types of particles with significantly different primary ionization, two separate plateaus will be obtained with the plateau corresponding to the more ionizing particles appearing first. Figure 5.21 shows such a plateau for a proportional counter detecting alpha and beta particles. The existence

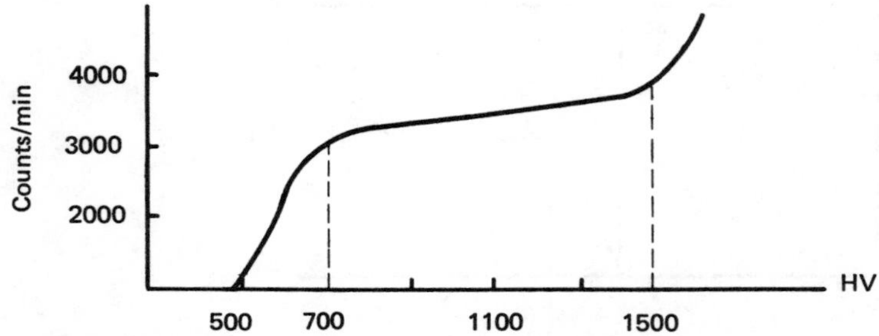

FIG. 5.20 The HV plateau used in Example 5.1.

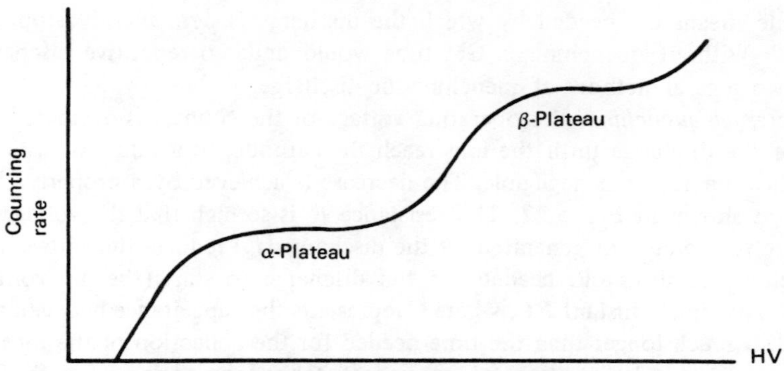

FIG. 5.21 Alpha and beta plateaus of a proportional counter.

of two plateaus is a consequence of the fact that in the proportional region, differentiation of the ionization produced by different types of particles is still possible (see region III in Fig. 5.4). In the GM region this distinction is lost, and for this reason GM counters have only one HV plateau regardless of the type of incident radiation (region IV of Fig. 5.4).

5.6 GEIGER-MÜLLER (GM) COUNTERS

5.6.1 Operation of a GM Counter and Quenching of the Discharge

A Geiger-Müller (GM) counter is a gas counter that operates in region IV of Fig. 5.3. Its construction and operation is in many ways similar to that of a proportional counter. The GM counter is usually cylindrical in shape, like most of the proportional counters. The electric field close to the central wire is so strong that $N\delta \approx 1$ (see Sec. 5.5.1) and the gas multiplication factor M is extremely high. In a GM counter, a single primary electron-ion pair triggers a great number of successive avalanches. Therefore, the output signal is independent of the primary ionization.

The operation of the GM counter is much more complicated than that of the proportional counter. When the electrons are accelerated in the strong field surrounding the wire they produce, in addition to a new avalanche of electrons, considerable excitation of the atoms and molecules of the gas. These excited atoms and molecules produce photons when they deexcite. The photons, in turn, produce photoelectrons in other parts of the counter. Thus the avalanche, which was originally located close to the wire, spreads quickly in most of the counter volume. During all this time, the electrons are continuously collected by the anode wire, while the much slower moving positive ions are still in the counter and form a positive sheath around the anode. When the electrons have been collected, this positive sheath, acting as an electrostatic screen, reduces the field to such an extent that the discharge should stop. However, this is not the case because the positive ions eject electrons when they finally strike the cathode, and since by that time the field has been restored to its original high value, a new avalanche starts and the process just described is repeated.

Clearly some means are needed by which the discharge is permanently stopped or "quenched." Without quenching, a GM tube would undergo repetitive discharging. There are two general methods of quenching the discharge.

In *external quenching*, the operating voltage of the counter is decreased, after the start of the discharge until the ions reach the cathode, to a value for which the gas multiplication factor is negligible. The decrease is achieved by a properly chosen RC circuit as shown in Fig. 5.22. The resistance R is so high that the voltage drop across it due to the current generated by the discharge (i_d) reduces the voltage of the counter below the threshold needed for the discharge to start (the net voltage is $V_0 - i_d R$). The time constant RC, where C represents the capacitance between anode and ground, is much longer than the time needed for the collection of the ions. As a result, the counter is inoperative for an unacceptably long period of time. Or, in other words, its dead time is too long.

The *self-quenching* method is accomplished by adding to the main gas of the counter a small amount of a polyatomic organic gas or a halogen gas.

The organic gas molecules, when ionized, lose their energy by dissociation rather than by photoelectric processes. Thus, the number of photoelectrons, which would spread and continue the avalanche, is greatly reduced. In addition to that, when the organic ions strike the surface of the cathode they dissociate instead of causing the ejection of new electrons. Therefore, new avalanches do not start.

GM counters using an organic gas as a quenching agent have a finite lifetime because of the dissociation of the organic molecules. Usually, the GM counters last for 10^8 to 10^9 counts. The lifetime of a GM detector increases considerably if a halogen gas is used as the quenching agent. The halogen molecules also dissociate during the quenching process but there is a certain degree of regeneration of the molecules which greatly extends the useful lifetime of the counter.

5.6.2 The Pulse Shape and the Dead Time of a GM Counter

The signal of a GM counter is formed in essentially the same way as the signal of a proportional counter and is given by the same equation, Eq. 5.24. For GM counters the signal is the result of the sum of the contributions from all the positive ion avalanches produced throughout the volume of the counter. The final pulse is similar in shape to that shown in Fig. 5.17, except that the pulse rises much slower. The shape and height of GM counter pulses are not very important because the pulse is only

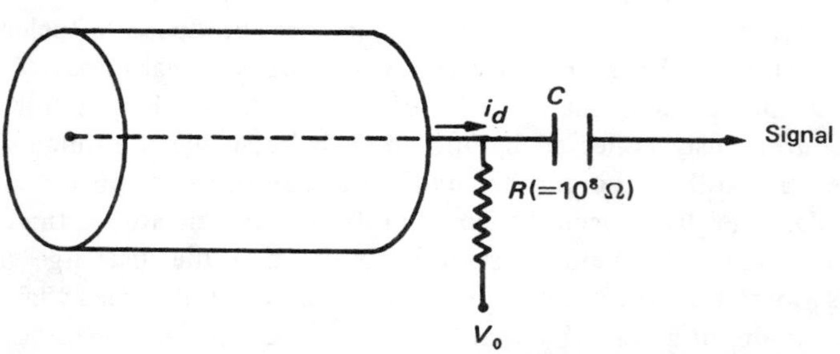

FIG. 5.22 The circuit used for external quenching of a GM counter.

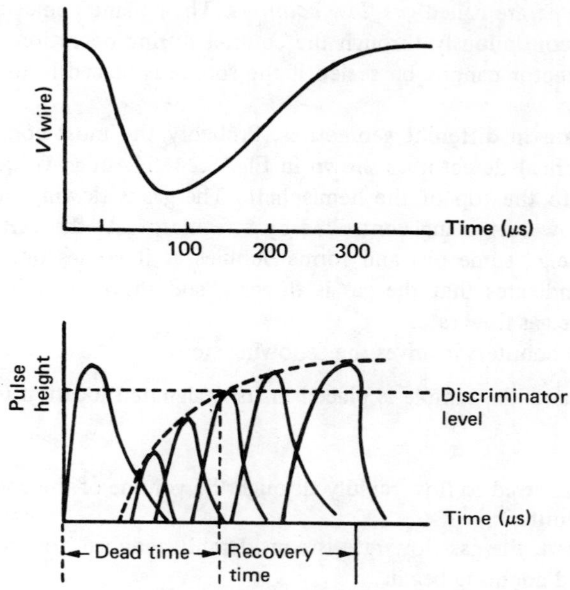

FIG. 5.23 Dead time and recovery time for a GM counter.

used to signal the presence of the particle and nothing else. However, how one pulse affects the formation of the next one is important.

As discussed in Sec. 5.6.1, during the formation of a pulse the electric field in the counter is greatly reduced because of the presence of the positive ions around the anode. If a particle arrives during that period, no pulse will be formed because the counter is insensitive. The insensitivity lasts for a certain time, called the dead time of the counter. Then, the detector slowly recovers with the pulse height growing exponentially during the *recovery* period. This is illustrated in Fig. 5.23, which shows the change of the voltage and pulse for a typical GM counter. Typical values of dead time are from 100 to 300 μs. If the dead time is 100 μs and the counting rate is 500 counts/s, there is going to be a 5 percent loss of counts due to dead time. Correction for dead time is described in Sec. 2.20.

5.7 GAS-FLOW COUNTERS

The gas counters described so far are all sealed. That is, the counter is a closed volume filled with a gas at a certain pressure. The radiation source is placed outside the detector; therefore, the particles have to penetrate the wall of the counter to be counted. In doing so, some particles may be absorbed by the wall and some may be backscattered; in the case of charged particles, these will all lose a certain fraction of their energy. To minimize these effects, most commercial gas counters have a thin window through which the radiation enters the counter. The window may still be too thick for some alpha and low-energy beta particles. For this reason, counters have been developed with the capability of having the source placed inside the chamber.

Gas counters of this type are called *gas-flow counters.* Their name comes from the fact that the gas flows continuously through the counter during operation. This is necessary because the detector cannot be sealed if the source is placed inside the chamber.

Gas-flow counters come in different geometries. Probably the most common one is that of the hemispherical detector as shown in Fig. 5.24. The high voltage is applied to a wire attached to the top of the hemisphere. The gas is flowing slowly through the counter, the flow rate being controlled by a regulator. At the exit, the gas goes through a liquid (e.g., some oil) and forms bubbles as it comes out. The formation of the bubbles indicates that the gas is flowing, and the rate of bubble formation gives an idea of the gas flow rate.

Counting with gas-flow counters involves the following steps:

1 The chamber is opened and the sample is placed in its designated location inside the chamber.
2 The chamber is closed.
3 Gas from the gas tank is allowed to flow rapidly through the volume of the counter and purge it (for a few minutes).
4 After the counter is purged, the gas flow rate is considerably reduced, to a couple of bubbles per second, and counting begins.

There are two advantages in placing the sample inside the detector:

1 The particles do not have to penetrate the window of the counter, where they might be absorbed, scattered out of the detector, or lose energy.
2 Close to 50 percent of the particles emitted by the source have a chance to be recorded in a hemispherical counter, or close to a 100 percent in a spherical one. If the source is placed outside the detector, there are always less than 50 percent of the particles entering the detector.

A hemispherical counter is also called a 2π counter, while a spherical counter with the source located at its center is called a 4π counter. Figure 5.24 shows a 2π counter.

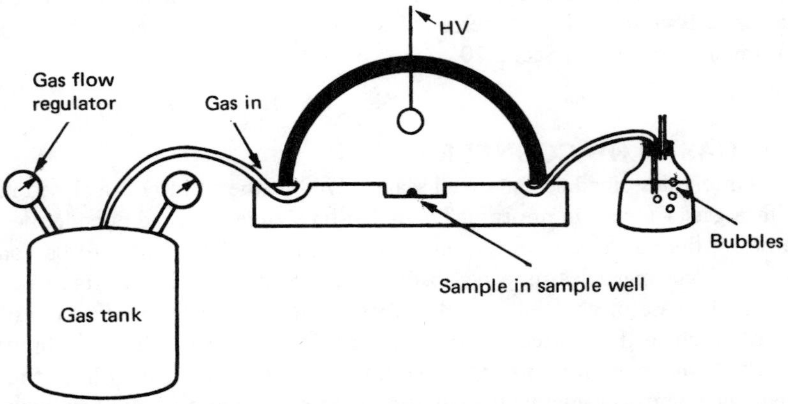

FIG. 5.24 A hemispherical (2π) gas-flow counter.

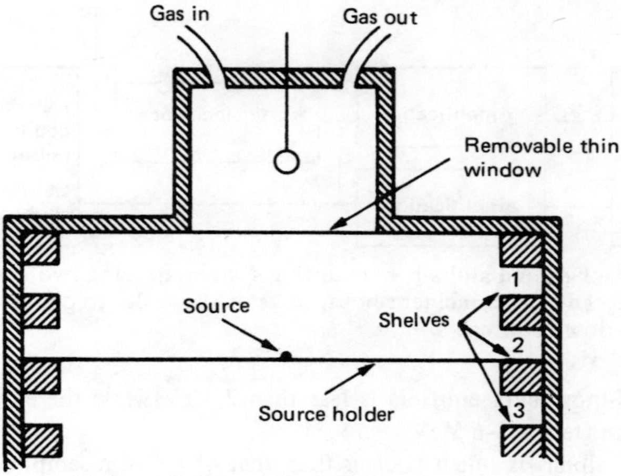

FIG. 5.25 A gas-flow counter with removable thin window and movable source holder.

Gas-flow counters may operate as proportional or GM counters. In fact, there are commercial models which may operate in one or the other region depending on the voltage applied and the gas used. In a proportional gas-flow counter, the gas is usually methane or mixture of argon and methane. In the GM region, the gas is a mixture of argon and isobutane.

In some gas-flow counter models, there is provision for placing a very thin window between the sample and the sensitive volume of the counter to reduce the effects of slight contamination of the sample well or of static charges that interfere with the measurement. In the counter of Fig. 5.24, the thin window will be placed on top of the sample well. A different arrangement is shown in Fig. 5.25.

Gas-flow counters are used as low-background alpha-beta detection systems. Requirements for low-background measurements arise in cases where the level of activity from the sample is very low, compared to background. Examples of such cases are samples that monitor contamination of water supplies or of air or ground.

There are commercially available systems that have a background counting rate of less than 1 count/min for betas and a considerably lower rate for alphas. Such a low background is achieved by shielding the counter properly (surrounding it with lead) and using electronic means to reject most of the background radiation. A system offered by one of the manufacturers uses two detectors. The first is the gas-flow counter and the second is a cosmic-ray detector (Fig. 5.26). The two detectors are operated in *anticoincidence* (see Sec. 10.8), which means that events due to particles going through both detectors (e.g., cosmic rays or other radiation from the environment) will not be counted. Only pulses produced by the activity of the sample in the gas-flow counter will be recorded.

Discrimination between alphas and betas can be achieved in many ways. The two methods most frequently used with gas-flow counters are based on range and energy differences. Before these methods are discussed, the reader should recall that

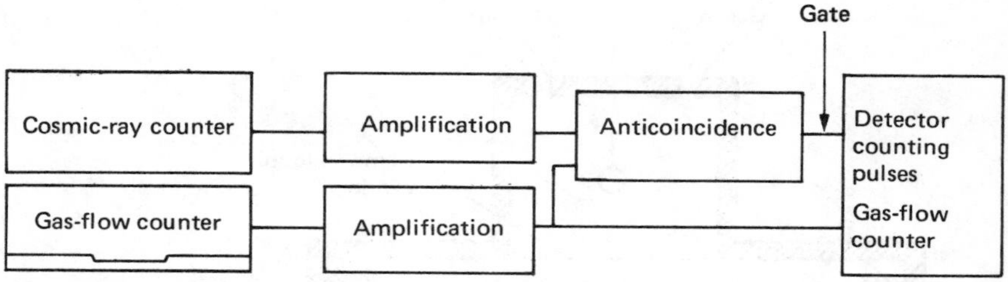

FIG. 5.26 A low-background alpha-beta counting system utilizing two counters and anticoincidence. The anticoincidence output gates the scaler to count only pulses from the gas-flow counter.

the maximum energy of most beta emitters is less than 2 MeV while the energy of alphas from most alpha emitters is 5–6 MeV.

Since the range of alphas is much shorter than that of betas, a sample can be analyzed for alpha and beta activity by counting it twice: once with a thin foil covering it to stop the alphas, and a second time without the foil to record alphas and betas.

Energy discrimination is based on the difference in pulse height produced by the two types of particles, alphas being more energetic produce higher pulses; thus a simple discriminator at an appropriate level can reject the beta pulses.

5.8 INTERNAL GAS COUNTING

An alternative to the gas-flow counter is *internal gas counting*, which is used with low-energy β-emitters. In internal gas counting, a gaseous form of the radioisotope is introduced into the counter (usually a proportional counter) along with the counting gas. As with gas-flow counters, by having the source inside the counter, losses in the window are avoided and an increase in efficiency is achieved by utilizing a 4π geometry.

Internal gas counting requires that corrections be made for wall and end effects and for the decrease in electric field intensity at the ends.[8–10] One way to reduce the end effect is to use a spherical proportional counter,[11] in which the anode wire is stretched along a diameter and the cathode is, of course, spherical. The electric field inside the sphere is

$$E = \frac{V}{\ln (b/a)} \frac{1}{r} \tag{5.2}$$

At a certain distance r from the anode, the electric field becomes stronger at the ends of the anode because b, the radius of the cathode, gets smaller. However, the supports of the wire tend to reduce the field. By properly adjusting the supports, one may make the field uniform. In cylindrical counters, corrections for end effects are applied by a length-compensation method.[8]

Internal gas counting is used for the production of standards. Using this technique, the National Bureau of Standards produced standards of ^{3}H, ^{14}C, ^{37}A, ^{85}K, ^{131m}Xe, and ^{133}Xe.

5.9 GENERAL COMMENTS ABOUT CONSTRUCTION OF GAS COUNTERS

This section summarizes the important characteristics of gas counters.

GEOMETRY

Parallel-plate counters are almost exclusively ionization chambers. The intense fields needed for gas multiplication can be produced only in cylindrical or spherical geometry.

In the cylindrical geometry, which is the most frequently used, the strong electric field exists close to the central wire. The wire is usually made of tungsten or platinum. It has a diameter of 25–100 μm (few mills of an inch); it must be uniform in radius, without any bends or kinks, and be placed concentrically with the outer cylinder. Of particular importance is the smoothness of the central wire. Any kinks or tiny specks of material attached to its surface amount to pointed tips where very high electric fields are generated. Such a high field is a source of spurious discharges that interfere with counting.

GASES AND PRESSURES USED

For ionization chambers, almost any gas or pressure may be used. Even atmospheric air has been used.

For proportional or GM counters, the noble gases—argon in particular—are normally used. A small percentage of additional gases is also used for quenching purposes. In proportional counters, methane is frequently added to the main gas. The so-called P-10 mixture, consisting of 90 percent argon and 10 percent methane, is extensively used. Another mixture is 4 percent isobutane and 96 percent helium. Several gas pressures have been used. As Figs. 5.15 and 5.16 show, the gas multiplication depends on the pressure. Usually the pressure is less than 1 atm. Of course, gas-flow counters operate at ambient pressure.

As discussed in Sec. 5.6.1, the quenching gas in a GM counter is either an organic polyatomic molecule such as ethyl alcohol, or a halogen such as bromine or chlorine. A typical mixture is 0.1 percent chlorine in neon. The gas pressure in a GM counter is, in most cases, less than 1 atm. The pressure affects the operating voltage.

COUNTER WINDOW

When the source is place outside the counter, it is very important for the radiation to enter the counter after traversing as thin a wall material as possible. Any material in the path of radiation may scatter, absorb, or cause energy loss. This is particularly critical in the measurement of alphas and low-energy betas, which have a very short range. It is not important for neutron and gamma counters.

All counters have walls as thin as possible (or practical), but in addition to that, many commercial designs have an area on the surface of the counter designated as the "window" consisting of a very thing material. In cylindrical counters, the window is usually the front end of the cylinder (the other end houses electrical connectors). There are some cylindrical counters with windows located on the cylindrical surface.

Materials and thickness of windows are:

1 Glass, down to 0.30–0.40 kg/m^2 (100 μm)
2 Aluminum, 0.25–0.30 kg/m^2 (100 μm)

3 Steel, 0.60–0.80 kg/m^2 (80 μm)
4 Mica, 0.01 kg/m^2 (3 μm)
5 Mylar (plain or aluminized), 0.01 kg/m^2
6 Special ultrathin membranes or foils, $\sim 10^{-3}$ kg/m^2

PROBLEMS

5.1 Sketch the HV plateau of a counter, if all the pulses out of the amplifier have exactly the same height.

5.2 How would the sketch of Prob. 5.1 change if there are two groups of pulses out of the amplifier (two groups, two different pulse heights)?

5.3 Sketch counting rate versus discriminator threshold, assuming that the electronic noise consists of pulses in the range $0 < V < 0.1$ V and all the pulses due to the source have height equal to 1.5 V.

5.4 In a cylindrical gas counter with a central wire radius equal to 25 μm (0.001″), outer radius 25 mm (~1 in), and 1000 V applied between anode and cathode, what is the distance from the center of the counter at which an electron gains enough energy in 1 mm of travel to ionize helium gas? (Take 23 eV as the ionization potential of helium.)

5.5 A GM counter with a mica window is to be used for measurement of ^{14}C activity. What should the thickness of the window be if it is required that at least 90 percent of the ^{14}C betas enter the counter?

5.6 What is the minimum pressure required to stop 6-MeV alphas inside the argon atmosphere of a spherical gas counter with a 25-mm radius? Assume the alpha source is located at the center of the counter.

5.7 What is the ratio of the saturation ionization currents for a chamber filled with He versus one filled with CH_4 (other things being equal)?

5.8 Show that the variance of M is equal to $\bar{M}^2$ if the probability distribution is given by Eq. 5.22.

5.9 Calculate the maximum value of the positive ion time given by Eq. 5.25 for a cylindrical counter with a cathode radius equal to 19 mm (~0.75 in) and a central anode wire with a radius of 25 μm (~0.001 in). The high voltage applied is 1000 V, the pressure of the gas is 13.3 kPa (10 cmHg), and the mobility of the ions is [13.34 (m/s·Pa)/(V/m)] [1000(cm/s·mmHg)/V/cm)].

5.10 The observed counting rate of a counter is 22,000 counts/min. What is the error in the true counting rate if the dead time is 300 μs and no dead-time correction is applied?

BIBLIOGRAPHY

Fenyves, E., and Haiman, O., *The Physical Principles of Nuclear Radiation Measurements*, Academic Press, New York, 1969.

Franzen, W., and Cochran, L. W., "Pulse Ionization Chambers and Proportional Counters," in A. H. Snell (ed.), *Nuclear Instruments and Their Uses*, Wiley, New York, 1962.

Gillespie, A. B., *Signal, Noise and Resolution in Nuclear Counter Amplifiers*, McGraw-Hill, New York, 1953.

Kowalski, E., *Nuclear Electronics*, Springer-Verlag, New York–Heidelberg–Berlin, 1970.

Price, W. J., *Nuclear Radiation Detection*, McGraw-Hill, New York, 1964.
Rossi, B. B., and Staub, H. H., *Ionization Chambers and Counters*, McGraw-Hill, New York, 1949.

REFERENCES

1 Diethorn, W., NYO-0628 (1956).
2 Kiser, R. W., *Appl. Sci. Res.* **8B**:183 (1960).
3 Williams, W., and Sara, R. I., *Intern. J. Appl. Rad. Isotopes* **13**:229 (1962).
4 Bennett, E. F., and Yule, T. J., ANL-7763 (1971).
5 Hanna, G. C., Kirkwood, H. W., and Pontecorvo, B., *Phys. Rev.* **75**:985 (1949).
6 Snyder, H. S., *Phys. Rev.* **72**:181 (1947).
7 Champion, P. J., *Nucl. Instr. Meth.* **112**:75 (1973).
8 Mann, W. B., Seliger, H. H., Marlow, W. F., and Medlock, R. W., *Rev. Sci. Instr.* **31**:690 (1960).
9 Garfunkel, S. B., Mann, W. B., Schima, F. J., and Unterweger, M. P., *Nucl. Instr. Meth.* **112**:59 (1973).
10 Bambynek, W., *Nucl. Instr. Meth.* **112**:103 (1973).
11 Benjamin, P. W., Kemsholl, C. D., and Redfearn, J., *Nucl. Instr. Meth.* **59**:77 (1968).

6

scintillation detectors

6.1 INTRODUCTION

Scintillators are materials—solids, liquids, gases—that produce sparks or scintillations of light when ionizing radiation passes through them. The first solid material to be used as a particle detector was a scintillator. It was used by Rutherford, in 1910, in his alpha-scattering experiments. In his experimental setup, alpha particles hit a zinc sulfide screen and produced scintillations, which were counted with or without the help of a microscope—a very inefficient process, inaccurate and time consuming. The method was abandoned for about 30 years and was remembered again when advanced electronics made possible amplification of the light produced in the scintillator.

The amount of light produced in the scintillator is very small. It must be amplified before it can be recorded as a pulse or in any other way. The amplification or multiplication of the scintillator's light is achieved with a device known as the *photomultiplier tube* (or *phototube*). Its name denotes its function: it accepts a small amount of light, amplifies it many times, and delivers a strong pulse at its output. Amplifications of the order of 10^6 are common for many commercial photomultiplier tubes. Apart from the phototube, a detection system that uses a scintillator is no different from any other (Fig. 6.1).

The operation of a scintillation counter may be divided into two broad steps:

1 Absorption of incident radiation energy by the scintillator and production of photons in the visible part of the electromagnetic spectrum
2 Amplification of the light by the photomultiplier tube and production of the output pulse

The sections that follow analyze these two steps in detail. The different types of scintillators are divided, for the present discussion, into three groups:

1 Inorganic scintillators
2 Organic scintillators
3 Gaseous scintillators

6.2 INORGANIC SCINTILLATORS (CRYSTAL SCINTILLATORS)

Most of the inorganic scintillators are crystals of the alkali metals, in particular alkali iodides, that contain a small concentration of an impurity. Examples are NaI(Tl), CsI(Tl), CaI(Na), LiI(Eu), and CaF_2(Eu). The element in parentheses is the impurity or activator. Although the activator has a relatively small concentration—e.g., thallium in NaI(Tl) is 10^{-3} on a per mole basis—it is the agent that is responsible for the luminescence of the crystal.

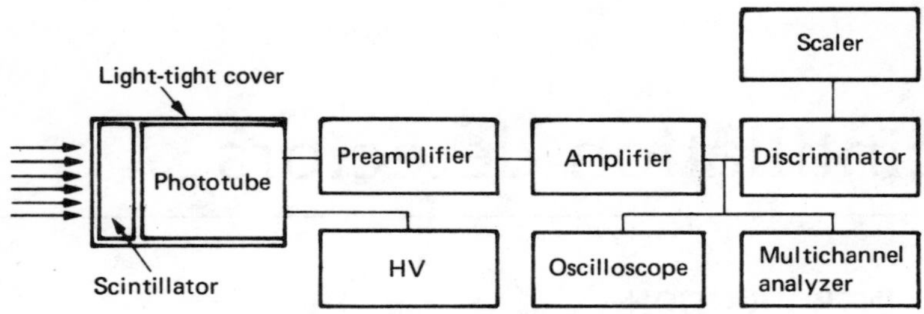

FIG. 6.1 A detection system using a scintillator.

6.2.1 The Mechanism of the Scintillation Process

The luminescence of inorganic scintillators can be understood in terms of the allowed and forbidden energy bands of a crystal. The electronic energy states of an atom are discrete energy levels, which in an energy level diagram are represented as discrete lines. In a crystal, the allowed energy states widen into bands (Fig. 6.2). In the ground state of the crystal, the uppermost allowed band that contains electrons is completely filled. This is called the *valence band.* The next allowed band is empty (in the ground state) and is called the *conduction band.* An electron may obtain enough energy from incident radiation to move from the valence to the conduction band. Once there, the electron is free to move anywhere in the lattice. The removed electron leaves behind a hole in the valence band, which can also move. Sometimes, the energy given to the electron is not sufficient to raise it to the conduction band. Instead, the electron remains electrostatically bound to the hole in the valence band. The electron-hole pair thus formed is called an *exciton.* In terms of energy states, the exciton corre-

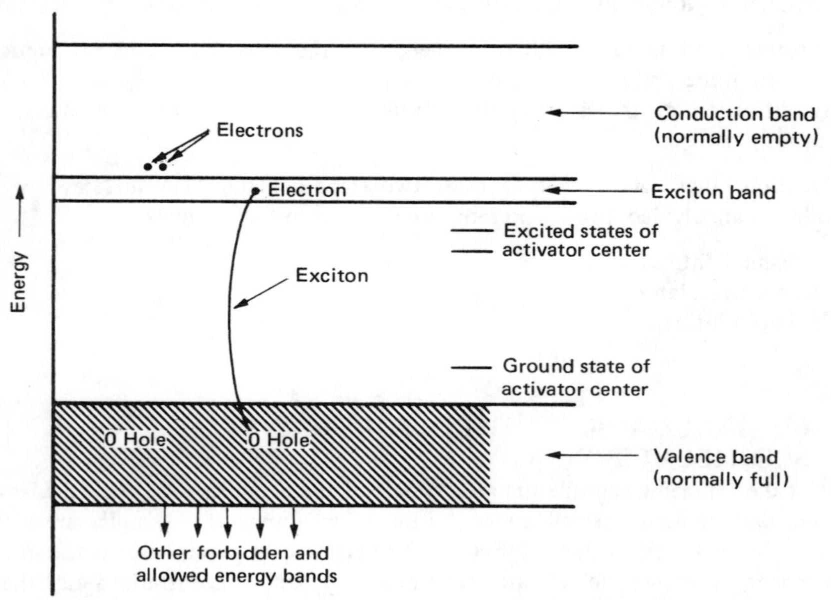

FIG. 6.2 Allowed and forbidden energy bands of a crystal.

corresponds to elevation of the electron to a state higher than the valence but lower than the conduction band. Thus, the exciton states form a thin band, with upper level coinciding with the lower level of the conduction band (Fig. 6.2). The width of the exciton band is of the order of 1 eV, while the gap between valence and conduction bands is of the order of 8 eV.

In addition to the exciton band, energy states may be created between valence and conduction bands because of crystal imperfections or impurities. Particularly important are the states created by the activator atoms such as thallium. The activator atom may exist in the ground state or in one of its excited states. Elevation to an excited state may be the result of a photon absorption or of the capture of an exciton or of the successive capture of an electron and a hole. The transition of the impurity atom from the excited to the ground state, if allowed, results in the emission of a photon in times of the order of 10^{-8} s. If this photon has a wavelength in the visible part of the electromagnetic spectrum, it contributes to a scintillation. Thus, production of a scintillation is the result of the occurrence of these events:

1 Ionizing radiation passes through the crystal.
2 Electrons are raised to the conduction band.
3 Holes are created in the valence band.
4 Excitons are formed.
5 Activation centers are raised to the excited states by absorbing electrons, holes, and excitons.
6 Deexcitation is followed by the emission of a photon.

The light emitted by a scintillator is primarily the result of transitions of the activator atoms, and not of the crystal. Since most of the incident energy goes to the lattice of the crystal—eventually becoming heat—the appearance of luminescence produced by the activator atoms means that energy is transferred from the host crystal to the impurity. For NaI(Tl) scintillators, about 12 percent of the incident energy appears as thallium luminescence.[1]

The magnitude of the light output and the wavelength of the emitted light are two of the most important properties of any scintillator. The light output affects the number of photoelectrons generated at the input of the photomultiplier tube (see Sec. 6.5), which in turn affects the pulse height produced at the output of the counting system. Information about the wavelength is necessary in order to match the scintillator with the proper photomultiplier tube. Emission spectra of NaI(Tl), CsI(Na), and CsI(Tl) are shown in Fig. 6.3. Also shown in Fig. 6.3 are the responses of two phototube cathode materials. Table 6.1 gives the most important properties of some inorganic scintillators.

The light output of the scintillators depends on temperature. Figure 6.4 shows the temperature response of NaI(Tl), Cs(Tl), and CsI(Na).

6.2.2 Time Dependence of Photon Emission

Since the photons are emitted as a result of decays of excited states, the time of their emission depends on the decay constants of the different states involved. Experiments show that the emission of light follows an exponential decay law of the form

$$N(t) = N_0 e^{-t/T} \tag{6.1}$$

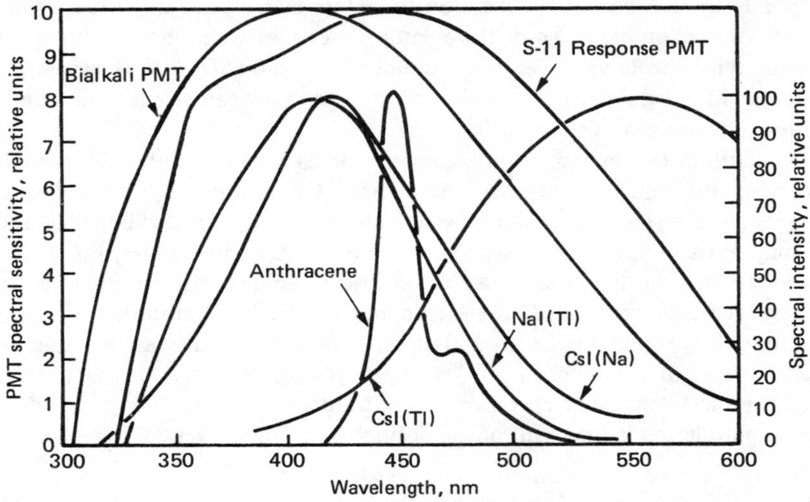

FIG. 6.3 Emission spectra of NaI(Tl), CsI(Tl), CsI(Na), and anthracene, com pared to the spectral response of two photocathode materials. (From HARSHAW Research Laboratory Report.)

where $N(t)$ = number of photons emitted at time t

T = decay time of the scintillator (see Table 6.1)

Most of the excited states in a scintillator have essentially the same lifetime T. There are, however, some states with longer lifetimes contributing a slow component in the decay of the scintillator known as *afterglow*. It is present to some extent in all inorganic scintillators and may be important in certain measurements where the integrated output of the phototube is used. Two scintillators with negligible afterglow are CaF_2(Eu) and $Bi_4Ge_3O_{12}$ (bismuth orthogermanate).

In a counting system using a scintillator, the light produced by the crystal is amplified by a photomultiplier tube and is transformed into an electric current having the exponential behavior given by Eq. 6.1. This current is fed into an RC circuit (see Sec. 10.2) as shown in Fig. 6.5, and a voltage pulse is produced of the form

$$V(t) = V_\infty(e^{-t/RC} - e^{-t/T}) \tag{6.2}$$

TABLE 6.1 Properties of Certain Inorganic Scintillators

Material	Wavelength of maximum emission (nm)	Scintillation efficiency (relative, %)	Decay time (μs)	Density (10^3 kg/m)
NaI(Tl)	410	100	0.23	3.67
CaF_2(Eu)	435	50	0.94	3.18
CsI(Na)	420	80	0.63	4.51
CsI(Tl)	565	45	1.80	4.51
$Bi_4Ge_3O_{12}$	480	8	0.30	7.13
$CdWO_4$	530	20	0.90	7.90
^{6}LiI(Eu)	470	30	0.94	3.49

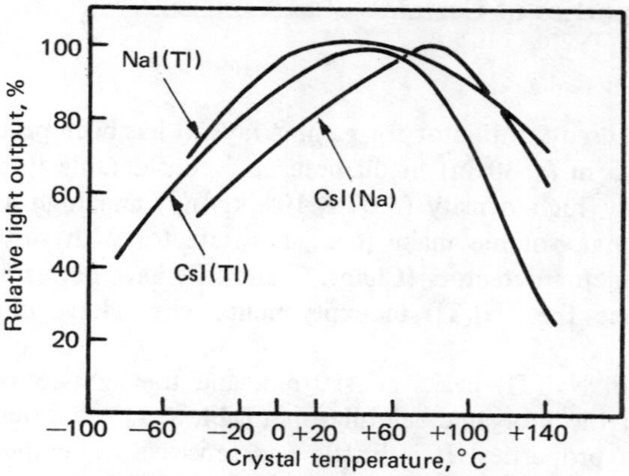

FIG. 6.4 Temperature dependence of light output of NaI(Tl), CsI(Tl), and CsI(Na). (From HARSHAW Research Laboratory Report.)

In practice, the value of RC is selected to be of the order of a few hundreds of microseconds. Thus, for short times–i.e., $T \ll RC$, which is the timespan of interest–Eq. 6.2 takes the form

$$V(t) = V_\infty(1 - e^{-t/T}) \tag{6.2a}$$

Notice that the rate at which the pulse rises (risetime) is determined by the decay time T. In certain measurements, e.g., coincidence-anticoincidence measurements (Chap. 10), the timing characteristics of the pulse are extremely important.

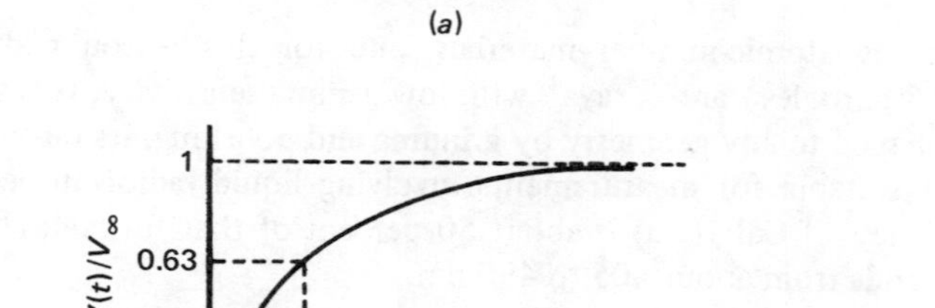

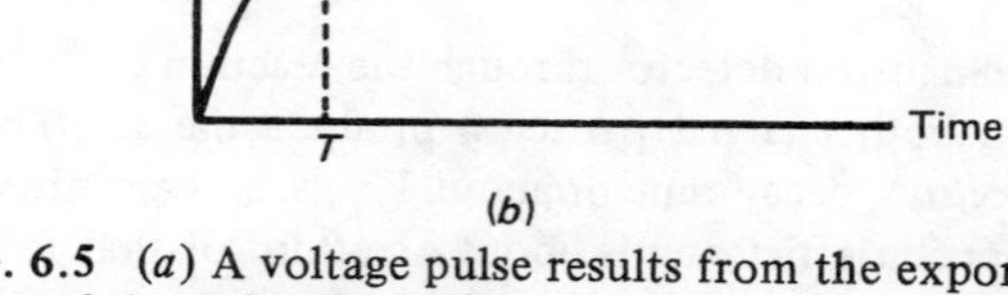

FIG. 6.5 (*a*) A voltage pulse results from the exponential current. (*b*) The shape of the pulse for $RC \gg T$.

6.2.3 Important Properties of Certain Inorganic Scintillators

NaI(Tl)

NaI(Tl) is the most commonly used scintillator for gamma rays. It has been produced in single crystals of up to 0.75 m (~30 in) in diameter and considerable thickness (0.25 m ≈ 10 in). Its relatively high density (3.67×10^3 kg/m^3) and high atomic number combined with the large volume make it a γ-ray detector with very high efficiency. Although semiconductor detectors (Chaps. 7 and 12) have better energy resolution, they cannot replace the NaI(Tl) in experiments where large detector volumes are needed.

The emission spectrum of NaI(Tl) peaks at 410 nm, and the light-conversion efficiency is the highest of all the inorganic scintillators (Table 6.1). As a material, NaI(Tl) has many undesirable properties. It is brittle and sensitive to temperature gradients and thermal shocks. It is also so hygroscopic that it should be kept encapsulated at all times. NaI always contains a small amount of potassium, which creates a certain background because of the radioactive ^{40}K.

CsI(Tl)

CsI(Tl) has a higher density (4.51×10^3 kg/m^3) and higher atomic number than NaI; therefore its efficiency for gamma detection is higher. The light-conversion efficiency of CsI(tl) is about 45 percent of that for NaI(Tl) at room temperature. At liquid nitrogen temperatures (77 K), pure CsI has a light output equal to that of NaI(Tl) at room temperature and a decay constant equal to 10^{-8} s.[2] The emission spectrum of CsI(Tl) extends from 420 to about 600 nm.

CsI is not hygroscopic. Being softer and more plastic than NaI, it can withstand severe shocks, acceleration, and vibration, as well as large temperature gradients and sudden temperature changes. These properties make it suitable for space experiments. Finally, CsI does not contain potassium.

CsI(Na)

The density and atomic number of CsI(Na) are the same as those of CsI(Tl). The light-conversion efficiency is about 85 percent of that for NaI(Tl). Its emission spectrum extends from 320 to 540 nm (see Fig. 6.3). CsI(Na) is slightly hygroscopic.

CaF_2(Eu)

CaF_2(Eu) consists of low-atomic-number materials, and for this reason makes an efficient detector for β particles[3] and x-rays[4] with low gamma sensitivity. It is similar to Pyrex and can be shaped to any geometry by grinding and polishing. Its insolubility and inertness make it suitable for measurements involving liquid radioisotopes. The light-conversion efficiency of CaF_2(Eu) is about 50 percent of that for NaI(Tl). The emission spectrum extends from about 405 to 490 nm.

LiI(Eu)

LiI(Eu) is an efficient thermal-neutron detector through the reaction $^6_3\text{Li}(n, \alpha)^3_1\text{H}$. The alpha particle and the triton, both charged particles, produce the scintillations. LiI has a density of 4.06×10^3 kg/m^3, decay time of about 1.1 μs, and emission spectrum peaking at 470 nm. Its conversion efficiency is about one-third of that for NaI. It is very hygroscopic and is subject to radiation damage as a result of exposure to neutrons.

OTHER INORGANIC SCINTILLATORS

Many other scintillators have been developed for special applications. Examples are $Bi_4Ge_3O_{12}$, $CdWO_4$, and more recently[5] $MF_2:UF_4:CeF_3$, where M stands for one of the following: Ca, Sr, Ba. This last scintillator containing 2 percent UF_4 and using Ce as the fluorescing agent has been used for detection of fission fragments.

6.3 ORGANIC SCINTILLATORS

The materials that are efficient *organic scintillators* belong to the class of aromatic compounds. They consist of planar molecules made up of benzenoid rings. Two examples are toluene and anthracene, having the structures shown in Fig. 6.6.

Organic scintillators are formed by combining appropriate compounds. They are classified as unitary, binary, ternary, and so on, depending on the number of compounds in the mixture. The substance with the highest concentration is called the *solvent.* The others are called *solutes.* A binary scintillator consists of a solvent and a solute, while a ternary scintillator is made of a solvent, a primary solute, and a secondary solute. Table 6.2 lists the most common compounds used.

6.3.1 The Mechanism of the Scintillation Process

The production of light in organic scintillators is the result of molecular transitions. Consider the energy-level diagram of Fig. 6.7, which shows how the potential energy of a molecule changes with interatomic distance. The ground state of the molecule is at point A, which coincides with the minimum of the potential energy. Ionizing radiation passing through the scintillator may give energy to the molecule and raise it to an excited state, i.e., the transition $A_0 \rightarrow A_1$ may occur. The position A_1 is not the point of minimum energy. The molecule will release energy through lattice vibrations (that energy is eventually dissipated as heat) and move to point B_1. The point B_1 is still an excited state and, in some cases, the molecule will undergo the transition $B_1 \rightarrow B_0$ accompanied by the emission of a photon with energy equal to $E_{B_1} - E_{B_0}$. This transition, if allowed, takes place at times of the order 10^{-8} s. It should be noted that the energy of the emitted photon $(E_{B_1} - E_{B_0})$ is less than the energy that caused the excitation $(E_{A_1} - E_{A_0})$. This difference is very important because otherwise the emission spectrum of the scintillator would completely coincide with its absorption spectrum and no scintillations would be produced. A more detailed description of the scintillation process is given in the references (see Birks and reference 6).

One of the important differences between inorganic and organic scintillators is in the response time, which is less than 10 ns for the latter (response time of inorganic scintillators is $\sim 1\ \mu s$, see Table 6.1) and makes them suitable for fast timing measurements (see Chap. 10). Table 6.3 lists important properties of some organic scintillators.

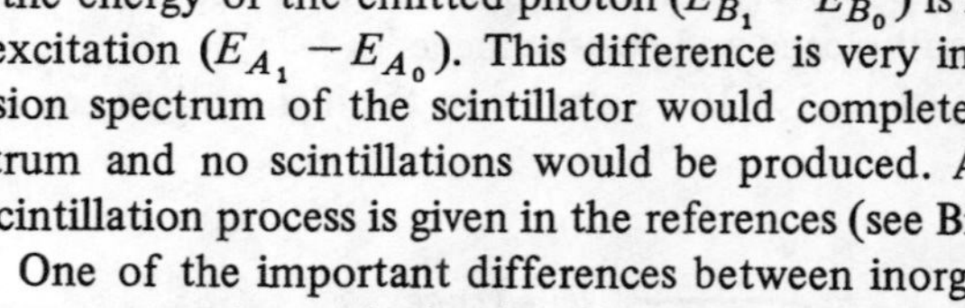

FIG. 6.6 Molecular structure of (*a*) toluene and (*b*) anthracene.

TABLE 6.2 Organic Scintillator Compounds†

Compound	Formula	Application‡
Benzene	C_6H_6	S
Toluene	$C_6H_5CH_3$	S
p-Xylene	$C_6H_4(CH_3)_2$	S
1,2,4-Trimethylbenzene (pseudocumene)	$C_6H_3(CH_3)_3$	S
Hexamethylbenzene	$C_6(CH_3)_6$	S
Styrene monomer	$C_6H_5C_2H_3$	S
Vinyltoluene monomer	$C_6H_4CH_3C_2H_3$	S
Naphthalene	$C_{10}H_8$	S′, C
Anthracene	$C_{14}H_{10}$	C
Biphenyl	$C_{12}H_{10}$	S′
p-Terphenyl	$C_{18}H_{14}$	C, PS
p-Quaterphenyl	$C_{24}H_{18}$	C
trans-Stilbene	$C_{14}H_{12}$	C
Diphenylacetylene	$C_{14}H_{10}$	C
1,1′,4,4′-Tetraphenylbutadiene	$C_{28}H_{22}$	SS
Diphenylstilbene	$C_{26}H_{20}$	SS
PPO (2,5-diphenyloxazole)	$C_{15}H_{11}NO$	PS
α-NPO [2-(1-Naphthyl)-5-phenyloxazole]	$C_{19}H_{13}NO$	PS
PBD [2-Phenyl,5-(4-biphenylyl)-1,3,4-oxadiazole]	$C_{20}H_{14}N_2O$	PS
BBO [2,5-Di(4-biphenylyl)-oxazole]	$C_{27}H_{19}NO$	SS
POPOP {1,4-Bis[2-(5-phenyloxazolyl)]-benzene}	$C_{24}H_{16}N_2O_2$	SS
TOPOT {1,4-Di-[2-(5-*p*-tolyloxazolyl)]-benzene}	$C_{26}H_{20}N_2O_2$	SS
DiMePOPOP {1,4-Di[2-(4-methyl-5-phenyloxazolyl)]-benzene}	$C_{26}H_{20}N_2O_2$	SS

†From [6].

‡S–primary solvent; S′–secondary solvent; PS–primary solute; SS–secondary solute; C–crystal scintillator.

6.3.2 Organic Crystal Scintillators

No activator is needed to enhance the luminescence of organic crystals. In fact, any impurities are undesirable because their presence reduces the light output and for this reason the material used to make the crystal is purified. Two of the most common organic crystal scintillators are anthracene and *trans*-stilbene.

Anthracene has a density of 1.25×10^3 kg/m^3 and the highest light conversion

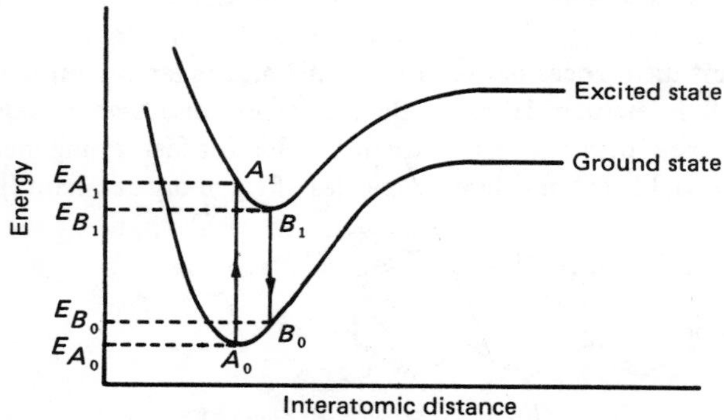

FIG. 6.7 A typical (simplified) energy diagram of a molecule.

TABLE 6.3 Properties of Certain Organic Scintillators

Material	Wavelength of maximum emission (nm)	Relative scintillation efficiency (%)	Decay time (ns)	Density (10^3 kg/m^3)
Anthracene	445	100	~30	1.25
trans-Stilbene	385	~60	4–8	1.16
NE 102	350–450	~65	2	1.06
NE 110	350–450	60	3	1.06
NE 213 (liquid)	350–450	~60	2	0.867
PILOT B	350–450	68	2	1.06
PILOT Y	350–450	64	~3	1.06

efficiency of all organic scintillators (see Table 6.3)—which is still only about one-third of the light conversion efficiency of NaI(Tl). Its decay time (~30 ns) is much shorter than that of inorganic crystals. Anthracene can be obtained in different shapes and sizes.

trans-Stilbene has a density of 1.15×10^3 kg/m^3 and a short decay time (4–8 ns). Its conversion efficiency is about half of that for anthracene. It can be obtained as a clear, colorless, single crystal with a size up to several millimeters. Stilbene crystals are sensitive to thermal and mechanical shock.

6.3.3 Organic Liquid Scintillators

The organic liquid scintillators consist of a mixture of a solvent with one or more solutes. Compounds that have been used successfully as solvents include xylene, toluene, and hexamethylbenzene (see Table 6.2). Satisfactory solutes include *p*-terphenyl, PBD, and POPOP.

In a binary scintillator, the incident radiation deposits almost all of its energy in the solvent but the luminescence is due almost entirely to the solute. Thus, as in the case of inorganic scintillators, an efficient energy transfer is taking place from the bulk of the phosphor to the material with the small concentration (activator in inorganic scintillators, solute in organic ones). If a second solute is added, it acts as a *wavelength shifter*, i.e., it increases the wavelength of the light emitted by the first solute, so that the emitted radiation is better matched with the characteristics of the cathode of the photomultiplier tube.

Liquid scintillators are very useful for measurements where a detector with large volume is needed to increase efficiency. Examples are counting of low-activity β-emitters (^{3}H and ^{14}C in particular), detection of cosmic rays, and measurement of the energy spectrum of neutrons in the MeV range (see Chap. 14) using the scintillator NE 213. The liquid scintillators are well suited for such measurements because they can be obtained and used in large quantities (kiloliters) and can form a detector of desirable size and shape by utilizing a proper container.

In certain cases, the radioisotope to be counted is dissolved in the scintillator, thus providing 4π geometry and, therefore, high detection efficiency. In others, an extra element or compound is added to the scintillator to enhance its detection efficiency without causing significant deterioration of the luminescence. Boron, cadmium, or gadolinium,[7-9] used as additives, cause an increase in neutron detection efficiency. On the other hand, fluorine-loaded scintillators consist of compounds in which fluorine has replaced hydrogen, thus producing a phosphor with a low neutron sensitivity.

6.3.4 Plastic Scintillators

The plastic scintillators may be considered as solid solutions of organic scintillators. They have properties similar to those of liquid organic scintillators (Table 6.3), but they have the added advantage, compared to liquids, that they do not need a container. Plastic scintillators can be machined into almost any desirable shape and size ranging from thin fibers to thin sheets. They are inert to water, air, and many chemicals, and for this reason they can be used in direct contact with the radioactive sample.

Plastic scintillators are also mixtures of a solvent and one or more solutes. The most frequently used solvents are polysterene and polyvinyltoluene. Satisfactory solutes include *p*-terphenyl and POPOP. The exact compositions of some plastic scintillators are given in reference 10.

Plastic scintillators have a density of about 10^3 kg/m^3. Their light output is lower than that of anthracene (Table 6.3). Their decay time is short, and the wavelength corresponding to the maximum intensity of their emission spectrum is between 350 and 450 nm. Trade names of commonly used plastic scintillators are Pilot B, Pilot Y, NE 102, and NE 110. The characteristics of these phosphors are discussed in references 11-13. Plastic scintillators loaded with tin and lead have been tried as x-ray detectors in the 5-100 keV range.[14,15] Thin plastic scintillator films (as thin as 20×10^{-5} kg/m^2 = 20 μg/cm^2) have proven to be useful detectors in time-of-flight measurements[16-18] (see Chap. 13).

6.4 GASEOUS SCINTILLATORS

Gaseous scintillators are mixtures of noble gases.[19,20] The scintillations are produced as a result of atomic transitions. Since the light emitted by noble gases belongs to the ultraviolet region, other gases, such as nitrogen, are added to the main gas to act as wavelength shifters. Thin layers of fluorescent materials used for coating the inner walls of the gas container achieve the same effect.

Gaseous scintillators exhibit the following features:

1 Very short decay time
2 Light output per MeV deposited in the gas depending very little on the charge and mass of the particle being detected
3 Very low efficiency for gamma detection

These properties make the gaseous scintillators suitable for the energy measurement of heavy charged particles (alphas, fission fragments, other heavy ions).

6.5 THE RELATIONSHIP BETWEEN PULSE HEIGHT AND ENERGY AND TYPE OF INCIDENT PARTICLE

To measure the energy of the incident particle with a scintillator, the relationship between the pulse height and the energy deposited in the scintillator must be known.

Since the pulse height is proportional to the output of the photomultiplier, which output is in turn proportional to the light produced by the scintillator, it is necessary to know the light conversion efficiency of the scintillator as a function of type and energy of incident radiation. The rest of this section presents experimental results for several cases of interest.

6.5.1 The Response of Inorganic Scintillators

PHOTONS

The response of NaI(Tl) to gammas is linear, except for energies below 400 keV where a slight nonlinearity is present. Experimental results are shown in Fig. 6.8.[21] More details about the NaI(Tl) response to gammas are given in Chap. 12.

CHARGED PARTICLES

For protons and deuterons, the response of the scintillator is proportional to the particle energy, at least for $E > 1$ MeV. For alpha particles, the proportionality begins at about 15 MeV (Fig. 6.9).[22] Theoretical aspects of the the response have been studied extensively.[23-26] Today, inorganic scintillators are seldom used for detection of charged particles.

NEUTRONS

Since neutrons are detected indirectly through charged particles produced as a result of nuclear reactions, to find the response to neutrons one looks at the response to alphas and protons. LiI(Eu), which is the crystal used for neutron detection, has essentially the same response as NaI(Tl) (Fig. 6.9).

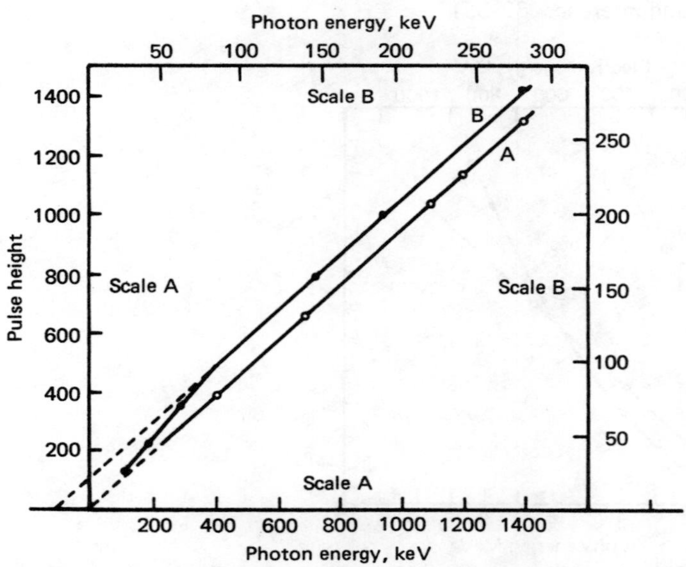

FIG. 6.8 Pulse height versus energy for a NaI(Tl) crystal. The region below 300 keV has been expanded in curve B to show the nonlinearity (from ref. 21).

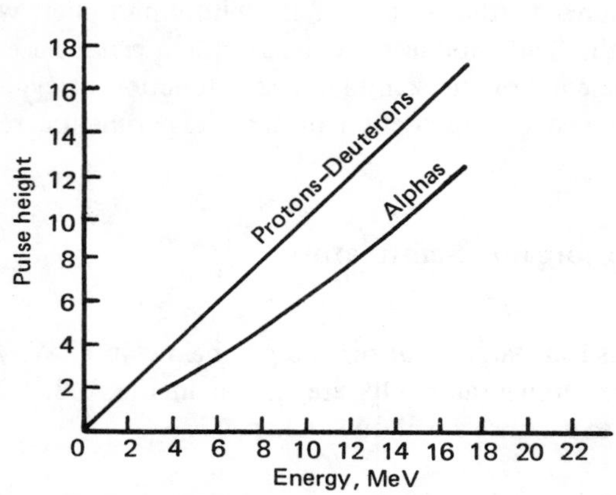

FIG. 6.9 Pulse height versus energy for a NaI(Tl) crystal resulting from charged particles (from ref. 22).

6.5.2 The Response of Organic Scintillators

CHARGED PARTICLES

Experiments have shown that organic crystal scintillators (e.g., anthracene) exhibit a direction-dependent response to alphas[27] and protons.[28] An adequate explanation of the direction-dependent characteristics of the response does not exist at present. The user should be aware of the phenomenon to avoid errors.

The response of plastic and liquid scintillators to electrons, protons, and alphas is shown in Figs. 6.10, 6.11, and 6.12.[29-31] Notice that the response is not linear, especially for heavier ions. The response has been studied theoretically by many investigators (Birks and references 32-35).

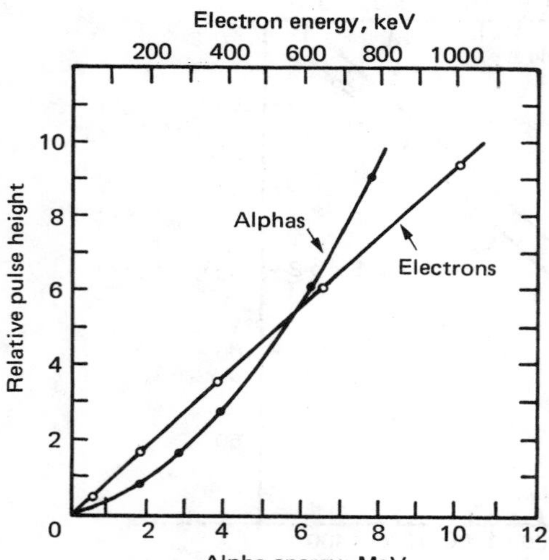

FIG. 6.10 Pulse height versus energy for a liquid scintillator resulting from alphas and electrons (from ref. 29).

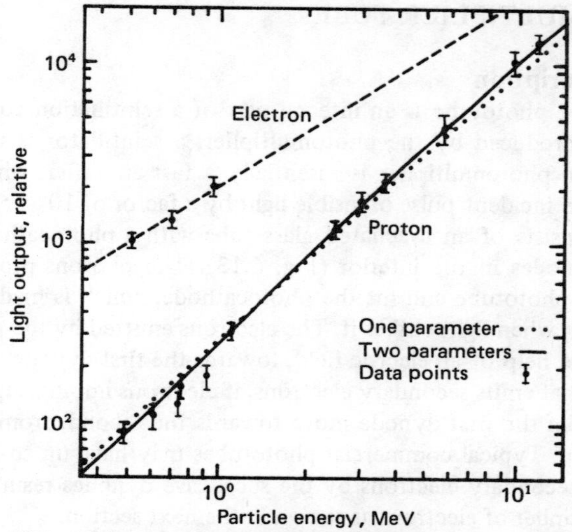

FIG. 6.11 Plastic scintillator (NE 102) response to electrons and protons (from ref. 30).

PHOTONS AND NEUTRONS

Organic scintillators are not normally used for detection of gammas because of their low efficiency. The liquid scintillator NE 213† is being used for γ detection in mixed neutron-gamma fields[36] because of its ability to discriminate against neutrons. Neutrons are detected by NE 213 through the proton-recoil method. More details about the use of NE 213 scintillator and its response function are given in Chaps. 12 and 14.

†NE 213 consists of xylene, activators, and POPOP as the wavelength shifter. Naphthalene is added to enhance the slow components of light emission. The composition of NE 213 is given as $CH_{1.21}$ and its density as 0.867×10^3 kg/cm³.

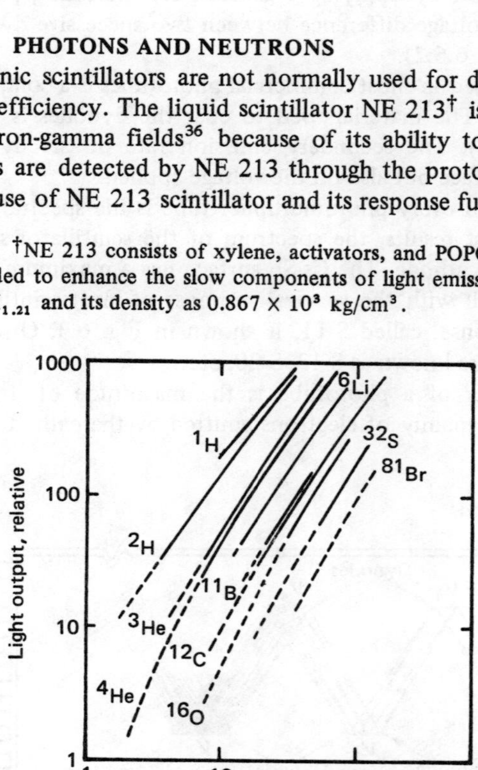

FIG. 6.12 Plastic scintillator (NE 102) response to heavy ions (from ref. 31).

6.6 THE PHOTOMULTIPLIER TUBE

6.6.1 General Description

The photomultiplier tube or phototube is an integral part of a scintillation counter. Without the amplification produced by the photomultiplier, a scintillator is uesless as a radiation detector. The photomultiplier is essentially a fast amplifier, which in times of 10^{-9} s amplifies an incident pulse of visible light by a factor of 10^6 or more.

A photomultiplier consists of an evacuated glass tube with a photocathode at its entrance and several dynodes in the interior (Fig. 6.13). The photons produced in the scintillator enter the phototube and hit the photocathode, which is made of a material that emits electrons when light strikes it. The electrons emitted by the photocathode are guided, with the help of an electric field, towards the first dynode, which is coated with a substance that emits secondary electrons, if electrons impinge upon it. The secondary electrons from the first dynode move towards the second, from there towards the third, and so on. Typical commercial phototubes may have up to 15 dynodes. The production of secondary electrons by the successive dynodes results in a final amplification of the number of electrons as shown in the next section.

The electrons produced in the phototube are directed from one dynode to the next by an electric field established by applying a successively increasing positive high voltage to each dynode. The voltage difference between two successive dynodes is of the order of 80–120 V (see Sec. 6.6.2).

The photocathode material used in most commercial phototubes is a compound of cesium and antimony (Cs-Sb). The material used to coat the dynodes is either Cs-Sb or silver-magnesium (Ag-Mg). The secondary emission rate of the dynodes depends not only on the type of surface but also on the voltage applied.

A very important parameter of every photomultiplier tube is the spectral sensitivity of its photocathode. For best results, the spectrum of the scintillator should match the sensitivity of the photocathode. The Cs-Sb surface has a maximum sensitivity at 440 nm, which agrees well with the spectral response of most scintillators (Tables 6.1 and 6.3). Such a response, called S-11, is shown in Fig. 6.3. Other responses of commercial phototubes are known as S-13, S-20, etc.

Another important parameter of a phototube is the magnitude of its *dark current*. The dark current consists mainly of electrons emitted by the cathode after

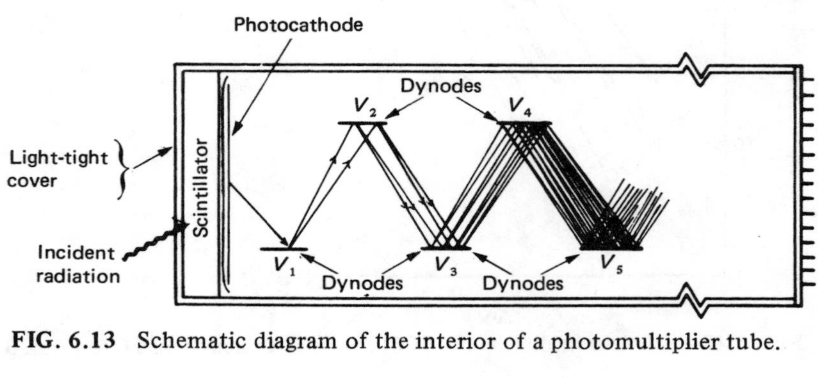

FIG. 6.13 Schematic diagram of the interior of a photomultiplier tube.

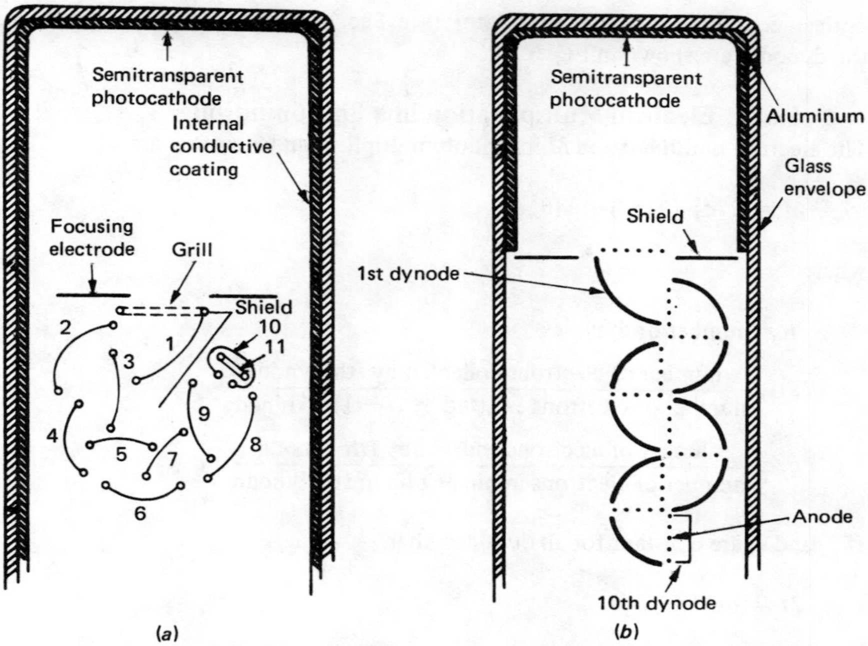

FIG. 6.14 Two dynode arrangements in commercial phototubes: (*a*) Model 6342 RCA, 1–10 are dynodes, 11 is anode; (*b*) Model 6292 DuMont.

thermal energy is absorbed. This process is called *thermionic emission*, and a 50-mm-diameter photocathode may release in the dark as many as 10^5 electrons/s at room temperature. Cooling of the cathode reduces this source of noise by a factor of about 2 per 10–15°C reduction in temperature. Thermionic emission may also take place from the dynodes and the glass wall of the tube, but this contribution is small. Electrons may be released from the photocathode as a result of its bombardment by positive ions coming from ionization of the residual gas in the tube. Finally, light emitted as a result of ion recombination may release electrons upon hitting the cathode or the dynodes. Obviously, the magnitude of the dark current is important in cases where the radiation source is very weak. Both the dark current and the spectral response should be considered when a phototube is to be purchased.

Recall that the electrons are guided from one dynode to the next by an electric field. If a magnetic field is present, it may deflect the electrons in such a way that not all of them hit the next dynode, and the amplification is reduced. Even the earth's weak magnetic field may sometimes cause this undesirable effect. The influence of the magnetic field may be minimized by surrounding the photomultiplier tube with a cylindrical sheet of metal, called *μ-metal*. The μ-metal is commercially available in various shapes and sizes.

Commercial photomultiplier tubes are made with a variety of geometrical arrangements of photocathode and dynodes. In general, the photocathode is deposited as a semitransparent layer on the inner surface of the end window of the phototube (Fig. 6.14). The external surface of the window is, in most phototubes, flat for easier optical coupling with the scintillator (see Sec. 6.7). Two different geometries for the dynodes are shown in Fig. 6.14.

6.6.2 Electron Multiplication in a Photomultiplier

The electron multiplication M in a photomultiplier can be written as:

$$M = (\theta_1 \epsilon_1)(\theta_2 \epsilon_2) \cdots (\theta_n \epsilon_n) \tag{6.3}$$

where

n = number of dynodes

$$\epsilon_i = \frac{\text{number of electrons collected by } i\text{th dynode}}{\text{number of electrons emitted by } (i-1)\text{th dynode}}$$

$$\theta_i = \frac{\text{number of electrons emitted by } i\text{th dynode}}{\text{number of electrons impinging upon } i\text{th dynode}}$$

If θ_i and ϵ_i are constant for all dynodes, then

$$M = (\theta \epsilon)^n \tag{6.4}$$

The quantity ϵ depends on the geometry. The quantity θ depends on the voltage between two successive dynodes and on the material of which the dynode is made. The dependence of θ on voltage is of the form

$$\theta = kV^a \tag{6.5}$$

where $V = V_i - V_{i-1}$ = potential difference between two successive dynodes, assumed the same for dynodes

k, a = constants (the value of a is about 0.7)

Using Eq. 6.5, the multiplication M becomes

$$M = \epsilon^n (kV^a)^n = CV^{an} \tag{6.6}$$

where $C = (\epsilon k)^n$ = constant, independent of the voltage.

Equation 6.6 indicates that the value of M increases with the voltage V and the number of stages n. The number of dynodes is limited, because as n increases, the charge density between two dynodes distorts the electric field and hinders the emission of electrons from the previous dynode with the lower voltage. In commercial photomultipliers, the number of dynodes is 10 or more. If one takes $n = 10$ and $\epsilon\theta = 4$, a typical value, the value of M becomes equal to 10^6.

To apply the electric field to the dynodes, a power supply provides a voltage adequate for all the dynodes. A voltage divider, usually an integral part of the preamplifier, distributes the voltage to the individual dynodes. When reference is made to *phototube voltage*, one means the total voltage applied. For example, if 900 V are applied to a phototube with 10 dynodes, the voltage between any two dynodes is 100 V.

6.7 ASSEMBLY OF A SCINTILLATION COUNTER AND THE ROLE OF LIGHT PIPES

A scintillation counter consists of the scintillator and the photomultiplier tube. It is extremely important that these two components be coupled in such a way that a maximum amount of light enters the phototube and strikes the photocathode. This section presents a brief discussion of the problems encountered during the assembly of a scintillation counter, with some of the methods used to solve them.

A solid scintillator is coupled to the photomultiplier through the end window of the tube (Fig. 6.15). During the transfer from the scintillator to the photocathode, light may be lost by leaving through the sides and front face of the scintillator, or by being reflected back to the scintillator when it hits the window of the phototube.

To avoid loss of light through the sides and front face, the scintillator is painted with a material that reflects toward the crystal the light that would otherwise escape. Examples of reflecting materials commercially available are alpha alumina and Al_2O_3.

To avoid reflection of light from the end window of the phototube, a transparent viscous fluid (such as Dow-Corning 200 Silicone fluid) is placed between the scintillator and the phototube (Fig. 6.15). The optical fluid minimizes reflection because it reduces the change of the index of refraction during the passage of light from the scintillator to the phototube. A sharp change in the index of refraction results in a small critical angle of reflection, which in turn increases total reflection.

In certain experiments, the scintillator has to be a certain distance away from the photocathode. Such is the case if the phototube should be protected from the radiation impinging upon the scintillator or from a magnetic field. Then a *light pipe* is interposed between the scintillator and the phototube. The light pipe is made of a material transparent to the light of the scintillator. Lucite, quartz, plexiglass, and glass have been used in many applications to form light pipes of different lengths and shapes. Light pipes of several feet—sometimes with bends—have been used with success. The optical coupling of the light pipe at both ends is accomplished by the same methods used to couple the scintillator directly to the phototube.

One of the major reasons for using scintillators is their availability in large sizes. In fact, commercially available scintillators are larger than the biggest commercial photomultipliers. In cases where the scintillator is too large, multiple phototubes are coupled to the same crystal. Figure 6.16 shows a NaI(Tl) crystal coupled to six photomultipliers.

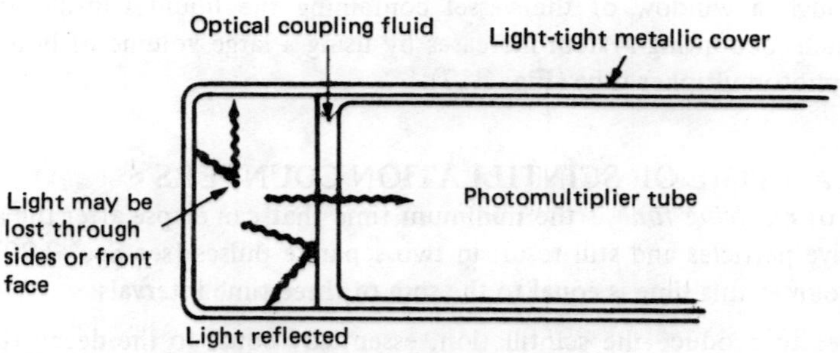

FIG. 6.15 Assembly of a scintillation counter.

FIG. 6.16 A special 30-in (0.762-m) diameter scintillator crystal coupled to six photomultiplier tubes (from HARSHAW Chemical Company).

When a liquid scintillator is used, the phototube is optically coupled to the scintillator through a window of the vessel containing the liquid scintillator. The efficiency of such a counting system increases by using a large volume of liquid and more than one photomultiplier tube (Fig. 6.17).

6.8 DEAD TIME OF SCINTILLATION COUNTERS

The *dead time* or *resolving time* is the minimum time that can elapse after the arrival of two successive particles and still result in two separate pulses (see Sec. 2.20). For a scintillation counter this time is equal to the sum of three time intervals:

1 Time it takes to produce the scintillation, essentially equal to the decay time of the scintillator (see Eq. 6.1 and Tables 6.1 and 6.3).

2 Time it takes for electron multiplication in the phototube, of the order of 20–40 ns.
3 Time it takes to amplify the signal and record it by a scaler. The recovery time of commercial scalers is of the order of 1 μs. The time taken for amplification and discrimination is negligible.

By adding the three above components, the resulting dead time of a scintillation counter is of the order of 1-5 μs. This is much shorter than the dead time of gas-filled counters, which is of the order of tens to hundreds of microseconds.

Scintillators are detectors with fast response. As seen in Tables 6.1 and 6.3, the risetime of the pulse is very short for all of them. Short risetime is important in measurements that depend on the time of arrival of the particle (see Chap. 10).

6.9 SOURCES OF BACKGROUND IN A SCINTILLATION COUNTER

One of the major sources of background in a scintillation counter is the dark current of the phototube (see Sec. 6.6.1). Other background sources are naturally occurring radioisotopes, cosmic rays, and phosphorescing substances.

The holder of a liquid scintillator may contain small amounts of naturally occurring isotopes. In particular, ^{40}K is always present (isotopic abundance of ^{40}K is 0.01 percent). Another isotope, ^{14}C, is a constituent of contemporary organic materials. Solvents, however, may be obtained from petroleum, consisting of hydrocarbons without ^{14}C.

The term *phosphorescence* refers to delayed emission of light as a result of deexcitation of atoms or molecules. Phosphorescent half-lives may extend to hours. This source of background may originate in phosphorescent substances contained in the glass of the phototube, the walls of the sample holder, or the sample itself.

Cosmic rays, which are highly energetic charged particles, produce background in all types of detectors, and scintillators are no exception. The effect of cosmic-ray background, as well as that of the other sources mentioned earlier, will be reduced if two counters are used in coincidence or anticoincidence.

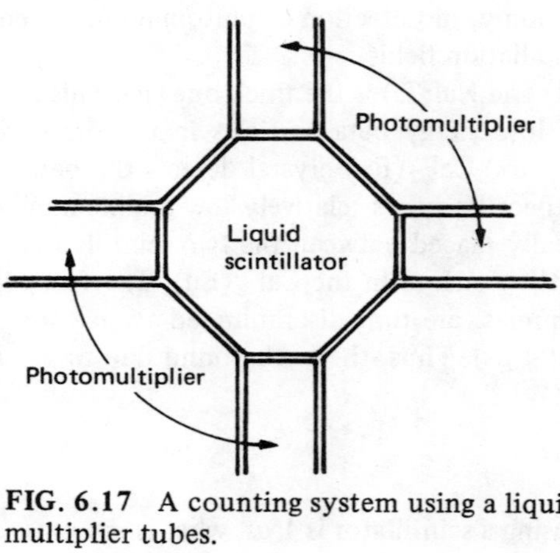

FIG. 6.17 A counting system using a liquid scintillator and four photomultiplier tubes.

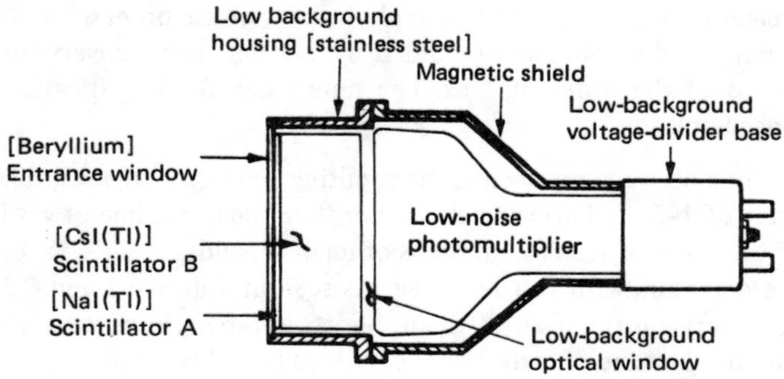

FIG. 6.18 A Phoswich detector (from HARSHAW Chemical Co.).

6.10 THE PHOSWICH DETECTOR

The phoswich detector is used for the detection of low-level radiation in the presence of considerable background. It consists of two different scintillators coupled together and mounted on a single photomultiplier tube.[1] By utilizing the difference in the decay constants of the two phosphors, differentiation between events taking place in the two detectors is possible. The combination of crystals used depends on the types of particles being present in the radiation field under investigation.

The basic structure of a phoswich detector is shown in Fig. 6.18. A thin scintillator (scintillator A) is coupled to a larger crystal (scintillator B), which in turn is coupled to the cathode of a single phototube. Two examples of scintillators used are these:

1 NaI(Tl) is the thin scintillator (A) and CsI(Tl) is the thick one (B). Pulses originating in the two crystals are differentiated based on the difference between the 0.25-μs decay constant of the NaI(Tl) and the 1-μs decay constant of the CsI(Tl). Slow pulses come from particles losing energy in the CsI(Tl) or in both crystals simultaneously. In a mixed low-energy–high-energy photon field, the relatively fast pulses of the NaI(Tl) will come from the soft component of the radiation. [Soft photons will not reach the CsI(Tl).] Phoswich detectors of this type have been used in x-ray and γ-ray astronomy, in detection of plutonium in the environment, and in other cases of mixed-radiation fields.

2 CaF_2(Eu) is the thin scintillator (A) and NaI(Tl) is the thick one (B). This combination is used for measurements of low-energy beta particles in the presence of a gamma background. The thin (0.1 mm) CaF_2(Eu) crystal detects the betas, but is essentially transparent to gammas because of its relatively low atomic number and thickness. A quartz window is usually placed between the two scintillators to stop the betas that did not deposit all their energy in the CaF_2(Eu). The fast pulses of the NaI(Tl), which are due to gammas, are time-discriminated against the slower pulses from the CaF_2(Eu) ($\tau = 0.94\ \mu$s). Thus, the background due to gammas is reduced.

PROBLEMS

6.1 If the dead time of a detection using a scintillator is 1 μs, what is the gross counting rate that will result in a loss of 2 percent of the counts?

6.2 A typical dead time for a scintillation detector is 5 μs. For a gas counter, the corresponding number is 200 μs. If a sample counted with a gas counter results in 8 percent loss of gross counts due to dead time, what is the corresponding loss in a scintillation counter that records the same gross counting rate?

6.3 A parallel beam of 1.5-MeV gammas strikes a 25-mm thick NaI crystal. What fraction of these gammas will have at least one interaction in the crystal ($\mu = 0.0047\ m^2/kg$)?

6.4 What is the range of 2-MeV electrons in a plastic scintillator? Assume that the composition of the scintillator is $C_{10}H_{11}$ ($\rho = 1.02 \times 10^3\ kg/m^3$).

6.5 Consider two electrons, one with kinetic energy 1 MeV, the other with 10 MeV. Which electron will lose more energy going through a 1-mm-thick plastic scintillator? Consider both ionization and radiation loss. Composition of the scintillator is given in Prob. 6.4. For radiation loss, use

$$Z_{eff} = \frac{N_H Z_H^2 + N_C Z_C^2}{N_H Z_H + N_C Z_C}$$

6.6 A phoswich detector consists of a 1-mm thick NaI(Tl) scintillator coupled to a 25-mm-thick CsI(Tl) scintillator. A 0.1-mm-thick beryllium window protects the NaI(Tl) crystal. If the detector is exposed to a thin parallel beam of 150-keV x-rays and 1.5-MeV α rays, what are the fractions of interactions of each type of photon in each scintillator?

BIBLIOGRAPHY

Birks, J. B., *The Theory and Practice of Scintillation Counting*, McMillan Co., New York, 1964.

Fenyves, E., and Haiman, O., *The Physical Principles of Nuclear Radiation Measurements*, Academic Press, New York, 1969.

Price, W. J., *Nuclear Radiation Detection*, McGraw-Hill, New York, 1964.

Snell, A. H. (ed.), *Nuclear Instruments and Their Uses*, Wiley, New York, 1962.

REFERENCES

1 Heath, R. L., Hofstadter, R., and Hughes, E. B., *Nucl. Instr. Meth.* **162**:431 (1979). (Review article listing 127 references.)

2 Aliaga-Kelly, D., and Nicoll, D. R., *Nucl. Instr. Meth.* **43**:110 (1966).

3 Colmenares, C., Shapiro, E. G., Barry, P. E., and Prevo, C. T., *Nucl. Instr. Meth.* **114**:277 (1974).

4 Campbell, M., Ledingham, K. W. D., Baillie, A. D., and Lynch, J. G., *Nucl. Instr. Meth.* **137**: 235 (1976).

5 Catalano, E., and Czirr, J. B., *Nucl. Instr. Meth.* **143**:61 (1977).

6 Brooks, F. D., *Nucl. Instr. Meth.* **162**:477 (1979). (Review article listing 274 references.)

7 Bollinger, L. M., and Thomas, G. E., *Rev. Sci. Instr.* **28**:489 (1957).

8 Hellstrom, J., and Beshai, S., *Nucl. Instr. Meth.* **101**:267 (1972).

9 Bergere, R., Beil, H., and Veyssiere, A., *Nucl. Phys.* **A121**:463 (1968).

10 Swank, R. K., *Ann. Rev. Nucl. Sci.* **4**:111 (1954).

11 Walker, J. K., *Nucl. Instr. Meth.* **68**:131 (1969).

12 Moszynski, M., and Bengtson, B., *Nucl. Instr. Meth.* **142**:417 (1977).

13 Moszynski, M., and Bengtson, B., *Nucl. Instr. Meth.* **158**:1 (1979).

14 Eriksson, L. A., Tsai, C. M., Cho, Z. H., and Hurlbut, C. R., *Nucl. Instr. Meth.* **122**:373 (1974).

15 Becker, J., Eriksson, L., Monberg, L. C., and Cho, Z. H., *Nucl. Instr. Meth.* **123**:199 (1975).

16 Muga, M. L., Burnsed, D. J., Steeger, W. E., and Taylor, H. E., *Nucl. Instr. Meth.* **83**:135 (1970).

17 Muga, M. L., *Nucl. Instr. Meth.* **95**:349 (1971).
18 Batra, R. K., and Shotter, A. C., *Nucl. Instr. Meth.* **124**:101 (1975).
19 Policarpo, A. J. P. L., Conde, C. A. N., and Alves, M. A. F., *Nucl. Instr. Meth.* **58**:151 (1968).
20 Morgan, G. L., and Walter, R. L., *Nucl. Instr. Meth.* **58**:277 (1968).
21 Engelkemeir, B., *Rev. Sci. Instr.* **27**:989 (1956).
22 Eby, F. S., and Jentschk, W. K., *Phys. Rev.* **96**:911 (1954).
23 Murray, R. B., and Meyer, A., *Phys. Rev.* **122**:815 (1961).
24 Meyer, A., and Murray, R. B., *Phys. Rev.* **128**:98 (1962).
25 Prescott, J. R., and Narayan, G. H.,
26 Hill, R., and Collinson, A. J. L., *Nucl. Instr. Meth.* **44**:245 (1966).
27 Brand, W., Dobrin, R., Jack, H., Aubert, R. L., and Roth, S., *Con. J. Phys.* **46**:537 (1968).
28 Brooks, F. D., and Jones, D. T. L., *Nucl. Instr. Meth.* **121**:69 (1974).
29 Flynn, K. F., Glendenin, C. E., Steinberg, E. P., and Wright, P. M., *Nucl. Instr. Meth.* **27**:13 (1964).
30 Craun, R. L., and Smith, D. L., *Nucl. Instr. Meth.* **80**:239 (1970).
31 Bechetti, F. D., Thorn, C. E., and Levine, M. S., *Nucl. Instr. Meth.* **138**:93 (1976).
32 Chou, C. N., *Phys. Rev.* **87**:376, 904 (1952).
33 Wright, G. T., *Phys. Rev.* **91**:1282 (1953).
34 Voltz, R., Lopes da Silva, J., Laustriat, G., and Coche, A., *J. Chem. Phys.* **45**:3306 (1966).
35 Voltz, R., du Pont, H., and Laustriat, G., *J. Physique* **29**:297 (1968).
36 Ingersoll, D. T., and Wehring, B. W., *Nucl. Instr. Meth.* **147**:551 (1977).

7

■ semiconductor detectors

7.1 INTRODUCTION

Semiconductor detectors are solid-state devices that operate essentially like ionization chambers. The charge carriers in semiconductors are not electrons and ions, as in the gas counters, but electrons and "holes."[1,2] At present, the most successful semiconductor detectors are made of silicon and germanium. Other materials have been tried, however, with some success, e.g., CdTe and HgI_2.

The most important advantage of the semiconductor detectors, compared to other types of radiation counters, is their superior *energy resolution*: the ability to resolve the energy of particles out of a polyenergetic energy spectrum (energy resolution and its importance are discussed in Chaps. 9, 12–14). Other advantages are:

1 Linear response (pulse height versus particle energy) over a wide energy range
2 Higher efficiency for a given size, because of the high density of the solid material used for construction
3 Possibility for special geometric configurations
4 Fast pulse risetime (relative to gas counters)
5 Ability to operate in vacuum
6 Insensitivity to magnetic fields

The characteristics of a semiconductor detector depend not only on the type of material used—e.g., Si or Ge—but also on the way the semiconductor is shaped and treated. The type, size, shape, and treatment of the crystal play a role in the operation and performance of a semiconductor detector.

This chapter first discusses the fundamentals of energy states in crystals, a subject necessary for understanding the creation and movement of electrons and holes in a solid. The properties of semiconductors are discussed next, with special emphasis given to the properties of silicon and germanium. The principle of construction and operation is accompanied by a description of the different types of detectors available in the market. Future prospects in this field are also discussed.

7.2 ELECTRICAL CLASSIFICATION OF SOLIDS

Solids are divided according to their electrical conductivity into three groups: conductors, insulators, and semiconductors. If a piece of solid material is placed in an electric field, whether or not current will flow depends on the type of material. If current flows, the material is a *conductor*. If current is zero at low temperatures but larger than zero at higher temperatures, the material is a *semiconductor*. If current is zero at all temperatures, the material is an *insulator*.

Conductivity and *electric current* mean motion of electrons, and according to the results of this simple experiment,

1 In conductors, electrons can move freely at any voltage different than zero.
2 In insulators, electrons cannot move under any voltage (except, of course, when the voltage is so high that an electrical discharge occurs).
3 In semiconductors, electrons cannot move at low temperatures (close to absolute zero) under any voltage. As the temperature of a semiconductor increases, however, electrons can move and electric current will flow at moderate voltages.

These properties can be explained through examining the electronic structure of crystals.

7.2.1 Electronic States in Solids and the Fermi Distribution Function

In a free atom the electrons are allowed to exist only in certain discrete energy states (Fig. 7.1*a*). In solids, the energy states widen into energy bands. Electrons can exist only in bands 1, 3, and 5, but not in bands 2 and 4 (Fig. 7.1*b*). An electron can move from band 1 to band 3 if:

1 The electron acquires the energy E_g necessary to cross the forbidden gap.
2 There is an empty energy state in band 3, which the jumping electron can occupy.†

The energy distribution of electronic states is described in terms of the following quantities:

$N(E)\,dE$ = number of electrons per unit volume with energy between E and $E + dE$

$S(E)\,dE$ = number of allowed electronic energy states, per unit volume, in the energy interval between E and $E + dE$

$P(E)$ = probability that a state of energy E is occupied = *fermi distribution function*

†This constraint is due to the Pauli principle, which forbids two or more electrons to be in the same state.

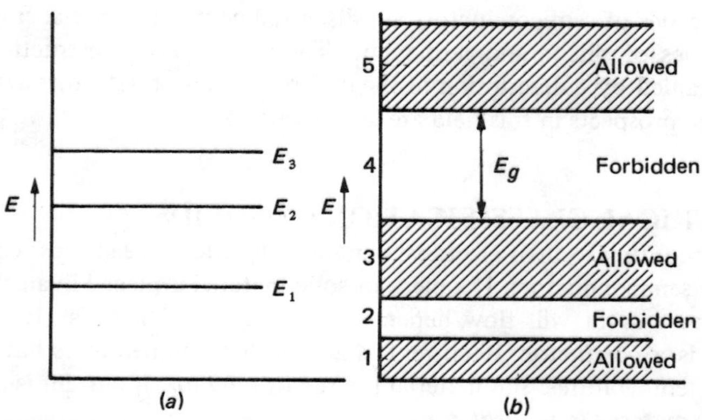

FIG. 7.1 (*a*) The atomic energy levels are discrete lines. (*b*) In a solid, the allowed energy states become energy bands.

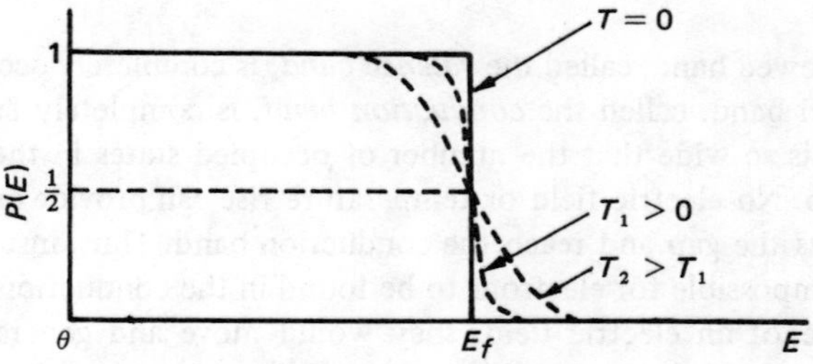

FIG. 7.2 The Fermi distribution function.

Then

$$N(E)\,dE = P(E)\,[S(E)\,dE] \tag{7.1}$$

The form of $P(E)$ is given by

$$P(E) = \frac{1}{1 + e^{(E-E_f)/kT}} \tag{7.2}$$

where E_f = Fermi energy
k = Boltzmann constant
T = temperature, Kelvin

The *Fermi energy* E_f is a constant that does not depend on temperature but it does depend on the purity of the solid. The function $P(E)$ is a universal function applying to all solids and having these properties (Fig. 7.2):

1 At $T = 0$,

$$P(E) = 1 \qquad E < E_f$$

$$P(E) = 0 \qquad E > E_f$$

2 At any T,

$$P(E_f) = \tfrac{1}{2}$$

3 For $T > 0$, the function $P(E)$ extends beyond E_f. If $E - E_f \gg kT$, $P(E)$ takes the form

$$P(E) = \frac{1}{1 + e^{(E-E_f)/kT}} \sim \frac{1}{e^{E-E_f/kT}} \approx \exp\left(-\frac{E - E_f}{kT}\right) \tag{7.3}$$

which resembles the classical Boltzmann distribution.

Notice that at $T = 0$ (Fig. 7.2), all the states are occupied for $E < E_f$ but all the states are empty for $E > E_f$.

7.2.2 Insulators

In insulators, the highest allowed band, called the *valence band*, is completely occupied (Fig. 7.3). The next allowed band, called the *conduction band*, is completely empty. As Fig. 7.3 shows, the gap is so wide that the number of occupied states in the conduction band is always zero. No electric field or temperature rise can provide enough energy for electrons to cross the gap and reach the conduction band. Thus, insulators *are* insulators because it is impossible for electrons to be found in the conduction band where, under the influence of an electric field, they would move and generate an electric current.

7.2.3 Conductors

In conductors, the conduction band is partially occupied (Fig. 7.4). An electron close to the top of the filled part of this band (point A, Fig. 7.4) will be able to move to the empty part (part B) under the influence of any electric field other than zero. Thus, because of the lack of a forbidden gap there is no threshold of electric field intensity below which electrons cannot move. Motion of the charge carriers and, consequently, conductivity are always possible for any voltage applied, no matter how small.

7.3 SEMICONDUCTORS

In semiconductors, the valence band is full and the conduction band is empty, but the energy gap between these two bands is very small. At very low temperatures, close to $T = 0$, the conductivity of the semiconductors is zero and the energy-band picture looks like that of an insulator (Fig. 7.3). As temperature increases, however, the "tail" of the Fermi distribution brings some electrons into the conduction band and conductivity increases (Fig. 7.5). That is, as temperature increases, some electrons obtain enough energy to cross over to the conduction band. Once there, they will move under

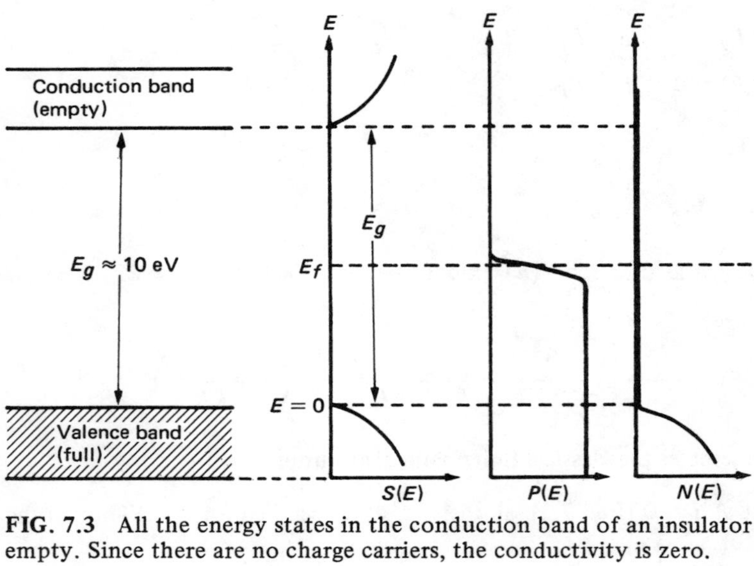

FIG. 7.3 All the energy states in the conduction band of an insulator are empty. Since there are no charge carriers, the conductivity is zero.

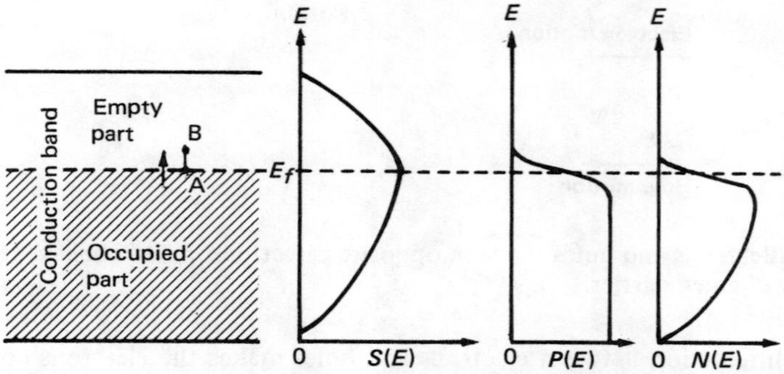

FIG. 7.4 In conductors, the conduction band is partially occupied. If an electric field is applied, the electrons move and conductivity is not zero.

the influence of an electric field for the same reason that electrons of conductors move.

When an electron moves to the conduction band, an empty state is left in the valence band. This is called a *hole*. A hole is the absence of an electron. When the electron moves in one direction, the hole moves in the opposite direction (Fig. 7.6). Holes are treated as particles with positive charges: $-(-e) = +e$. They contribute to the conductivity in the same way electrons do (see Sec. 7.3.2). In a pure and electrically neutral semiconductor, the number of electrons is always equal to the number of holes.

Heat–i.e., temperature increase–is not the only way energy may be given to an electron. Absorption of radiation or collision with an energetic charged particle may produce the same effect. The interaction of ionizing radiation with a semiconductor is a complex process and there is no agreement upon a common model explaining it. One simplified model is the following.

An energetic incident charged particle collides with electrons of the semiconductor and lifts them, not only from the valence to the conduction band but also from deeper lying occupied bands to the conduction band, as shown in Fig. 7.7*a*. Electrons appear in normally empty bands and holes appear in normally fully occupied bands. However, this configuration does not last long. In times of the order

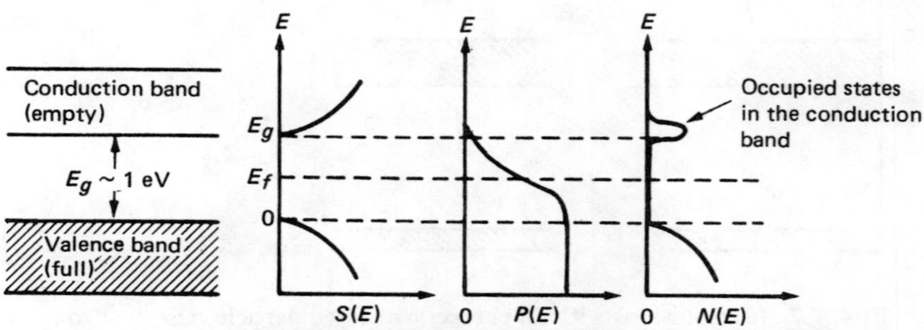

FIG. 7.5 In semiconductors, the energy gap is relatively narrow. As temperature increases, some electrons have enough energy to be able to move to the conduction band and conductivity appears.

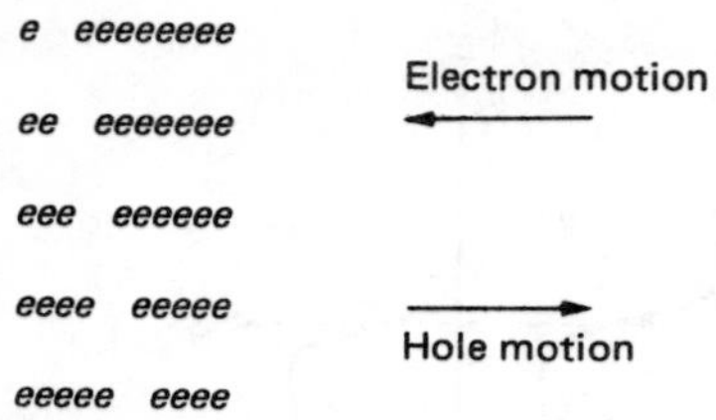

FIG. 7.6 Electrons and holes move in opposite directions. A hole behaves like a positively charged carrier.

of 10^{-12} s, the interaction between electrons and holes makes the electrons concentrate at the bottom of the lowest lying unoccupied (conduction) band. The holes, on the other hand, concentrate near the top of the highest full (valence) band. During this deexcitation process, many more electrons and holes are generated. Because of this multistep process, the average energy necessary for the creation of one electron-hole pair is much larger than the energy gap E_g. For example, for silicon at room tempera-

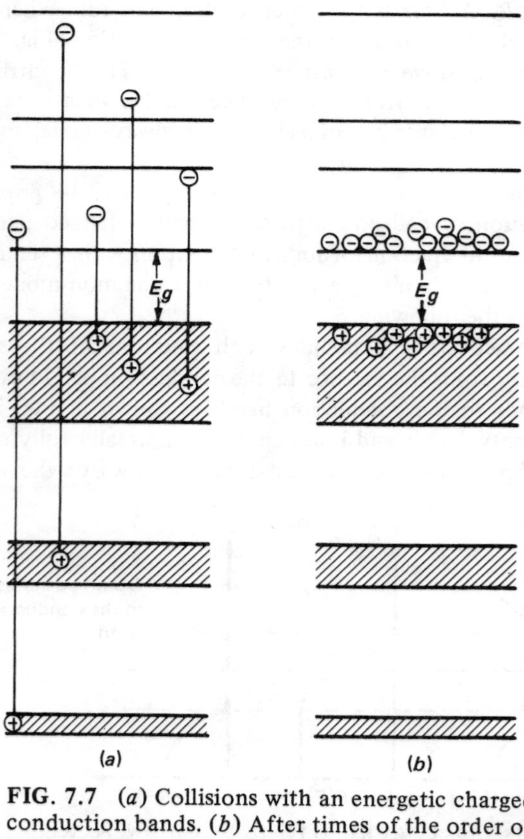

FIG. 7.7 (*a*) Collisions with an energetic charged particle raise electrons to the conduction bands. (*b*) After times of the order of 10^{-12} s, electrons and holes tend to deexcite to the upper part of the valence band and lower part of conduc- part, respectively.

ture the energy gap $E_g = 1.106$ eV, and the average energy for the production of one electron-hole pair is 3.66 eV.

In the absence of an electric field, the final step of the deexcitation process is the recombination of electrons and holes and the return of the crystal to its neutral state.

7.3.1 The Change of the Energy Gap with Temperature

The value of the energy gap E_g (Fig. 7.5) is not constant, but it changes with temperature as shown in Fig. 7.8. For silicon and germanium, E_g initially increases linearly as temperature decreases; but at very low temperatures, E_g reaches a constant value.

The average energy needed to create an electron-hole pair follows a similar change with temperature (Fig. 7.9).

7.3.2 Conductivity of Semiconductors

Conductivity σ is the inverse of resistivity and is defined by

$$j = \sigma E \tag{7.4}$$

where j = current density (A/m²)
σ = conductivity (A/V·m)
E = electric field (V/m)

Another expression for the current density is

$$j = eNv \tag{7.5}$$

where N = number of charge carriers/m³
v = speed of carriers

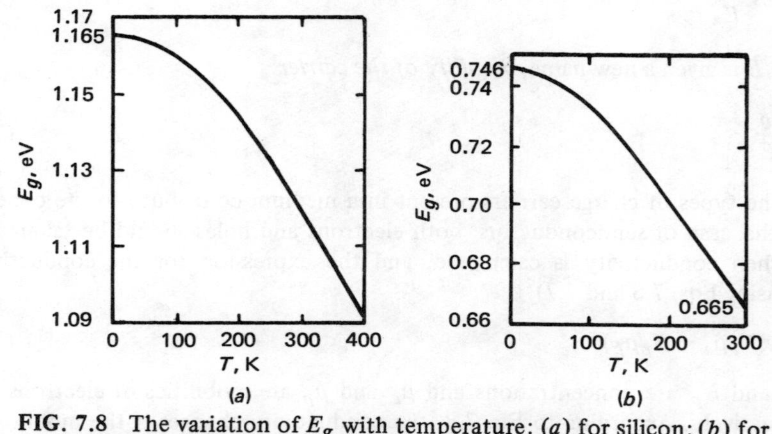

FIG. 7.8 The variation of E_g with temperature: (*a*) for silicon; (*b*) for germanium (from Chap. 1.1.1 of Bertolini-Coche).

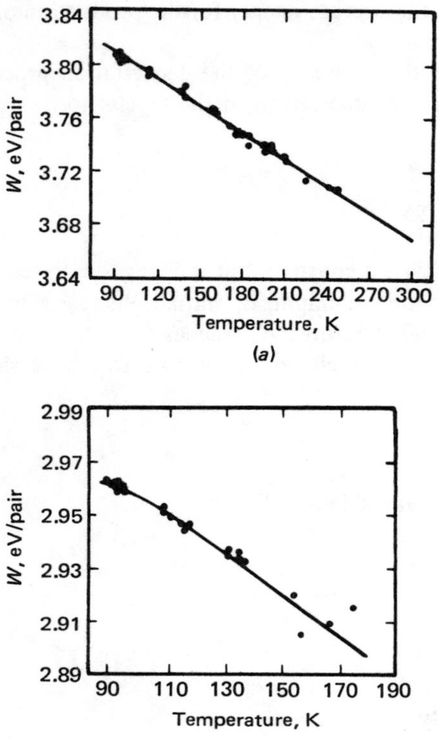

FIG. 7.9 Energy needed to produce an electron-hole pair in (*a*) silicon and (*b*) germanium, as a function of temperature (from ref. 3).

Using Eq. 7.4 and 7.5, one obtains the following equation:

$$\sigma = eN\frac{v}{E} \tag{7.6}$$

The ratio v/E is given a new name, *mobility of the carrier*:

$$\mu = \frac{v}{E} \tag{7.7}$$

All the types of charge carriers present in a medium contribute to the conductivity. In the case of semiconductors, both electrons and holes should be taken into account when conductivity is calculated, and the expression for the conductivity becomes (using Eqs. 7.6 and 7.7)

$$\sigma = e(N_e\mu_e + N_p\mu_p) \tag{7.8}$$

where N_e and N_p are concentrations and μ_e and μ_p are mobilities of electrons and holes, respectively. According to Eq. 7.8, the conductivity changes if the mobility of the carriers or their concentration or both change.

The mobilities of electrons and holes are independent of the electric field over a wide range of carrier velocities, but they change with temperature. If the temperature decreases, the mobility of both carriers increases, a fact that tends to increase the conductivity at low temperatures. The mobility of electrons and holes in pure germanium as a function of temperature is shown in Fig. 7.10.[4] The mobility changes at $\mu \sim T^{-\alpha}$ with $\alpha \approx 1.5$, for $T < 80\,\text{K}$. For $T > 80\,\text{K}$, the value of α is somewhat larger. It is worth noting that for $T < 80\,\text{K}$, $\mu_n \approx \mu_p$.

The motion of the carriers in a semiconductor is also affected by the presence of impurities and defects of the crystal. A small amount of impurities is always present, although impurities are usually introduced deliberately to make the properties of the crystal more appropriate for radiation detection (see Sec. 7.3.3). Crystal defects are present too. Even if one starts with a perfect crystal, defects are produced by the incident particles (this is called *radiation damage*). In the language of energy bands, impurities and defects represent new energy levels that may trap the carriers. Trapping is, of course, undesirable because it means loss of part of the charge generated by the incident particle.

For semiconductors, the probability that an electron will move from the valence to the conduction level is proportional to the factor (Eq. 7.3)

$$\exp\left(-\frac{E_g}{2kT}\right) \tag{7.9}$$

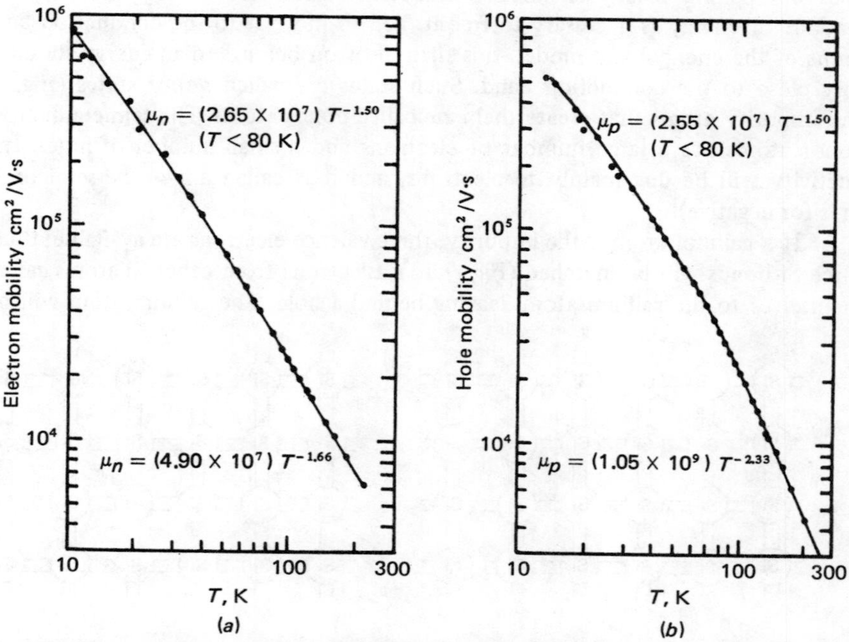

FIG. 7.10 (*a*) Electron mobility versus temperature for *n*-type semiconductors. (*b*) Hole mobility versus temperature for *p*-type semiconductors (from ref. 4).

(E_f is located in the middle of the gap; thus $E - E_f = E_g/2$.) Because of the exponential form of Eq. 7.9, there are always some electrons in the conduction band. These electrons produce a leakage current. Obviously, a successful detector should have as low a leakage current as possible to be able to detect the ionization produced by the incident radiation. The leakage current decreases with temperature, and for two different materials it will be smaller for the material with the larger energy gap.

7.3.3 Extrinsic and Intrinsic Semiconductors—The Role of Impurities

The properties of a pure semiconductor change if impurities are introduced. With impurities present, new states are created and the semiconductor obtains extra electrons or extra holes, which increase the conductivity of the material.

Actually, pure semiconductors are not available. All materials contain some impurities and for this reason they are called *impure* or *extrinsic*, in contrast to a pure semiconductor, which is called *intrinsic.* In most cases, controlled amounts of impurities are introduced purposely by a process called *doping*, which increases the conductivity of the material by orders of magnitude.

Doping works in the following way. Consider silicon (Si), which has four valence electrons. In a pure Si crystal, every valence electron makes a covalent bond with a neighboring atom (Fig. 7.11*a*). Assume now that one of the atoms is replaced by an atom or arsenic (As), which has five valence electrons (Fig. 7.11*b*). Four of the valence electrons form covalent bonds with four neighboring Si atoms, but the fifth electron does not belong to any chemical bond. It is bound very weakly and only a small amount of energy is necessary to free it, i.e., to move it to the conduction band. In terms of the energy-band model, this fifth electron belongs to an energy state located very close to the conduction band. Such states are called *donor* states (Fig. 7.12), and impurity atoms that create them are called *donor* atoms. The semiconductor with donor atoms has a large number of electrons and a small number of holes. Its conductivity will be due mainly to electrons, and it is called an *n-type* semiconductor (n is for negative).

If a gallium atom is the impurity, three valence electrons are available; thus only three Si bonds will be matched (Fig. 7.13). Electrons from other Si atoms can attach themselves to the gallium atom, leaving behind a hole. The gallium atom will behave

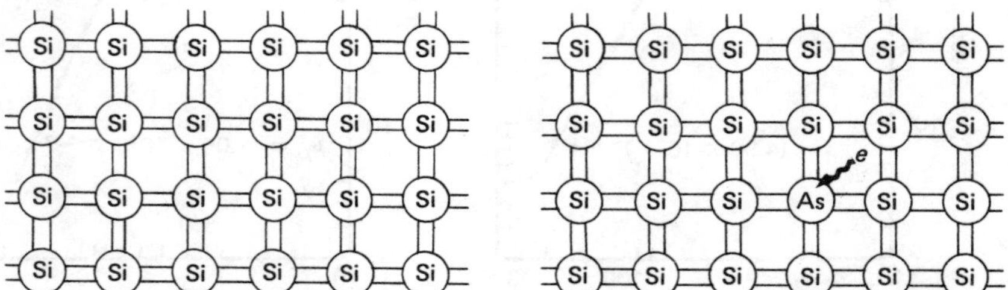

FIG. 7.11 (*a*) Pure (intrinsic) silicon. (*b*) Silicon doped with arsenic. The fifth electron of the arsenic atom is not tightly bound, and little energy is needed to move it to the conduction band.

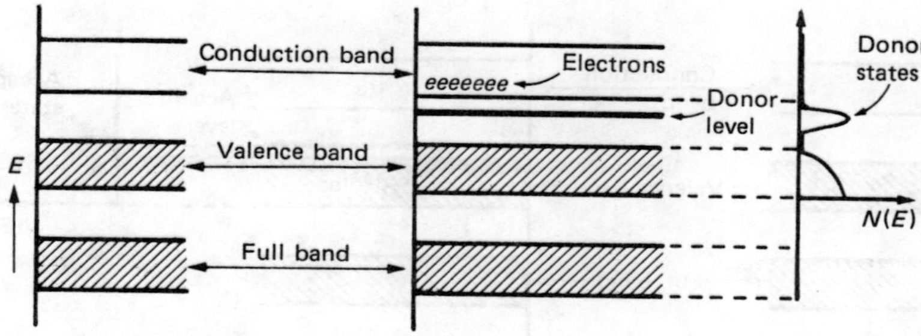

FIG. 7.12 (*a*) Intrinsic and (*b*) n-type semiconductor. New electron states are created (donor states) close to the conduction band.

like a negative ion after it accepts the extra electron. In terms of the energy-band theory, the presence of the gallium atom creates new states very close to the valence band (Fig. 7.14). These are called *acceptor* states. The impurity is called an *acceptor* atom. For every electron that moves to the acceptor states, a hole is left behind. The acceptor impurity atoms create holes. The charge carriers are essentially positive, and the semiconductor is called *p-type.*

Interstitial atoms can act as donors or acceptors. Lithium, as an interstitial in either silicon or germanium, creates donor states very close to the conduction band. Copper and nickel introduce donor states midway between the valence and conduction bands. Gold may act as either an acceptor or donor, depending on its position on the lattice.

For every atom of n-type or p-type impurity, an electron or hole is located at the donor or acceptor state, respectively. The material is still neutral, but when conductivity appears:

Electrons are the major carriers for n-type.
Holes are the major carriers for p-type.

Since the addition of impurities creates new states that facilitate the movement of the carriers, it should be expected that the conductivity of a semiconductor in-

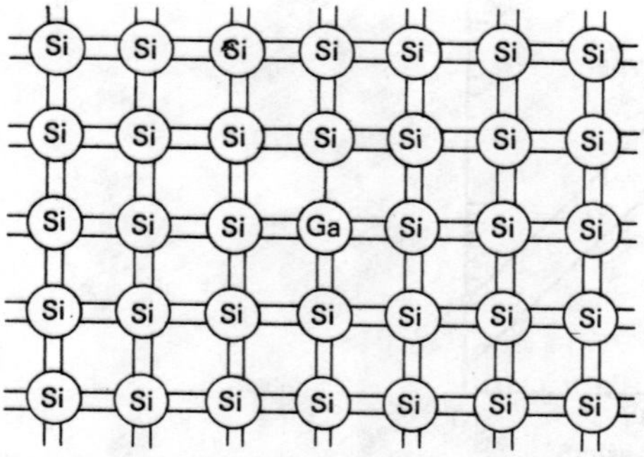

FIG. 7.13 Silicon doped with gallium. One of the covalent bonds is not matched.

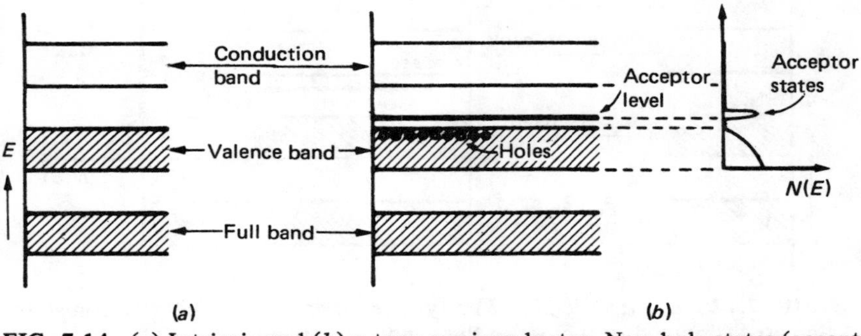

FIG. 7.14 (*a*) Intrinsic and (*b*) p-type semiconductor. New hole states (acceptor states) are created close to the top of the valence band.

creases with impurity concentration. Figures 7.15 and 7.16 show how the resistivity of germanium and silicon changes with impurity concentration.

The energy gap E_g depends on temperature, as shown in Fig. 7.8, and on the number of impurities and defects of the crystal. With increasing temperatures, if E_g is small as in germanium, the electrical conduction is dominated by electron-hole pairs created by thermal excitation and not by the presence of the impurity atoms. Therefore, at high enough temperatures any semiconductor can be considered as intrinsic.

Table 7.1 presents the most important physical and electrical properties of silicon and germanium, the two most widely used semiconductors.

7.4 THE p–n JUNCTION

As stated in the introduction to this chapter, semiconductor detectors operate like ionization counters. In ionization counters (see Chap. 5), the charges produced by the incident radiation are collected with the help of an electric field from an external

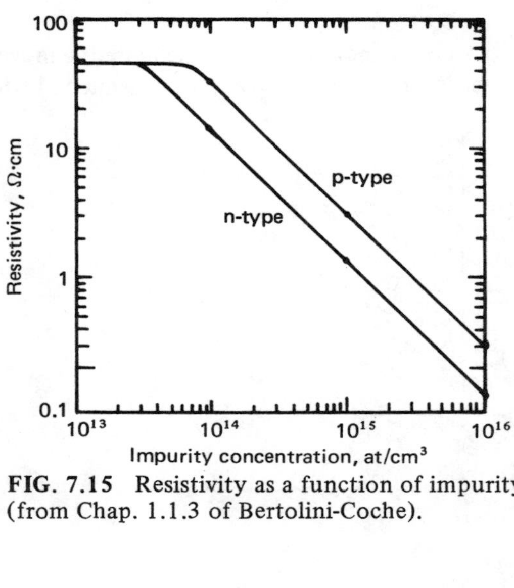

FIG. 7.15 Resistivity as a function of impurity concentration in germanium (from Chap. 1.1.3 of Bertolini-Coche).

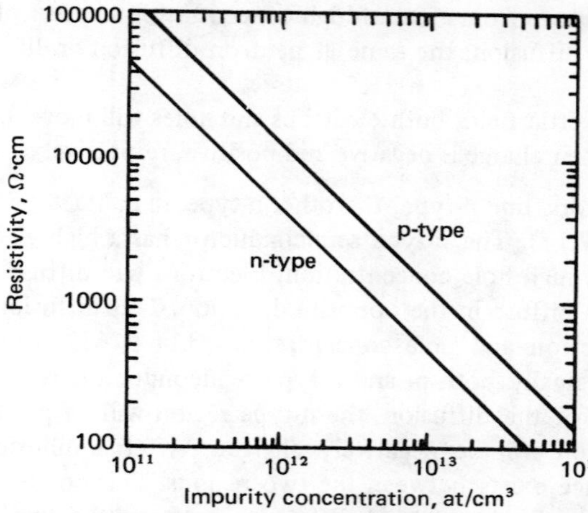

FIG. 7.16 Resistivity as a function of impurity concentration in silicon (from Chap. 1.1.3 of Bertolini-Coche).

voltage. In semiconductor detectors, the electric field is established by a process more complicated than in gas counters, a process that depends on the properties of n- and p-type semiconductors. The phenomena involved will be better understood with a brief discussion of the so-called p–n junction.

An n-type semiconductor has an excess of electron carriers. A p-type has excess holes. If a p-type and an n-type semiconductor join together, electrons and holes move because of two reasons:

TABLE 7.1 Properties of Si and Ge (from Fenyres and Haiman and Ref. 2)

Property	Si	Ge
Atomic number	14	32
Atomic weight	28.1	72.6
Density (300 K)	2.33×10^3 kg/m³	5.33×10^3 kg/m³
Energy gap (E_g), 300 K	1.106 eV	0.67 eV
Energy gap (E_g), 0 K	1.165 eV	0.75 eV
Average energy per electron-hole pair, 77 K	3.7 eV	2.96 eV
Average energy per electron-hole pair 300 K	3.65 eV	–
Diffusion voltage (V_0)	0.7 V	0.4 V
Atomic concentration	5×10^{28} m⁻³	4.5×10^{28} m⁻³
Intrinsic carrier concentration (300 K)	1.5×10^{16} m⁻³	2.4×10^{19} m⁻³
Intrinsic resistivity (300 K)	2.3×10^3 Ω·m	0.47 Ω·m
Intrinsic resistivity (77 K)	∞	5×10^2 Ω·m
Electron mobility (300 K)	0.1350 m²/V·s	0.3900 m²/V·s
Hole mobility (300 K)	0.0480 m²/V·s	0.1900 m²/V·s
Electron mobility (77 K)	4.0–7.0 m²/V·s	3.5–5.5 m²/V·s
Hole mobility (77 K)	2.0–3.5 m²/V·s	4.0–7.0 m²/V·s
Dielectric constant	12	16

1 Both electrons and holes will move from areas of high concentration to areas of low concentration. This is simply diffusion, the same as neutron diffusion or diffusion of gas molecules.
2 Under the influence of an electric field, both electrons and holes will move, but in opposite directions because their charge is negative and positive, respectively.

Consider two semiconductors, one p-type, the other n-type, in contact, without an external electric field (Fig. 7.17). The n-type semiconductor has a high electron concentration; the p-type has a high hole concentration. Electrons will diffuse from the n- to the p-type; holes will diffuse in the opposite direction. This diffusion will produce an equilibrium of electron and hole concentrations, but it will upset the original charge equilibrium. Originally both p- and n-type semiconductors were electrically neutral, but as a result of the diffusion, the n-type region will be positively charged, while the p-type region will be negatively charged. After equilibrium is established, a potential difference exists between the two regions. This combination of p- and n-type semiconductor with a potential difference between the two types constitutes a p–n junction.

The potential V_0 (Fig. 7.17) depends on electron-hole concentrations and is of the order of 0.5 V. If an external voltage V_{ext} is applied with the positive pole connected to the n side, the total potential across the junction becomes $V_0 + V_b$. This is called *reverse bias.* Such external voltage tends to make the motion of both electrons and holes more difficult. In the region of the changing potential, there is an electric field $E = -\partial V/\partial x$. The length X_0 of the region where the potential and the electric field exist increases with reverse bias. Calculation shows that

$$X_0 \approx \sqrt{\mu_p \rho (V_0 + V_b)} \qquad \text{for p-type semiconductor} \tag{7.10}$$

and

$$X_0 \approx \sqrt{\mu_n \rho (V_0 + V_b)} \qquad \text{for n-type semiconductor} \tag{7.10a}$$

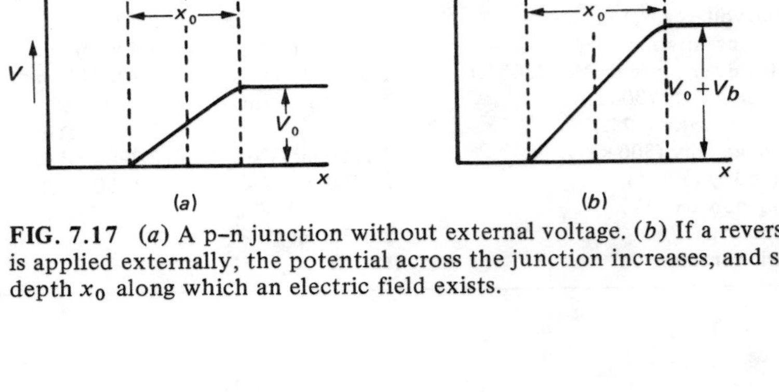

FIG. 7.17 (*a*) A p–n junction without external voltage. (*b*) If a reverse voltage is applied externally, the potential across the junction increases, and so does the depth x_0 along which an electric field exists.

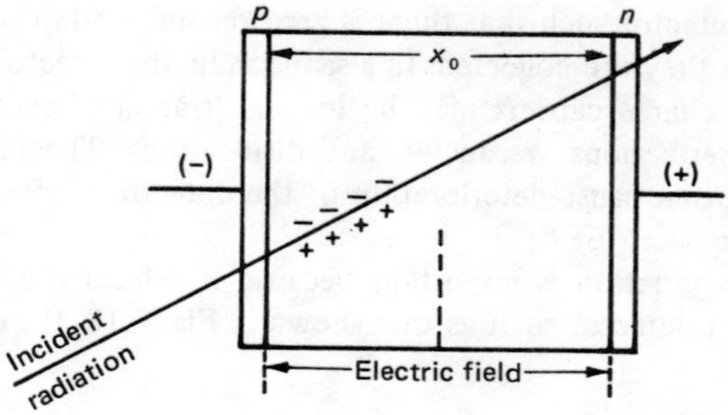

FIG. 7.18 A p–n junction with reverse bias operating as a detector.

where ρ ($\Omega \cdot$m) is the resistivity of the crystal. Application of a negative potential on the n side will have the opposite effect. The total potential difference will be $V_0 - V_b$. This is called *forward bias.* For a successful detector, reverse bias is applied. Since, usually, $V_b \gg V_0$, $X_0 \sim \sqrt{V_b}$.

In practice, a p–n junction is not made by bringing two pieces of semiconductor into contact. Instead, one starts with a semiconductor of one type (say n-type) and then transforms one end of it into the other type (p-type).

7.4.1 The p–n Junction Operating as a Detector

The operation of a semiconductor detector is based, essentially, on the properties of the p–n junction with reverse bias (Fig. 7.18). Radiation incident upon the junction produces electron-hole pairs as it passes through it. For example, if a 5-MeV alpha particle impinges upon the detector and deposits all its energy there, it will create about

$$\frac{5 \times 10^6 \text{ eV}}{3 \text{ eV/pair}} \sim 1.7 \times 10^6 \text{ electron-hole pairs}$$

Electrons and holes are swept away under the influence of the electric field and, with proper electronics, the charge collected produces a pulse that can be recorded.

The performance of a semiconductor detector depends on the region of the p–n junction where the electric field exists (region of width X_0, Fig. 7.18). Electrons and holes produced in that region find themselves in an environment similar to what electrons and ions see in a plate ionization chamber (see Sec. 5.4). There are some differences, however, between these two types of detectors.

In a gas counter, the electron mobility is thousands of times bigger than that of the ions. In semiconductors, the electron mobility is only about two to three times bigger than that of the holes. The time it takes to collect all the charge produced in a gas counter is of the order of milliseconds. In semiconductors, the sensitive region of the counter is only a few millimeters and the speed of electrons and holes is such that the charge carriers can traverse the sensitive region and be collected in times of the order of 10^{-7} s.

It is always the objective in either an ionization or a semiconductor detector to collect all the charges produced by the incident particle. This is achieved by establishing an electric field in the detector such that there is zero recombination of electrons and ions (or holes) before they are collected. In a semiconductor detector, even if recombination is zero, some charge carriers may be lost in "trapping" centers of the crystal such as lattice imperfections, vacancies, and dislocations. The incident radiation creates crystal defects that cause deterioration of the detector performance and, thus, reduce its lifetime (see Sec. 7.6).

The capacitance of a p-n junction is important because it affects the energy resolution of the detector. For a detector such as that shown in Fig. 7.18, the *capacitance C* is given by

$$C = \epsilon \frac{A}{4\pi X_0} \tag{7.11}$$

where ϵ = dielectric constant of the material
A = surface area of the detector
X_0 = depletion depth (detector thickness)

Combining Eqs. 7.10 and 7.11,

$$C \sim \frac{1}{\sqrt{V_b}}$$

To summarize, a material that will be used for the construction of a detector should have certain properties, the most important of which are the following:

1 *High resistivity.* This is essential, since otherwise current will flow under the influence of the electric field, and the charge produced by the particles will result in a pulse that may be masked by the steadily flowing current.
2 *High carrier mobility.* Electrons and holes should be able to move quickly and be collected before they have a chance to recombine or be trapped. This is in conflict with (1) because in high-resistivity materials carrier mobility is low. Semiconductor materials doped with impurities have proven to have the proper resistivity-carrier mobility combination.
3 *Capability of supporting strong electric fields.* This property is related to (1). Its importance stems from the fact that the stronger the field, the better and faster the charge collection becomes. Also, as the electric field increases, so does the depth of the sensitive region (Eq. 7.10) for certain detectors.
4 *Perfect crystal lattice.* Apart from externally injected impurities, the semiconductor detector material should consist of a perfect crystal lattice without any defects, missing atoms, or interstitial atoms. Any such defect may act as a "trap" for the moving charges.

7.5 THE DIFFERENT TYPES OF SEMICONDUCTOR DETECTORS

The several types of semiconductor detectors that exist today (1981) differ from one another because of the material used for their construction or the method by which that material is treated. The rest of this section describes briefly the method of construction and the characteristics of the most successful detectors—made of silicon or germanium—and two promising ones made of CdTe and HgI_2.

7.5.1 Surface-Barrier Detectors

Silicon of high purity, usually n-type, is cut, ground, polished, and etched until a thin wafer with a high-grade surface is obtained. The silicon is then left exposed to air or to another oxidizing agent for several days. As a result of surface oxidization, surface energy states are produced that induce a high density of holes and form, essentially, a p-type layer on the surface (Fig. 7.19). A very thin layer of gold evaporated on the surface serves as the electrical contact that will lead the signal to the preamplifier. In Fig. 7.19, X_0 is the depth of the sensitive region, t is the total silicon thickness,

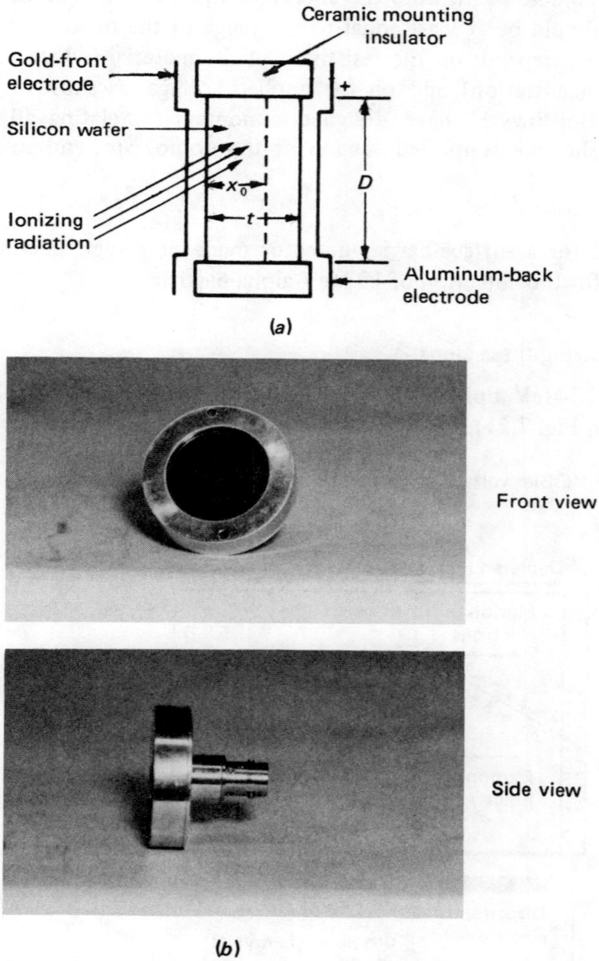

FIG. 7.19 A typical surface barrier detector: (*a*) a schematic representation; (*b*) photograph of a commercial detector (from ORTEC catalogue).

and D is the diameter of the detector. The size of the detector is the length (or depth) X_0.

7.5.2 Diffused-Junction Detectors

Silicon of high purity, normally p-type, is the basic material for this detector type. As with surface-barrier detectors, the silicon piece has the shape of a thin wafer. A thin layer of n-type silicon is formed on the front face of the wafer by applying a phosphorus compound to the surface and then heating the assembly to temperatures as high as 800–1000°C for less than an hour. The phosphorus diffuses into the silicon and "dopes" it with donors (Fig. 7.20). The n-type silicon in front and the p-type behind it form the p–n junction.

Both surface-barrier and diffused-junction detectors are used for the detection of charged particles. To be able to measure the energy of the incident radiation, the size X_0 of the detector should be at least equal to the range of the incident particle in silicon. The value of X_0 depends on the resistivity of the material (which in turn depends on impurity concentration) and on the applied voltage, as shown by Eq. 7.10. Blankenship and Borkowski[5] have designed a nomogram relating all these quantities. Figure 7.21 shows a simplified version of the nomogram, and Example 7.1 explains its use.

EXAMPLE 7.1

What is the bias needed for a surface-barrier detector made of p-type silicon with resistivity 15 Ω·m, used for the detection of 10-MeV alpha particles?

ANSWER

The bias is found by following these steps:

1 Find the range of a 10-MeV alpha particle in silicon. From Sec. 4.6, one obtains $R = 65\ \mu m$ (point A in Fig. 7.21).

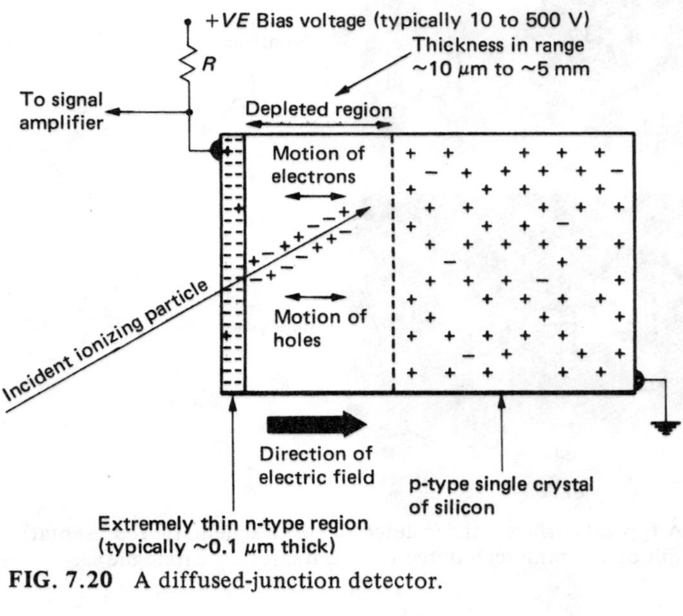

FIG. 7.20 A diffused-junction detector.

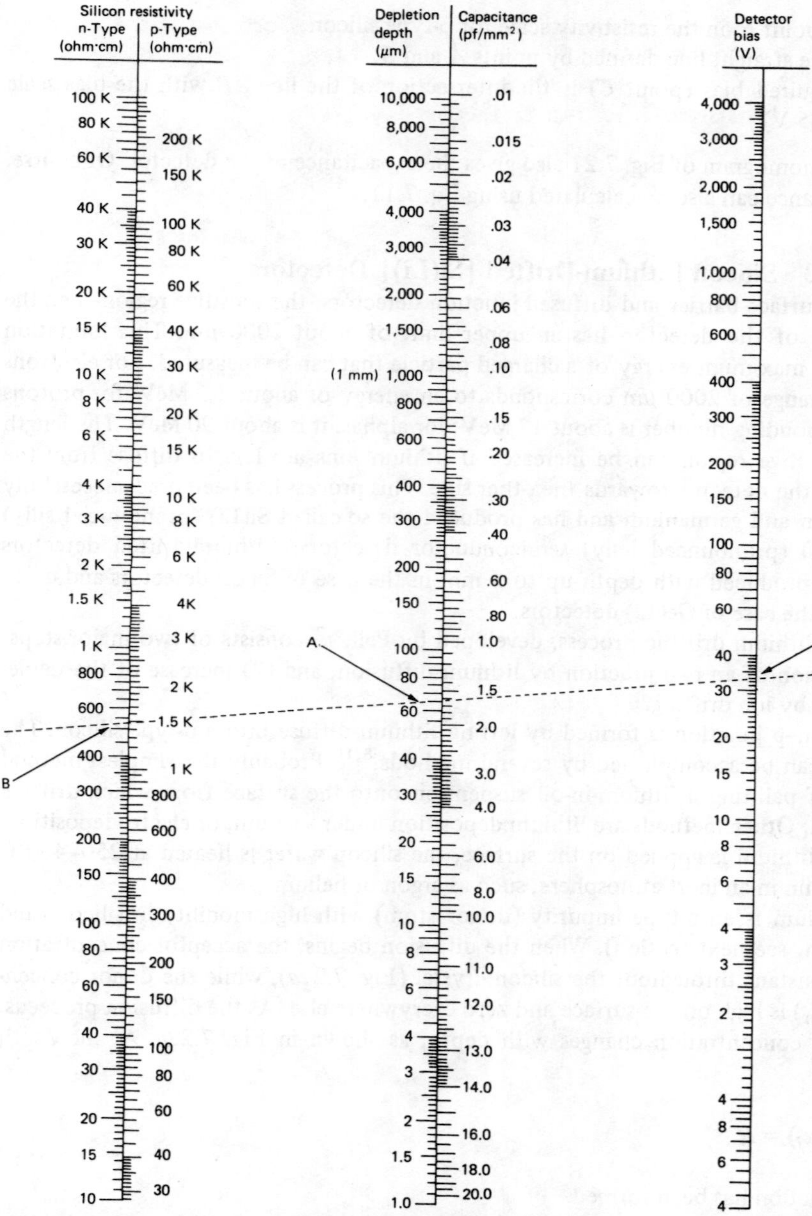

FIG. 7.21 The Blankenship-Borkowski[5] nomogram that relates resistivity, detector thickness, and detector bias. The detector capacitance as a function of detector thickness is also given.

2 Define point B on the resistivity scale for p-type silicon.
3 Draw the straight line defined by points A and B.
4 The required bias (point C) is the intersection of the line AB with the bias scale ($V_b \simeq 35$ V).

The nomogram of Fig. 7.21 also gives the capacitance of the detector. Of course, the capacitance can also be calculated using Eq. 7.11.

7.5.3 Silicon Lithium-Drifted [Si(Li)] Detectors

For both surface-barrier and diffused-junction detectors, the sensitive region–i.e., the actual size of the detector–has an upper limit of about 2000 μm. This limitation affects the maximum energy of a charged particle that can be measured. For electrons in Si, the range of 2000 μm corresponds to an energy of about 1.2 MeV; for protons the corresponding number is about 17 MeV; for alphas, it is about 90 MeV. The length of the sensitive region can be increased if lithium ions are left to diffuse from the surface of the detector towards the other side. This process has been used successfully with silicon and germanium and has produced the so-called Si(Li) (pronounced silly) and Ge(Li) (pronounced jelly) semiconductor detectors. Lithium-drifted detectors have been produced with depth up to 5 mm in the case of Si(Li) detectors and up to 12 mm in the case of Ge(Li) detectors.

The lithium drifting process, developed by Pell,[6,7] consists of two major steps: (1) formation of an n-p junction by lithium diffusion, and (2) increase of the depletion depth by ion drifting.

The n-p junction is formed by letting lithium diffuse into a p-type silicon. The diffusion can be accomplished by several methods.[8,11] Probably the simplest method consists of painting a lithium-in-oil suspension onto the surface from which drifting is to begin. Other methods are lithium deposition under vacuum, or electrodeposition. After the lithium is applied on the surface, the silicon wafer is heated at 250–400°C for 3–10 min in an inert atmosphere, such as argon or helium.

Lithium is an n-type impurity (donor atom) with high mobility in silicon (and germanium, see next section). When the diffusion begins, the acceptor concentration (N_p) is constant throughout the silicon crystal (Fig. 7.22*a*), while the donor concentration (N_n) is high on the surface and zero everywhere else. As the diffusion proceeds, the donor concentration changes with depth, as shown in Fig. 7.22*a*. At the depth x_j where

$$N_n(x_j) = N_p$$

an n-p junction has been formed.

After the diffusion is completed, the crystal is left to cool, the excess lithium is removed, and ohmic contacts are put on the n and p sides of the junction. The contact on the p side is usually formed by evaporating aluminum or gold doped with boron. The contact on the n side can be formed by using pure gold or antimony-doped gold.

Drifting is accomplished by heating the junction to 120–150°C while applying a reverse bias that may range from 25 V up to about 1000 V. As shown in Eq. 7.12,

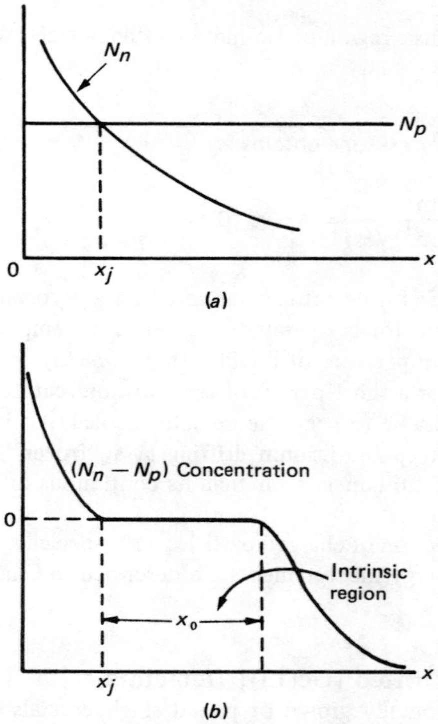

FIG. 7.22 (*a*) During the diffusion of lithium the donor concentration changes with depth as shown. (*b*) During drifting (at elevated temperature and under reverse bias), an almost intrinsic region is formed with thickness x_0.

the higher the temperature and the voltage are, the faster the drifting proceeds. Depending on the special method used, the semiconductor may be under vacuum or in air or be placed in a liquid bath (e.g., silicon oil or fluorocarbon). The electric field established by the reverse bias tends to move the n-type atoms (lithium) toward the p-side of the junction. As a result, the concentration of lithium atoms becomes lower for $x < x_j$ (Fig. 7.22*a*) and higher for $x > x_j$. For $x < x_j$, N_n cannot become less than N_p because then a local electric field would appear pushing the lithium atoms toward the n-side. Similarly, for $x > x_j$, N_n cannot increase very much because the local electric field works against such a concentration. Thus, a region is created that looks like an intrinsic semiconductor because $N_n \approx N_p$. For long drifting times, the thickness of the intrinsic region $X_0(t)$ as a function of time is given by

$$X_0(t) = \sqrt{2V\mu_{\text{Li}}t} \tag{7.12}$$

where V = applied voltage

μ_{Li} = mobility of Li ions in silicon at the drifting temperature

The mobility of lithium, which increases with temperature,[12] has a value of about 5×10^{-14} $m^2/V{\cdot}s$ at $T = 150°C$. Drifting is a long process. Depending on the desired thickness, drifting may take days and sometimes weeks.

EXAMPLE 7.2

How long will it take to obtain an intrinsic region of 1.5 mm in a silicon wafer drifted at 150°C under a reverse bias of 500 V?

ANSWER

Using Eq. 7.12 with $\mu_{Li} = 5 \times 10^{-14}$ m²/V·s, one obtains

$$t = \frac{X_0^2(t)}{2V\mu} = \frac{(1.5 \times 10^{-3})^2 \text{ m}^2}{2(500 \text{ V})(5 \times 10^{-14} \text{ m}^2/\text{V}\cdot\text{s})} = 4.5 \times 10^4 \text{ s} = 12.5 \text{ h}$$

After drifting is completed, the Si(Li) detector is mounted on a cryostat since the best results are obtained if the detector is operated at a very low temperature. Usually this temperature is 77 K, the temperature of liquid nitrogen. Si(Li) detectors may be stored at room temperature for a short period of time without catastrophic results, but for longer periods it is advisable to keep the detector cooled at all times. The low temperature is necessary to keep the lithium drifting at a "frozen" stage. At room temperature, the mobility of lithium is such that its continuous diffusion and precipitation[12] will ruin the detector.

Si(Li) detectors are used for detection of charged particles and especially x-rays. Their characteristics with respect to energy measurements are described in Chaps. 12 and 13.

7.5.4 Germanium Lithium-Drifted [Ge(Li)] Detectors

Ge(Li) detectors are made from horizontally grown or pulled single crystals of germanium. As the crystal is grown, it is doped with acceptor impurities such as indium, gallium, or boron, and becomes a p-type semiconductor. Germanium crystals may be cut to length and shaped by a variety of means, including the use of diamond wheels or band saws. In these mechanical operations, great care must be taken not to fracture the brittle material.

Lithium drifting in germanium follows the same approach as in silicon. The deposition and diffusion of lithium is accomplished by one of the methods discussed in the previous section. The ohmic contacts are made by electrolytic deposition of gold,[13] by using gallium-indium[14] or mercury-indium,[15] or by ion implantation.[16] The drifting process itself takes place at a lower temperature (<60°C) than for silicon, with the germanium diode in air[17] or immersed in a liquid maintained at its boiling point.[18]

As with silicon, the mobility of lithium in germanium increases with temperature, having a value of about 10^{-13} m²/V·s at 50°C. The drift rate, as Eq. 7.12 shows, can be accelerated by increasing the voltage or the temperature, but there are advantages in having a relatively slow rate. It takes about four weeks to achieve a drifting depth of 10 to 15 mm. (It should be noted that Eq. 7.12 is not exact for lithium drifting in germanium, because of the presence of secondary effects like precipitation of lithium in the crystal.)

After the drifting process has been completed, the detector is mounted on a crystal and is always kept at a low temperature ($\sim$77 K). Keeping the Ge(Li) detector at a low temperature is much more critical than for a Si(Li) detector. The mobility of the lithium atoms in germanium is so high at room temperature that the detector will be ruined if brought to room temperature even for a short period of time.†

A Ge(Li) detector is always operated at liquid nitrogen temperatures. This is necessary to minimize the electronic noise and thus achieve the best results. The energy gap in germanium is so small (~0.67 eV) that considerable thermal excitation of carriers is possible at room temperature. The result of the excitation is larger leakage current and, generally, increased electronic noise. Thus, even if one could find a way to achieve stability of drifting at room temperature, a Ge(Li) detector would still have to be operated at low temperatures.

Ge(Li) detectors are fabricated in many geometries. The most common are the following:

1 *Planar detector.* This consists of a round or square piece of germanium (Fig. 7.23) with lithium diffusing from one surface toward the other. The maximum length of the intrinsic region is about 15 mm.
2 *The U-type detector.* This geometry is shown in Fig. 7.24. Lithium is drifted from three sides into a block of p-type germanium. This geometry produces large depletion depths.
3 *The wrap-around detector.* This geometry, shown in Fig. 7.25, produces large detector volumes. Lithium is drifted from five sides toward the center of a trapezoidal-shaped block of p-type germanium.
4 *Coaxial detector.* Lithium is drifted from the outer surface toward the axis of a cylindrical piece of p-type germanium (Fig. 7.26). There are two variations, the wrap-around coax (Fig. 7.26*a*) and the true coax (Fig. 7.26*b*).

Ge(Li) detectors are used for detection of gammas. Their characteristics with respect to γ energy measurements are discussed in Chap. 12.

7.5.5 Totally Depleted Detectors

A totally depleted detector is not a new kind of semiconductor counter. *Total depletion* means elimination of any insensitive region, e.g., elimination of the dead regions at the front and back of the detector shown in Fig. 7.20.

Totally depleted detectors have been obtained by selecting the proper resistivity and reverse bias, or in lithium-drifted detectors by letting the lithium drift from one end of the crystal all the way to the other.

†If this happens, the detector may be redrifted, but at a considerable financial cost.

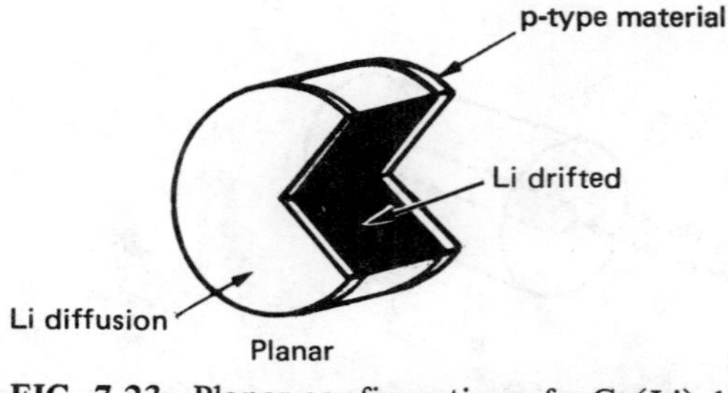

FIG. 7.23 Planar configuration of a Ge(Li) detector.

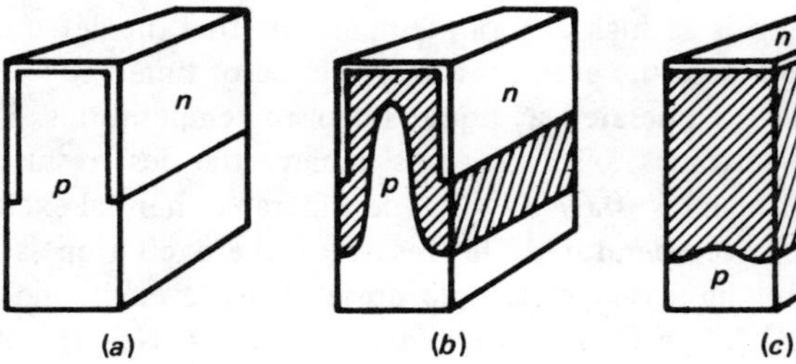

FIG. 7.24 The U-type configuration of a Ge(Li) detector. (*a*) Lithium diffused on three sides. (*b*) Partial lithium drift. (*c*) Completed detector with *n*-layers removed.

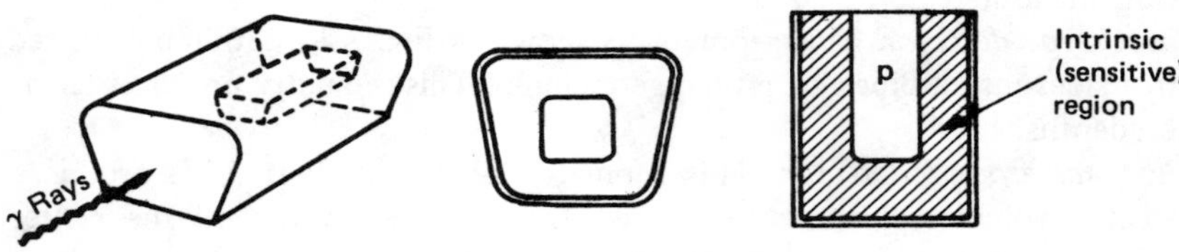

FIG. 7.25 The wrap-around configuration of a Ge(Li) detector.

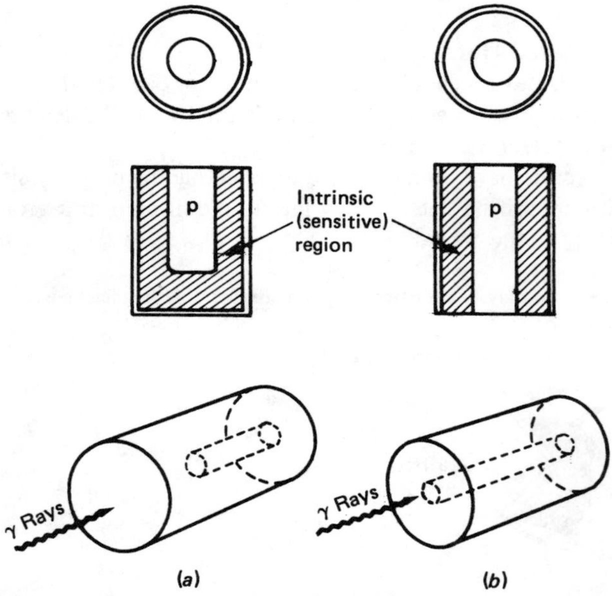

FIG. 7.26 The coaxial Ge(Li) detector: (*a*) wrap-around coax; (*b*) true coax.

The advantages of the totally depleted detector are:

1 The radiation may enter the detector on either side. Therefore, such a device may be used as a transmission counter where the signal simply records the passage of the particle.
2 There is no series resistance between the sensitive region and the electrodes.
3 Capacitance of the detector does not depend on the applied voltage.
4 The thickness of the detector is independent of the applied voltage.

7.5.6 Hyperpure Germanium (HPGe) Detectors

The production of high-purity germanium (HPGe) with an impurity concentration of 10^{16} atoms/m^3 has made possible the construction of detectors without lithium drifting.[19–21] The detector is simply formed by applying a reverse bias across a piece of germanium. The sensitive depth of the detector (depletion layer) depends on the impurity concentration and the applied voltage, as shown in Fig. 7.27.

The fabrication of HPGe detectors follows the same procedure as with Ge(Li) detectors, except that the lithium-drifting process is not needed. A very important step of the fabrication of HPGe detectors is the construction of the ohmic contacts. The n-side contact (on the front surface) is formed by letting lithium diffuse into the crystal or by depositing gold or palladium on it. A recent breakthrough reported by one of the manufacturers[22,23] makes it possible to eliminate the front metal contact. A good contact on the back side has been made by using metals or by implanting boron.[21]

HPGe detectors are made in planar or coaxial geometry. Coaxial detectors do not have the undrifted central core of Ge(Li) detectors (Fig. 7.26). Therefore, the central part along the axis of the crystal is removed and a contact is made on the inside of this central hole. HPGe detectors can be made with either n- or p-type germanium. High-purity germanium blocks up to 60 mm in diameter have been produced, making feasible the construction of coaxial detectors with a volume up to

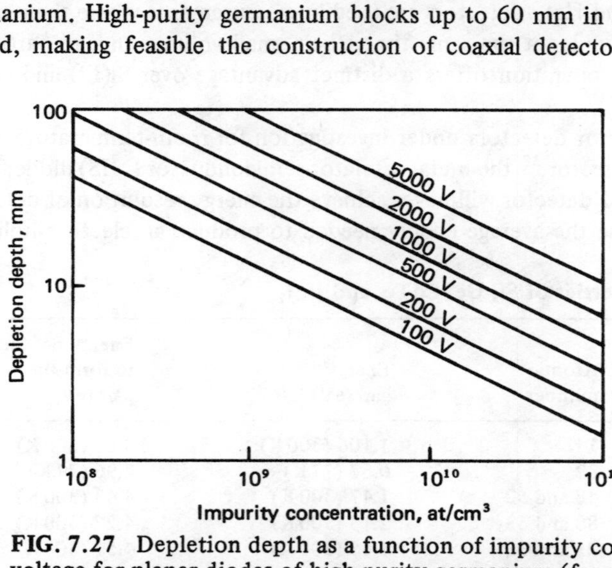

FIG. 7.27 Depletion depth as a function of impurity concentration and applied voltage for planar diodes of high-purity germanium (from ref. 42).

2×10^5 mm^3 (200 cm^3). Planar detectors with a thickness up to 20 mm and coaxial ones with a volume up to 5×10^4 mm^3 (50 cm^3) have been constructed.[12]

The major advantage of HPGe detectors is that they can be stored at room temperature because of the absence of lithium drifting. Room-temperature storage is particularly helpful when there is a need to ship the detector.

7.5.7 CdTe, HgI_2, and Other Detectors

The major disadvantage of lithium-drifted detectors is the requirement for continuous cooling. In the case of HPGe detectors, the requirement for cooling during operation is also a disadvantage. Cooling requires a cryostat, which makes the counter bulky and thus impossible to use in cases where only a small space is available; another disadvantage is the cost of continuously buying liquid nitrogen. There is a great incentive, therefore, to develop semiconductor detectors that can be stored and operated at room temperature. Two materials that have been studied and show great promise for the construction of such detectors are CdTe and HgI_2.[24–33] A comprehensive review of the state-of-the-art (until 1978) for both materials can be found in reference 34.

Successful detectors using CdTe or HgI_2 have been constructed with thickness up to 0.7 mm and area 100 mm^2 (as of 1978.)[33] These detectors are small in size, compared to Si(Li) or Ge(Li) detectors, but the required detector volume depends on the application. For CdTe and HgI_2, the favored applications are those that require a small detector volume: monitoring in space,[35] measurement of activity in nuclear power plants,[36] medical portable scanning,[37] or medical imaging devices.[38] Although the detector volume is small, efficiency is considerable because of the high atomic number of the elements involved (Table 7.2). The energy needed for the production of an electron-hole pair is larger for CdTe and HgI_2 than it is for Si and Ge; as a result, the energy resolution of the former is inferior to that of the latter (see also Chap. 12). But CdTe and HgI_2 detectors are used in measurements where their energy resolution is adequate while, at the same time, their small volume and, in particular, their room-temperature operation offers a distinct advantage over Si(Li) and Ge(Li) detectors.

Other semiconductor detectors under investigation for room-temperature operation are the diamond detector,[38] the metal-insulator-semiconductor (MIS) diode,[39] and GaAs.[40,41] The diamond detector will never achieve the energy resolution of detectors described earlier because the average energy needed to produce an electron-hole pair

TABLE 7.2 Properties of Si, Ge, CdTe, and HgI_2

Material	Atomic number	Energy gap (eV)	Energy needed to form the pair (eV)
Si	14	1.106 (300 K)	3.65 (300 K)
Ge	32	0.67 (77 K)	2.96 (77 K)
CdTe	48 and 52	1.47 (300 K)	4.43 (300 K)
HgI_2	80 and 53	2.13 (300 K)	4.22 (300 K)
GaAs	21 and 33	1.45 (300 K)	4.51 (300 K)

TABLE 7.3 Particle Fluence that Causes Significant Radiation Damage

	Heavy ions	Alphas	Fast neutrons
Junction detectors	10^{12} particles/m^2	10^{14} α/m^2	
Si(Li)		10^{12} α/m^2	10^{14} n/m^2
Ge(Li) or HPGe			10^{13}–10^{14} n/m^2

is about 13 eV. It looks promising, however, for detection of charged particles at high temperature in adverse chemical environment.

The MIS diode consists of three layers: a metal, an insulator, and a semiconductor. One example is metal-silicon oxide-silicon.[39] The thickness of the insulators (100–250 mm) seems to be a critical parameter for the operation of the MIS diode. The presence of the insulator reduces the leakage current and makes operation at room temperature possible.

Finally, GaAs is mentioned as one of the semiconducting materials with some promise for operation at room temperature. The energy gap and the average energy needed for the production of an electron-hole pair have values close to those of CdTe (Table 7.2). The atomic numbers of the elements involved, however, are lower; thus the efficiency of a GaAs detector, for the same volume, will be lower than that for CdTe.

7.6 RADIATION DAMAGE TO SEMICONDUCTOR DETECTORS

The fabrication and operation of a semiconductor detector is based on the premise that one starts with a perfect crystal containing a known amount of impurities. Even if this is true at the beginning, a semiconductor detector will suffer damage after being exposed to radiation. The principal type of radiation damage is caused by the collision of an incident particle with an atom. As a result of the collision, the atom may be displaced into an interstitial position, thus creating an interstitial-vacancy pair known as the *Frenkel defect.* A recoiling atom may have enough energy to displace other atoms; therefore on incident particle may produce many Frenkel defects.

Crystal defects affect the performance of the detector because they may act as trapping centers for electrons and holes or they may create new donor or acceptor states. New trapping centers and new energy states change the charge collection efficiency, the leakage current, the pulse risetime, the energy resolution, and other properties of the detector. The changes are gradual, but the final result is shortening of the detector lifetime.

Electrons and photons cause negligible radiation damage compared to charged particles and neutrons. Heavier and more energetic charged particles cause more damage than lighter and less energetic particles.[2,21] Also, the damage is not the same for all detector types. Table 7.3 gives the fluences that cause considerable radiation damage for different detectors and bombarding particles.

Ge(Li) detectors are not affected by gammas, but they are damaged by the neutrons in a mixed n-γ field. Recovery is possible, however, after annealing and redrifting. HPGe detectors suffer approximately the same damage as Ge(Li) detectors.

PROBLEMS

7.1 What is the probability that an electron energy state in Ge will be occupied at temperature $T = 300$ K if the energy state is greater than the Fermi energy by 2 eV?

7.2 Repeat Prob. 7.1 for $T = 77$ K.

7.3 The energy gap for diamond is 7 eV. What temperature will provide thermal energy (kT) equal to that amount?

7.4 What should be the maximum thickness of the gold layer covering the front face of a surface barrier detector used for the measurement of 10-MeV alphas, if the energy loss of the alphas traversing the layer should be less than 0.1 percent of their kinetic energy?

7.5 Repeat Prob. 7.4 for 6-MeV electrons.

7.6 The thickness of the gold layer covering the front face of a semiconductor detector may be measured by detecting particles entering the detector at two different angles. Calculate that thickness if alphas that enter in a direction perpendicular to the front face register as having energy 4.98 MeV, but those that enter at a 45° angle register as having energy 4.92 MeV.

7.7 What is the average distance traveled in Si by a 50-KeV gamma before it has an interaction? What is the corresponding distance in Ge?

7.8 Lithium has been drifted in germanium at 50°C under a reverse bias of 500 V for 2 weeks. What is your estimate of the drifting depth? ($\mu_{Li} = 1.5 \times 10^{-13}$ m²/V·s.)

7.9 A parallel beam of 0.5-MeV gammas is normally incident upon 2-mm thick crystals of Si, Ge, CdTe, and HgI_2. What fraction of photons will interact at least once in each crystal?

BIBLIOGRAPHY

Bertolini, G., and Coche, A., *Semiconductor Detectors*, North-Holland Publishing Co., Amsterdam, 1968.

Brown, W. L., Higinbotham, W. A., Miller, G. L., and Chace, R. L. (eds.), "Semiconductor Nuclear Particle Detectors and Circuits," proceedings of a conference conducted by the Committee of Nuclear Science of the National Academy of Sciences, NAS Publication 1593, Washington, D.C. (1969).

Dearnaley, G., and Northrop, D. C., *Semiconductor Counters for Nuclear Radiations*, E & F.N. Spon. Ltd., London, 1964.

Fenyves, E., and Haiman, O., *The Physical Principles of Nuclear Radiation Measurements*, Academic Press, New York, 1969.

Price, W. J., *Nuclear Radiation Detection*, McGraw-Hill, New York, 1964.

REFERENCES

1 McKenzie, J. M., *Nucl. Instr. Meth.* **162**:49 (1979).
2 Ewan, G. T., *Nucl. Instr. Meth.* **162**:75 (1979).
3 Pehl, R. H., Goulding, F. S., Landis, D. A., and Lenzlinger, M., *Nucl. Instr. Meth.* **59**:45 (1968).
4 Delaet, L. H., Schoenmaekers, W. K., and Guislain, H. J., *Nucl. Instr. Meth.* **101**:11 (1972).
5 Friedland, S. S., Mayer, J. W., and Wiggins, J. S., *IRE Trans. Nucl. Sci.* **NS-7**(2–3):181 (1960).
6 Pell, E. M., *J. Appl. Phys.* **31**:291 (1960).
7 Mayer, J. W., *J. Appl. Phys.* **33**:2894 (1962).

8 Elliott, J. H., *Nucl. Instr. Meth.* **12**:60 (1961).
9 Baily, N. A., and Mayer, J. W., *Radiology* 76:116 (1961).
10 Siffert, P., and Coche, A., *Compt. Rend.* **256**:3277 (1963).
11 Dearnaley, G., and Lewis, J. C., *Nucl. Instr. Meth.* **25**:237 (1964).
12 Siffert, P., and Coche, A., "Behavior of Lithium in Silicon and Germanium," in G. Bertolini and A. Coche (eds.), *Semiconductor Detectors*, North-Holland Publishing Co., Amsterdam, 1968.
13 Janarek, F. J., Helenberg, H. W., and Mann, H. M., *Rev. Sci. Instr.* **36**:1501 (1965).
14 Mooney, J. B., *Nucl. Instr. Meth.* **50**:242 (1967).
15 Hansen, W. L., and Jarrett, B. V., *Nucl. Instr. Meth.* **31**:301 (1964).
16 Meyer, O., and Haushahn, G., *Nucl. Instr. Meth.* **56**:177 (1967).
17 Ewan, G. T., and Tavendale, A. J., *Can. J. Phys.* **42**:3386 (1964).
18 Cappellani, E., Fumagulli, W., and Restelli, G., *Nucl. Instr. Meth.* **37**:352 (1965).
19 Hall, R. N., and Soltys, T. J., *IEEE Trans. Nucl. Sci.* **NS-18**:160 (1971).
20 Hansen, W. L., *Nucl. Instr. Meth.* **94**:377 (1971).
21 Pehl, R. H., *Physics Today* **30**:11, 50 (1977).
22 ORTEC Catalogue and brochures (1977 and 1979).
23 Baedecker, P. A., Powe, J. J., and Steinnes, E., *Journal Radiol. Chem.* **40**:115 (1977).
24 Siffert, I., Gonidec, J. P., and Cornet, A., *Nucl. Instr. Meth.* **115**:13 (1974).
25 Eichinger, P., Halder, N., and Kemmer, J., *Nucl. Instr. Meth.* **117**:305 (1974).
26 Jones, L. T., and Woollam, P. B., *Nucl. Instr. Meth.* **124**:591 (1975).
27 Iwanczyk, J., and Dabrowski, A. J., *Nucl. Instr. Meth.* **134**:505 (1976).
28 Siffert, P., *Nucl. Instr. Meth.* **150**:1 (1978).
29 Dabrowski, A. J., Iwanczyk, J., and Szymczyk, W. M., *Nucl. Instr. Meth.* **150**:25 (1978).
30 Schieber, M., Beinglass, I., Dishon, G., Holzer, A., and Yaron, G., *Nucl. Instr. Meth.* **150**:71 (1978).
31 Shalev, S., *Nucl. Instr. Meth.* **150**:79 (1978).
32 Caine, S., Holzer, A., Beinglass, I., Dishon, G., Lowenthal, E., and Schieber, M., *Nucl. Instr. Meth.* **150**:83 (1978).
33 Whited, R. C., and Schieber, M. M., *Nucl. Instr. Meth.* **162**:113 (1979).
34 "International Workshop on Mercuric Iodide and Cadmium Telluride Nuclear Detectors," *Nucl. Instr. Meth.* **150**:1–112 (1978).
35 Lyons, R. B., *Rev. Phys. Appl.* **12**:385 (1977).
36 Jones, L. T., *Rev. Phys. Appl.* **12**:379 (1977).
37 Vogel, J., Ullman, J., and Entine, G., *Rev. Phys. Appl.* **12**:375 (1977).
38 Canali, C., Gutti, E., Kozlov, S. F., Manfredi, P. F., Manfredotti, C., Nava, F., and Quirini, A., *Nucl. Instr. Meth.* **160**:73 (1979).
39 Konova, A. A., *Nucl. Instr. Meth.* **160**:115 (1979).
40 Eberhardt, J. E., Ryan, R. D., and Tavendale, A. J., *Appl. Phys. Letters* **17**:427 (1970).
41 Hesse, K., Gramann, W., and Höppner, D., *Nucl. Instr. Meth.* **101**:39 (1972).
42 Pehl, R. H., *Physics Today* **30**(11):53 (1977).

8

relative and absolute measurements

8.1 INTRODUCTION

An *absolute measurement* is one in which the exact number of particles emitted or the exact number of events taking place is determined, for example:

1 Determination of the activity of a radioactive source, i.e., measurement of the number of particles emitted by the source per second
2 Determination of the neutron flux (neutrons per square meter per second) at a certain point in a reactor
3 Measurement of the number of neutrons emitted per fission
4 Measurement of the first cross section for a nuclear interaction.

A *relative measurement* is one in which the exact number of particles emitted or the exact number of events taking place is not determined. Instead, a "relative" number of particles or events is measured, a number that has a fixed, but not necessarily known, relationship to the exact number, for example:

1 Determination of the G-M plateau. The relative change of the number of particles counted versus HV is measured. The exact number of particles emitted by the source is not determined; in fact, it is not needed.
2 Determination of half-life by counting the decaying activity of an isotope. The relative change of the number of atoms versus time is measured. The exact number of nuclei decaying per second is not needed.
3 Measurement of the fission cross section for ^{239}Pu, based on the known fission cross section for ^{235}U.
4 Determination of the variation of the neutron flux along the axis of a cylindrical reactor. The relative change of the flux from point to point along the axis of the reactor is measured, and not the exact number of neutrons per square meter per second.

Relative measurements are, in most cases, easier than absolute measurements. For this reason, investigators tend to perform the very minimum of absolute measurements and use their results in subsequent relative measurements. One of the most characteristic examples is the determination of the value of nuclear cross sections. Absolute measurements have been performed for very few cross sections. After certain cross sections have been measured, most of the others may be determined relative to the known ones.

This chapter discusses the factors that should be taken into account in performing relative and absolute measurements. Assume that there is a source of particles placed a certain distance away from a detector (Fig. 8.1) and that the detector is

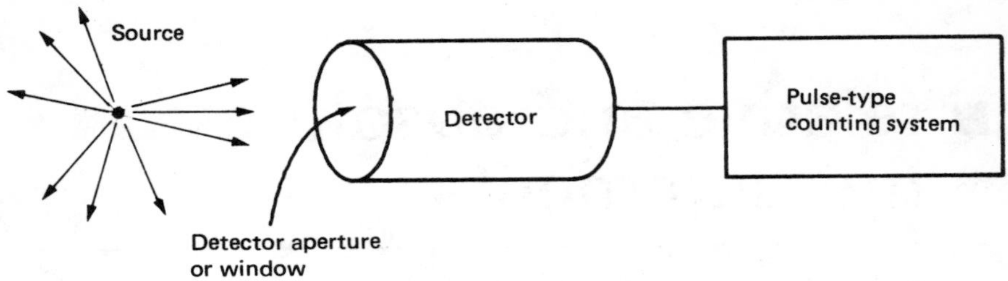

FIG 8.1 A point isotropic source counted by a pulse-type counting system.

connected to a pulse-type counting system. The source may be located outside the detector as shown in Fig. 8.1, or it may be inside the detector (e.g., liquid-scintillation counting and internal-gas counting), and may be isotropic (e.g., particles emitted with equal probability in all directions) or anisotropic (e.g., a parallel beam of particles). Both cases will be examined.

Let

S = number of particles per second emitted by the source
r = number of particles per second recorded by the scaler

It is assumed that the counting rate r has been corrected for dead time and background, if such corrections are necessary. The measured rate r is related to S by

$$r = f_1 f_2 f_3 \cdots f_n S \tag{8.1}$$

where the f factors represent the effects of the experimental setup on the measurement. These factors may be grouped into three categories, to be discussed in detail in the following sections.

1 *Geometry effects.* The term *geometry* refers to size and shape of source (point, disc, rectangular), size and shape of detector aperture (cylindrical, rectangular, etc.), and distance between source and detector.
2 *Source effects.* The size and, in particular, the way the source is made may have an effect on the measurement. Whether the source is a solid material or a thin deposit evaporated on a metal foil may make a difference. The effect of source thickness is different on charged particles, gammas, and neutrons.
3 *Detector effects.* The detector may affect the measurement in two ways. First, the size and thickness of the detector window (Fig. 8.1) determines how many particles enter the detector and how much energy they lose, as they traverse the window. Second, particles entering the detector will not necessarily be counted. The fraction of particles that is recorded depends on the efficiency of the detector (see Sec. 8.4.2).

8.2 GEOMETRY EFFECTS

The geometry may affect the measurement in two ways. First, the medium between the source and the detector may scatter and may also absorb some particles. Second,

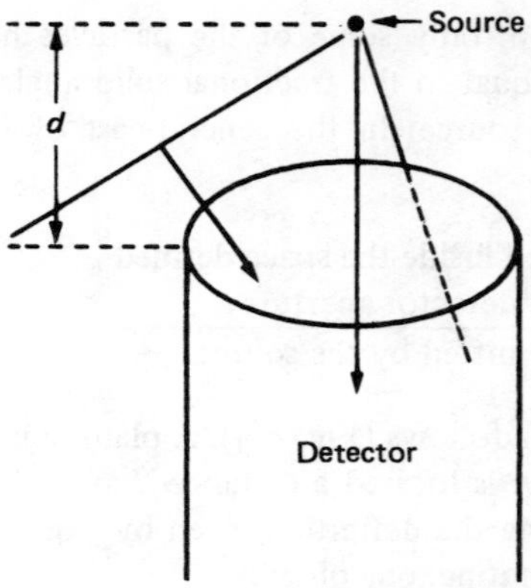

FIG. 8.2 The medium between the source and the detector may scatter and/or absorb particles emitted by the source.

the size and shape of the source and the detector and the distance between them determines what fraction of particles will enter the detector and have a chance to be counted.

8.2.1 The Effect of the Medium between Source and Detector

Consider a source and a detector separated by a distance d (Fig. 8.2). Normally, the medium between the source and detector is air, a medium of low density. For measurements of photons and neutrons, the air has no effect. If the source emits charged particles, however, all the particles suffer some energy loss, and some of them may be scattered in or out of the detector (Fig. 8.2). If this effect is important for the measurement, it can be eliminated by placing the source and the detector inside an evacuated chamber. If the use of an evacuated chamber is precluded by the conditions of the measurement, then appropriate corrections should be applied to the results.

8.2.2 The Solid Angle–General Definition

To illustrate the concept of solid angle, consider a point isotropic source at a certain distance from a detector as shown in Fig. 8.3. Since the particles are emitted by the

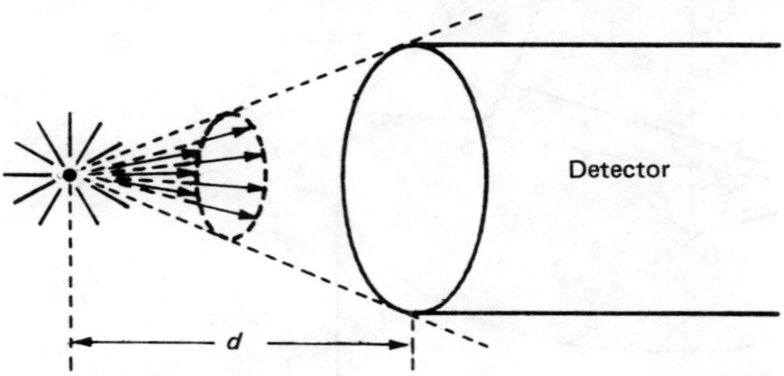

FIG. 8.3 The fraction of particles emitted by a pont isotropic source and entering the detector is defined by the solid angle subtended by the detector at the location of the source.

source with equal probability in every direction, only some of the particles have a chance to enter the detector. That portion is equal to the fractional solid angle subtended by the detector at the location of the source. In the general case of an extended source, the *solid angle* Ω is defined by

$$\Omega = \frac{\text{number of particles per second emitted inside the space defined by the contours of the source and the detector aperture}}{\text{number of particles per second emitted by the source}} \quad (8.2)$$

The mathematical expression for Ω is derived as follows (Fig. 8.4). A plane source of area A_s emitting S_0 particles/m$^2\cdot$s, isotropically, is located a distance d away from a detector with an aperture equal to A_d. Applying the definition given by Eq. 8.2 for the two differential areas dA_s and dA_d and integrating, one obtains†

$$\Omega = \frac{\int_{A_s}\int_{A_d} (S_0\, dA_s/4\pi r^2)\, dA_d(\hat{\mathbf{n}}\cdot\mathbf{r}/r)}{S_0 A_s} \quad (8.3)$$

†Equation 8.3 applies to isotropic sources. Nonisotropic sources, seldom encountered in practice, need special treatment.

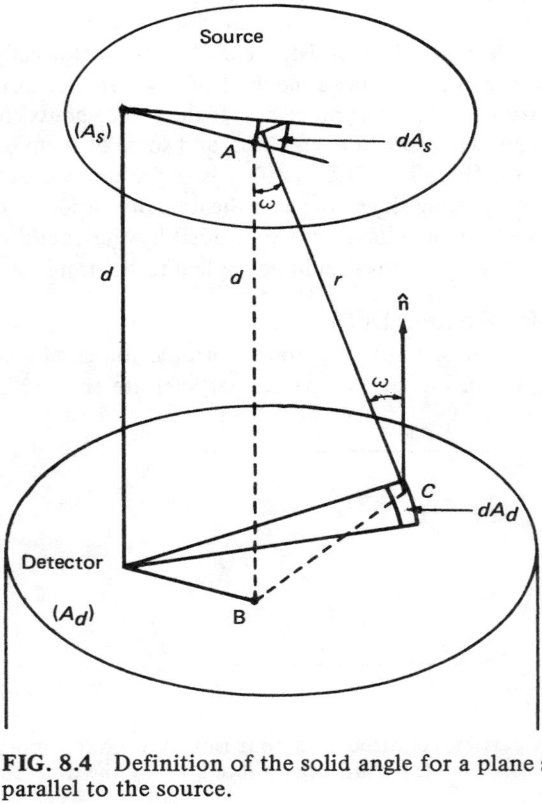

FIG. 8.4 Definition of the solid angle for a plane source and a plane detector parallel to the source.

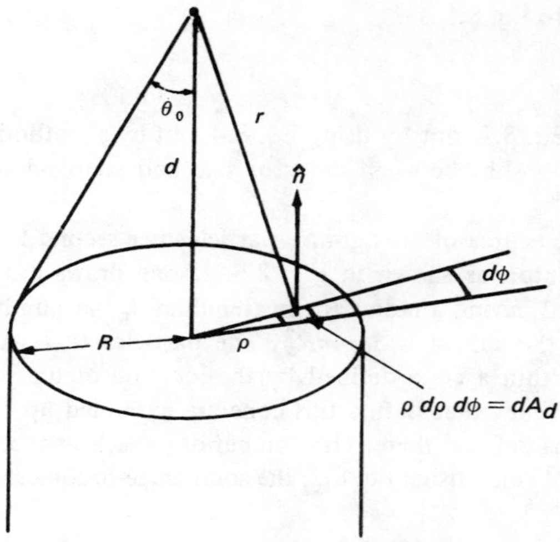

FIG. 8.5 The solid angle between a point isotropic source and a detector with a circular aperture.

where $\hat{\mathbf{n}}$ is a unit vector normal to the surface of the detector aperture. Since $\hat{\mathbf{n}}\cdot\mathbf{r}/r = \cos\omega$, Eq. 8.3 takes the form

$$\Omega = \frac{1}{4\pi A_s}\int_{A_s} dA_s \int_{A_d} dA_d \frac{\cos\omega}{r^2} \tag{8.4}$$

Equation 8.4 is valid for any shape of source and detector. In practice one deals with plane sources and detectors having regular shapes, examples of which are given in the following sections.

As stated earlier, Ω is equal to the fractional solid angle ($0 \leqslant \Omega \leqslant 1$). In radiation measurements, it is called either *solid angle* or *geometry factor*. In this text it will be called the solid angle.

8.2.3 The Solid Angle for a Point Isotropic Source and a Detector with a Circular Aperture

The most frequently encountered case of obtaining a solid angle is that of a point isotropic source at a certain distance away from a detector with a circular aperture (Fig. 8.5). In Eq. 8.4, $\cos\omega = d/r$, and the integration gives

$$\Omega = \frac{1}{2}\left(1 - \frac{d}{\sqrt{d^2 + R^2}}\right) \tag{8.5}$$

From Fig. 8.5,

$$\cos\theta_0 = \frac{d}{\sqrt{d^2 + R^2}} \tag{8.6}$$

Therefore, an equation equivalent to Eq. 8.5 is

$$\Omega = \tfrac{1}{2}(1 - \cos\theta_0) \tag{8.7}$$

It is instructive to rederive Eq. 8.7, not by using Eq. 8.4, but by a method that gives more insight into the relationship between detector size and source-detector distance.

Consider the point isotropic source of strength S_0 particles per second located a distance d away from the detector, as shown in Fig. 8.6. If one draws a sphere centered at the source position and having a radius R_s greater than d, the number of particles/m²·s on the surface of the sphere is $S_0/4\pi R_s^2$. The particles that will hit the detector are those emitted within a cone defined by the location of the source and the detector aperture. If the lines that define this cone are extended up to the surface of the sphere, an area A_s is defined there. The number of particles per second entering the detector is $A_s(S_0/4\pi R^2)$ and, using Eq. 8.2, the solid angle becomes

$$\Omega = \frac{A_s(S_0/4\pi R^2)}{S_0} = \frac{A_s}{4\pi R^2}$$

The area A_s is given by (Fig. 8.7)

$$A_s = \int dA = \int (R_s\, d\theta)(R_s \sin\theta\, d\phi) = R_s^2 \int_0^{2\pi} d\phi \int_0^{\theta_0} d\phi \sin\theta$$

$$= 2\pi R_s^2(1 - \cos\theta_0)$$

Therefore, the expression for the solid angle becomes

$$\Omega = \frac{A_s}{4\pi R_s^2} = \frac{2\pi R_s^2(1 - \cos\theta_0)}{4\pi R_s^2} = \frac{1}{2}(1 - \cos\theta_0) \tag{8.7}$$

which is, of course, Eq. 8.7.

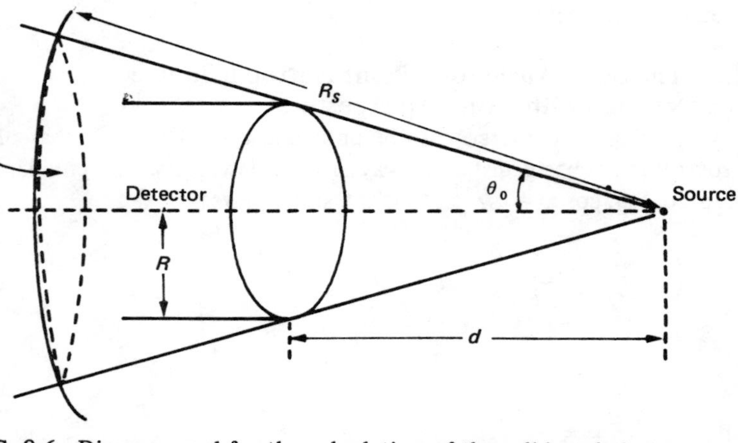

FIG. 8.6 Diagram used for the calculation of the solid angle between a point isotropic source and a detector with a circular aperture.

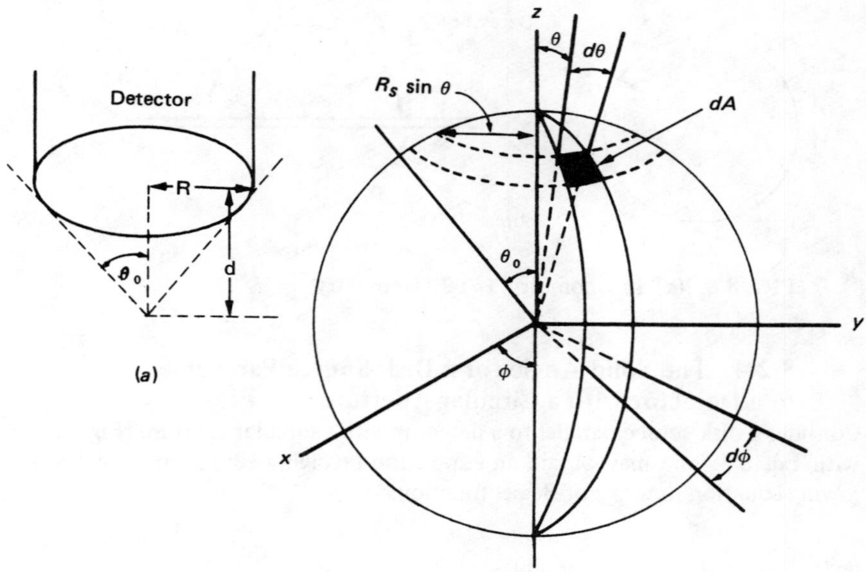

(a)

(b)

FIG. 8.7 (*a*) The detector is at distance *d* from the source. (*b*) The source is assumed to be at the center of the sphere. The cone defined by the angle θ_0 determines the area A_s (differential area dA) on the surface of the sphere.

If $R \ll d$, Eq. 8.5 takes the form (after expanding the square root (Eq. 8.6) and keeping only the first two terms)

$$\Omega = \frac{R^2}{4d^2} = \frac{\pi R^2}{4\pi d^2} = \frac{\text{detector aperture}}{4\pi d^2} \tag{8.8}$$

Equation 8.8 is valid even for a noncylindrical detector *if the source-detector distance is much larger than any of the linear dimensions of the detector aperture.*

EXAMPLE 8.1

A typical Geiger-Muller counter is a cylindrical detector with an aperture 50 mm in diameter. What is the solid angle if a point isotropic source is located 0.10 m away from the detector?

ANSWER

Using Eq. 8.5 with $d = 0.10$ m and $R = 25$ mm,

$$\Omega = \frac{1}{2}\left[1 - \frac{0.10}{\sqrt{0.10^2 + (25 \times 10^{-3})^2}}\right] = 0.015$$

If $\Omega = 1$, the setup is called a 4π geometry because the detector sees the full 4π solid angle around the source. A spherical detector represents such a case (Fig. 8.8*a*). If $\Omega = \frac{1}{2}$, the setup is called a 2π geometry. Then, half of the particles emitted by the source enter the detector (Fig. 8.8*b*).

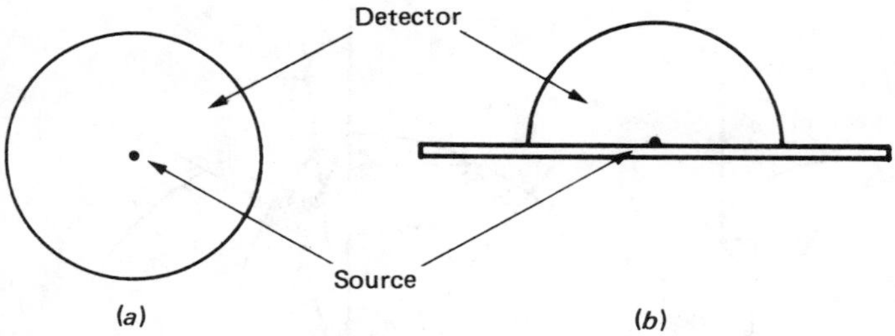

FIG. 8.8 (*a*) 4π Geometry. (*b*) 2π Geometry.

8.2.4 The Solid Angle for a Disk Source Parallel to a Detector with a Circular Aperture

Consider a disk source parallel to a detector with a circular aperture (Fig. 8.9). Starting with Eq. 8.4, one may obtain an expression involving elliptic integrals[1,2] or the following equation in terms of Bessel functions[3]:

$$\Omega = s \int_0^\infty dx\, e^{-xz} \frac{J_1(x)}{x} J_1(xs) \tag{8.9}$$

where $s = R_d/R_s$, $z = d/R_s$, and $J_1(x)$ = Bessell function of the first kind. If R_d/d and R_s/d are less than 1, the following algebraic expression is obtained for the solid angle (see Prob. 8.1):

$$\Omega = \frac{\omega^2}{4}\left\{1 - \frac{3}{4}(\psi^2 + \omega^2) + \frac{15}{8}\left(\frac{\psi^4 + \omega^4}{3} + \psi^2\omega^2\right) - \frac{35}{16}\left[\frac{\psi^6 + \omega^6}{4} + \frac{3}{2}\psi^2\omega^2(\psi^2 + \omega^2)\right]\right\} \tag{8.10}$$

where $\psi = R_s/d$
$\omega = R_d/d$

The accuracy of Eq. 8.10 increases as ψ and ω decrease. If $\psi < 0.2$ and $\omega < 0.5$, the error is less than 1 percent.

8.2.5 The Solid Angle for a Point Isotropic Source and a Detector with a Rectangular Aperture

Consider the geometry of Fig. 8.10 with a point isotropic source located a distance d away from a detector having a rectangular aperture with area equal to ab. The solid angle is given by[4]

$$\Omega = \frac{1}{4} \arctan \frac{ab}{d\sqrt{a^2 + b^2 + d^2}} \tag{8.11}$$

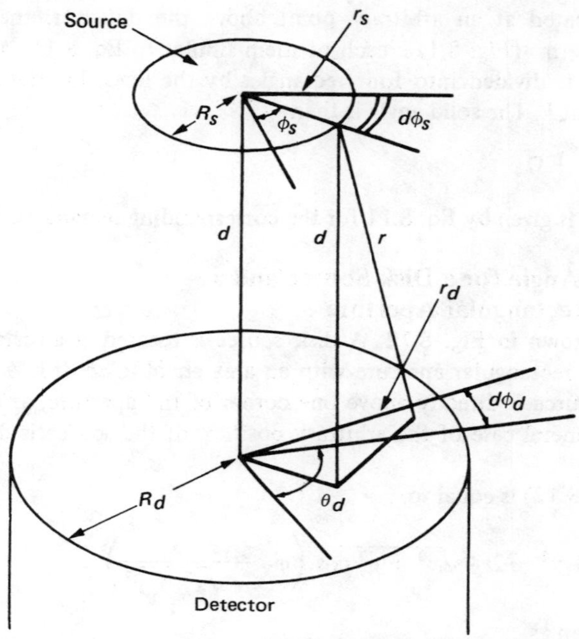

FIG. 8.9 A disk source and a detector with a circular aperture.

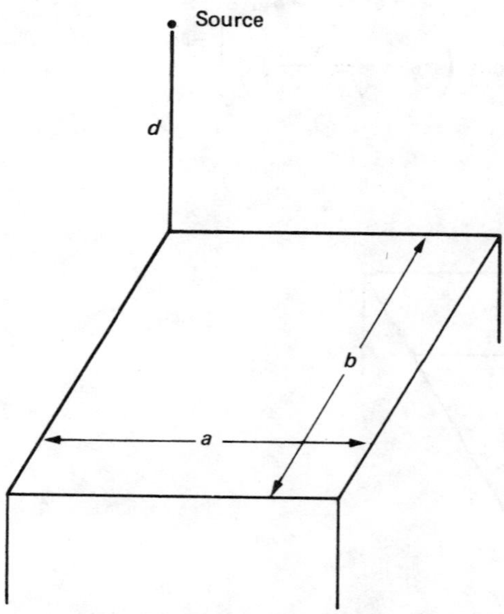

FIG. 8.10 The solid angle between a point isotropic source and a detector with a rectangular aperture. Source is located directly above one corner of the detector.

If the source is located at an arbitrary point above the detector, the solid angle is the sum of four terms (Fig. 8.11), each of them similar to Eq. 8.11. As Fig. 8.11 shows, the detector is divided into four rectangles by the lines that determine the coordinates of the point P. The solid angle is then

$$\Omega = \Omega_1 + \Omega_2 + \Omega_3 + \Omega_4$$

where Ω_i for $i = 1, \ldots, 4$ is given by Eq. 8.11 for the corresponding rectangles.

8.2.6 The Solid Angle for a Disk Source and a Detector with a Rectangular Aperture

Consider the geometry shown in Fig. 8.12. A disk source is located at a distance d above a detector having a rectangular aperture with an area equal to ab. It is assumed that the center of the source is directly above one corner of the aperture, as shown in Fig. 8.12. The more general case of the arbitrary position of the source is derived from the present example.

The distance r (Fig. 8.12) is equal to

$$r^2 = d^2 + r_S^2 + x^2 + y^2 - 2r_S\sqrt{x^2+y^2}\cos\left(\phi_S - \frac{x}{\sqrt{x^2+y^2}}\right)$$

Equation 8.4 is then written as

$$\Omega = \frac{1}{4\pi A_S d^2}\int_0^{R_S} dr_S\, r_S \int_0^{2\pi} d\phi_S \int_0^a dx \int_0^b dy \left[1 + \frac{r_S^2}{d^2} + \frac{x^2}{d^2} + \frac{y^2}{d^2} - 2\frac{r_S}{d}\sqrt{\frac{x^2+y^2}{d^2}}\cos\left(\phi_S - \frac{x}{\sqrt{x^2+y^2}}\right)\right]^{-3/2} \tag{8.12}$$

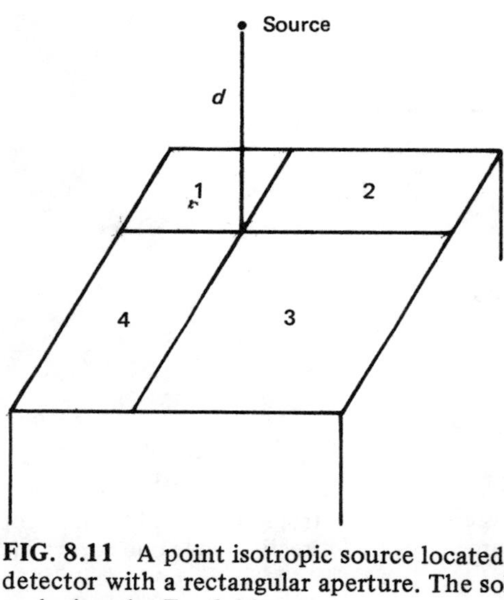

FIG. 8.11 A point isotropic source located at an arbitrary point above a detector with a rectangular aperture. The solid angle is equal to four terms, each given by Eq. 8.11.

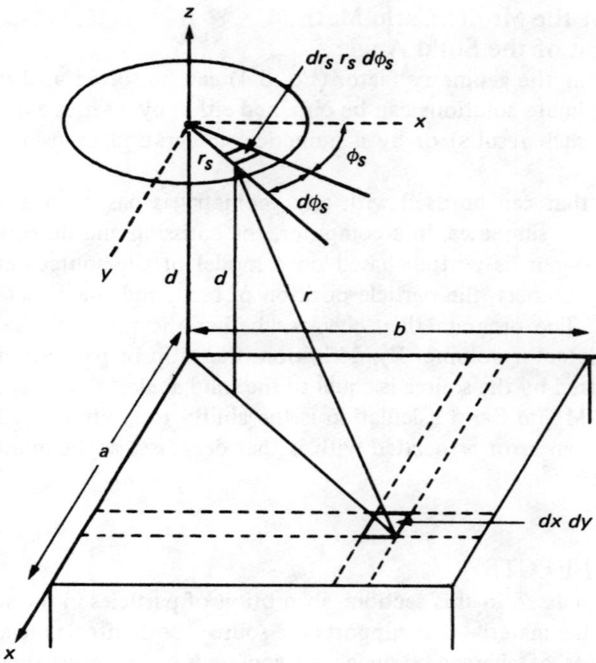

FIG. 8.12 A disk source and a detector with a rectangular aperture.

As in Sec. 8.2.4, if the ratios R_s/d, a/d, and b/d are less than 1, the expression in the bracket may be expanded in a series. If only the first four terms are kept, the result of the integration is

$$\Omega = \frac{\omega_1\omega_2}{4\pi}\left[1 - \frac{3}{4}\psi^2 - \frac{1}{2}(\omega_1^2 + \omega_2^2) + \frac{1}{8}(\psi^4 + 3\omega_1^4 + 3\omega_2^4)\right.$$
$$+ \frac{1}{3}\psi^2(\omega_1^2 + \omega_2^2) - \frac{35}{16}\psi^6 + \frac{5}{12}\omega_1^2\omega_2^2 - \frac{35}{32}\psi^4(\omega_1^2 + \omega_2^2)$$
$$- \frac{7}{48}\psi^2(9\omega_1^4 + 9\omega_2^4 + 10\omega_1^2\omega_2^2) - \frac{7}{16}\omega_1^2\omega_2^2(\omega_1^2 + \omega_2^2)$$
$$\left. + \frac{5}{16}(\omega_1^6 + \omega_2^6)\right] \tag{8.13}$$

where $\omega_1 = a/d$
$\omega_2 = b/d$
$\psi = R_s/d$

If the source is located at an arbitrary point above the detector, the solid angle is the sum of four terms as shown in Fig. 8.11.

8.2.7 The Use of the Monte Carlo Method for the Calculation of the Solid Angle

The basic equation defining the geometry factor (Eq. 8.4) can be solved analytically in very few cases. Approximate solutions can be obtained either by a series expansion (Eqs. 8.10 and 8.13 are such results) or by a numerical integration or using other approximations.[1,5,6]

A general method that can be used with any geometry is based on a Monte Carlo calculation,[7–10] which simulates, in a computer, the emission and detection of particles. A computer program is written based on a model of the source-detector geometry. Using random numbers, the particle position of birth and the direction of emission are determined. The program then checks whether the randomly selected direction intersects the detector volume. By definition, the ratio of particles hitting the detector to those emitted by the source is equal to the solid angle.

The advantage of a Monte Carlo calculation is the ability to study complicated geometries. The result has an error associated with it that decreases as the number of particles studied increases.

8.3 SOURCE EFFECTS

Two source effects are discussed in this section: absorption of particles in the source, and the effect of the backing material that supports the source. Both effects are always important in measurements of charged particles. In some cases, however, they may also be significant in x-ray or thermal-neutron measurements.

8.3.1 Source Self-Absorption Factor (f_a)

Radioactive substances are deposited on a backing material in thin deposits. But no matter how thin, the deposit has a finite thickness and may cause absorption of some particles emitted by the source. Consider the source of thickness t shown in Fig. 8.13. Particle 1 traverses the source deposit and enters the detector. Particle 2 is absorbed inside the source so that it will not be counted. Therefore, source self-absorption will produce a decrease of the counting rate r.

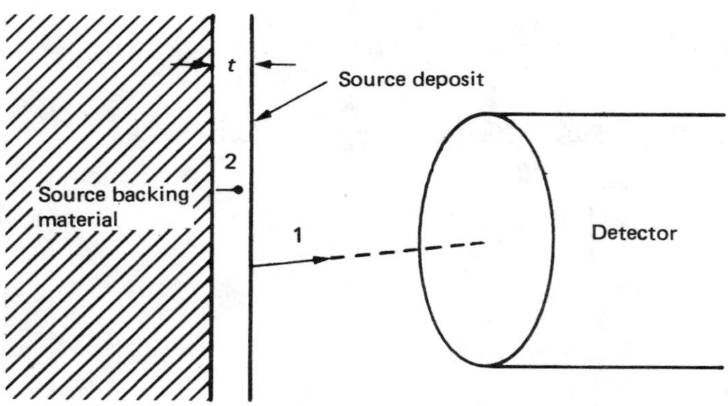

FIG. 8.13 Source self-absorption. Particles may be absorbed in the source deposit.

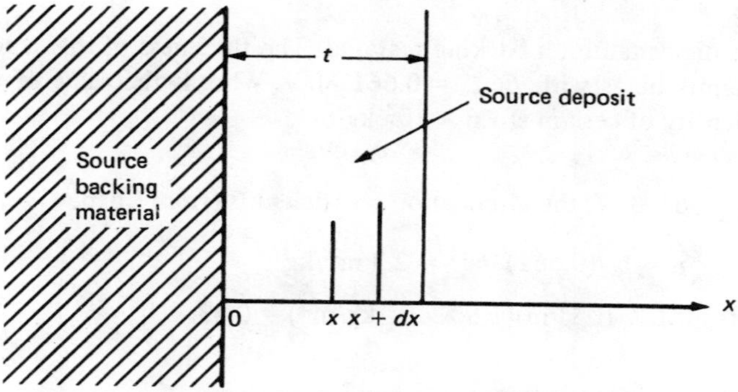

FIG. 8.14 Diagram used for the calculation of the source self-absorption factor for betas.

Source self-absorption may be reduced to an insignificant amount but it cannot be eliminated completely. It is always important for charged particles and generally more crucial for heavier particles (p, α, d, heavy ions) than for electrons.

Source self-absorption, in addition to altering the number of particles leaving the source, may also change the energy of the particles escaping from it. Particle 1 in Fig. 8.13 successfully leaves the deposit, but it loses some energy as it goes through the deposit. This energy loss is important when the energy of the particle is measured.

An approximate correction for self-absorption can be calculated if the source emits particles following a known attenuation law. As an example, consider a source with thickness t (Fig. 8.14) that has a uniform deposit of a radioisotope emitting β particles. Assume that the source gives S betas per second in the direction of the positive x axis. If self-absorption is absent, S betas per second will leave the source (towards positive x). Because of the source thickness, betas produces *in* dx, around x, have to successfully penetrate the thickness $(t-x)$ to escape. The probability of escape is $e^{-\mu(t-x)}$, where μ is the attenuation coefficient for the betas in the material of which the deposit is made. The total number of betas escaping is

$$\int_0^t \frac{dx}{t} S e^{-\mu(t-x)} = \frac{S}{t\mu}(1-e^{-\mu t})$$

A self-absorption factor f_a is defined by

$$f_a = \frac{\text{number of particles leaving source with self-absorption}}{\text{number of particles leaving source without self-absorption}}$$

Using the result obtained above,

$$f_a = \frac{(S/t\mu)(1-e^{-\mu t})}{S} = \frac{1}{t\mu}(1-e^{-\mu t}) \qquad 0 \leqslant f_a \leqslant 1 \tag{8.14}$$

EXAMPLE 8.2

Assume that ^{137}Cs is deposited on a backing material. The thickness of the deposit is $t = 0.1$ mm. ^{137}Cs emits betas with $E_{max} = 0.661$ MeV. What is the value of f_a for such a source? The density of cesium is 1.6×10^3 kg/m^3.

ANSWER

For betas of $E_{max} = 0.661$ MeV, the attenuation coefficient is (from Chap. 4)

$$\mu = 1.7E_{max}^{-1.14} \qquad \mu = 1.7(0.661)^{-1.14} = 2.7 \text{ m}^2/\text{kg}$$

$$\mu t = (2.7 \text{ m}^2/\text{kg})(0.1 \times 10^{-3} \text{ m})(1.6 \times 10^3 \text{ kg/m}^3) = 0.43$$

Using Eq. 8.14,

$$f_a = \frac{1}{0.43}(1 - e^{-0.43}) = 0.81$$

Therefore, only 81 percent of the betas escape this source. Or, if this effect is not taken into account, the source strength will be underestimated by 19 percent.†

If the source emits monoenergetic charged particles, essentially all the particles leave the source deposit as long as $t < R$, where R = range of the particles. In practice, the sources for monoenergetic charged particles are such that $t \ll R$, in which case, $f_a \approx 1$. Then the only effect of the source deposit is an energy loss for the particles that traverse it (see also Chap. 13).

8.3.2 Source Backscattering Factor (f_b)

A source cannot be placed in midair. It is always deposited on a material that is called *source backing* or *source support*. The source backing is usually a very thin material, but no matter how thin, it may backscatter particles emitted in a direction away from the detector (Fig. 8.15). To understand the effect of backscattering, assume that the

†A similar calculation of f_a may be repeated for an x-ray or a neutron source. For x-rays the probability of escape is $e^{-\mu t}$; for neutrons it is $e^{-\Sigma t}$.

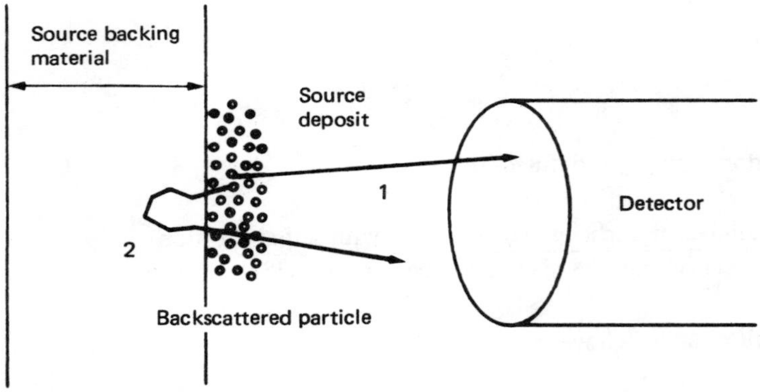

FIG. 8.15 The source backing material backscatters particles and necessitates the use of a backscattering factor f_b.

solid angle in Fig. 8.12 is $\Omega = 10^{-2}$. Also assume that all the particles entering the detector are counted, self-absorption is zero, and there is no other medium that might absorb or scatter the particles except the source backing.

Particle 1 in Fig. 8.15 is emitted towards the detector. Particle 2 is emitted in the opposite direction. Without the source backing, particle 2 would not turn back. With the backing material present, there is a possibility that particle 2 will have scattering interactions there, have its direction of motion changed, and enter the detector. If the counting rate is $r = 100$ counts per minute and there is no backscattering of particles toward the detector, the strength of the source will be correctly determined as

$$S = \frac{r}{\Omega} = \frac{105}{10^{-2}} = 10{,}000 \text{ counts/min}$$

If the source backing backscatters 5 percent of the particles, the counting rate will become 105 counts per minute, even though it is still the same source as before. If source backscattering is not taken into account, the source strength will be erroneously determined as

$$S = \frac{r}{\Omega} = \frac{105}{10^{-2}} = 10{,}500 \text{ counts/min}$$

To correct properly for this effect, a source backscattering factor (f_b) is defined by

$$f_b = \frac{\text{number of particles counted with source backing}}{\text{number of particles counted without source backing}} \tag{8.15}$$

From the definition it is obvious that

$$2 > f_b \geqslant 1$$

In the example discussed above, $f_b = 1.05$ and the correct strength of the source is

$$S = \frac{r}{\Omega f_b} = \frac{105}{10^{-2} \times 1.05} = 10{,}000 \text{ counts/min}$$

The backscattering factor is important, in most cases, only for charged particles. It depends on three variables:

1 Thickness (b) of the backing material
2 Particle kinetic energy (T)
3 Atomic number of the backing material (Z)

The dependence of f_b on thickness b is shown in Fig. 8.16. As $b \to 0$, $f_b \to 1$, which should be expected. For large thicknesses, f_b reaches a saturation value, which should also be expected. Since charged particles have a definite range, there is a maximum distance they can travel in the backing material, be backscattered, and

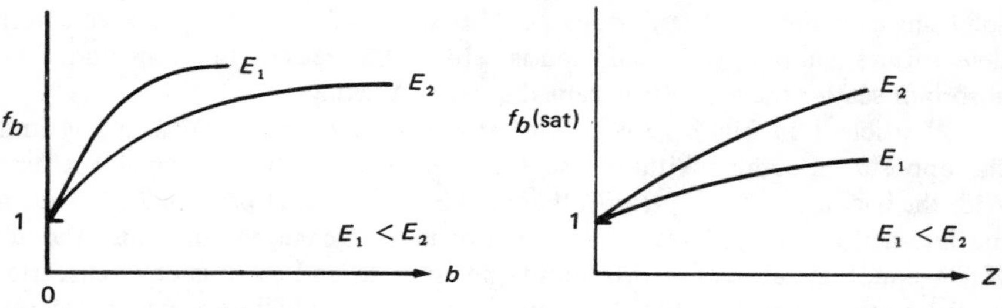

(a) (b)

FIG. 8.16 (*a*) The backscattering factor f_b as a function of thickness b of the backing material. (*b*) The saturation backscattering factor as a function of the atomic number Z of the backing material.

traverse the material again in the opposite direction. Therefore an upper limit for that thickness is $b = R/2$, where R is the range of the particles. Experiments have shown that

$$b_s = b\ (\text{saturation}) \approx 0.2R$$

The dependence of the saturation backscattering factor of electrons on the kinetic energy and the atomic number of the backing material is given by the following empirical equation,[11,12] based on a least-squares fit of experimental results.

$$f_b(\text{sat}) = 1 + \frac{b_1 \exp(-b_2 Z^{-b_3})}{1 + (b_4 + b_5 Z^{-b_6})\gamma^{(b_7 - b_8/Z)}} \tag{8.16}$$

where the constants b_i for $i = 1, 2, \ldots, 8$ have these values:

$b_1 = 1.15 \pm 0.06$ $b_5 = 15.7 \pm 3.1$ $\alpha = \dfrac{T}{mc^2}$

$b_2 = 8.35 \pm 0.25$ $b_6 = 1.59 \pm 0.07$

$b_3 = 0.525 \pm 0.02$ $b_7 = 0.56 \pm 0.02$

$b_4 = 0.0185 \pm 0.0019$ $b_8 = 4.42 \pm 0.18$

Figure 8.17 shows the change of f_b(sat) versus kinetic energy T for four elements.

A backscattering correction should be applied to alpha counting in 2π counters (Fig. 8.8). It has been determined[13-15] that the number of backscattered alphas is between 0 and 5 percent, depending on the energy of the alphas, the uniformity of the source, and the atomic number of the material forming the base of the counter.

Correction for source backscattering is accomplished in two ways.

1 An extremely thin backing material is used for which $f_b \approx 1$. In general, a low-Z material is used, e.g., plastic, if possible.
2 A thick backing material is used, for which the saturation backscattering factor should be employed for correction of the data.

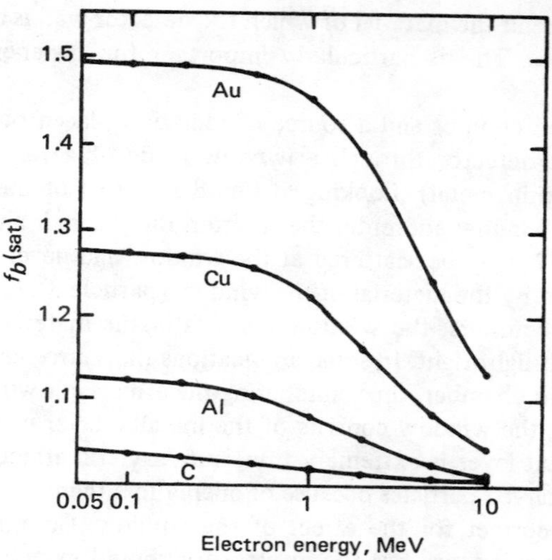

FIG. 8.17 Backscattering factor as a function of energy for C, Al, Cu, and Au. Curves were obtained using Eq. 8.16.

For accurate results, the backscattering factor should be measured for the actual geometry of the experiment.

8.4 DETECTOR EFFECTS

The detector may affect the measurement in two ways. First, if the source is located outside the detector (which is usually the case), the particles may be scattered or absorbed by the detector window. Second, some particles may enter the detector and not produce a signal, or they may produce a signal lower than the discriminator threshold.

8.4.1 Scattering and Absorption Due to the Window of the Detector

In most measurements the source is located outside the detector (Fig. 8.18). The radiation must penetrate the walls of the counter and enter it in order to be counted.

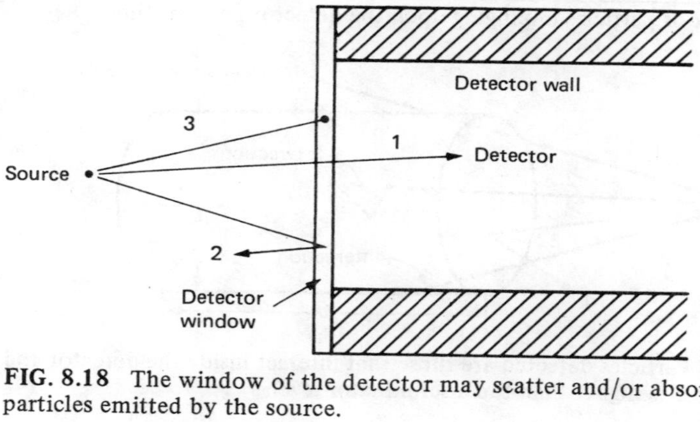

FIG. 8.18 The window of the detector may scatter and/or absorb some of the particles emitted by the source.

Interactions between the radiation and the material of which the detector wall is made may scatter and/or absorb particles. This is particularly important for low-energy β particles.

Figure 8.18 shows a gas-filled counter and a source of radiation placed outside it. Usually the particles enter the detector through a *window* made of a very thin material (such as glass, mica, or thin metal). Looking at Fig. 8.18, most of the particles, like particle 1, traverse the window and enter the counter. But, there is a possibility that a particle, like particle 2, may be scattered at the window and never enter the counter. Or, it may be absorbed by the material of the window (particle 3).

In the case of scintillation counters, the window consists of the material that covers the scintillator and makes it light-tight. In some applications the source and the scintillator are placed in a light-tight chamber, thus eliminating the effects of a window.

In semiconductor detectors, the window consists of the metallic layer covering the front face of the detector. That layer is extremely thin, but may still affect measurements of alphas and heavier charged particles because of energy loss there.

There is no direct way to correct for the effect of the window. Commercial detectors are made with very thin windows, but the investigator should examine the importance of the window effect for the particular measurement performed. If there is a need for an energy-loss correction, it is applied separately to the energy spectrum. If, however, there is a need to correct for the number of particles stopped by the window, that correction is incorporated into the detector efficiency.

8.4.2 Detector Efficiency (ϵ)

It is not certain that a particle will be counted when it enters a detector. It may, depending on the type and energy of the particle and type and size of detector, go through without having an interaction (particle 1 in Fig. 8.19); it may produce a signal so small it is impossible to record with the available electronic instruments (particle 3); or, it may be prevented from entering the detector by the window (particle 4). In Fig. 8.19, the particle with the best chance of being detected is particle 2.

The quantity that gives the fraction of particles being detected is called the *detector efficiency* ϵ, given by

$$\epsilon = \frac{\text{number of particles recorded per unit time}}{\text{number of particles impinging upon the detector per unit time}} \tag{8.17}$$

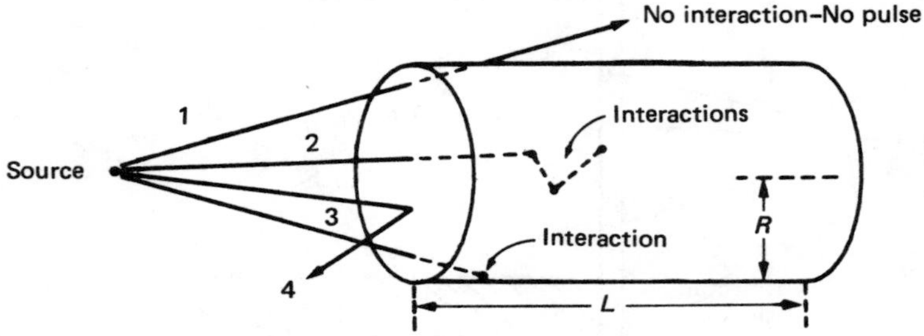

FIG. 8.19 Particles detected are those that interact inside the detector and produce a pulse higher than the discriminator level.

The detector efficiency depends upon†

1 Density and size of detector material
2 Type and energy of radiation
3 Electronics

EFFECT OF DENSITY AND SIZE OF DETECTOR MATERIAL

The efficiency of a detector will increase if the probability of an interaction between the incident radiation and the material of which the detector is made increases. That probability increases with detector size. But larger size is of limited usefulness because the background increases proportionally with the size of the detector, and because in some cases it is practically impossible to make large detectors. (Semiconductor detectors are a prime example.)

The probability of interaction per unit distance traveled is proportional to the density of the material. The density of solids and liquids is about a thousand times greater than the density of gases at normal pressure and temperature. Therefore, other things being equal, detectors made of solid or liquid material are more efficient than those using gas.

EFFECT OF TYPE AND ENERGY OF RADIATION

Charged particles moving through matter will always have Coulomb interactions with the electrons and nuclei of that medium. Since the probability of interaction is almost a certainty, the efficiency for charged particles will be close to 100 percent. Indeed, detectors for charged particles have an efficiency that is practically 100 percent, regardless of their size or the density of the material of which they are made. For charged particles, the detector efficiency is practically independent of particle energy except for very low energies, when the particles may be stopped by the detector window.

Charged particles have a definite range. Therefore, it is possible to make a detector with a length L such that all the particles will stop and deposit their energy in the counter. Obviously, the length L should be greater than R, where R is the range of the particles in the material of which the detector is made.

Photons and neutrons traversing a medium show an exponential attenuation (see Chap. 4), which means that there is always a nonzero probability for a photon or a neutron to traverse any thickness of material without an interaction. As a result of this property, detectors for photons on neutrons have efficiency less than 100 percent regardless of detector size and energy of the particle.

EFFECT OF ELECTRONICS

The electronics of a detector affects the counter efficiency indirectly. If a particle interacts in the detector and produces a signal, that particle will be recorded only if the signal is recorded. The signal will be registered if it is higher than the discriminator level, which is, of course, determined by the electronic noise of the counting system. Thus, the counting efficiency may increase if the level of electronic noise is decreased.

†In gamma spectroscopy, several other efficiencies are being used in addition to this one (see Chap. 12).

As an example, consider a counting system with electronic noise such that the discriminator level is at 1 mV. In this case, only pulses higher than 1 mV will be counted; therefore, particles which produce pulses lower than 1 mV will not be recorded. Assume next that the preamplifier or the amplifier or both are replaced by quieter ones, and the new noise level is such that the discriminator level can be set at 0.8 mV. Now, pulses as low as 0.8 mV will be registered, more particles will be recorded, and hence the efficiency of the counting system increases.

If electronics is included in the discussion, it is the efficiency of the *system* (detector plus electronics) that is considered rather than the efficiency of the counter.

8.4.3 Determination of Detector Efficiency

The efficiency of a detector can be determined either by measurement or by calculation. Many methods have been used for the measurement of detection efficiency,[16–18] but the simplest and probably the most accurate is the method of using a calibrated source, i.e., a source of known strength. In Fig. 8.15, assume that the source is a monoenergetic point isotropic source emitting S particles per second. If the true net counting rate is r counts per second, the solid angle is Ω, and the efficiency is ϵ, the equation giving the efficiency is

$$\epsilon = \frac{r}{\Omega F S} \tag{8.18}$$

where $F = f_a f_b \cdots$ is a combination of all the correction factors that may have to be applied to the results.

Accurate absolute measurements rely on measured rather than calculated efficiencies. Nevertheless, an efficiency calculation is instructive because it brings forward the parameters which are important for this concept. For this reason, two cases of efficiency calculation for a photon detector are presented below.

Consider first a parallel beam of photons of energy E impinging upon a detector of thickness L (Fig. 8.20). The probability that a photon will have at least one interaction in the detector is $1 - e^{-\mu(E)L}$, where $\mu(E)$ is the total linear attenuation coefficient of photons with energy E in the material of which the detector is made. If one interaction is enough to produce a detectable pulse, the efficiency is

$$\epsilon(E) = 1 - e^{-\mu(E)L} \tag{8.19}$$

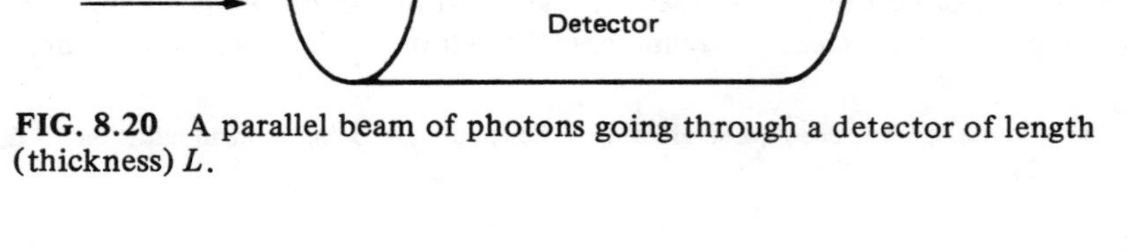

FIG. 8.20 A parallel beam of photons going through a detector of length (thickness) L.

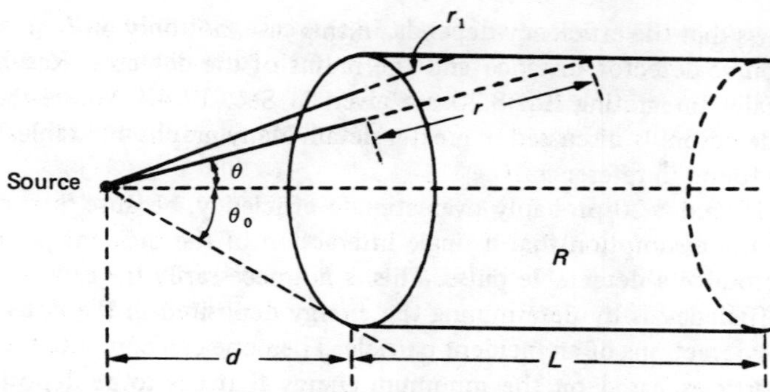

FIG. 8.21 A point isotropic photon source at a distance d away from a cylindrical detector.

Equation 8.19 shows the dependence of $\epsilon(E)$ on

1 The size L of the detector
2 The photon energy (through μ)
3 The density of the material (through μ)

EXAMPLE 8.3

What is the efficiency of a 50-mm long NaI(Tl) crystal for a parallel beam of (a) 2-MeV gammas or (b) 0.5-MeV gammas?

ANSWER

(a) From the table in App. D, the total mass attenuation coefficient for 2-MeV gammas in NaI(Tl) is $\mu = 0.00412\ \text{m}^2/\text{kg}$. The density of the scintillator is $3.67 \times 10^3\ \text{kg/m}^3$. Therefore, Eq. 8.19 gives

$$\begin{aligned}\epsilon &= 1 - \exp\,[-0.00412\ \text{kg/m}^3(3.67 \times 10^3\ \text{kg/m}^3)0.05\ \text{m}] = 1 - \exp\,(-0.756)\\ &= 0.53 = 53\%\end{aligned}$$

(b) For 0.50-MeV gammas, $\mu = 0.0091\ \text{m}^2/\text{kg}$. Therefore,

$$\begin{aligned}\epsilon &= 1 - \exp\,[-0.0091\ \text{m}^2/\text{kg}(3.67 \times 10^3\ \text{kg/m}^3)0.05\ \text{m}] = 1 - \exp\,(-1.67)\\ &= 0.081 = 81\%\end{aligned}$$

The next case to consider is that of a point isotropic monoenergetic source, at a distance d away from a cylindrical detector of length L (see Fig. 8.21). For photons emitted at an angle θ, measured from the axis of the detector, the probability of interaction is $1 - \exp\,[-\mu(E)r(\theta)]$ and the probability of emission between angles θ and $\theta + d\theta$ is $\frac{1}{2}\sin\theta\, d\theta$. Assuming, as before, that one interaction is enough to produce a detectable pulse, the efficiency is given by

$$\epsilon(E) = \frac{\displaystyle\int_0^{\theta_0} S\{1 - \exp\,[-\mu(E)r(\theta)]\}\tfrac{1}{2}\sin\theta\, d\theta}{\displaystyle(S/2)\int_0^{\theta_0} \sin\theta\, d\theta} \tag{8.20}$$

Equation 8.20 shows that the efficiency depends, in this case, not only on L, μ, and E, but also on the source-detector distance and the radius of the detector. Results obtained by numerically integrating Eq. 8.20 are given in Sec. 12.4.1, where the efficiency of gamma detectors is discussed in greater detail. Many graphs and tables based on Eq. 8.20 can be found in reference 20.

Equations 8.19 and 8.20 probably overestimate efficiency, because their derivation was based on the assumption that a single interaction of the incident photon in the detector will produce a detectable pulse. This is not necessarily the case. A better way to calculate efficiency is by determining the energy deposited in the detector as a result of all the interactions of an incident particle. Then one can compute the number of recorded particles based on the minimum energy that has to be deposited in the detector in order that a pulse higher than the discriminator level may be produced. The Monte Carlo method, which is ideal for such calculations, has been used by many investigators[10,21] for that purpose.

Efficiencies of neutron detectors are calculated by methods similar to those used for gammas. Neutrons are detected indirectly through gammas or charged particles produced by reactions of nuclei with neutrons. Thus, the neutron detector efficiency is essentially the product of the probability of a neutron interaction, with the probability to detect the products of that interaction (see Chap. 14).

8.5 RELATIONSHIP BETWEEN COUNTING RATE AND SOURCE STRENGTH

Equation 8.18 rewritten in terms of the true net counting rate r gives the relationship between r and the source strength:

$$r = \Omega F \epsilon S \tag{8.21}$$

In terms of gross counts G obtained over time t_G and background count B obtained over time t_B, the true net counting rate is Eq. 2.113)

$$r = \frac{G/t_G}{1-(G/t_G)\tau} - \frac{B}{t_B} \tag{8.22}$$

where τ is the counter dead time. Usually the objective of the measurement is to obtain the source strength S using a detection system of known Ω, F, and ϵ. Combining Eqs. 8.21 and 8.22, the source strength becomes

$$S = \frac{r}{\Omega F \epsilon} = \frac{1}{\Omega F \epsilon}\left[\frac{G/t_G}{1-(G/t_G)\tau} - \frac{B}{t_B}\right] \tag{8.23}$$

The error of S is due to errors in the values of Ω, F, ϵ, and the statistical error of r. In many cases encountered in practice, the predominant error is that of r. Then one obtains, from Eq. 8.23,

$$\frac{\sigma_S}{S} = \frac{\sigma_r}{r}$$

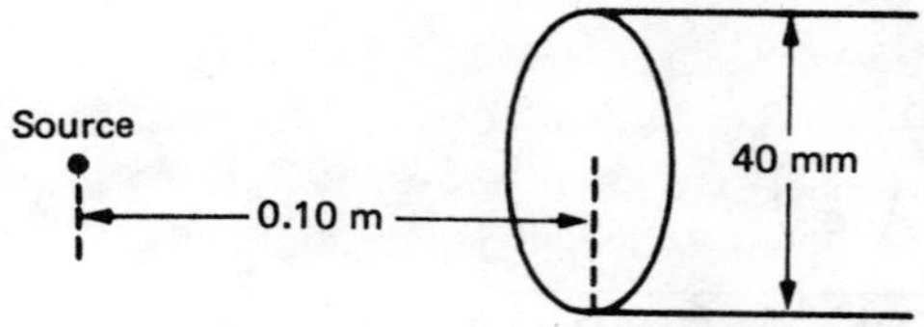

FIG. 8.22 Geometry assumed in Example 8.4.

That is, the percent error of S is equal to the percent error of the true net counting rate r.

EXAMPLE 8.4
The geometric setup shown in Fig. 8.22 was used for the measurement of the strength of the radioactive source. The following data were obtained.

$G = 6000 \quad B = 400 \quad \tau = 100\ \mu s \quad \epsilon = 0.60 \pm 0.005$

$t_G = 10\text{ min} \quad t_B = 10\text{ min} \quad F = 1 \pm 0.001$

What is the strength S and its standard error?

ANSWER
The true net counting rate r is

$$r = \frac{6000/10}{1-(6000 \times 100 \times 10^{-6}/10 \times 60)} - \frac{400}{10} = \frac{6000/10}{.999} - 40$$
$$= 600.6 - 40 \approx 540 \text{ counts/min}$$

The standard error of r is (Eq. 2.109)

$$\sigma_r = \sqrt{\left[\frac{1}{1-(G_\tau/t_G)}\right]^4 \frac{G}{t_G^2} + \frac{B}{t_B^2}} = \sqrt{\left(\frac{1}{.999}\right)^5 \frac{6000}{100} + \frac{400}{100}} = 8 \text{ counts/min}$$

Therefore

$$\frac{\sigma_r}{r} = \frac{8}{540} = 1.5\%$$

The solid angle is

$$\Omega = \frac{1}{2}\left(1 - \frac{d}{\sqrt{d^2 + r^2}}\right) = 0.0097$$

Using Eq. 8.23,

$$S = \frac{r}{\Omega F \epsilon} = \frac{540}{(0.0097)(0.60)} = 92{,}783 \text{ counts/min}$$

The standard error of S is (Sec. 2.15.1)

$$\frac{\sigma S}{S} = \sqrt{\left(\frac{\sigma r}{r}\right)^2 + \left(\frac{\sigma F}{F}\right)^2 + \left(\frac{\sigma \epsilon}{\epsilon}\right)^2}$$

$$= \sqrt{(1.5 \times 10^{-2})^2 + \left(\frac{0.001}{1}\right)^2 + \left(\frac{0.005}{0.60}\right)^2} = 1.7 \times 10^{-2} = 5.7\%$$

PROBLEMS

8.1 Show that if $R_a/d < 1$ and $R_s/d < 1$, the solid angle between two parallel disks with radii R_a and R_s a distance d apart is given to a good approximation by

$$\Omega = \frac{\omega^2}{4}\left\{1 - \frac{3}{4}(\psi^2 + \omega^2) + \frac{15}{8}\left(\frac{\psi^4 + \omega^4}{3} + \psi^2\omega^2\right) - \frac{35}{16}\left[\frac{\psi^6 + \omega^6}{4} + \frac{3}{2}\psi^2\omega^2(\psi^2 + \omega^2)\right]\right\}$$

where $\psi = R_s/d$ and $\omega = R_a/d$.

8.2 Show that an approximate expression for the solid angle between two nonparallel disks is

$$\Omega = \frac{\omega^2}{4}\left\{1 - \frac{3}{4}[(\psi^2 + \omega^2(1 + \sin^2\theta)] + \frac{15}{8}\left[\frac{\omega^4 + \psi^4}{3} + \psi^2\omega^2 + \omega^2\left(1 + \frac{1}{4}\omega^2 + \frac{2}{3}\psi^2\right)\sin^2\theta\right]\right\}$$

where θ is the angle between the planes of the two discs, and ψ and ω are defined as in Prob. 8.1.

8.3 Show that the solid angle between a disk source and a detector with a rectangular aperture is given, approximately, by Eq. 8.13 under the conditions given in Sec. 8.2.6.

8.4 A 1-mCi point isotropic gamma source is located 0.10 m away from a 60° spherical shell of a NaI detector, as shown in the figure below. Assuming that all the pulses at the output of the photomultiplier tube are counted, what is the counting rate of the scaler? The gamma energy is 1.25 MeV.

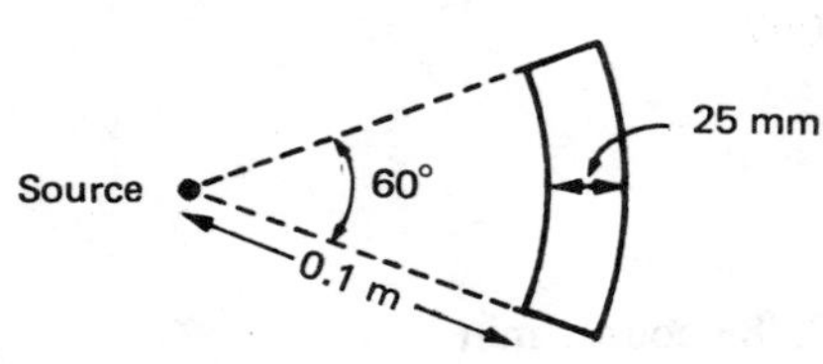

8.5 Calculate the counting rate in particles/s for the case shown in the figure below. The source has the shape of a ring and emits 10^6 particles/s isotopically. The background is zero. The detector efficiency is 80 percent, and $F = 1$.

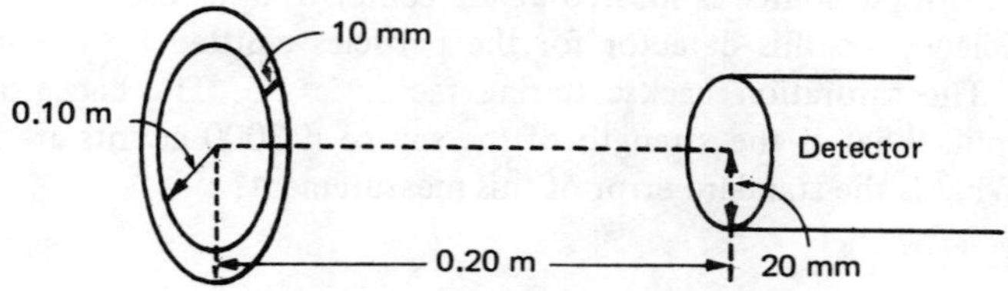

8.6 Calculate the self-absorption factor for a ^{14}C source which has a thickness of 10 $\mu g/cm^2$ (10^{-5} kg/m^2); E_{max} = 156 keV.

8.7 An attempt was made to measure the backscattering factor by placing foils of continuously increasing thickness behind the source and observing the change in the counting rate. The foils were of the same material as the source backing. The results of the measurements are given in the table below. Calculate the saturation backscattering factor and the source backscattering factor.

Thickness behind source (mm)	Counting rate (counts/min)
0.1 (Source backing only)	3015
0.15	3155
0.2	3365
0.25	3400
0.3	3420
0.35	3430
0.4	3430

8.8 What is the counting rate in a detector with a rectangular aperture measuring 1 mm × 40 mm, if a 1-mCi gamma-ray point isotropic source is 0.10 m away? The efficiency of the detector for these gammas is 65 percent.

8.9 A radioactive source emits electrons isotropically at the rate of 10^4 electrons/s. A plastic scintillator having the shape of a cylindrical disk with a 25-mm radius is located 120 mm away from the source. The efficiency of the detector for these electrons is 95 percent. The backscattering factor is 1.02, and the source self-absorption factor is 0.98. Dead time of the counting system is 5 μs. How long should one count, under these conditions, to obtain the strength of the source with a standard error of 5 percent? Background is negligible. The only error involved is that due to counting statistics.

8.10 How would the result of Prob. 8.9 change if the backscattering factor was known with an error of ±1 percent, the efficiency with an error of ±0.5 percent, and the source self-absorption factor with an error of ±1 percent?

8.11 Calculate the strength of a point isotropic radioactive source if it is given that the gross counting rate is 200 counts/min, the background counting rate is 25 counts/min, the counter efficiency is 0.90, the source detector distance 0.15 m, and the detector aperture has a radius of 20 mm ($F = 1$). What is the standard error of the result if the error of the gross counting rate is known with an accuracy of ±5 percent and the background with ±3 percent? Dead time is 1 μs.

8.12 A point isotropic source is located at the center of a hemispherical 2π counter. The efficiency of this detector for the particles emitted by the source is 85 percent. The saturation backscattering factor is 1.5. The background 25 ± 1 counts/min. What is the strength of the source if 3000 counts are recorded in 1 min? What is the standard error of this measurement?

REFERENCES

1 Masket, A. V., *Rev. Sci. Instr.* **28**:191 (1957).
2 Ruffle, M. P., *Nucl. Instr. Meth.* **52**:354 (1967).
3 Ruby, L., and Rechen, J. B., *Nucl. Instr. Meth.* **58**:345 (1968).
4 Gotoh, H., and Yagi, H., *Nucl. Instr. Meth.* **96**:485 (1971).
5 Gardner, R. P., and Verghese, K., *Nucl. Instr. Meth.* **93**:163 (1971).
6 Verghese, K., Gardner, R. P., and Felder, R. M., *Nucl. Instr. Meth.* **101**:391 (1972).
7 Williams, I. R., *Nucl. Instr. Meth.* **44**:160 (1966).
8 Green, M. V., Aamodt, R. L., and Johnston, G. S., *Nucl. Instr. Meth.* **117**:409 (1974).
9 Wielopolski, L., *Nucl. Instr. Meth.* **143**:577 (1977).
10 Beam, G. B., Wielopolski, L., Gardner, R. P., and Verghese, K., *Nucl. Instr. Meth.* **154**:501 (1978).
11 Tabata, T., Ito, R., and Okabe, S., *Nucl. Instr. Meth.* **94**:509 (1971).
12 Kuzminikh, V. A., and Vorobiev, S. A., *Nucl. Instr. Meth.* **129**:561 (1975).
13 Deruytter, A. J., *Nucl. Instr. Meth.* **15**:164 (1962).
14 Walker, D. H., *Int. J. Appl. Rad. Isot.* **16**:183 (1965).
15 Hutchinson, J. M. R., Nass, C. R., Walker, D. H., and Mann, W. B., *Int. J. Appl. Rad. Isot.* **19**:517 (1968).
16 Waibel, E., *Nucl. Instr. Meth.* **74**:236 (1969).
17 Waibel, E., *Nucl. Intr. Meth.* **86**:29 (1970).
18 Waibel, E., *Nucl. Instr. Meth.* **131**:133 (1975).
19 Bell, P. R., in K. Siegbahn (ed.), *Beta and Gamma-Ray Spectrometry*, Interscience Publishers, New York, 1955, Chap. 5.
20 Heath, R. L., "Scintillation Spectrometry–Gamma-Ray Spectrum Catalogue," IDO-16880-1 (1964).
21 Nakamura, T., *Nucl. Instr. Meth.* **86**:163 (1970).

9

■ introduction to spectroscopy

9.1 INTRODUCTION

Spectroscopy is the aspect of radiation measurements that deals with measuring the energy distribution of particles emitted by a radioactive source or produced by a nuclear reaction.

This introduction to spectroscopy is complemented by Chap. 11, which discusses methods of analysis of spectroscopic data, and Chaps. 12–14, which present details on spectroscopy of photons, charged particles, and neutrons. This chapter discussed the following broad subjects:

1 Definition of differential and integral spectra
2 Energy resolution of the detector
3 The function of a multichannel analyzer (MCA)

9.2 DEFINITION OF ENERGY SPECTRA

A particle energy spectrum is a function giving the distribution of particles in terms of their energy. There are two kinds of energy spectra, differential and integral.

The *differential energy spectrum*, the most commonly studied distribution, is also known as an energy spectrum. It is a function $n(E)$ with the following meaning:

$$n(E)\,dE = \text{number of particles with energies between } E \text{ and } E + dE$$

or $$n(E) = \text{number of particles per unit energy interval}$$

The quantity $n(E)\,dE$ is represented by the cross-hatched area of Fig. 9.1.

The *integral energy spectrum* is a function $N(E)$, where $N(E)$ is the number of particles with energy greater than or equal to E. The quantity $N(E)$ is represented by the hatched area of Fig. 9.1. The integral energy spectrum $N(E)$ and the differential energy spectrum $n(E)$ are related by

$$N(E) = \int_E^{\infty} n(E)\,dE \tag{9.1}$$

The two examples that follow illustrate the relationship between a differential spectrum and an integral spectrum.

EXAMPLE 9.1

Consider a monoenergetic source emitting particles with energy E_0. The differential energy spectrum $n(E)$ is shown in Fig. 9.2. Since there are no particles with energy

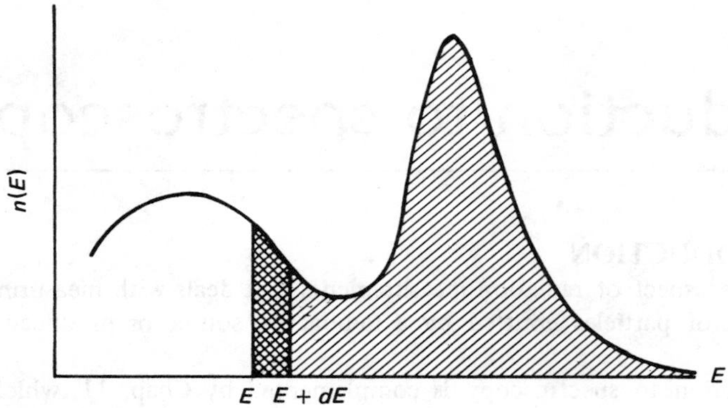

FIG. 9.1 A differential energy spectrum. The quantity $n(E)\, dE$ is equal to the number of particles between E and $E + dE$ (cross-hatched area).

different than E_0, the value of $n(E)$ is equal to zero for any energy other than $E = E_0$.

The corresponding integral spectrum $N(E)$ is shown in Fig. 9.3. It indicates that there are no particles with $E > E_0$. Furthermore, the value of $N(E)$ is constant for $E \leqslant E_0$ since all the particles have energy E_0 and only those particles exist. In other words:

$N(E_0)$ = number of particles with energy greater than or equal to $E_0 = N(E_1)$
= number of particles with energies greater than or equal to E_1 (Fig. 9.3)

EXAMPLE 9.2

Consider the energy spectrum shown in Fig. 9.4. According to this spectrum there are 10 particles per MeV at 11, 12, and 13 MeV. The total number of particles is 30. The integral spectrum is shown in Fig. 9.5. Its value at different energies are:

$N(14) = 0$ no particles above $E = 14$ MeV

$N(13) = 10$ 10 particles at $E = 13$ MeV and above

$N(12) = 20$ 20 particles at $E = 12$ MeV and above

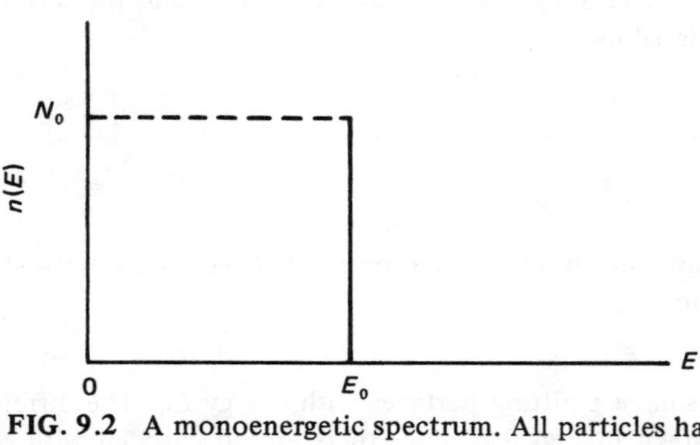

FIG. 9.2 A monoenergetic spectrum. All particles have energy E_0.

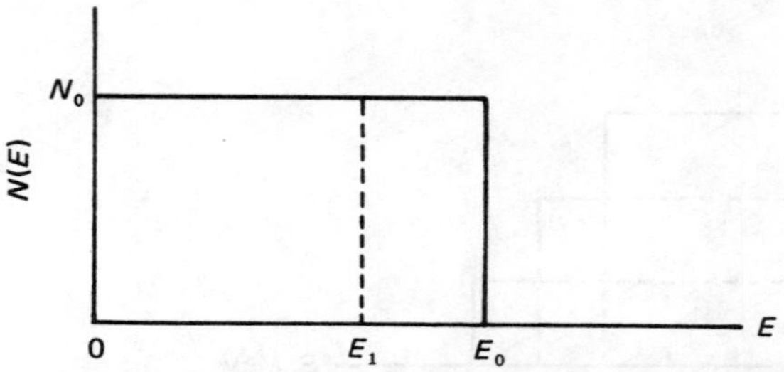

FIG. 9.3 The integral spectrum of a monoenergetic source.

$N(11) = 30$ 30 particles at $E = 11$ MeV and above

$N(10) = 30$ 30 particles at $E = 10$ MeV and above

$N(0) = 30$ 30 particles above $E = 0$

The determination of energy spectra is based on the measurement of pulse-height spectra, as shown in the following sections. Therefore, the definitions of differential and integral spectra given in this section in terms of energy could be expressed equivalently in terms of pulse height. The relationship between particle energy and pulse height is discussed in Sec. 9.5.

9.3 MEASUREMENT OF AN INTEGRAL SPECTRUM WITH A SINGLE-CHANNEL ANALYZER (SCA)

Measurement of an integral spectrum means to count all particles that have energy greater than or equal to a certain energy E or, equivalently, to record all particles that produce pulse height greater than or equal to a certain pulse height V. A device is needed that can sort out pulses according to height. Such a device is a ***single-channel analyzer*** (SCA) operating as a discriminator (integral mode). If the discriminator is set at V_0 volts, all pulses with height less than V_0 will be rejected, while all pulses with heights above V_0 will be recorded. Therefore, a single discriminator can measure an integral energy spectrum. The measurement proceeds as follows.

Consider the differential pulse spectrum shown in Fig. 9.6 for which all the pulses have exactly the same height V_0. To record this spectrum, one starts with the

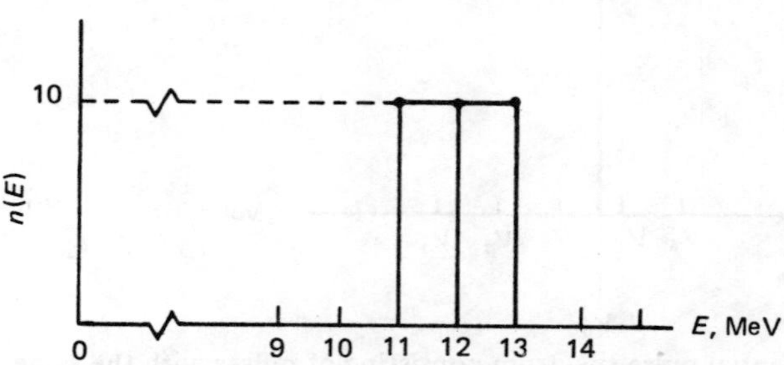

FIG. 9.4 The energy spectrum considered in Example 9.2.

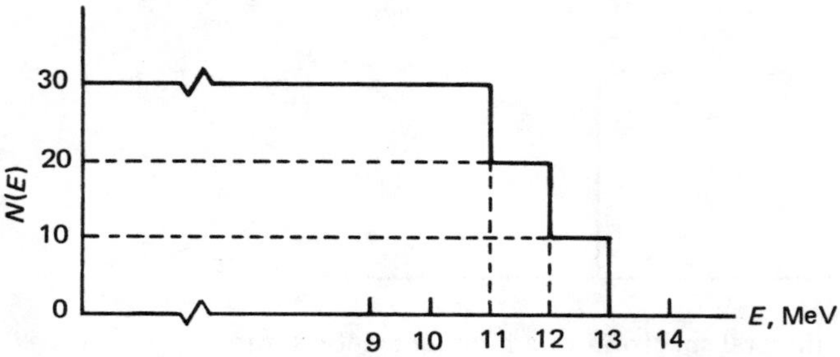

FIG. 9.5 The integral spectrum corresponding to that of Fig. 9.4.

discriminator threshold set very high (higher than V_0) and then lowers the threshold by a certain amount ΔV (or ΔE) in successive steps. Table 9.1 shows the results of this measurement, where $N(V)$ is the number of pulses higher than or equal to V. A plot of these results is shown in Fig. 9.7.

9.4 MEASUREMENT OF A DIFFERENTIAL SPECTRUM WITH AN SCA

Measurement of a differential energy spectrum amounts to the determination of the number of particles within a certain energy interval ΔE for several values of energy; or, equivalently, it amounts to the determination of the number of pulses within a certain interval ΔV, for several pulse heights. An SCA operating in the differential mode is the device that is used for such a measurement.

If the lower threshold of the SCA is set at V_1 (or E_1) and the window has a width ΔV (or ΔE), then only pulses with height between V_1 and $V_1 + \Delta V$ are recorded. All pulses outside this range are rejected. To measure the pulse spectrum of Fig. 9.6, one starts by setting the lower threshold at V_1, where $V_1 > V_0$, with a certain window ΔV (e.g., $\Delta V = 0.1$ V) and then keeps lowering the lower threshold of the SCA. Table 9.2 shows the results of the measurement where $n(V)\,\Delta V$ is the number of pulses with height between V and $V + \Delta V$. Figure 9.8 shows these results.

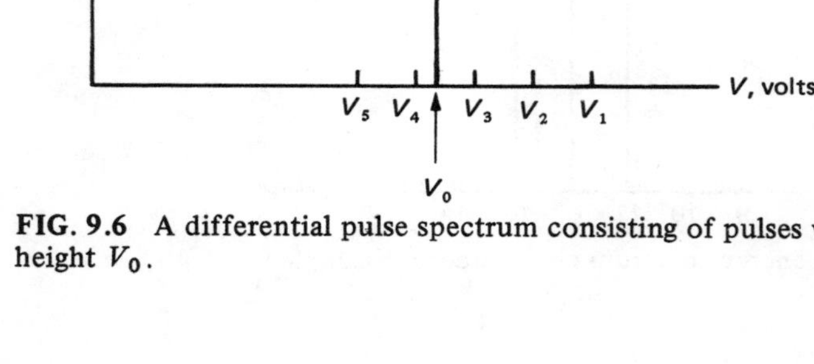

FIG. 9.6 A differential pulse spectrum consisting of pulses with the same height V_0.

TABLE 9.1 Measurement of Integral Spectrum

Discriminator Threshold	$N(V)$
$V_1 > V_0$	0
V_2	0
V_3	0
$V_4 < V_0$	N_0
V_5	N_0
V_6	N_0
$V_i < V_0$	N_0

It is assumed that the width is $\Delta V = V_i - V_{i+1}$, where V_i are the successive settings of the lower threshold of the SCA. It is important to note that one never measures the value of $n(V)$, but only the product $n(V)\,\Delta V$.

9.5 THE RELATIONSHIP BETWEEN PULSE HEIGHT DISTRIBUTION AND ENERGY SPECTRUM

To determine the energy spectrum of particles emitted by a source, one measures, with the help of a detector and appropriate electronics, the pulse-height distribution produced by these particles. Fundamental requirements for the detector and the electronics are as follows:

1 The particle should deposit all its energy or a known constant fraction of it in the detector.

2 The voltage pulse produced by the detector should be proportional to the particle energy dissipated in it, or a known relationship should exist between energy dissipated and pulse height.

3 The electronic amplification should be the same for all pulse heights.

Since the relationship between pulse-height distribution and energy spectrum depends on these three requirements, it is important to discuss them in some detail.

Charged particles deposit all their energy in the detector, as long as their range is

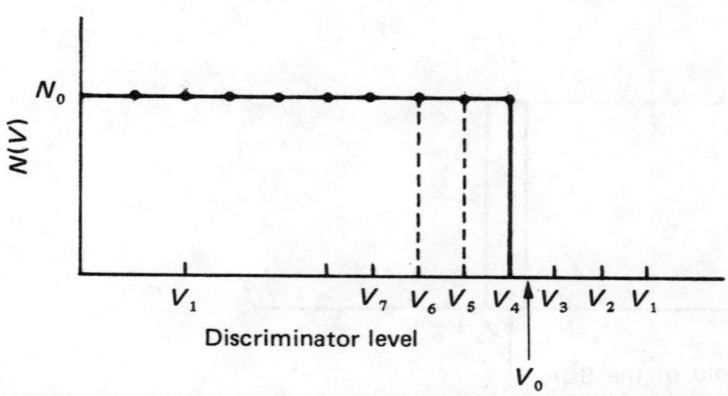

FIG. 9.7 The integral spectrum corresponding to the pulse spectrum of Fig. 9.6.

TABLE 9.2 Measurement of a Differential Spectrum

SCA threshold	$n(V)\,\Delta V$
$V_1 > V_0$	0
V_2	0
V_3	0
$V_4 < V_0$	N_0
V_5	0
$V_i < V_5$	0

smaller than the size of the detector. Gammas do not necessarily deposit all their energy in the detector, regardless of detector size. Neutrons are detected indirectly through other particles produced by nuclear reactions. The energy deposited in the detector depends not only on the energy of the neutron but also on the energy of the reaction products.

The events that transform the particle energy into a voltage pulse are statistical in nature. As a result, even if all the particles deposit exactly the same energy in the detector, the output pulses will not be the same but they will have a certain distribution.

The state of commercial electronics is such that the amplification is essentially the same for all pulse heights (see also Sec. 10.11).

As a result of incomplete energy deposition and the statistical nature of the events that take place in the detector, the shape of the pulse-height distribution is different from that of the source energy spectrum. In other words, two spectra are involved in every measurement:

1 The *source spectrum* is the energy spectrum of particles emitted by the source.
2 The *measured spectrum* is the measured pulse-height spectrum (or distribution).

Figure 9.9 shows as an example the source and the measured spectrum of a monoenergetic gamma source. The objective of the measurement is to obtain the spectrum of Fig. 9.9*a*, but the observer actually measures the distribution shown by Fig. 9.9*b*. The task of the observer is, therefore, to apply appropriate corrections to the measured spectrum to finally obtain the source spectrum.

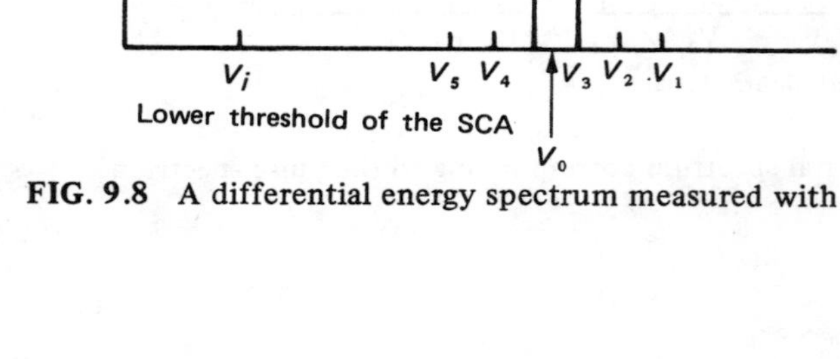

FIG. 9.8 A differential energy spectrum measured with an SCA.

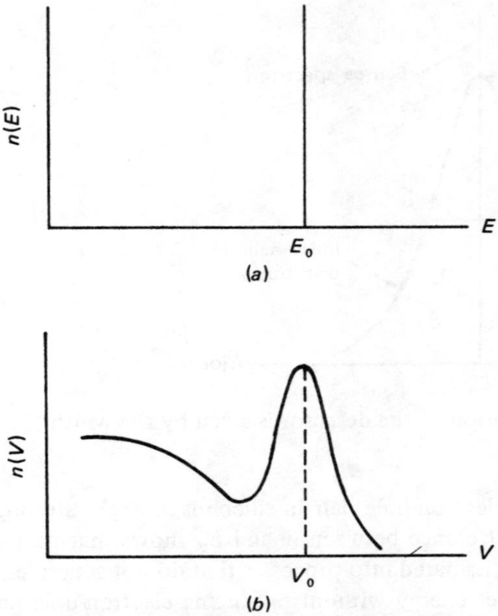

FIG. 9.9 (*a*) The source energy spectrum of a monoenergetic gamma source. (*b*) The pulse height distribution obtained with a NaI(Tl) scintillation counter.

9.6 ENERGY RESOLUTION

The performance of a detector used for energy measurements is characterized by the width of the pulse distribution of a monoenergetic source (Fig. 9.10). The *width*, measured at half of the maximum of the bell-shaped curve, is indicated by Γ or by FWHM (full width half maximum). The ability of a detector to identify particles of different energies, called the *energy resolution*, is given either in terms of Γ or in terms of the ratio $R(E_0)$, where

$$R(E_0) = \frac{\Gamma}{E_0} \tag{9.2}$$

The width Γ is given in energy units, while the ratio $R(E_0)$ is given as a percentage. The three most important factors affecting the energy resolution are:

1 Statistical fluctuations in the number of charge carriers produced in the detector
2 Electronic noise in detector itself, the preamplifier, and the amplifier
3 Incomplete collection of the charge produced in the detector

9.6.1 The Effect of Statistical Fluctuations: The Fano Factor

To discuss the effect of the statistical fluctuations on energy resolution, consider a monoenergetic source of charged particles being detected by a silicon semiconductor detector. (The discussion would apply to a gas-filled counter as well.) The average

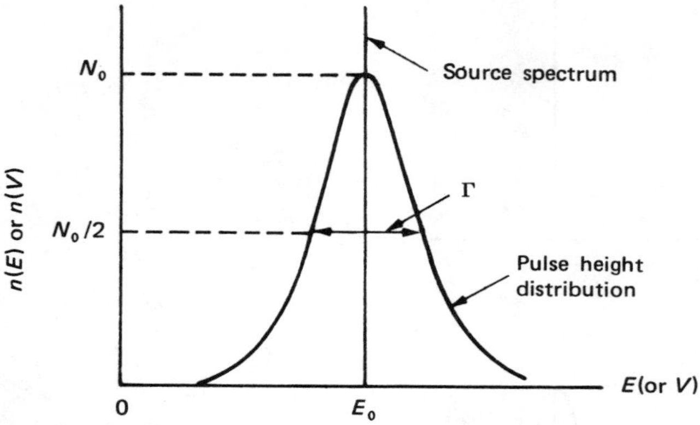

FIG. 9.10 The energy resolution of the detector is given by the width Γ or the ratio Γ/E_0.

energy w needed to produce one electron-hole pair in silicon is 3.66 eV, although the energy gap (E_g) is 1.1 eV. This difference between w and E_g shows that part of the energy of the incident particles is dissipated into processes that do not generate charge carriers. Any process that consumes energy without producing electron-hole pairs is, of course, useless to the generation of the detector signal. If the energy deposited in the detector is E, the average number of charge carriers is E/w. If the process of electron-hole generation were purely statistical, Poisson statistics would apply and the standard deviation of the number of pairs would be

$$\sigma = \sqrt{\frac{E}{w}} \tag{9.3}$$

Experience has shown that the fluctuations are smaller than what Eq. 9.3 gives. The observed statistical fluctuations are expressed in terms of the *Fano factor F*,[1] where

$$F = \frac{(\text{standard deviation of the number of pairs produced})^2}{\text{number of pairs produced}}$$

or, using Eq. 9.3,

$$\sigma = \sqrt{\frac{FE}{w}} \tag{9.4}$$

The two extreme values of F are 0 and 1.

$F = 0$ means that there are no statistical fluctuations in the number of pairs produced. That would be the case if all the energy was used for production of charge carriers. $F = 1$ means that the number of pairs produced is governed by Poisson statistics.

Fano factors have been calculated and also measured.[2–5] For semiconductor detectors, F values as low as 0.06 have been reported.[6] For gas-filled counters, reported F values lie between 0.2 and 0.5. Values of $F < 1$ mean that the generation of electron-hole pairs does not exactly follow Poisson statistics. Since Poisson statistics applies to outcomes that are independent, it seems that the ionization events in a counter are interdependent.

The width Γ of a Gaussian distribution, such as that shown in Fig. 9.10, is related to the standard deviation σ by

$$\Gamma_f = 2\sqrt{2 \ln 2}\, w\sigma \approx 2.355 w\sigma \tag{9.5}$$

Combining Eqs. 9.4 and 9.5,

$$\Gamma_f = 2\sqrt{2(\ln 2) wFE} \tag{9.6}$$

Equation 9.5 shows that the width Γ_f, which is due to the statistical fluctuations, is roughtly proportional to the square root of the energy (the Fano factor is a weak function of energy).

To compare the contribution of the statistical fluctuations to the resolution of different types of detectors at a certain energy, one can use Eqs. 9.2 and 9.6 and write for detectors 1 and 2

$$\frac{R_1}{R_2} = \frac{\Gamma_1/E}{\Gamma_2/E} = \sqrt{\frac{w_1 F_1}{w_2 F_2}} \tag{9.7}$$

It can be seen from Eq. 9.7 that the resolution is better for the detector with the smaller average energy needed for the creation of a charge carrier pair (and smaller Fano factor). Thus, the energy resolution of a semiconductor detector ($w = 3$ eV, $F < 0.1$) should be expected to be much better than the resolution of a gas-filled counter ($w \approx 30$ eV, $F \approx 0.2$), and indeed it is (see Chaps. 12 and 13).

9.6.2 The Effect of Electronic Noise on Energy Resolution

The electronic noise consists of a small voltage variation around the zero line (Fig. 9.11), with average voltage $\bar{v}_n \neq 0$. To see the effect of the noise on the energy resolution, consider pulses of constant height V. In the absence of noise, the FWHM of the distribution of these pulses is zero. If noise is present, the pulses will be superimposed on the noise with the results that the pulses are not of equal height any more (Fig. 9.12), and that the pulses form a Gaussian distribution centered at V and having a width equal to $\Gamma_n = 2\sqrt{2 \ln 2}\, \sigma_n$. The width Γ_n is due to the noise only and has nothing to do with statistical effects in the detector.

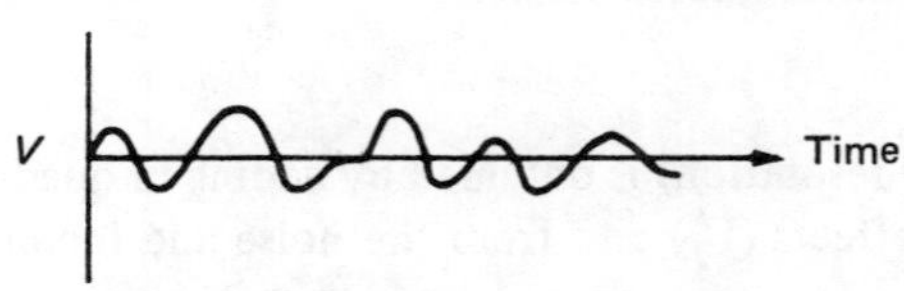

FIG. 9.11 The electronic noise.

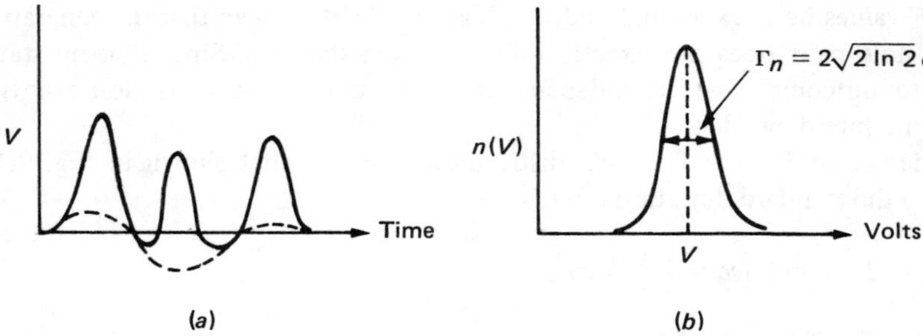

FIG. 9.12 (*a*) The pulses are superimposed on the noise, as a result of which (*b*) they show a distribution with a width which depends on the standard deviation of the noise.

The *signal-to-noise ratio* is frequently the quantity used to indicate the magnitude of the noise. It is defined by:

$$\text{Signal-to-noise ratio} = \frac{\text{mean pulse height}}{\text{noise standard deviation}} = \frac{V}{\sigma_n}$$

Or, one can write

$$\frac{V}{\sigma_n} = 2\sqrt{2 \ln 2}\,\frac{V}{\Gamma_n} = \frac{2\sqrt{2 \ln 2}}{R} \tag{9.8}$$

where R is given by Eq. 9.2. This last equation may be rewritten as

$$R = \frac{2\sqrt{2 \ln 2}}{V/\sigma_n} \tag{9.8a}$$

to show that the higher the signal-to-noise ratio is, the better the resolution becomes (other things being equal, of course).

9.6.3 The Effect of Incomplete Charge Collection

The effect of incomplete charge collection in gas counters is small compared to the effect of the statistical fluctuations. In semiconductor detectors, incomplete charge collection is due to trapping of carriers. The amount of charge trapped is approximately proportional to the energy deposited in the detector, which in turn is proportional to the energy of the incident particles.[7] For this reason, the resolution is affected by trapping effects more at high energy than at low energy. As discussed in Chap. 7, trapping effects depend on the material of which the detector is made and on radiation damage suffered by the semiconductor. Usually, the effect of incomplete charge collection is included in the statistical fluctuations.

9.6.4 The Total Width Γ

The total width Γ (or the total energy resolution) is obtained by adding in quadrature the contributions from the statistical effects (Γ_f) and from the noise and incomplete charge collection (Γ_n). Thus,

$$\Gamma = \sqrt{\Gamma_f^2 + \Gamma_n^2} \tag{9.9}$$

For gas and scintillator counters, the main contribution comes from the statistical fluctuations. For semiconductor detectors at low energies, measurements have shown that $\Gamma_n \gtrsim \Gamma_f$. At higher energies this is reversed, since Γ_n is essentially independent of energy while Γ_f increases with it (see Eq. 9.6).

9.7 DETERMINATION OF THE ENERGY RESOLUTION–THE RESPONSE FUNCTION

Depending on the type and energy of the incident particle and the type of the detector, a monoenergetic source produces a pulse-height distribution that may be a Gaussian (Fig. 9.10) or a more complicated function (Fig. 9.9). In either case, one concludes that although all the particles start at the source with the same energy, there is a probability that they may be recorded within a range of energies. That probability is given by the *response function* or *energy resolution function* $R(E, E')$ of the detection system, defined as:

$R(E, E')\,dE$ = probability that a particle emitted by the source with energy E' will be recorded with energy between E and $E + dE$

One measures, of course, a pulse-height distribution, but the energy calibration of the system provides a one-to-one correspondence between energy and pulse height. If one defines

$S(E)\,dE$ = source spectrum = number of particles emitted by the source with energy between E and $E + dE$

and

$M(E)\,dE$ = measured spectrum = number of particles recorded as having energy between E and $E + dE$

then the three functions $R(E, E')$, $S(E)$, and $M(E)$ are related by

$$M(E) = \int_0^\infty R(E, E')S(E')\,dE' \tag{9.10}$$

Equation 9.10 is an integral equation with the source spectrum $S(E)$ being the unknown. The procedure by which $S(E)$ is obtained, after $R(E, E')$ and $M(E)$ have been determined, is called *unfolding* of the measured spectrum. Methods of unfolding are discussed in Chaps. 11–14.

To determine the response function of a detection system at energy E, the energy spectrum of a monoenergetic source emitting particles with that energy is recorded. Since the resolution changes with energy, the measurement is repeated using several sources spanning the energy range of interest. The response function can also be calculated, as shown in Chap. 12–14. Figure 9-13 shows response functions for several commonly encountered cases.

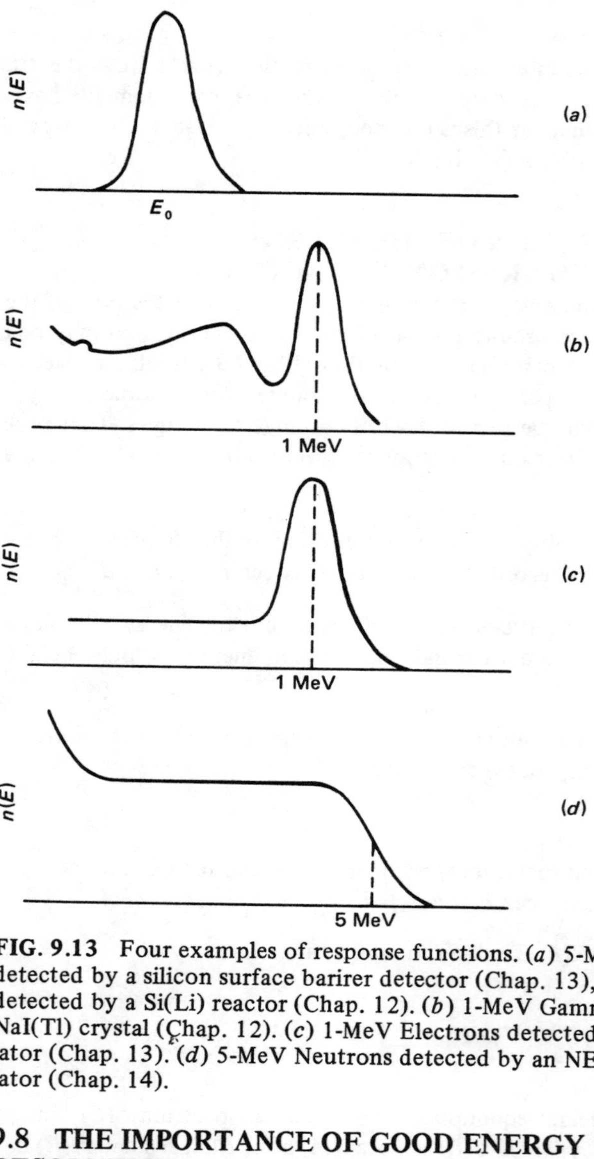

FIG. 9.13 Four examples of response functions. (*a*) 5-MeV Alpha particles detected by a silicon surface barirer detector (Chap. 13), or 20-keV x-rays detected by a Si(Li) reactor (Chap. 12). (*b*) 1-MeV Gamma ray detected by a NaI(Tl) crystal (Chap. 12). (*c*) 1-MeV Electrons detected by a plastic scintillator (Chap. 13). (*d*) 5-MeV Neutrons detected by an NE-213 organic scintillator (Chap. 14).

9.8 THE IMPORTANCE OF GOOD ENERGY RESOLUTION

The importance of good energy resolution becomes obvious if the energy spectrum to be measured consists of several energies. Consider as an example the source spectrum of Fig. 9.14, consisting of two energies E_1 and E_2. Assume that this spectrum is measured with a system having energy resolution equal to Γ,† and examine the following cases.

†Γ may be different at E_1 and E_2. However, the difference is very small since E_1 and E_2 are close. For the present discussion, the same Γ will be used at E_1 and E_2.

CASE I: $E_2 - E_1 > 2\Gamma$

The measured spectrum is shown in Fig. 9.15 for this case. The system can *resolve* the two peaks–i.e., the two peaks can be identified as two separate energies.

CASE II: $E_2 - E_1 = 2\Gamma$

This case is shown in Fig. 9.16. The peaks can still be resolved.

CASE III: $E_2 - E_1 = \Gamma$

This case is shown in Fig. 9.17. The solid line shows how the measured spectrum will look as the sum of the two peaks (dashed lines).

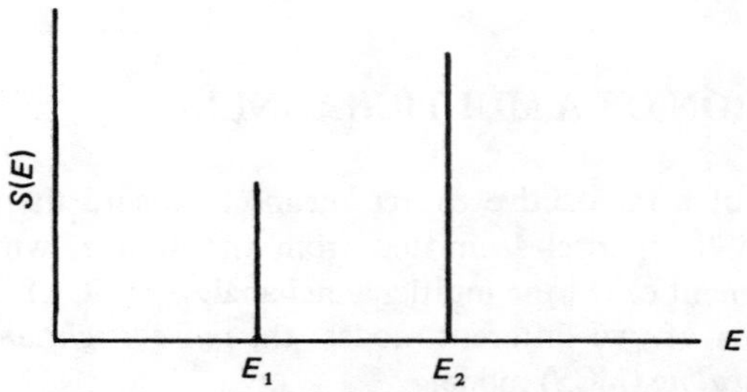

FIG. 9.14 Source spectrum consisting of two distinct energies.

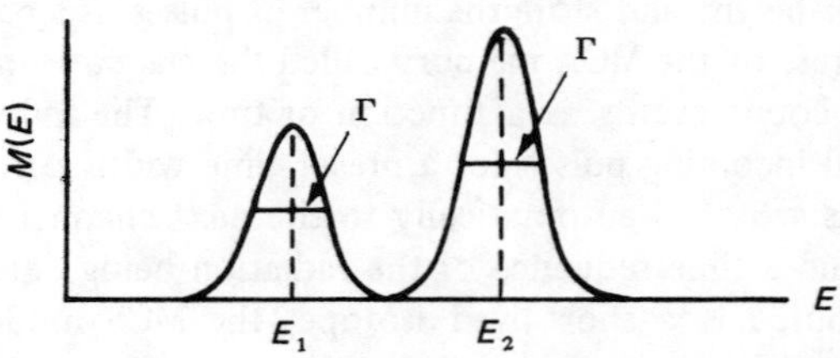

FIG. 9.15 Measured spectrum for Case I: $2\Gamma < E_2 - E_1$.

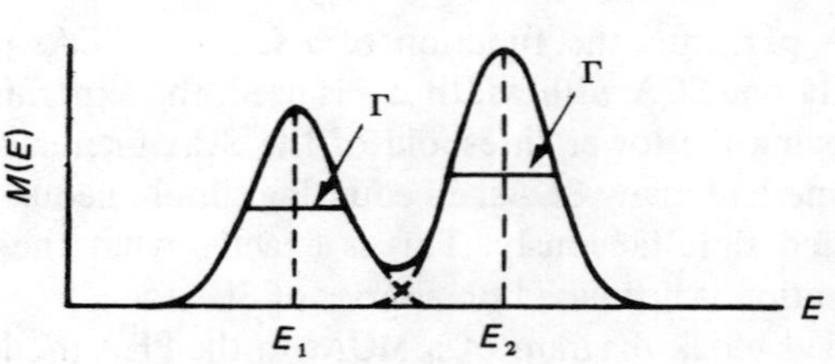

FIG. 9.16 Measured spectrum for case II: $2\Gamma = E_2 - E_1$.

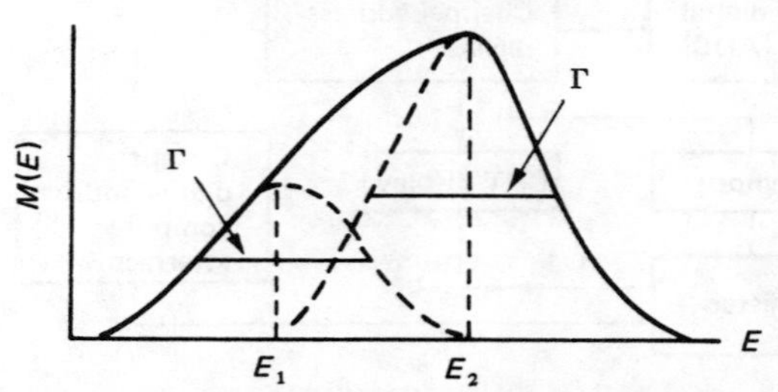

FIG. 9.17 Measured spectrum for Case III: $\Gamma = E_2 - E_1$.

It is obvious that it is difficult to identify two distinct peaks if $E_2 - E_1 = \Gamma$, and the situation will be worse if $E_2 - E_1 < \Gamma$.

The three cases examined above intend to show how important good energy resolution is for the measurement of spectra with many energy peaks. If the response function of the detector is not known and the measured spectrum shows no well-identified peaks, the following criterion is used for the energy resolution required to identify the peaks of about equal magnitude:

> To be able to resolve two energy peaks at E_1 and E_2, the resolution of the system should be such that $\Gamma \leqslant |E_1 - E_2|$.

9.9 BRIEF DESCRIPTION OF A MULTICHANNEL ANALYZER (MCA)

To measure an energy spectrum of a radioactive source means to record the pulse-height distribution produced by the particles emitted from the source, which is achieved with the use of an instrument called the multichannel analyzer (MCA). Multichannel analyzers are used in either of two different modes: the *pulse-height analysis* (PHA) mode or the *multichannel scaling* (MCS) mode.

The PHA mode is the tradiational function of a MCA and is used to sort out incoming pulses according to their height and store the number of pulses of a particular height in a corresponding address of the MCA memory called the *channel number.*

The MCS mode is used to count events as a function of time. The individual channels of the memory count all incoming pulses for a preset time width Δt. After time Δt, the counting operation is switched automatically to the next channel in the memory, thus producing in the end a time sequence of the radiation being detected. For example, if the radiation source is a short-lived isotope, the MCS mode will provide the exponential decay curve that can be used for the measurement of the half-life of this isotope.

In the PHA mode, an MCA performs the function of a series of SCAs placed adjacent to one another. When only one SCA with width ΔE is used, the experimenter has to sweep the spectrum by moving the lower threshold of the SCA manually (see Sec. 9.4). On the other hand, if one had many SCAs, all counting simultaneously, the whole spectrum would be recorded simultaneously. This is exactly what the MCA does, although its principle of operation is not based on a series of SCAs.

Figure 9.18 shows a simplified block diagram of a MCA. In the PHA mode, the

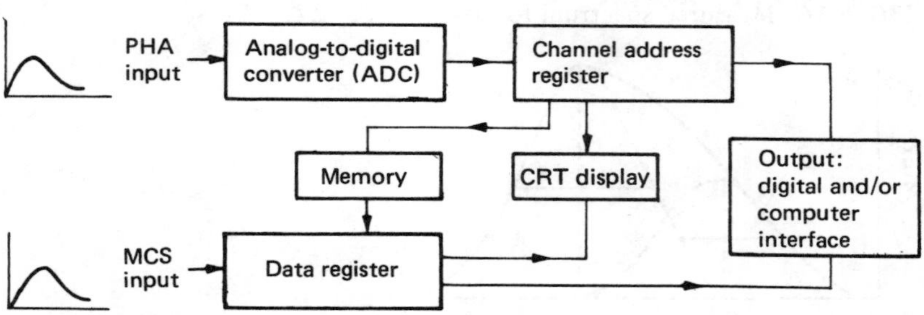

FIG. 9.18 A simplified block diagram of an MCA.

incoming pulse enters into a unit called the *analog-to-digital converter* (ADC). The ADC *digitizes* the pulse amplitude: it produces a number proportional to the height of the pulse, a number which determines the channel where the pulse will be stored. The size of the ADC, given in terms of channels, defines the absolute resolution of the system. Actually, the ADC determines the number of discrete parts into which the pulse height can be subdivided. Commercial ADCs have at the present time a size up to 8192 channels, with the full scale adjustable in steps of 256, 512, 1024, etc., channels.

The number of discrete parts (channels) into which the input pulse range (0 to +10 V) is divided is called the *conversion gain.* The conversion gain is set by a stepwise control knob located on the front of the instrument. As an example, if the conversion gain is set at 2048 channels, it means that the maximum pulse height (10 V) is divided into that many parts. Therefore, the resolution of the MCA at this setting is:

$$10 \text{ V}/2048 = 4.88 \text{ mV/channel}$$

More details about the operation and characteristics of ADCs are given in Sec. 10.12.

The memory of the MCA is a data-storage unit arranged in a series of channels. Every channel is capable of storing up to $2^{20} - 1$ data (pulses), in most cases. Commercial MCAs have memories with up to 8192 channels. Normally, the MCA provides for selection and use of the full memory, of only half of it, or of one-fourth of it. Transfer of data from one fraction of the memory to another is also possible.

In the PHA mode, the first channel of the region used is called *channel zero* and records, in almost all late model MCAs, the live time of the analysis, in seconds. If the full memory or the first half or first quarter of the memory is used, channel zero is the address 0000. If the second half of a 4096 memory is used, channel zero is address 2048; if the second quarter is used, channel zero is address 1024; and so on.

How does one determine the size of the MCA memory needed for a specific experiment? The decision is made based on the requirements for the PHA mode. One equation frequently used is

$$\text{Number of channels} = h \frac{\text{energy range of interest (keV)}}{\Gamma \text{ (keV)}} \tag{9.12}$$

where Γ is the FWHM of the detector used. The factor h is equal to the number of channels at or above the FWHM of the peak. Its value is between 3 and 5.

As an example, assume that the energy range of interest is 0 to 2.0 MeV and consider a NaI(Tl) and a Ge(Li) detector. The resolution of the NaI(Tl) detector is about 50 keV. Therefore, the minimum number of channels is ($h = 5$)

$$5 \left(\frac{2000}{50} \right) \approx 200 \text{ channels}$$

The resolution of a Ge(Li) detector is about 2 keV. Now, the number of channels is

$$5 \left(\frac{2000}{2} \right) \approx 5000 \text{ channels}$$

The user should remember that the ADC, not the memory, determines the absolute resolution of an MCA. An MCA with an ADC of 1000 channels and a memory of 2000 channels has an actual resolution of only 1000 channels.

The major disadvantage of every MCA is the dependence of its dead time on the channel where the pulse is stored. The equation for dead time is of the form

$$\tau = a_1 + a_2 C + a_3 \tag{9.13}$$

where a_1, a_2, and a_3 are constants given by the manufacturers and C is the channel number. Typical values are $a_1 \approx 2.5$ μs, $a_2 = 0.01$ to 0.02 μs/channel and $a_3 = 0$ to 7 μs. Thus, if a pulse is stored in channel 2000, the dead time is

$$\tau \approx 2.5 + 0.02(2000) + 7 \approx 47 \ \mu\text{s}$$

The dead-time correction should be applied separately for every channel, using the method shown in Sec. 2.20.

Commercial MCAs have a meter that shows, during counting, the percent of dead time. They also have timers that determine the counting period in *live time* or *clock time.* In clock time mode, the counting continues for as long as the clock is set up. In live time mode, an automatic correction for dead time is performed. In this case, the percent dead time indication can be used to determine the approximate amount of actual time the counting will take. For example, if the clock is set to count for 5 min (in live mode) and the dead time indicator shows 25 percent, the approximate actual time of this measurement is going to be,

$$\text{Actual time} = \frac{\text{live time}}{1 - (\text{dead time fraction})} = \frac{300 \text{ s}}{1 - 0.25} = 400 \text{ s}$$

Modern MCAs can do much more than store the number of pulses in the memory. They are, essentially, hard-wired minicomputers that may be able to integrate the area under a spectrum, find the energy of a peak and, in some cases, select the isotope that emits a photon of certain energy. Also marketed today are MCAs that are connected with minicomputers of their own, have their own library of programs, and can analyze quantitatively any spectrum. Most of the controls are on a push-button console located in front of a large cathode-ray tube.

9.10 CALIBRATION OF A MULTICHANNEL ANALYZER

The calibration of an MCA follows these steps:

1. *Determination of range of energies involved.* Assume this is $0 \leqslant E \leqslant E_m$ (MeV).

2. *Determination of preamplifier-amplifier setting.* Using a source which emits particles of known energy, one observes the signal generated on the screen of the oscilloscope. It should be kept in mind that the maximum possible signal at the output of the amplifier is 10 V. In energy spectrum measurements, one should try to stay in the range 0–9 V.

Assume that the particle energy E_1 results in pulse height V_1. Is this amplification proper for obtaining a pulse height $V_m \leqslant 10$ V for energy E_m? To find this out, the observer should use the fact that pulse height and particle energy are proportional. Therefore:

$$\frac{V_m}{E_m} = \frac{V_1}{E_1} \rightarrow V_m = \frac{E_m}{E_1} V_1$$

If $V_m < 10$ V, then the amplification setting is proper. If $V_m \geqslant 10$ V, the amplification should be reduced. (If $V_m < 2$ V, amplification should be increased. It is good practice, but not necessary, to use the full range of allowed voltage pulses.) The maximum pulse V_m can be changed by changing the amplifier setting.

3. *Determination of MCA settings.* One first decides the part of the MCA memory to be used. Assume that the MCA has a 1024-channel memory and it has been decided to use 256 channels, one-fourth of the memory. Also assume that a spectrum of a known source with energy E_1 is recorded and that the peak is registered in channel C_1. Will the energy E_m be registered in $C_m < 256$, or will it be out of scale?

The channel number and energy are almost proportional,† i.e., $E_i \sim C_i$. Therefore

$$\frac{C_m}{E_m} \approx \frac{C_1}{E_1} \rightarrow C_m \approx \frac{E_m}{E_1} C_1$$

If $C_m \leqslant 256$, the setting is proper and may be used. If $C_m > 256$, a new setting should be employed. This can be done in one of two ways or a combination of the two:

1 The fraction of the memory selected may be changed. One may use 526 channels of 1024, instead of 256.

2 The conversion gain may be changed. In the example discussed here, if a peak is recorded in channel 300 with conversion gain of 1024, that same peak will be recorded in channel 150 if the conversion gain is switched to 512.

There are analyzer models that do not allow change of conversion gain. For such an MCA, if C_m is greater than the total memory of the instrument, one should return to step 2 and decrease V_m by reducing the gain of the amplifier.

4. *Determination of the energy-channel relationship.* Calibration of the MCA means finding the expression that relates particle energy to the channel where a particular energy is stored. That equation is written in the form

$$E = a_1 + a_2 C + a_3 C^2 \tag{9.14}$$

where C = channel number and $a_1, a_2, a_3, \ldots$ are constants.

The constants a_1, a_2, a_3, ... are determined by recording spectra of sources with known energy. In principle, one needs as many energies as there are constants.

†The correction equation is $E = a + bC$, but a is small and for this argument it may be neglected; proper evaluation of a and b is given in step 4 of the calibration procedure.

In practice, a large number of sources is recorded with energies covering the whole range of interest and the constants are then determined by a least-squares fitting process (see Chap. 11).

Most detection systems are essentially linear, which means that Eq. 9.14 takes the form

$$E = a_1 + a_2 C \tag{9.15}$$

EXAMPLE 9.3

Obtain the calibration constants for an MCA based on the spectrum shown in Fig. 9.19. The peaks correspond to the following three energies:

$E_1 = 0.662$ MeV $C_1 = 160$

$E_2 = 1.173$ MeV $C_2 = 282.5$

$E_3 = 1.332$ MeV $C_3 = 320$

ANSWER

Plotting energy versus channel on linear graph paper, one obtains the line shown in Fig. 9.20, which indicates that the linear equation, Eq. 9.15, applies, and one can determine the constants a_1 and a_2 from the slope and the zero intercept of the straight line. From Fig. 9.19, the value of a_2 is

$$a_2 = \frac{950 - 400}{230 - 97.5} = 4.15 \text{ keV/channel}$$

The constant a_1 is equal to the zero-intercept of the line.† In the present case, it is almost zero. Based on these results, the calibration equation of this MCA is $E = 4.15C$.

5. *Calculation of the energy resolution.* By definition, the energy resolution is $R = \Gamma/E$, where Γ is the FWHM of the peak of energy E. Therefore, using Eq. 9.15,

$$R = \frac{\Gamma}{E} = \frac{(a_1 + a_2 C_R) - (a_1 - a_2 C_L)}{E} = \frac{a_2(C_R - C_L)}{E} \tag{9.16}$$

†Most commercial MCAs have a hand-screw adjustment that makes a_1 equal to zero.

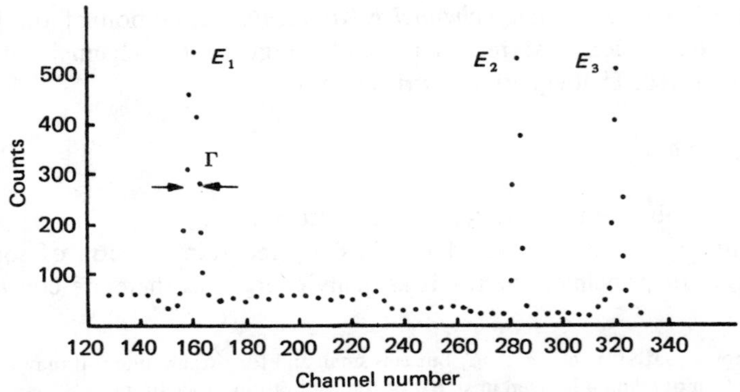

FIG. 9.19 A gamma spectrum used for calibration of an MCA.

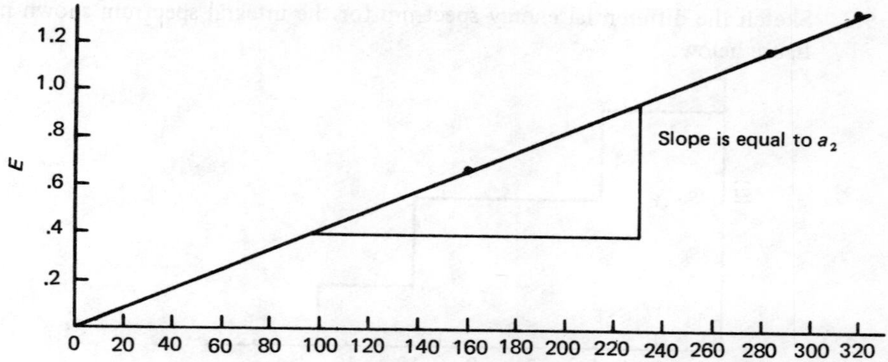

FIG. 9.20 Plot of energy versus channel number. In this case, the relationship is linear.

where C_R and C_L are the channel numbers on either side of the peak at half of its maximum. If a_1 is zero, the resolution is given by

$$R = \frac{a_2(C_R - C_L)}{a_2 C_{peak}} = \frac{C_R - C_L}{C_{peak}} \tag{9.17}$$

For peak E_1 (Fig. 9.18),

$$C_L = 158 \qquad C_{peak} = 160 \qquad C_R = 162$$

Therefore

$$R = \frac{162 - 158}{160} = 2.5\%$$

or $\quad \Gamma = a_2(C_R - C_L) = 4.15(4) = 16.6 \text{ keV}$

PROBLEMS

9.1 Sketch the integral spectrum for the diffferential spectrum shown in the figure below.

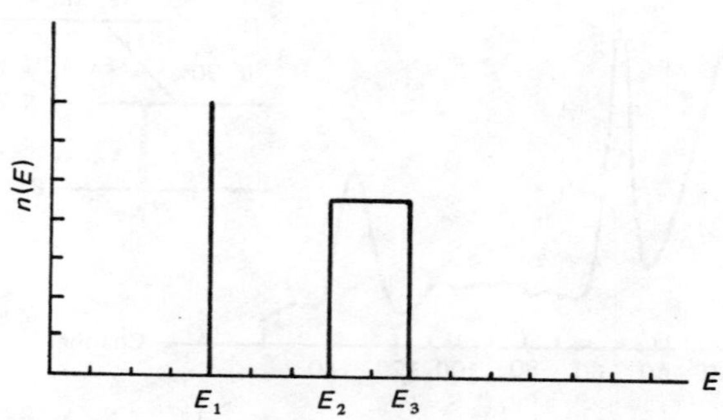

9.2 Sketch the differential energy spectrum for the integral spectrum shown in the figure below.

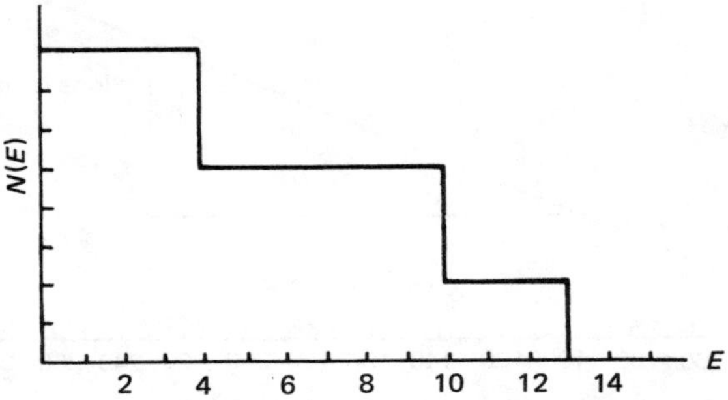

9.3 Sketch the integral spectrum for the differential spectrum shown in the figure below.

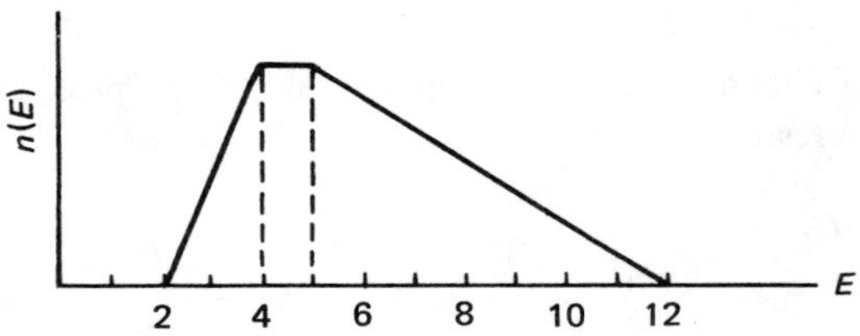

9.4 If the energy resolution of a NaI(Tl) scintillator system is 11 percent at 600 keV, what is the width Γ of a peak at that energy?

9.5 What is the maximum energy resolution ncesssary to resolve two peaks at 720 keV and 755 keV?

9.6 Prove that if a detection system is known to be linear the calibration constants are given by

$$a_1 = \frac{E_2 C_1 - E_1 C_2}{C_1 - C_2} \qquad a_2 = \frac{E_1 - E_2}{C_1 - C_2}$$

where E_1 and E_2 are two energies recorded in channels C_1 and C_2, respectively.

9.7 Shown in the figure below is the spectrum of ^{22}Na, with its decay scheme. Determine the calibration constants of the MCA which recorded this spectrum, based on the two peaks of the ^{22}Na spectrum.

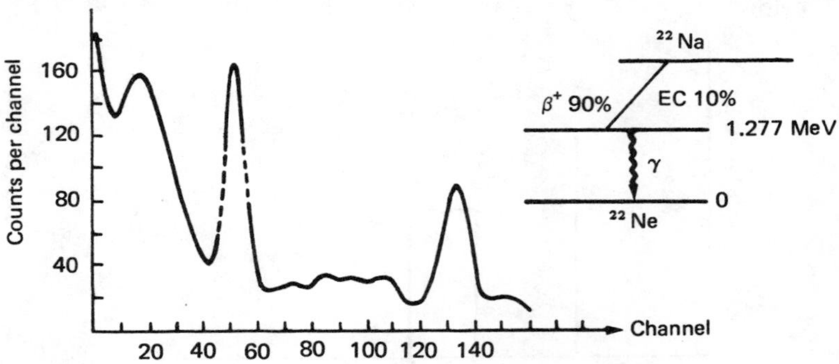

9.8 In Prob. 9.7, the channel number cannot be read exactly. What is the uncertainty of the calibration constants a_1 and a_2 if the uncertainty in reading the channel is one channel for either peak?

9.9 Assume that the energy resolution of a scintillation counter is 9 percent and that of a semiconductor detector is 1 percent at energies around 900 keV. If a source emits gammas at 0.870 MeV and 0.980 MeV, which of these peaks can be resolved with a scintillator and which ones with a semiconductor detector?

9.10 Consider the two peaks shown in the accompanying figure. How does the peak at E_2 affect the width of the peak at E_1 and vice versa? What is the width Γ for either peak?

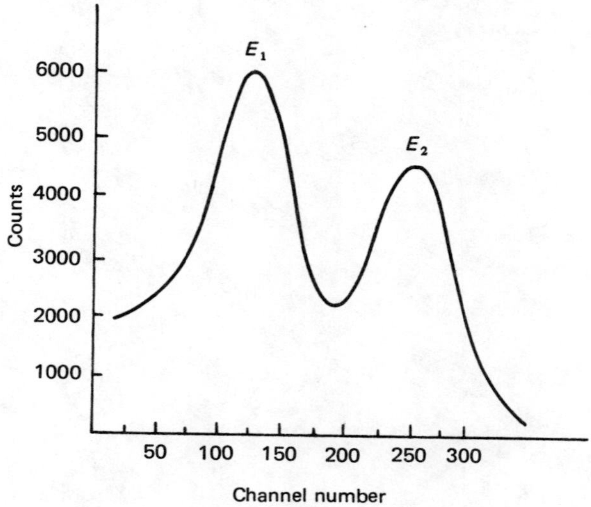

REFERENCES

1 Fano, U., *Phys. Rev.* 72:26 (1947).
2 van Roosbroeck, W., *Phys. Rev.* **139A**:1702 (1965).
3 Goulding, F. S., *Nucl. Instr. Meth.* **43**:1 (1966).
4 Mann, H. M., Bilger, H. R., and Sherman, I. S., *IEEE* **NS-13**(3):352 (1966).
5 Deshpande, R. Y., *Nucl. Instr. Meth.* **57**:125 (1967).
6 Pehl, R. H., and Goulding, F. S., *Nucl. Instr. Meth.* **81**:329 (1970).
7 Ewan, G. T., *Nucl. Instr. Meth.* **162**:75 (1979).

10

■ electronics

10.1 INTRODUCTION

This chapter presents a brief and general description of electronic units used in radiation measurements. The subject is approached from the viewpoint of "input-output"—i.e., the input and output signals of every component unit or instrument are presented with a minimum of discussion on circuitry. The objective is to make the reader aware of the capabilities and limitations of the different types of units and, at the same time, create the capacity to choose the right component for a specific counting system.

Details about construction and operation of electronic components and systems are given in books specializing on that subject. A few such texts are listed in the bibliography at the end of the chapter. Also, the vendors of nuclear instruments provide manuals for their products with useful information about their operation.

10.2 RESISTANCE, CAPACITANCE, INDUCTANCE, AND IMPEDANCE

To understand what factors affect the formation, transmission, amplification, and detection of a detector signal, it is important to comprehend the function of resistance, capacitance, inductance, and impedance, which are the basic constituents of any electronic circuit. For this reason, a brief review of these concepts is offered.

The *resistance R* is a measure of how difficult (or easy) it is for a current to flow through a conductor. The resistance is defined by Ohm's law as the ratio of a voltage to current flowing through a conductor (Fig. 10.1*a*). The resistance is measured in ohms (Ω). If a potential difference of 1 V generates a current of 1 A, the resistance is 1 Ω; that is,

$$R = \frac{V}{i} \tag{10.1}$$

Capacitance C is the ability to store electrical charge. A capacitor usually consists of two conductors separated by an insulator or a dielectric (Fig. 10.1*b*). Every conductor, however, has a certain capacitance. The capacitance is measured in farads (F). If a charge of 1 coulomb produces a potential difference of 1 V between the two conductors forming the capacitor, then its capacitance is 1 farad; that is,

$$C = \frac{q}{V} \tag{10.2}$$

If the voltage across the capacitor is constant, no current flows through it; i.e., a capacitor acts as an open circuit to dc voltage. If, however, the voltage changes, a current flows through the capacitor equal to

$$i = \frac{dq}{dt} = C\frac{dV}{dt} \tag{10.3}$$

Inductance refers to the property of conductors to try to resist a change in a magnetic field. If the current flowing through a conductor changes with time (in which case the magnetic field produced by the current also changes), a potential difference is induced that opposes the change. The induced potential difference V_L is given by (Fig. 10.1*c*)

$$V_L = -L\frac{di}{dt} \tag{10.4}$$

where L is called the inductance of the conductor and is measured in henrys (H). If a current change of 1 A/s induces a potential difference of 1 V, the inductance is 1 H. An inductor is usually indicated as a coil (Fig. 10.1*c*), but any conductor has a certain inductance. No pure inductor exists because there is always some ohmic resistance and some capacitance in the wires making the coils.

Capacitance and inductance are important for time-varying signals. To be able to introduce and discuss the pertinent concepts, consider a sinusoidal current signal applied to an RC circuit as shown in Fig. 10.2. If the current is given by

$$i(t) = i_m \sin \omega t \tag{10.5}$$

then the instantaneous voltage across the resistor is

$$V_R(t) = iR = i_m R \sin \omega t \tag{10.6}$$

which is a sine wave with the same phase as the input signal. The voltage across the capacitor is obtained by integrating Eq. 10.3:

$$V_C(t) = \frac{1}{C}\int i\,dt = -\frac{i_m}{\omega C}\cos \omega t = \frac{i_m}{\omega C}\sin\left(\omega t - \frac{\pi}{2}\right) \tag{10.7}$$

Thus, V_C is also sinusoidal but follows the current with a phase difference equal to $\pi/2$. By analogy with Ohm's law (Eq. 10.1), the ratio

$$\frac{(V_C)_{\max}}{i_{\max}} = \frac{V_{C,m}}{i_m} = \frac{1}{\omega C} = R_C \tag{10.8}$$

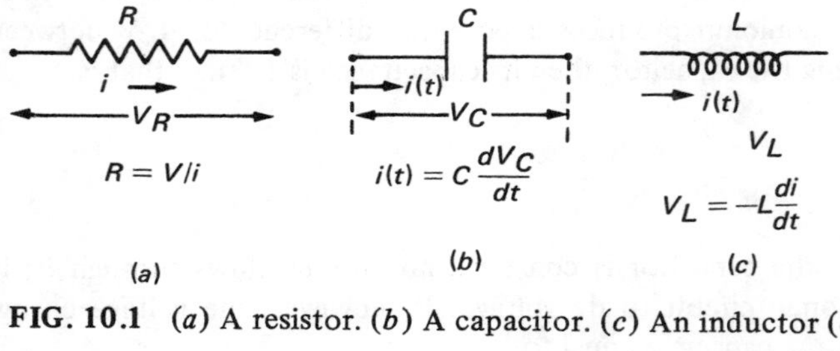

FIG. 10.1 (*a*) A resistor. (*b*) A capacitor. (*c*) An inductor (coil).

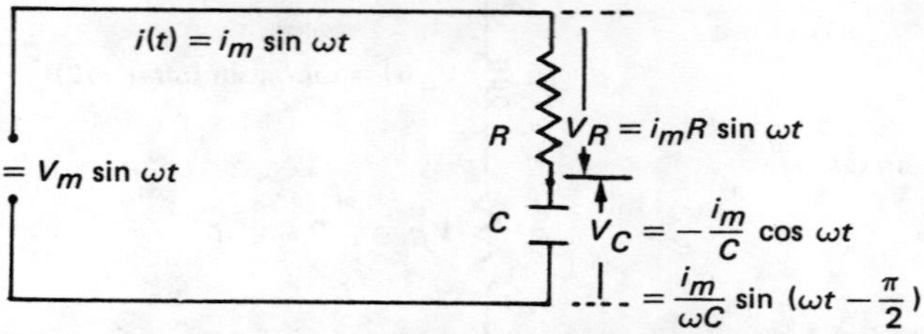

FIG. 10.2 The interaction of an RC circuit with a sinusoidal input.

is called the *capacitive resistance* or *reactance* to the change induced by the source signal. Capacitive reactance is measured in ohms.

At every time t, the instantaneous potentials $V_S(t)$, $V_R(t)$, and $V_C(t)$ satisfy the equation

$$V_S(t) = V_R(t) + V_C(t) \tag{10.9}$$

The peak potentials, however, are not additive because they do not all occur at the same time. The correct addition is performed by considering the peak potentials $V_{R,0}$ and $V_{C,0}$ as two vectors forming a 90° angle (Fig. 10.3) because their instantaneous values have a phase difference equal to $\pi/2$ (see Eqs. 10.6 and 10.7). Thus,

$$V_{S,m} = \sqrt{V_{R,m}^2 + V_{C,m}^2} \tag{10.10}$$

By analogy with Ohm's law, one writes

$$\frac{(V_S)_{\max}}{i_{\max}} = \frac{V_{S,m}}{i_m} = Z = \sqrt{R^2 + R_C^2} \tag{10.11}$$

where Z is the impedance of the circuit. Impedance should be viewed as a generalized resistance which may consist of a pure resistance R or a capacitive reactance RC or a mixture. Impedance is also measured in ohms.

Consider now an LR circuit as shown in Fig. 10.4. It will be left as an exercise to the reader to show that the potential across the inductor is

$$V_L(t) = \omega L i_m \sin\left(\omega t + \frac{\pi}{2}\right) \tag{10.12}$$

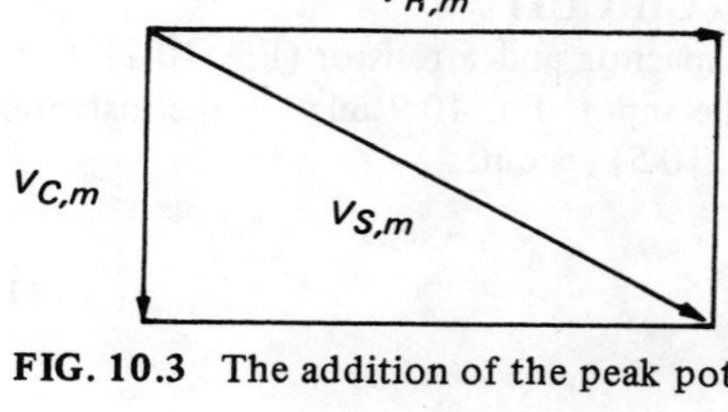

FIG. 10.3 The addition of the peak potentials in an RC circuit.

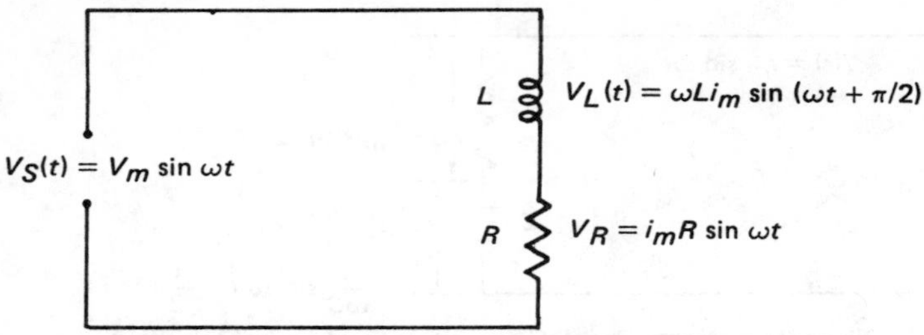

FIG. 10.4 The interaction of an LR circuit with a sinusoidal input.

The inductive reactance is

$$R_L = \frac{(V_L)_{\max}}{i_{\max}} = \omega L \tag{10.13}$$

and the impedance is

$$Z = \sqrt{R^2 + R_L^2} \tag{10.14}$$

Finally, the peak voltages are related by

$$V_{S,0}^2 = \sqrt{V_R^2 + V_L^2} \tag{10.15}$$

If a circuit contains all three elements in series (R, L, C), it is easy to show that the total impedance is given by

$$Z = \sqrt{R^2 + (R_L - R_C)^2} \tag{10.16}$$

Every electronic component has a characteristic impedance. When a signal is transmitted from a unit with a high-impedance output to a low-impedance input, there is going to be a loss in the signal unless an impedance-matching device is used to couple the two units. Manufacturers of preamplifiers and amplifiers quote the impedance of the input and output for their products. Coaxial cables have an impedance between 90 and 100 Ω.

10.3 A DIFFERENTIATING CIRCUIT

A *differentiating circuit* consists of a capacitor and a resistor (Fig. 10.2). If a time-dependent voltage $V_i(t)$ is applied at the input, Eq. 10.9 relating the instantaneous values of the three voltages involved (Fig. 10.5) becomes

$$\frac{q(t)}{C} + Ri = \frac{q(t)}{C} + R\frac{dq(t)}{dt} = V_i(t) \tag{10.17}$$

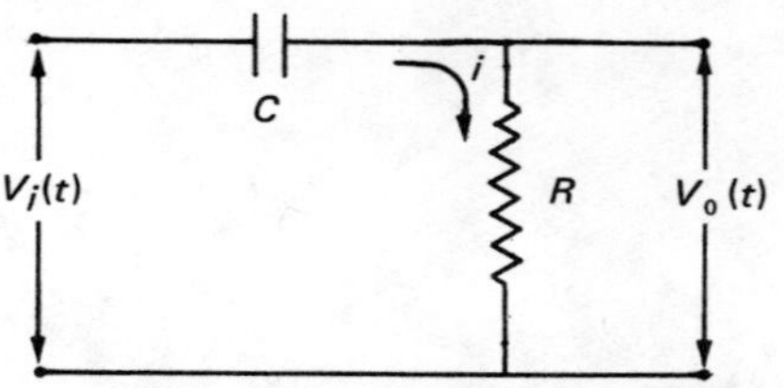

FIG. 10.5 A CR shaping circuit (differentiator).

where $q(t)$ is the charge of the capacitor at time t. If the input signal is a step function (Fig. 10.6), the output voltage is given (after solving Eq. 10.17) by

$$V_0(t) = R\frac{dq(t)}{dt} = V_{i,m}e^{-t/RC} \tag{10.18}$$

Figure 10.7 shows the output voltage if the input signal is a rectangular pulse of height $V_{i,m}$ and duration T. Notice that if $RC \ll T$, the output signal represents the derivative of the input. Indeed, from Eq. 10.17, if $RC \ll T$, then

$$\frac{V_i(t)}{R} = \frac{q}{RC} + \frac{dq}{dt} \approx \frac{q}{RC}$$

or

$$\frac{1}{R}\frac{dV_i}{dt} = \frac{1}{RC}\frac{dq}{dt} = \frac{1}{RC}i$$

Thus,

$$V_0(t) = iR = R\frac{dq}{dt} = RC\frac{dV_i}{dt} \tag{10.19}$$

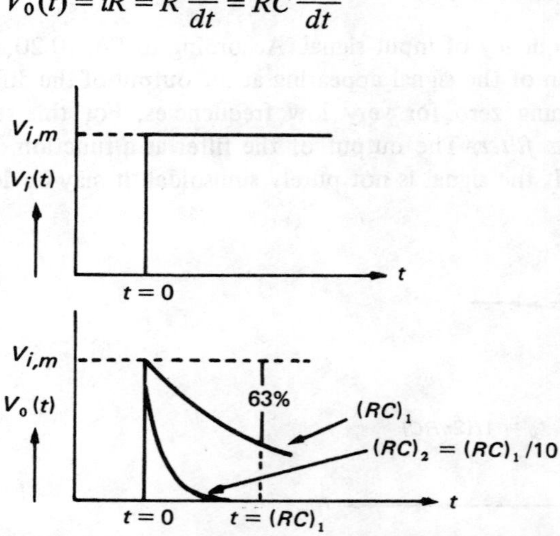

FIG. 10.6 The output signal of a CR shaping circuit for a step input.

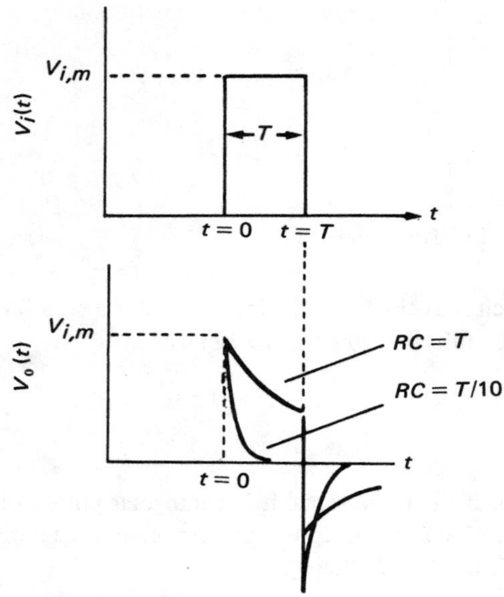

FIG. 10.7 The response of a CR circuit to a rectangular pulse.

and for this reason, this circuit is called a *differentiator.* If $V_i(t)$ is the pulse from a detector, the effect of differentiation is to force the pulse to decay faster.

As shown in Sec. 10.2, for a sinusoidal signal the peak value of the potential across the resistor of an RC circuit is related to the peak of the input signal by

$$\frac{V_{o,m}}{V_{i,m}} = \frac{R}{\sqrt{R^2 + R_C^2}} = \frac{R}{\sqrt{R^2 + 1/\omega^2 C^2}} \tag{10.20}$$

where $\omega = 2\pi f$ and f is the frequency of input signal. According to Eq. 10.20, as the frequency decreases the fraction of the signal appearing at the output of the differentiator also decreases, approaching zero for very low frequencies. For this reason, this circuit is called a *high-pass filter.* The output of the filter as a function of frequency is shown in Fig. 10.8. If the signal is not purely sinusoidal, it may be decom-

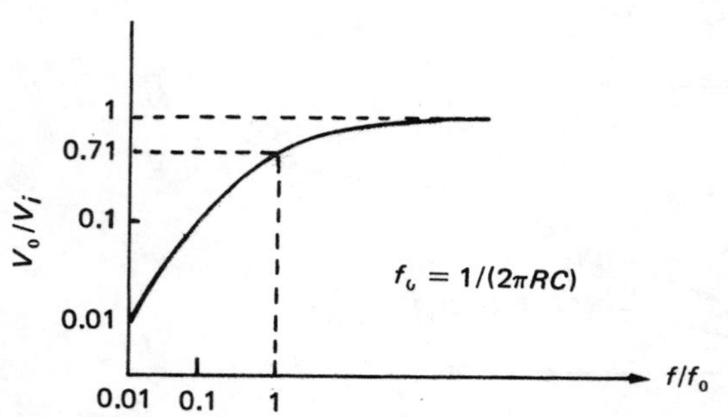

FIG. 10.8 The output of a differentiator (high-pass) filter as a function of input frequency.

posed into a series of sine components with frequencies that are multiples of a fundamental one (this is called Fourier analysis). Going through the high-pass filter, the lower frequencies will be attenuated more than the higher ones.

10.4 AN INTEGRATING CIRCUIT

An *integrating circuit* also consists of a resistor and a capacitor, but now the output signal is taken across the capacitor (Fig. 10.9). Equation 10.17 applies in such a case too, and the output signal as a result of a step input is given by

$$V_0(t) = \frac{q(t)}{C} = V_{i,m}(1 - e^{-t/RC}) \tag{10.21}$$

Input and output signals are shown in Fig. 10.10. Figure 10.11 shows the output voltage if the input signal is a rectangular pulse of height $V_{i,m}$ and duration T. If $RC \gg T$, the output signal looks like the integral of the input. Indeed, from Eq. 10.17,

$$\frac{V_i(t)}{R} = \frac{q(t)}{CR} + \frac{dq(t)}{dt}$$

which gives, if $RC \gg T$,

$$\frac{V_i(t)}{R} \approx \frac{dq}{dt}$$

Then, from Eq. 10.21,

$$V_0(t) = \frac{q(t)}{C} = \frac{1}{RC}\int V_i(t)\,dt \tag{10.22}$$

and for this reason, this circuit is called an *integrator.*

For a sinusoidal input signal, the peak value of the voltage across the capacitor is related to the peak value of the input signal by

$$\frac{V_{0,m}}{V_{i,m}} = \frac{R_C}{\sqrt{R^2 + R_C^2}} = \frac{1/\omega C}{\sqrt{R^2 + 1/\omega^2 C^2}} \tag{10.23}$$

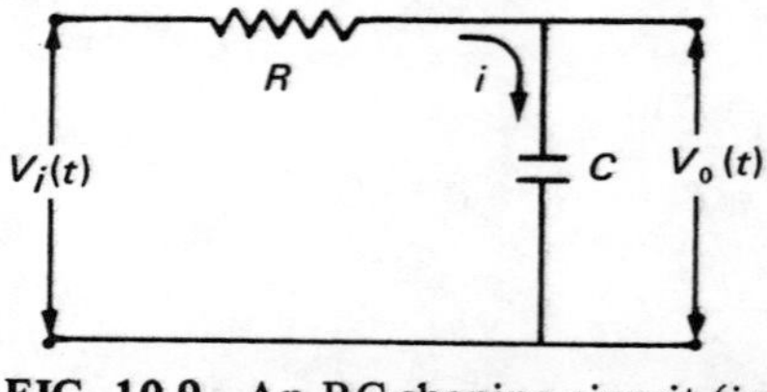

FIG. 10.9 An RC shaping circuit (integrator).

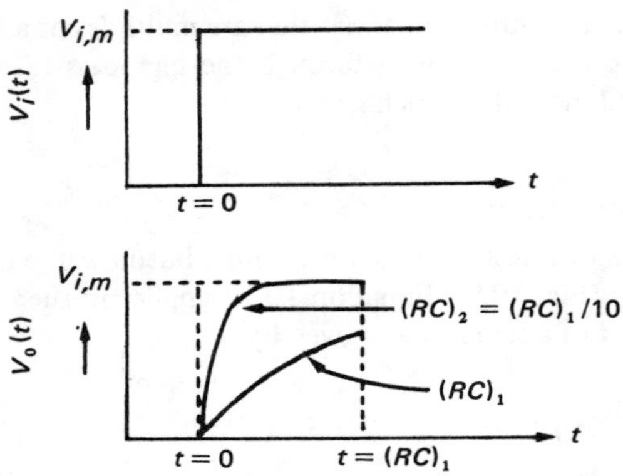

FIG. 10.10 The output signal of an RC shaping circuit for a step input.

As the frequency increases, the ratio given by Eq. 10.23 decreases, and for this reason this circuit is called a *low-pass filter.* Going through the filter, lower frequencies fare better than the higher ones. The output of the filter as a function of frequency is shown in Fig. 10.12.

10.5 DELAY LINES

Any signal transmitted through a coaxial cable is delayed by a time $T = \sqrt{LC}$ seconds per unit length, where L is the inductance per unit length and C is the capacitance per unit length. For ordinary coaxial cables, the delay is about 5 ns/m. For larger delays, the central conductor of the cable is spiraled to increase the inductance per unit length.

Commercial delay lines are a little more complicated than a simple cable. They are used not only to delay a signal, but also to produce a rectangular pulse for subsequent pulse shaping or for triggering another electronic unit (e.g., a scaler). The formation of the rectangular pulse is achieved by reflecting the delayed signal at the end

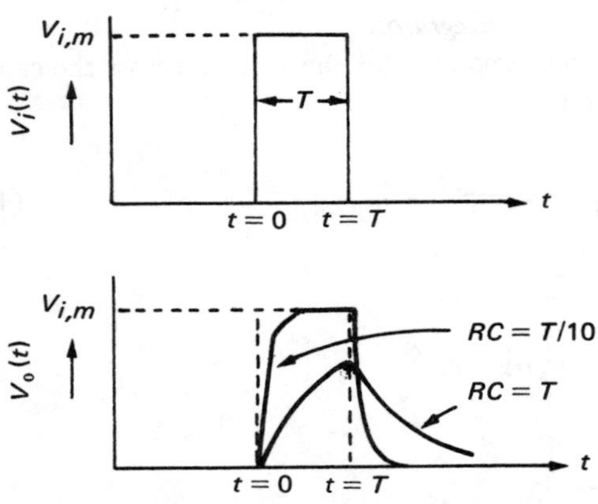

FIG. 10.11 The response of an RC circuit to a rectangular pulse.

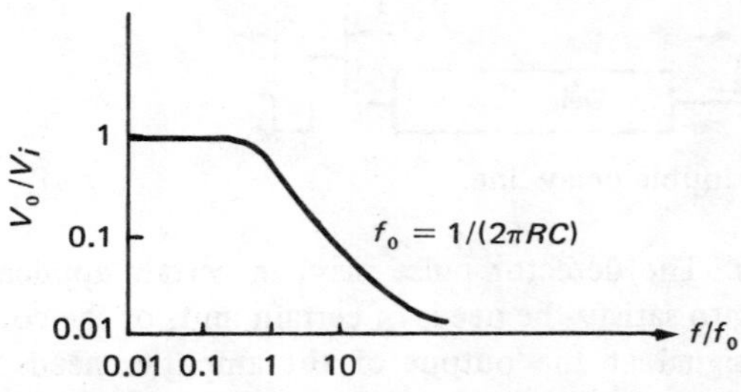

FIG. 10.12 The output of an integrator (low-pass) filter as a function of input frequency.

of the delay line, bringing it back to the input and adding it to the original signal (Fig. 10.13). A double delay line produces the double rectangular pulse shown in Fig. 10.14.

10.6 PULSE SHAPING

The pulse produced at the output of a radiation detector has to be modified or *shaped* for better performance of the counting system. There are three reasons that necessitate pulse shaping:

1 *To prevent overlap.* Each pulse should last for as short a period of time as possible, and then its effect should be abolished so that the system may be ready for the next pulse. Without pulse shaping, the detector signal lasts so long that pulses overlap. If only the number of particles is counted, pulse overlap leads to loss of counts (dead time loss). In spectroscopy measurements, pulse overlap worsens the resolution.

2 *To improve the signal-to-noise ratio.* Noise created in the detector and the early amplification stages accompanies the detector signal. Appropriate pulse shaping can enhance the signal while at the same time reduce the noise. Thus, the signal-to-noise ratio will improve, which in turn leads to better energy resolution.

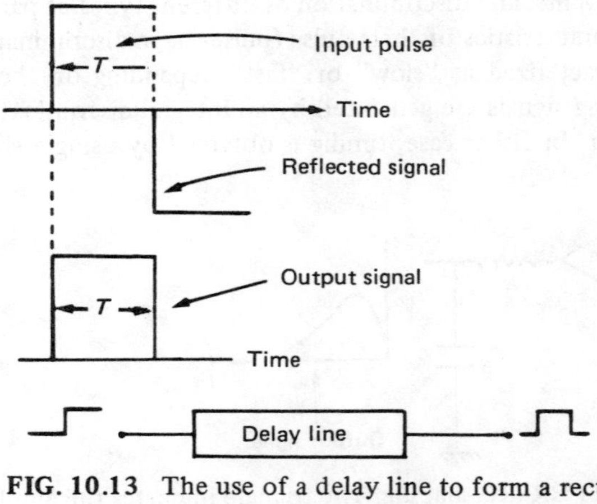

FIG. 10.13 The use of a delay line to form a rectangular pulse.

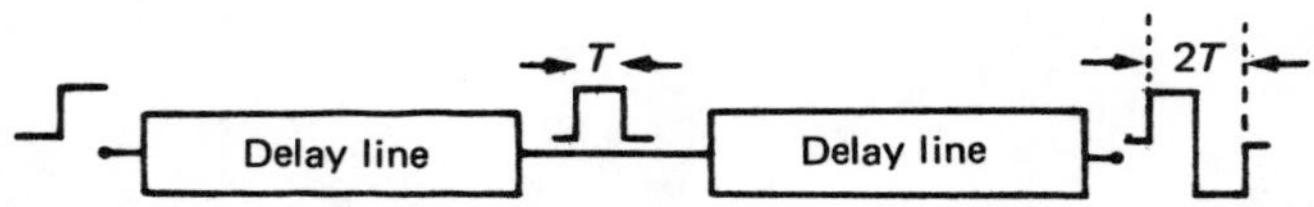

FIG. 10.14 The effect of a double delay line.

3 *For special pulse manipulation.* The detector pulse may, in certain applications, need special pulse pulse shaping to satisfy the needs of certain units of the counting system. As an example, the signal at the output of the amplifier needs to be stretched before it is recorded in the memory of a multichannel analyzer (see Sec. 10.12).

The pulse-shaping methods used today are based on combinations of RC circuits and delay lines. For example, the use of a CR–RC circuit combination produces the pulse shown in Fig. 10.15. The exact shape and size of the output pulse depends on the relative magnitudes of the time constants C_1R_1 and C_2R_2. The use of the CR-RC circuit combination provides, in addition to pulse shaping, a better signal-to-noise ratio by acting as high-pass and low-pass filter for undesired frequencies.

If one adds more RC integrating circuits, the pulse will approach a Gaussian shape (Fig. 10.16).

If one applies a CR–RC–CR combination, the result is a doubly differentiated pulse as shown in Fig. 10.17. Commercial amplifiers usually provide either singly or doubly differential pulses. In all cases, the final pulse is the result of repeated application of CR–CR circuits. Figure 10.18 shows such pulses produced by the application of many RC–CR circuits, called unipolar and bipolar pulses, respectively. Pulse shaping using a delay line and an RC circuit is shown in Fig. 10.19.

10.7 TIMING

The term *timing* refers to the determination of the time of arrival of a pulse. Timing experiments are used in measurement of the time-development of an event (e.g., measurement of the decay of a radioactive species), measurement of true coincident events out of a large group of events, and discrimination of different types of particles based on the different time characteristics of their pulse (pulse-shape discrimination).

Timing methods are characterized as "slow" or "fast" depending on the way the signal is derived. Slow timing signals are generated by an integral discriminator or a timing single-channel analyzer. In either case, timing is obtained by using a shaped

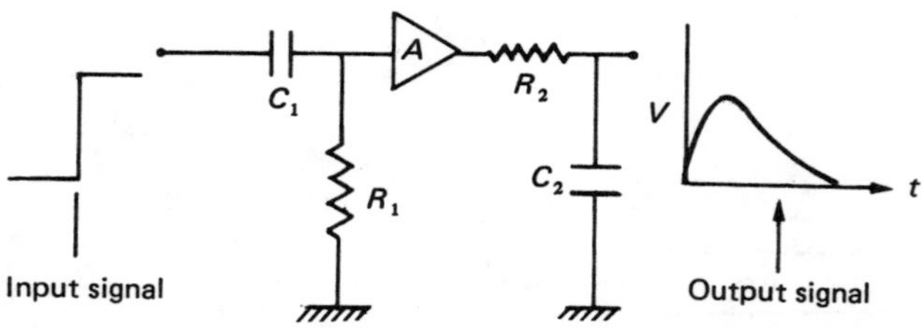

FIG. 10.15 An example of CR–RC shaping. The triangle indicates the amplification unit (A) that isolates the two shaping circuits.

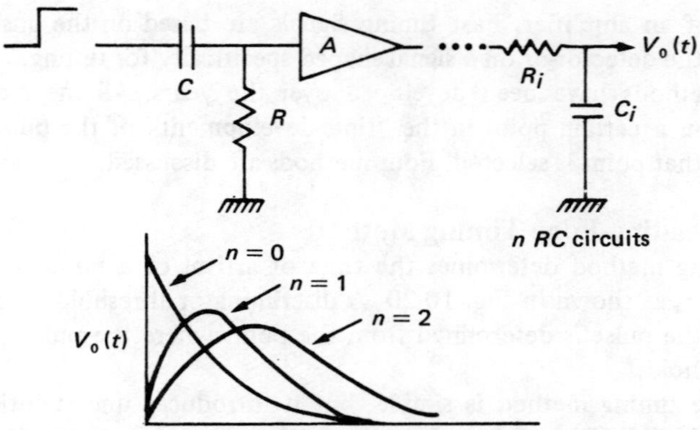

FIG. 10.16 The output pulse after using many shaping circuits. The triangle indicates the amplification unit that isolates any two consecutive shaping circuits.

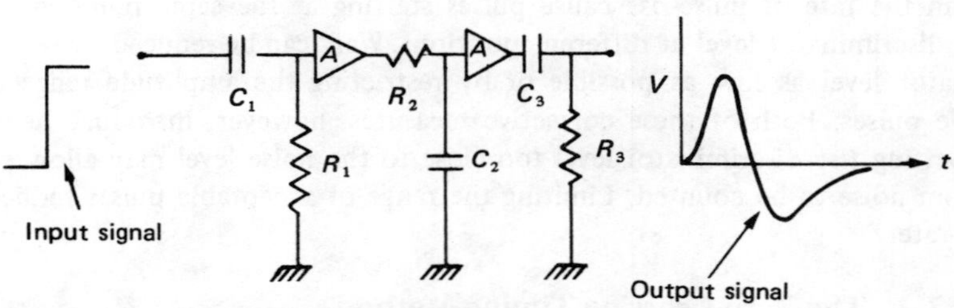

FIG. 10.17 A doubly differentiated pulse.

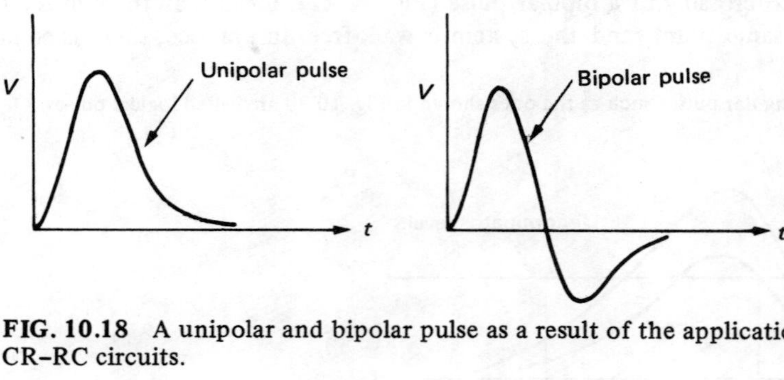

FIG. 10.18 A unipolar and bipolar pulse as a result of the application of many CR–RC circuits.

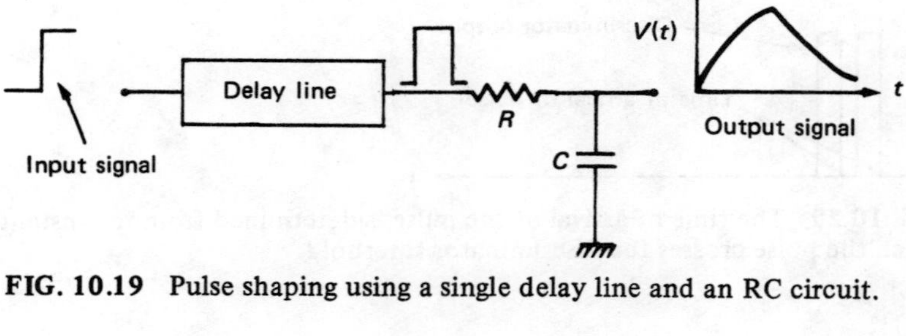

FIG. 10.19 Pulse shaping using a single delay line and an RC circuit.

signal at the output of an amplifier. Fast timing signals are based on the unshaped pulse at the output of the detector or on a signal shaped specifically for timing.

Many timing methods have been developed over the years. All the methods pick the time based on a certain point in the "time-development" of the pulse, but they differ in the way that point is selected. Four methods are discussed.

10.7.1 The Leading-Edge Timing Method

The leading-edge timing method determines the time of arrival of a pulse with the help of a discriminator, as shown in Fig. 10.20. A discriminator threshold is set and the time of arrival of the pulse is determined from the point where the pulse crosses the discriminator threshold.†

The leading-edge timing method is simple, but it introduces uncertainties because of "jitter" and "walk" (Fig. 10.21). *Jitter* is another name for electronic noise. The timing uncertainty due to jitter depends on the amplitude of the noise and the slope of the signal close to the discriminator threshold. *Walk* originates when differences in the rate of pulse-rise cause pulses starting at the same point in time to cross the discriminator level at different positions. Walk can be reduced by setting the discriminator level as low as possible or by restricting the amplitude range of the acceptable pulses. Both of these corrective measures, however, introduce new difficulties. Setting the discriminator level too close to the noise level may allow part of the random noise to be counted. Limiting the range of acceptable pulses reduces the counting rate.

10.7.2 The Zero-Crossing Timing Method

The zero-crossing method reduces the errors due to jitter and walk by picking the time from the zero crossing of a bipolar pulse (Fig. 10.22). Ideally, all the pulses cross the zero at the same point, and the system is walk-free. In practice, there is some walk

†Rectangular pulses such as the ones shown in Fig. 10.20 are called logical pulses.

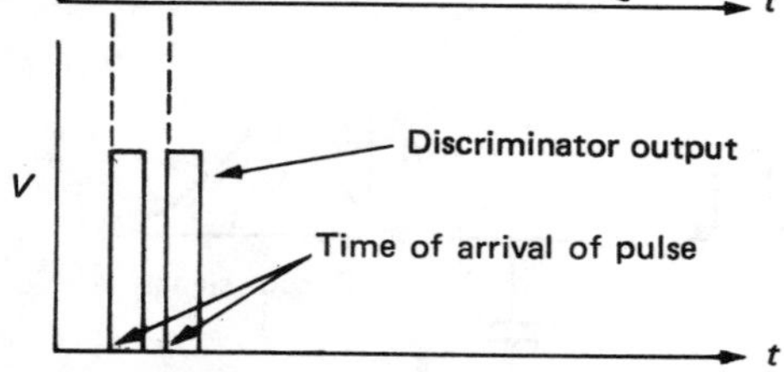

FIG. 10.20 The time of arrival of the pulse is determined from the instant at which the pulse crosses the discriminator threshold.

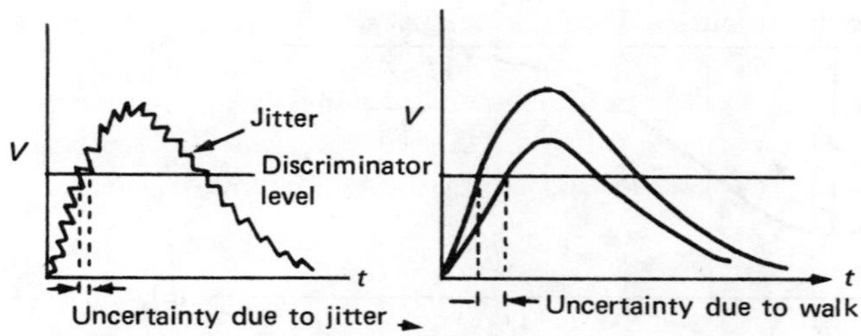

FIG. 10.21 Timing uncertainty due to jitter and walk.

because the position of zero crossing depends on pulse risetime.† The dependence on pulse risetime is particularly important for Ge(Li) detectors because the pulses produced by Ge(Li) detectors exhibit considerable variations in their time characteristics. To reduce the uncertainties still present with the zero-crossing method, the constant-fraction method and the ARC method have been developed specifically for Ge(Li) detectors.

10.7.3 The Constant-Fraction Timing Method

The principle of constant-fraction timing is shown in Fig. 10.23. First the original pulse A is attenuated by a factor *f* equal to the fraction of the pulse height on which the timing will be based (B). The original pulse is inverted and delayed (C) for a time longer than its rise time. Finally the signals C and B are added to give D. The time pick-off, taken as the zero-crossing point, is thus defined by the preselected fraction of the pulse height and is independent of the pulse amplitude. It can be shown that pulses with the same risetime always give the same zero-crossing time.

10.7.4 The ARC Timing Method

The amplitude and risetime compensation (ARC) method[1] is a variation of the constant-fraction technique, from which it differs by the amount of pulse delay

†Pulse risetime is taken as the time it takes the pulse to increase from 10 percent to 90 percent of its value.

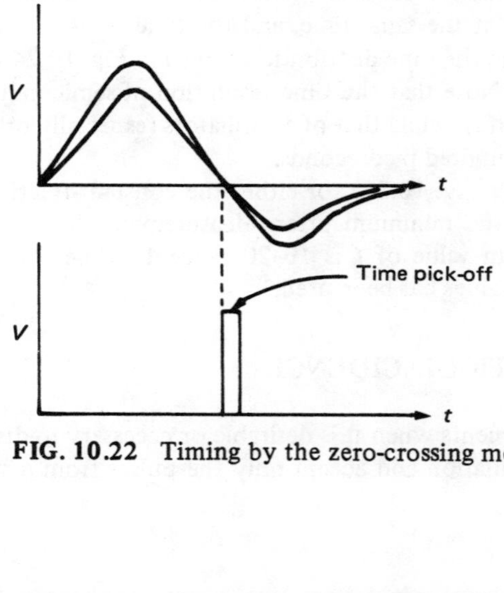

FIG. 10.22 Timing by the zero-crossing method.

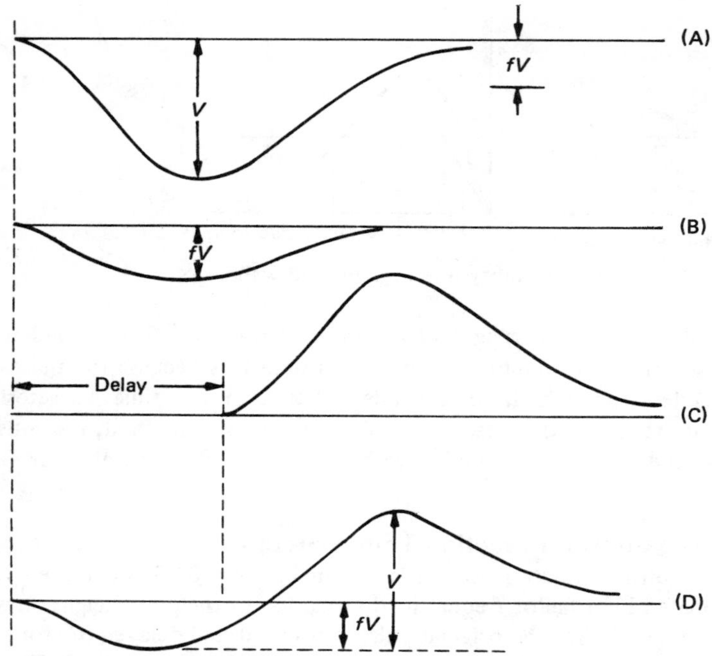

FIG. 10.23 Timing by the constant-fraction technique.

employed. The delay used by the constant-fraction method (Fig. 10.23) is longer than the risetime of the pulse, while that of the ARC method is shorter than the risetime of the pulse. The ARC method is the one most frequently used with Ge(Li) detectors. A comparison between the timing spectrum obtained with the ARC method versus that obtained with the leading-edge method is shown in Fig. 10.24. This timing spectrum was obtained by counting in coincidence the two annihilation gammas from a positron-emitting source (e.g., ^{22}Na) using a Ge(Li) and a NaI detector. The detection of one 0.511-MeV photon by the NaI detector defined the zero time with respect to which pulses generated by the Ge(Li) detector were timed. Since both annihilation gammas arrived at the two detectors at the same time, and the time response of the NaI is known to be less than 1 ns, then the time distribution shown in Fig. 10.24 must be due to the Ge(Li) detector alone. Note that the time resolution of semiconductor detectors is of the order of nanoseconds, while that of scintillators (especially organic scintillators) is of the order of a few hundred picoseconds.

The value of the fraction f (see Fig. 10.23) for either the constant-fraction or the ARC method is chosen to give the minimum jitter. Measurements have shown that for scintillators,[2,3] the optimum value of f is 10–20 percent, while for semiconductor detectors, a wide range of values has been used.[4,5]

10.8 COINCIDENCE–ANTICOINCIDENCE MEASUREMENTS

There are times in radiation measurements when it is desirable or necessary to discard the pulses due to certain types of radiation and accept only the pulses from a single type of particle or from a particle or particles coming from a specific direction. Here are two examples of such measurements.

1. *Detection of pair-production events.* When pair production occurs, two 0.511-MeV gammas are emitted back-to-back. To insure that only annihilation photons are counted, two detectors are placed 180° apart, and only events that register simultaneously (coincident events) in both detectors are recorded.

2. *Detection of Internal Conversion Electrons.* Radioisotopes emitting internal conversion (IC) electrons also emit gammas and x-rays. The use of a single detector to count electrons will record not only IC electrons but also Compton electrons produced in the detector by the gammas. To eliminate the Compton electrons, one can utilize the x-rays that are emitted simultaneously with the IC electrons. Thus, a second detector is added for x-rays and the counting system is required to record only events that are coincident in the two detectors. This technique excludes the detection of Compton electrons.

Elimination of undesirable events is achieved by using a coincidence (or anti-

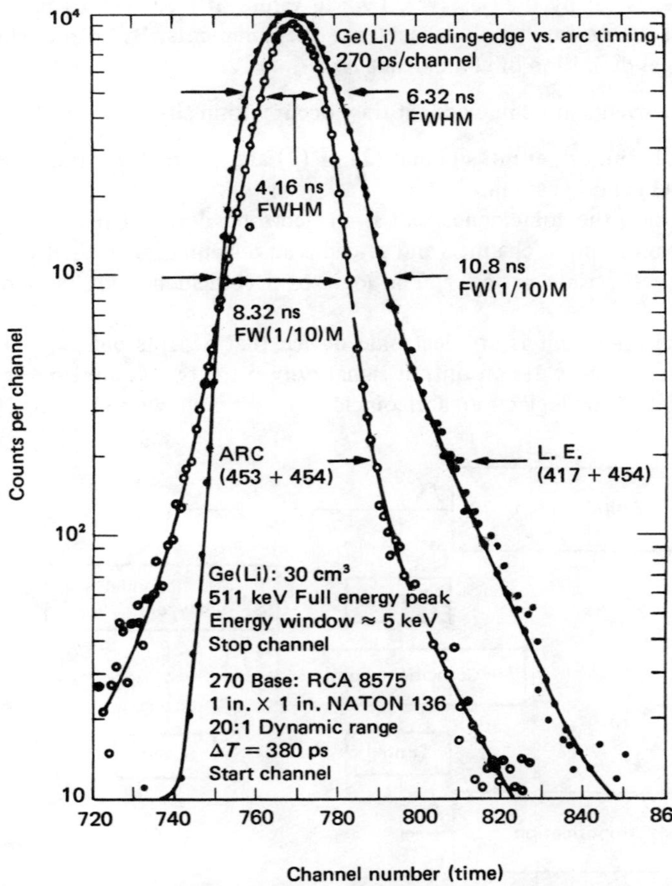

FIG. 10.24 A Ge(Li) timing spectrum taken with the ARC and the leading-edge methods (from ORTEC).

coincidence) unit. Consider the counting system shown in Fig. 10.25. The source emits particles detected by detectors 1 and 2. After amplification, the detector signals are fed into a timing circuit, which in turn generates a pulse signifying the time of occurrence in the detector of the corresponding event 1 or 2. The timing signals are fed into a coincidence unit, so constructed that it produces an output signal only when the two timing pulses are coincident. If the objective is to count only the number of coincident events, the output of the coincidence unit is fed into a scaler. If, on the other hand, the objective is to measure the energy spectrum of particles counted by detector 1 in coincidence with particles counted by detector 2, the output signal of the coincidence unit is used to "gate" a multichannel analyzer (MCA) that accepts the energy pulses from detector 1. The gating signal permits the MCA to store only those pulses from detector 1 that are coincident with events in detector 2.

In theory, a true coincidence is the result of the arrival of two pulses at exactly the same time. In practice this "exact coincidence" seldom occurs, and for this reason a coincidence unit is designed to register as a coincident event pulses arriving within a finite but short time interval τ. The interval τ, called the *resolving time* or the *width* of the coincidence, is set by the observer. Typical values of τ are 1-5 μs for "slow" coincidence and 1-10 ns for "fast" coincidence measurements. By introducing the width τ, the practical definition of coincidence is:

Two or more events are coincident if they occur within the time period τ.

According to this definition, events (1) and (2) or (2) and (3) in Fig. 10.26 are coincident, but events (1) and (3) are not.

As stated earlier, the coincidence unit is an electronic device that accepts pulses (events) in two or more input channels and provides an output signal *only* if the input pulses arrive within the time period τ. The logic of a coincidence unit is shown in Fig. 10.27.

An anticoincidence unit is an electronic device that accepts pulses (events) in two input channels and provides an output signal *only* if the two events *do not* arrive with the time period τ. The logic of an anticoincidence unit is shown in Fig. 10.28.

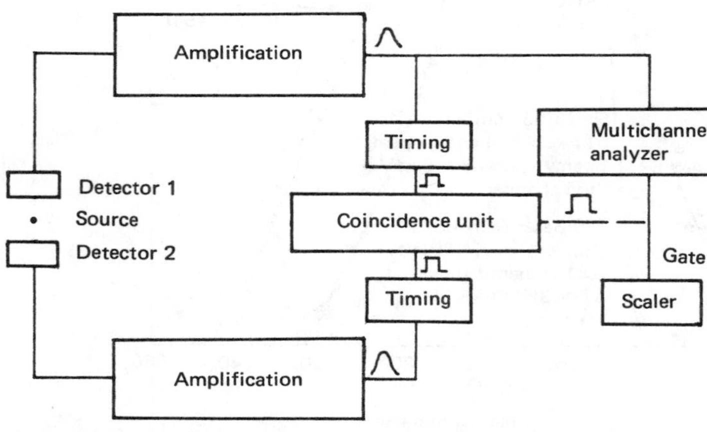

FIG. 10.25 A simple coincidence measurement setup.

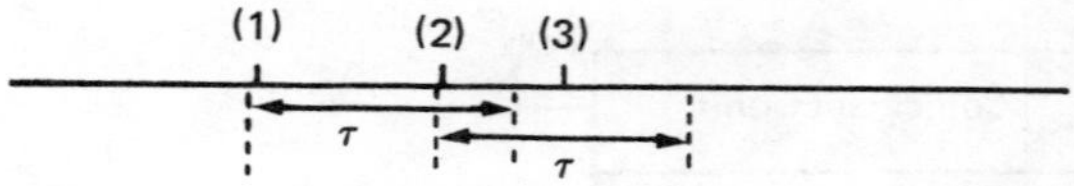

FIG. 10.26 Events [(1) and (2)] or [(2) and (3)] are coincident. Events [(1) and (3)] are not.

For a successful coincidence or anticoincidence measurement, the detector signals should not be delayed by any factors other than the time of arrival of the particles at the detector. If it is known that it takes longer to generate the signal in one detector than in another, the signal from the fast detector should be delayed accordingly to compensate for this difference. This compensation is accomplished by passing the signal through a delay line before it enters the coincidence unit. A delay line is always needed if the detectors used in the coincident measurement are not identical. The value of the relative delay needed is determined as follows.

Consider what is probably the simplest type of coincidence circuit, the additive type shown in Fig. 10.29. The coincidence unit is summing the input pulses. When two pulses overlap, their sum exceeds a discriminator threshold and the unit produces an output pulse. If the width of the input pulse is T, the resolving time is essentially $\tau = 2T$. Assume now that a system has been set up as shown in Fig. 10.25, with the addition of a delay line in channel 2 between the timing and the coincidence units. If one measures the number of coincidences as a function of the delay between the two signals, the result will be the delay or resolving-time curve shown in Fig. 10.30. The proper relative delay is the value corresponding to the center of the flat region. The ideal (rectangular) curve will be obtained if the time jitter is zero.

Since the pulses from the two detectors arrive randomly, a certain number of accidental (or chance) coincidences will always be recorded. Let

r_1 = counting rate of detector 1

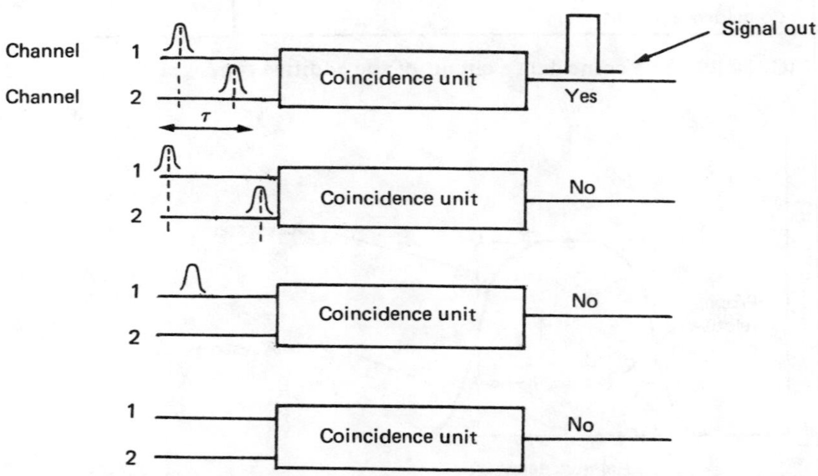

FIG. 10.27 The logic of a coincidence unit with two input channels.

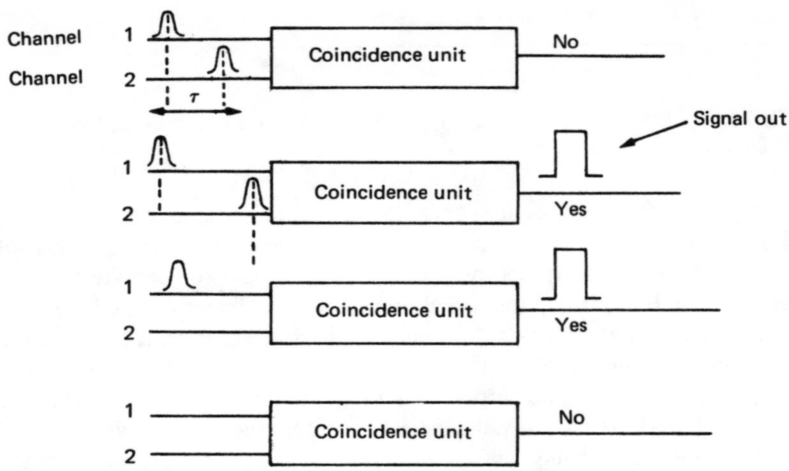

FIG. 10.28 The logic of an anticoincidence unit.

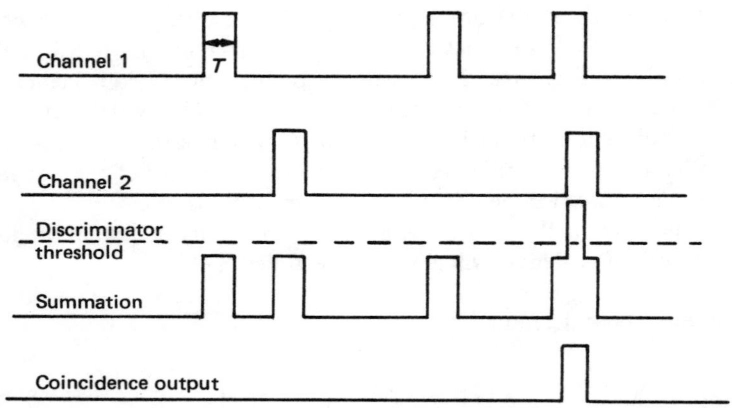

FIG. 10.29 The coincidence circuit of the additive type.

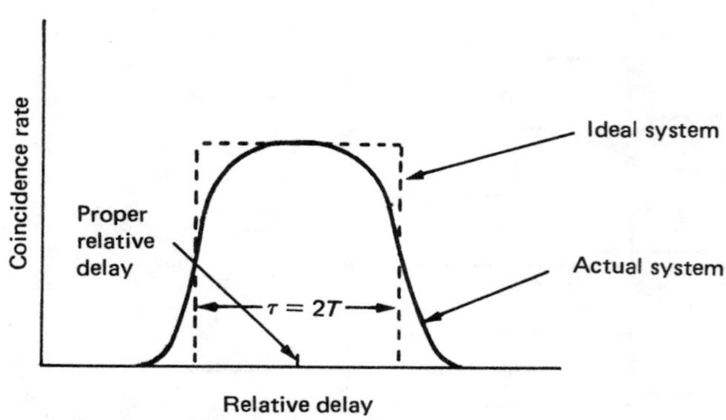

FIG. 10.30 The delay or resolving-time curve.

r_2 = counting rate of detector 2

r_a = accidental coincidence rate

Consider a single pulse in channel 1. If a pulse occurs in channel 2 within the time period τ, then a coincidence will be registered. Since the number of pulses in channel 2 during time τ is $r_2 \tau$, the rate of accidental coincidences is

$$r_a = r_1 r_2 \tau \tag{10.24}$$

Equation 10.24 gives accidental coincidences of first order. Corrections for multiple coincidences of higher order have also been calculated.[6,7]

If S is the strength of the source, ϵ_1 and ϵ_2 the efficiencies, Ω_1 and Ω_2 the solid angle factors, and F_1 and F_2 any other factors that affect the measurement of particles counted by detectors 1 and 2 (see Chap. 8), then the true coincidence rate r_t is given by

$$r_t = S\epsilon_1 \epsilon_2 \Omega_1 \Omega_2 F_1 F_2 \tag{10.25}$$

and from Eq. 10.24 the accidental coincidence rate is

$$r_a = S^2 \epsilon_1 \epsilon_2 \Omega_1 \Omega_2 F_1 F_2 \tau \tag{10.26}$$

The *figure of merit* in a coincidence experiment is the ratio

$$Q = \frac{r_t}{r_a} = \frac{1}{S\tau} \tag{10.27}$$

which should be as high as possible. Equation 10.27 shows that this ratio improves when S and τ decrease. Unfortunately, the values for both of these quantities have constraints. The value of τ is limited by the performance of the detector and by the electronics. The source strength S has to be of a certain value for meaningful counting statistics to be obtained in a reasonable time. It is interesting to note that when the source strength increases, both true and accidental coincidence rates increase but the ratio Q (Eq. 10.27) decreases, because $r_t \propto S$ but $r_a \propto S^2$.

Another coincidence technique involves the use of a *time-to-amplitude converter* (TAC). A TAC is an electronic unit that converts the time difference between two pulses into a voltage pulse between 0 and 10 V. The height of the pulse is proportional to the time difference between the two events. The time spectrum of the two detectors is stored directly in the MCA. A "time" window is set around the coincidence peak (Fig. 10.31). A second window of equal width is set outside the peak to record accidental coincidences only.

The advantages of using a TAC are

1 No resolving curve need be taken.
2 No resolving time is involved.
3 The number of channels and the range of time intervals analyzed may be changed over a wide range by simply changing the conversion gain of the MCA.

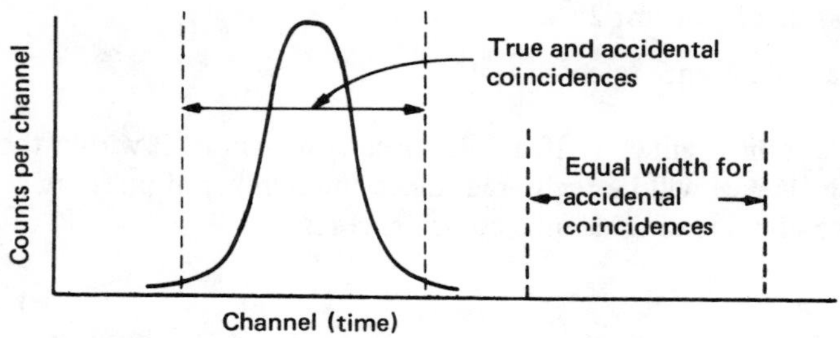

FIG. 10.31 An MCA spectrum taken with a TAC.

Most TACs cannot distinguish the sequence of events–i.e., they cannot tell if a pulse from channel 1 precedes a pulse from channel 2 and vice versa. To avoid this ambiguity and also to create a measurable difference between the pulses, the signal from one detector is usually shifted by a fixed delay.

Figure 10.32 shows what is now a common counting system for γ-γ coincidence measurements using a NaI(Tl) detector and a Ge(Li) detector.

10.9 PULSE-SHAPE DISCRIMINATION (PSD)

Pulse-shape discrimination (PSD) is the name given to a process that differentiates pulses produced by different types of particles in the same detector. Although PSD

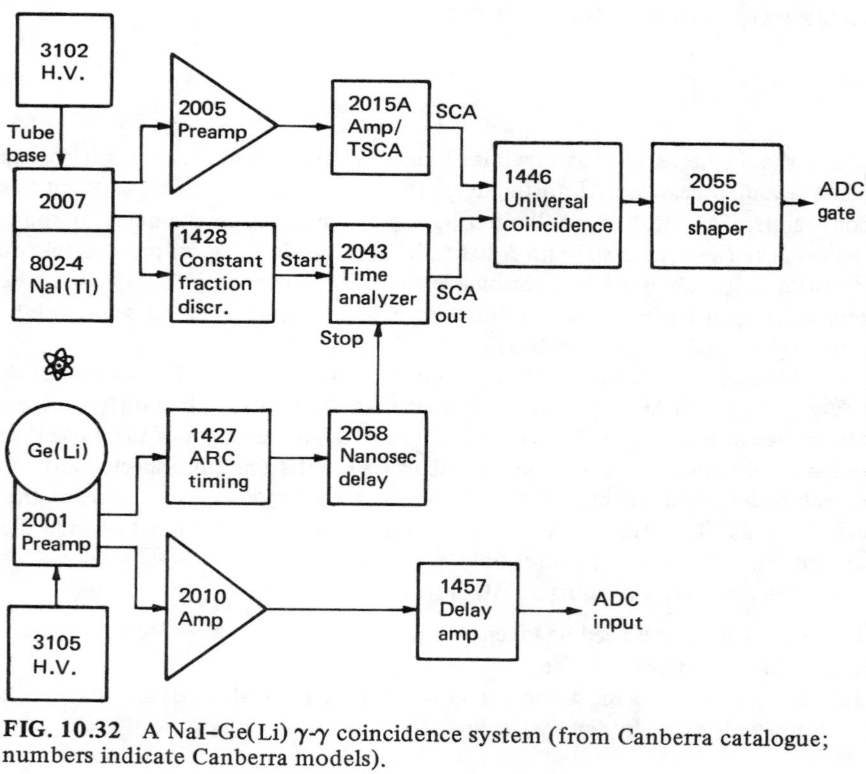

FIG. 10.32 A NaI–Ge(Li) γ-γ coincidence system (from Canberra catalogue; numbers indicate Canberra models).

has found many applications, its most common use is to dscriminate between pulses generated by neutrons and gammas in organic scintillators (see also Chap. 14), and it is this type of PSD that will be discussed.

Measurement of the amount of light produced in organic scintillators by neutrons and gammas shows that both the differential and integral light intensities are different as functions of time. Figure 10.33, presenting the results of Kuchnir and Lynch,[8] illustrates this point. It is obvious that the pulses from neutrons and gammas have different time characteristics, and it is this property that is used as the basis for PSD.

Many different methods have been proposed and used for successful PSD.[9-14] One method doubly differentiates the detector pulse, either using CR circuits or a delay line, and bases the PSD on the time interval between the beginning of the pulse and the zero crossing point. This time interval, which is essentially independent of the pulse amplitude but depends on the pulse shape, is usually converted into a pulse by means of a TAC. The pulse from the TAC may be used to gate the counting system. Figure 10.34 shows a block diagram for such a counting system. The result of n-γ discrimination is usually a spectrum that resembles Fig. 10.35. Actually, the γ peak

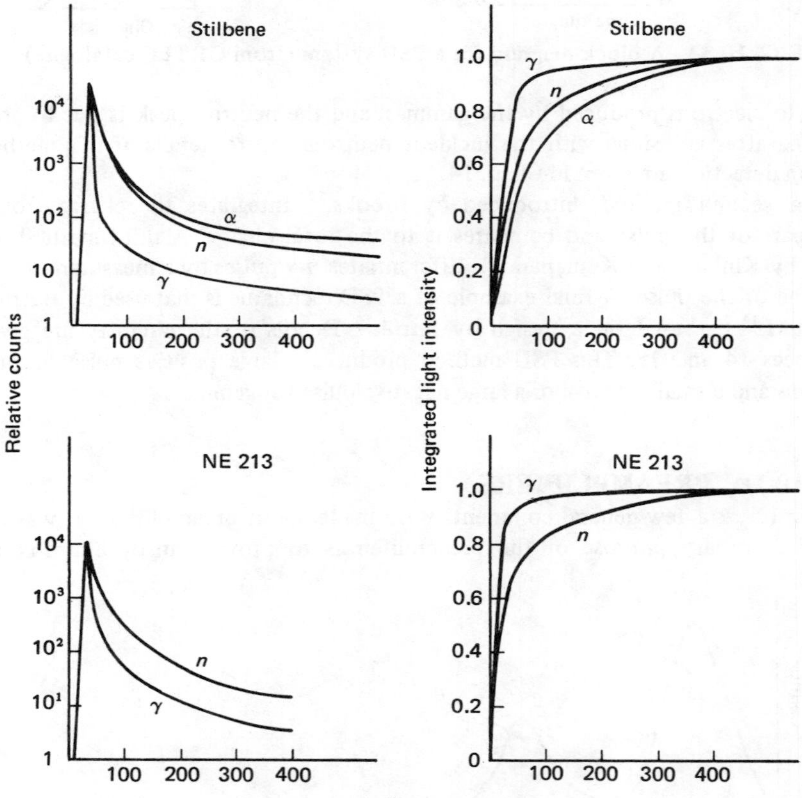

FIG. 10.33 The light produced by neutrons and gammas in stilbene and NE 213. Light intensity is shown at left; integrated light intensity is shown at right. (Based on data of reference 8.)

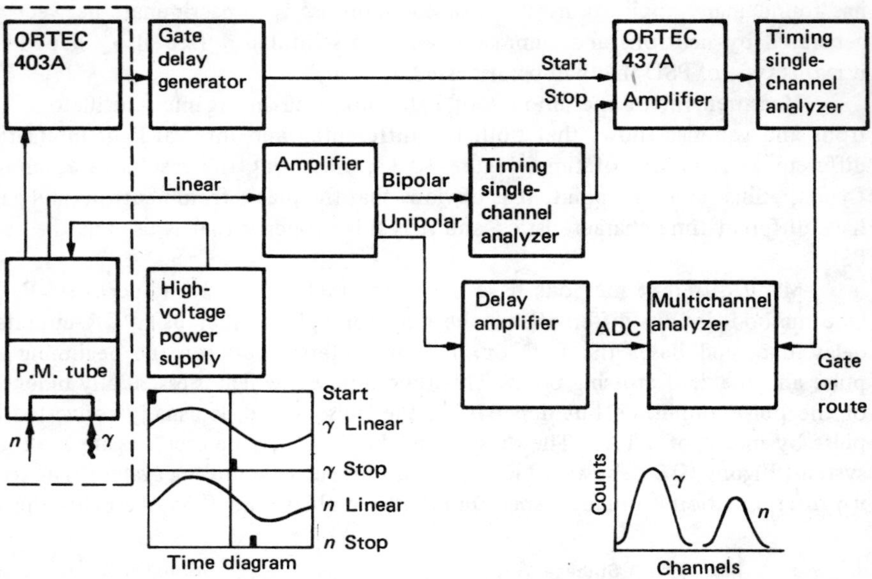

FIG. 10.34 A block diagram for a PSD system (from ORTEC catalogue).

is due to electrons produced by the gammas, and the neutron peak is due to protons recoiling after collisions with the incident neutrons. More details of this method of neutron detection are given in Chap. 14.

A second method, introduced by Brooks,[12] integrates the charge from the early part of the pulse and compares it to the total charge. A third method, introduced by Kinbara and Kumahara,[13] differentiates n-γ pulses by a measurement of the rise time of the pulse. A final example of a PSD technique is that used by Burrus and Verbinski,[14,15] based on a design by Forté.[16] Details of the circuitry are given in references 14 and 15. This PSD method produces a large positive pulse output for neutrons and a small positive or a large negative pulse for gammas.

10.10 PREAMPLIFIERS

In Sec. 1.5.5, a few general comments were made about preamplifiers. It was stated that the primary purpose of the preamplifier is to provide an optimum coupling

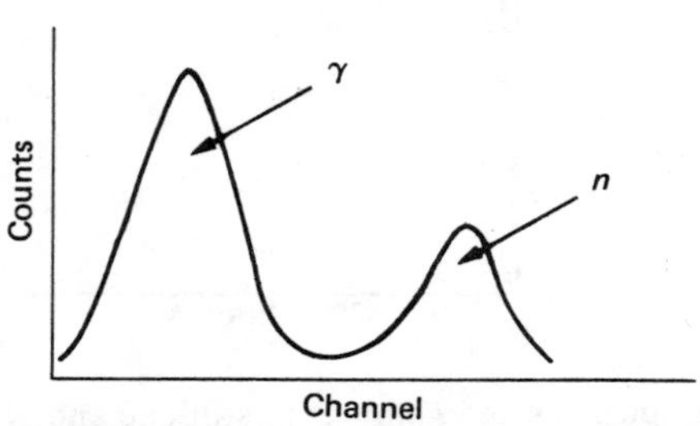

FIG. 10.35 The result of γ-n discrimination using PSD.

between the detector and the rest of the counting system. A secondary purpose of the preamplifier is to minimize any sources of noise, which will be transmitted along with the pulse and thus degrades the energy resolution of the system. This second objective, low noise, is particularly important with semiconductor detectors, which are the counters offering the best energy resolution.

There are three basic types of preamplifiers: charge-sensitive, current-sensitive, and voltage-sensitive. The voltage-sensitive preamplifier is not used in spectroscopy because its gain depends on the detector capacitance, which in turn depends on the detector bias. The charge-sensitive preamplifier is the most commonly used in spectroscopic measurements and the only type used with semiconductor detectors.

To understand the basic features of a charge-sensitive preamplifier, consider the basic circuit associated with a semiconductor detector, shown in Fig. 10.36. The HV bias applied to the detector is usually connected to the first stage of the charge-sensitive preamplifier. In Fig. 10.36, C_f is the feedback capacitor (~1 pF) and R_f is the feedback resistor (~1000 MΩ). The triangle with the letter A indicates the first stage of the preamplifier, which today is usually a field-effect transistor (FET). The FET is a p–n junction with reverse bias, exhibiting extremely low noise. The type of coupling shown in Fig. 10.36 is called dc coupling. There is an ac coupling, too, in which the detector is coupled to the FET through a coupling capacitor (see Nicholson, p. 180).

The detector sees the FET stage as a large capacitor of magnitude AC_f. As long as $AC_f \gg C_i$, where C_i is the total input capacitance consisting of the detector capacitance C_d, the cable capacitance, etc., the voltage at the output of the preamplifier is equal to

$$V_0 = \frac{Q}{C_f} \tag{10.28}$$

where Q, the charge produced in the detector, is given by

$$Q = \frac{Ee}{w} \tag{10.29}$$

where E = energy of the paarticle
e = electronic charge = 1.6×10^{-19} coulombs
w = average energy required to produce one electron-hole pair

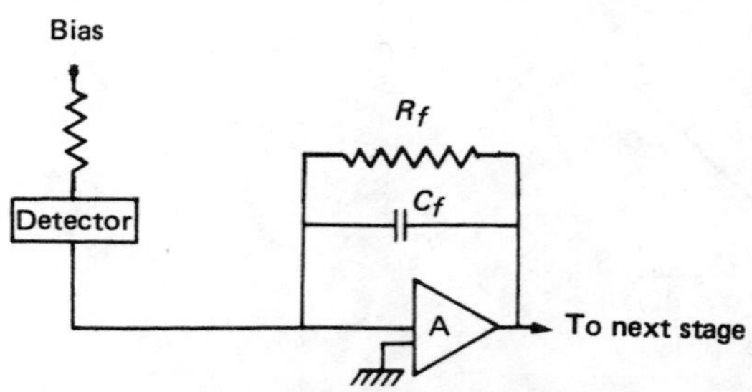

FIG. 10.36 The first stage of a charge-sensitive preamplifier dc-coupled to the bias circuit.

The major components of C_i are the detector capacitance C_D and that of the cables between the detector and the preamplifier. Both of these components are controlled by the user.

The noise of the charge sensitive preamplifier depends on three parameters: the noise of the input FET, the input capacitance C_i, and the resistance connected to the input. The noise can be determined by injecting a charge Q, equivalent to E, into the preamplifier and measuring the amplitude of the generated pulse. Commercial preamplifiers are provided with a test input for that purpose. In general, the noise expressed as the width (keV) of a Gaussian distribution increases as input capacitance increases (Fig. 10.37).

The output pulse of the preamplifier has a fast risetime (on the order of nanoseconds) followed by a slow exponential decay, ~100 μs (Fig. 10.38). The useful information in the pulse is its amplitude and its risetime. The risetime is particularly important when the signal is going to be used for timing. The observer should be aware that the risetime increases with external capacitance. The preamplifier pulse is shaped in the amplifier by the methods described in Sec. 10.6.

The *sensitivity* (or *gain*) of a charge-sensitive preamplifier is expressed by the ratio V/E, where V is given by Eq. 10.38. For a 1-MeV particle in a germanium detector, the sensitivity is (using $C_i \approx 5$ pF)

$$\frac{V}{E} = \frac{Q}{EC_i} = \frac{Ee}{EC_i w} = \frac{e}{C_i w} = \frac{1.6 \times 10^{-19}}{(5 \times 10^{-12})(3 \times 10^{-6})} = 10 \text{ mV/MeV}$$

A current-sensitive preamplifier is used to transform fast current pulses produced by a photomultiplier into a voltage pulse. The current-sensitive preamplifier is an amplifying instrument. The sensitivity (or gain) of such a unit is expressed as V_{out}/I_{in}, i.e., in mV/mA with typical values of the order of 500 mV/mA. The risetime of the pulse is ~1 ns.

10.11 AMPLIFIERS

As explained in Sec. 1.5.6, the amplifier plays the two roles of amplifying and shaping the signal. The need for amplification is obvious. The output signal of the preamplifier, being in the range of a few millivolts, cannot travel very far or be manipulated in any substantial way without losing the information it carries or being itself lost in the noise.

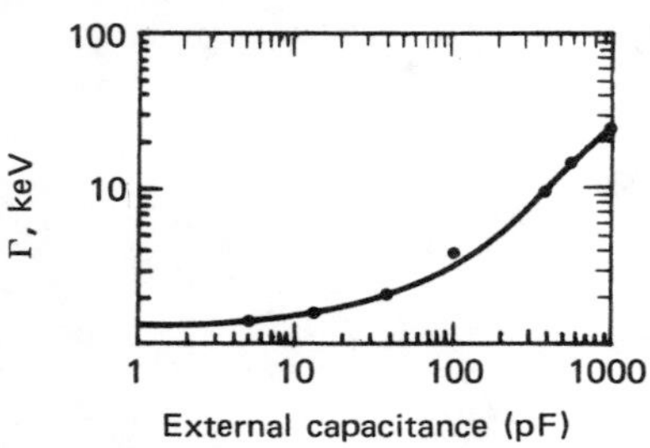

FIG. 10.37 The dependence of noise on external capacitance for a typical charge-sensitive preamplifier (from reference 25).

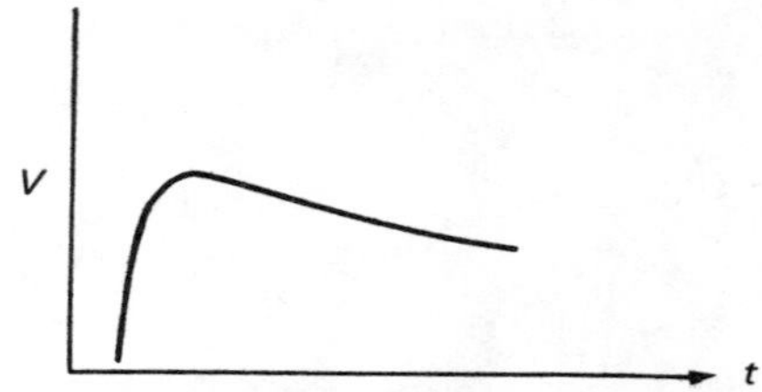

FIG. 10.38 Typical pulse from a charge-sensitive preamplifier.

Commercial amplifiers consisting of many amplification stages increase the amplitude of the input signal by as many as 2000 times, in certain models.

The need for shaping the signal was explained in Sec. 10.6. The type of shaping that is applied depends on the requirements of the measurement. For spectroscopy measurements where good energy resolution is the important parameter, pulse shaping should decrease the signal-to-noise ratio. For timing measurements, depending on the method of time pick-off, the signal may be singly or doubly differentiated or be shaped by a single or double delay line.

For a good measurement, the amplifier should satisfy many requirements.[19-23] Not all types of measurements, however, require the same level of performance. For example, of one measures only the number of particles and not their energy, the precision and stability of the amplification process can be relatively poor. It is in spectroscopy measurements, particularly measurements using semiconductor detectors, that the requirements for precision and stability are extremely stringent. Since the energy resolution of Ge(Li) detectors is of the order of 0.1 percent, the dispersion of the pulses due to the amplification process should be much less, about 0.01 percent.

An ideal spectroscopy amplifier should have a constant amplification for pulses of all amplitudes without distorting any of them. Unfortunately, some pulse distortion is always present because of electronic noise, gain drift due to temperature, pulse pile-up, and limitations on the linearity of the amplifier.

The effect of electronic noise on energy resolution was discussed in Sec. 9.6.2. Random electronic noise added to equal pulses makes them unequal (see Fig. 9.12). Gain drift of an amplifier is caused by small changes in the characteristics of resistors, capacitors, transistors, etc., as a result of temperature changes. The value of the gain drift, always quoted by the manufacturer of the instrument, is for commercial amplifiers of the order of 0.005 percent per °C or less.

Since the time of arrival of pulses is random, it is inevitable that a pulse may arrive at a time when the previous one did not fully decay. Then the incoming pulse "piles up" on the tail of the earlier one and appears to have a height different than its true one. Pulse pile-up depends on the counting rate.

The linearity of an amplifier is expressed as differential and integral.

Differential nonlinearity is a measure of the change in amplifier gain as a function of amplifier output signal. Referring to Fig. 10.39, the differential nonlinearity in percent is given by

$$\frac{(\Delta V_0/\Delta V_i)A}{(\Delta V_0/\Delta V_i)B}(100) \tag{10.30}$$

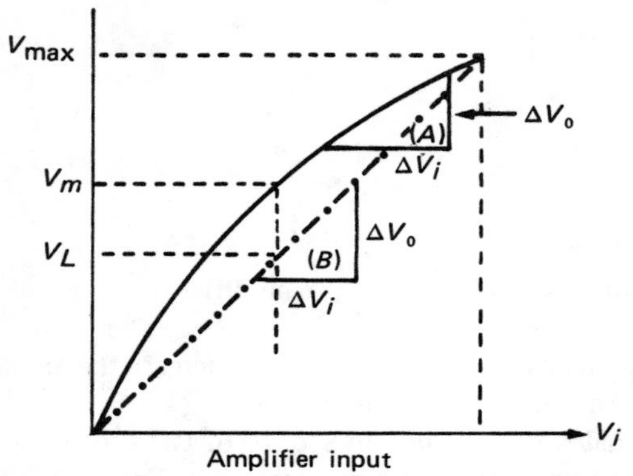

FIG. 10.39 Diagram used for the definition of differential and integral linearity of an amplifier. The output signal of a perfect amplifier plotted versus input signal should give the straight line shown (— • — • — •).

In Eq. 10.30, the numerator is the slope of the amplifier gain curve at the point where the nonlinearity is measured, and the denominator is the slope of the straight line as shown in Fig. 10.39.

Integral nonlinearity is defined as the maximum vertical deviation between the straight line shown in Fig. 10.39 and the actual amplifier gain curve, divided by the maximum rated output of the amplifier. Referring to Fig. 10.39, the integral nonlinearity in percent is given by

$$\frac{V_m - V_L}{V_{max}}(100) \tag{10.31}$$

The integral nonlinearity is one of the specifications of commercial amplifiers and has a value of about 0.05 percent or less over the range 0 to 10 V.

There are many types of commercial amplifiers designed to fit the specific needs of spectroscopic or timing measurements. Companies like Canberra, EG&G (Ortec), and Tennelec offer a wide selection of such instruments.

10.12 ANALOG-TO-DIGITAL CONVERTERS (ADCs)

As discussed in Sec. 9.9, the backbone of an MCA is the analog-to-digital converter (ADC), the unit that digitizes the input pulse height and assigns it to a specific channel. Many types of ADCs have been developed, but the most frequently used is the Wilkinson type.[24]

The Wilkinson-type ADC operates as shown in Fig. 10.40. When a pulse enters the MCA, two events are initiated:

1 A capacitor starts charging.
2 An input gate prevents the acceptance of another pulse until this one is fully processed and registered.

The capacitor keeps charging until the peak of the pulse is reached. At that point in time, two new events are initiated:

1 The voltage on the charged capacitor is discharged by a constant current.
2 An oscillator-clock starts to be counted. The clock stops being counted when the capacitor is fully discharged.

The number of oscillations during this time—called *rundown* or *conversion time*—is proportional to the pulse height and constitutes the information that determines the channel number of that pulse.

A variation of this method is shown in Fig. 10.41. The steps followed in this case are:

1 The input pulse is stretched in such a way that its flat portion is proportional to its height.
2 At the moment the pulse reaches its maximum (time t_1), a linear ramp generator is triggered producing a voltage C.
3 At the same moment (t_1), a gate signal is produced and an oscillator-clock is turned on.

When the voltage ramp signal reaches the flat part of the stretched pulse (P), the gate signal turns the clock off. Thus, the time interval $(t_2 - t_1)$ and, therefore, the number of oscillations during $(t_2 - t_1)$ are again proportional to the height of the pulse. This second method of ADC operation (Fig. 10.41) is not favored because it is difficult to keep the pulse height constant for the time interval $(t_2 - t_1)$.

Figure 10.40 shows, in addition to the principle of operation of the Wilkinson

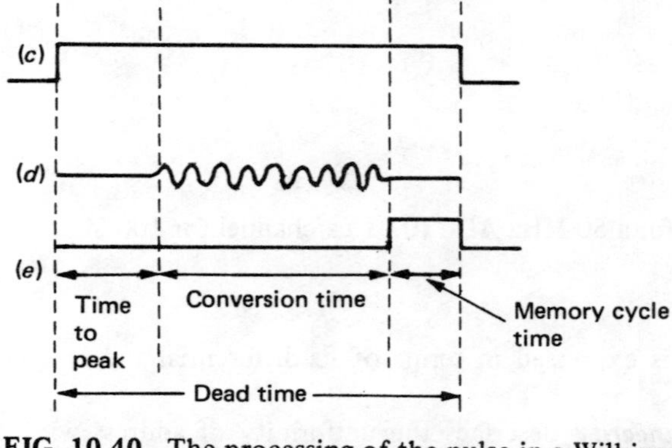

FIG. 10.40 The processing of the pulse in a Wilkinson ADC. (*a*) Input pulse. (*b*) Charge capacitor voltage. (*c*) Gate stops other pulses. (*d*) Oscillator clock. (*e*) Pulse storage.

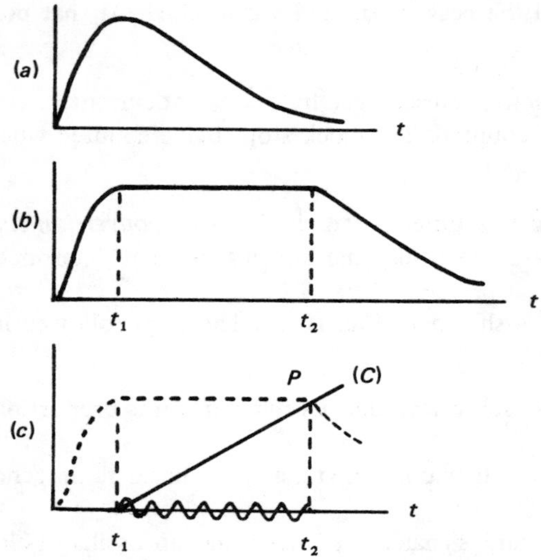

FIG. 10.41 The processing of the pulse by the ADC. (*a*) Input pulse. (*b*) Input pulse stretched; flat part proportional to pulse height. (*c*) Ramp voltage and oscillator clock start at t_1.

ADC, the reason for the dependence of the MCA dead time on the channel number. The dead time consists of three components:

1 Pulse risetime component
2 Conversion time component
3 Memory cycle time component (time it takes to store the digitized signal)

Of the three components, the second is the most important because it depends on the channel number. One can reduce the size of the conversion time by using a clock with higher frequency. Today's ADCs use quartz-stabilized clocks with a frequency of 50 to 200 MHz. Based on the three components given above, the dead time τ is given by

$$\tau(c) = a_1 + a_2 c + a_3 \tag{10.32}$$

Typical values of these constants are:

$a_1 \approx 2.5\ \mu s$

$a_2 = 0.02\ \mu s$/channel for a 50-MHz ADC (0.01 μs/channel for 100 MHz)

$a_3 \approx 0–7\ \mu s$

The accuracy of the ADC is expressed in terms of its differential and integral nonlinearity.

The *differential nonlinearity* describes the uniformity of address widths over the entire range of the ADC. To make this point better understood, assume that a 1000-channel ADC is used to process pulses with maximum height of 10 V. Then the average address width is $10/1000 = 1$ mV/channel. The ideal ADC should provide a conversion of 1 mV/channel at any channel. Any deviation between this width and the actual one is expressed by the differential nonlinearity. Mathematically, if

$$\overline{\Delta V} = \text{average width}$$

$$\Delta V_{max} = \text{maximum width}$$

$$\Delta V_{min} = \text{minimum width}$$

then the differential nonlinearity is given by the equation

$$\% \text{ Differential nonlinearity} = \frac{\Delta V_{max} - \Delta V_{min}}{\overline{\Delta V}} (100) \tag{10.33}$$

Commercial ADCs have differential nonlinearity of the order of ±0.5 percent to ±1 percent.

The *integral nonlinearity* is defined as the maximum deviation of any address (ADC channel) from its nominal position, determined by a linear plot of address (ADC channel) versus input pulse amplitude (Fig. 10.42). The maximum pulse height V_{max} corresponds to the maximum address N_{max}. If N is the address number with the maximum deviation between the actual and nominal pulse heights, the integral nonlinearity is given by the equation

$$\% \text{ Integral nonlinearity} = \frac{V_{nom} - V_{act}}{V_{max}} (100) \tag{10.34}$$

Modern commercial ADCs have integral nonlinearity of the order of ±0.05 percent over 98–99 percent of the full range.

The integral nonlinearity affects the centroid position of energy peaks, which in turn affects the calibration of the system as well as the identification of unknown energy peaks.

10.13 MULTIPARAMETER ANALYZERS

The MCA is an instrument that stores events by a single parameter, pulse height. There are many experiments, however, when the need arises for the study of events in terms of more than one parameter. Such requirements occur in:

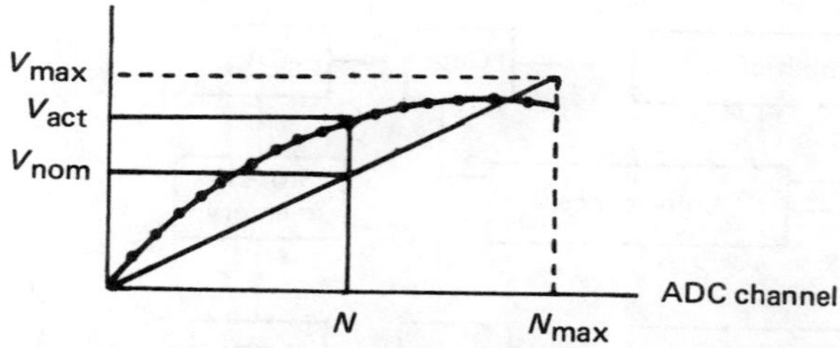

FIG. 10.42 The definition of integral nonlinearity is based on a linear plot of ADC channel versus pulse amplitude.

1 Coincidence measurements where the energy spectrum from both detectors need be analyzed
2 Simultaneous measurement of energy and mass distribution of fission fragments
3 Study of energy and angular dependence of nuclear reactions involving many particles, etc.

The "direct" method of multiparameter analysis would be to use such an arrangement that all parameters but one are limited to a narrow range (by using a single-channel analyzer) and the remaining parameter is recorded by an MCA. After an adequate number of events has been recorded, the value of one of the fixed parameters is changed and the measurement is repeated. This process continues until all values of all parameters are covered. Obviously, such an approach is cumbersome and time consuming.

A more efficient way of performing the measurement is by storing the information simultaneously for more than one parameter. For example, consider a coincidence measurement involving two detectors (Fig. 10.43). The detector signals are fed into a coincidence unit, which then is used to gate the corresponding ADCs. The amplified detector pulses that are coincident are thus digitized by the ADCs, and the information is stored in the memory of the system. Any event that reaches the memory is defined like a point in a two-dimensional space. For example, if a pulse from ADC_1 has the value 65 (i.e., ADC channel 65) and one from ADC_2 has the value 18, the event is registered as 6518 (assuming 100 channels are available for each parameter). The final results of the measurement may be displayed on a cathode-ray tube or may be written onto a magnetic tape for analysis by a computer. The results of a dual-parameter system such as that shown in Fig. 10.43 constitutes a three-dimensional histogram as shown by Fig. 10.44.

The greatest difficulty with multiparameter measurements is to secure sufficient memory capacity to register all possible events. The necessary storage increases exponentially with the number of parameters. For a k-parameter measurement with N channels per parameter, the capacity of the memory should be N^k. Thus a two-parameter system with 100 channels per parameter needs 10^4 memory locations. If both parameters are registered in 1000 channels, the requirements are 10^6 locations.

Most ways to remedy the difficulty associated with the need for a large memory rely on the use of large digital computers. One approach is to store the data in a magnetic tape and have them analyzed later by a large digital computer. A second approach is to use the computer on-line with the measurement.

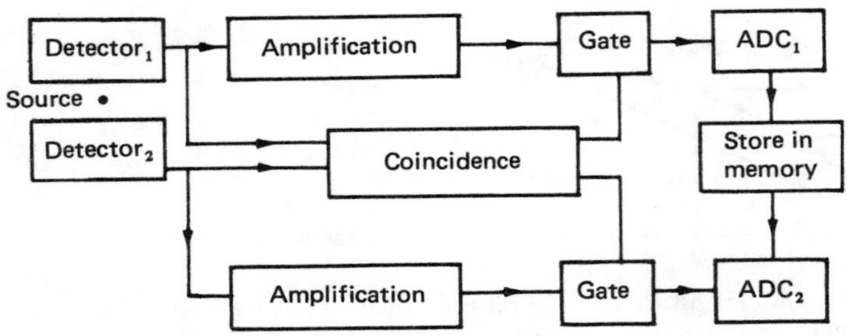

FIG. 10.43 A two-parameter measurement.

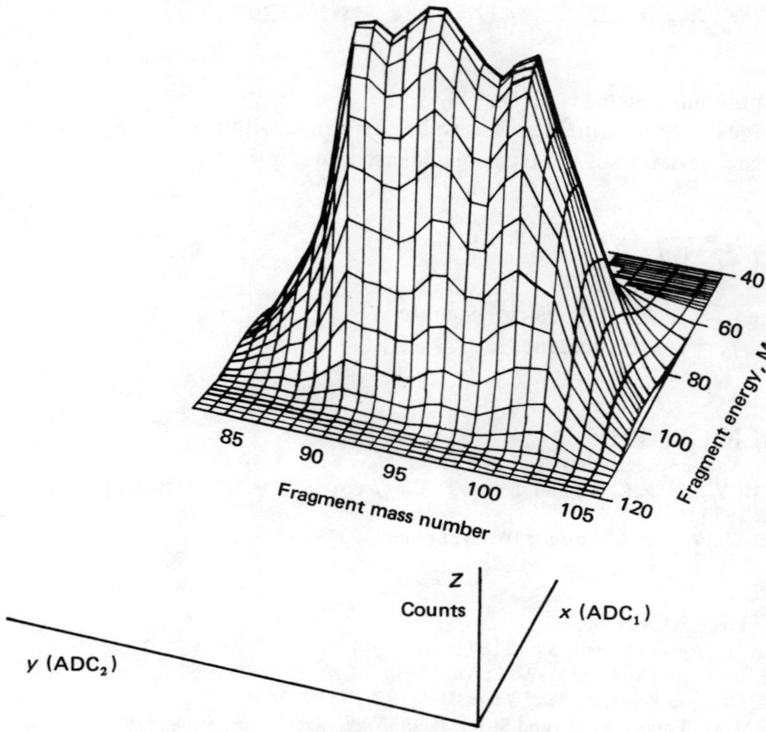

FIG. 10.44 A sample of the output from a two-parameter system (from reference 26).

PROBLEMS

10.1 Prove that the output signal of a differentiating circuit is, for a step input, equal to

$$V_0(t) = V_i e^{-t/RC}$$

10.2 Show that the output signal of a differentiating circuit is given by

$$V_0(t) = \frac{V}{\tau} RC(1 - e^{-t/RC})$$

when the input signal is given by $V_i(t) = Vt/\tau$.

10.3 Show that the output signal of an integrating circuit is, for a step input, equal to

$$V_0(t) = V_i(1 - e^{-t/RC})$$

10.4 Show that the output signal of an integrating circuit is given by

$$V_0(t) = V \frac{1}{1 - \tau/RC} (e^{-t/RC} - e^{-t/\tau})$$

when the input signal is $V_i(t) = V_0(1 - e^{-t/\tau})$.

10.5 A coincidence measurement has to be performed within a time T. Show that the standard deviation of the true coincidence rate is given by

$$\sigma_t = \frac{1}{\sqrt{T}} (\sqrt{r_a + r_t} + \sqrt{r_a})$$

where r_a = accidental coincidence rate
r_t = true coincidence rate

BIBLIOGRAPHY

Kowalski, E., *Nuclear Electronics*, Springer-Verlag, New York, 1970.

Malmstadt, H. V., Enke, C. G., and Toren, E. C., *Electronics for Scientists*, W.A. Benjamin, New York, 1963.

Nicholson, P. W., *Nuclear Electronics*, Wiley, London, 1974.

REFERENCES

1. Chase, R. L., *Rev. Sci. Instr.* **39**:1318 (1968).
2. Gedcke, D. A., and McDonald, W. J., *Nucl. Instr. Meth.* **55**:377 (1967).
3. Maier, M. R., and Sperr, P., *Nucl. Instr. Meth.* **87**:13 (1970).
4. Strauss, M. G., Larsen, R. N., and Sifter, L. L., *Nucl. Instr. Meth.* **46**:45 (1967).
5. Graham, R. L., Mackenzie, I. K., and Ewan, G. T., *IEEE Trans. NS 13* **1**:72 (1966).
6. Viencent, C. H., *Nucl. Instr. Meth.* **127**:421 (1975).
7. Smith, D., *Nucl. Instr. Meth.* **152**:505 (1978).
8. Kuchnir, F. T., and Lynch, F. J., *IEEE Trans. NS 15* **3**:107 (1968).
9. Alexander, T. K., and Goulding, F. S., *Nucl. Instr. Meth.* **13**:244 (1961).
10. Heistek, L. J., and Van Der Zwan, L., *Nucl. Instr. Meth.* **80**:213 (1970).
11. McBeth, G. W., Lutkin, J. E., and Winyard, R. A., *Nucl. Instr. Meth.* **93**:99 (1971).
12. Brooks, F. D., *Nucl. Instr. Meth.* **4**:151 (1959).
13. Kinbara, S., and Kumahara, T., *Nucl. Instr. Meth.* **70**:173 (1969).
14. Burrus, W. R., and Verbinski, V. V., *Nucl. Instr. Meth.* **67**:181 (1969).
15. Verbinski, V. V., Burrus, W. R., Love, T. A., Zobel, W., and Hill, N. W., *Nucl. Instr. Meth.* **65**:8 (1968).
16. Forte, M., Konsta, A., and Moranzana, C., *Electronic Methods for Discrimination of Scintillation Shapes* (IAEA Conf. Nucl. Electr. Belgrade, 1961) NE-59.
17. Nowlin, C. H., and Blankenship, J. L., *Rev. Sci. Instr.* **36**:1830 (1965).
18. Brenner, R., *Rev. Sci. Instr.* **40**:1011 (1969).
19. Fairstein, E., and Hahn, J., *Nucleonics 23*, no. 7:56 (1965).
20. Fairstein, E., and Hahn, J., *Nucleonics 23*, no. 9:81 (1965).
21. Fairstein, E., and Hahn, J., *Nucleonics 23*, no. 11:50 (1965).
22. Fairstein, E., and Hahn, J., *Nucleonics 24*, no. 1:54 (1966).
23. Fairstein, E., and Hahn, J., *Nucleonics 24*, no. 3:68 (1966).
24. Wilkinson, D. H., *Phil. Soc.* **46**:508 (1950).
25. Ortec catalogue (1976).
26. Diorio, G., and Wehring, B. W., *Nucl. Instr. Meth.* **147**:487 (1977).

11

data analysis methods

11.1 INTRODUCTION

Raw† experimental data seldom give the answer to the problem that is the objective of the measurement. In most cases, additional calculations or analysis of the raw data are necessary. The analysis of the raw data may consist of a simple division of the counts recorded in a scaler by the counting time to obtain counting rates, may require fitting an analytical function to the data, or may necessitate unfolding of a measured spectrum.

Whatever the analysis of the data may entail, there are some general methods helpful to the analyst. The objective of this chapter is to present a brief introduction to these general methods and principles of data analysis.

11.2 CURVE FITTING

The results of most experiments consist of a finite number of values (and their errors) of a dependent variable y measured as a function of the independent variable x (Fig. 11.1). The objective of the measurement of $y = y(x)$ may be one of the following:

1 To find how y changes with x
2 To prove that $y = y(x)$ follows a theoretically derived function
3 To use the finite number of measurements of $y(x)$ for the evaluation of the same function at intermediate points or at values of x beyond those measured

These objectives could be immediately achieved if the function $y(x)$ were known. Since it is not, the observer tries to determine it with the help of the experimental data. The task of obtaining an analytic function that represents $y(x)$ is called *curve fitting.*

The first step in curve fitting is to plot the data (y versus x) on linear graph paper (Fig. 11.1). A smooth curve is then drawn, following as closely as possible the general trend of the data and trying to have an equal number of points on either side of the curve. The experimental points always have an error associated with them, so the smooth curve is not expected to pass through all the measured (x, y) points. Obviously, there is no guarantee that the smooth curve so drawn is the "true" one. Criteria that may help the observer draw a curve with a certain degree of confidence are then needed. Such criteria exist and are described in Sec. 11.4, Least-Squares Fitting.

After the data are plotted and a smooth curve is drawn, the observer has to answer two questions.

†Raw data consists of the numbers obtained by the measuring device, e.g., a scaler, a clock, or a voltmeter.

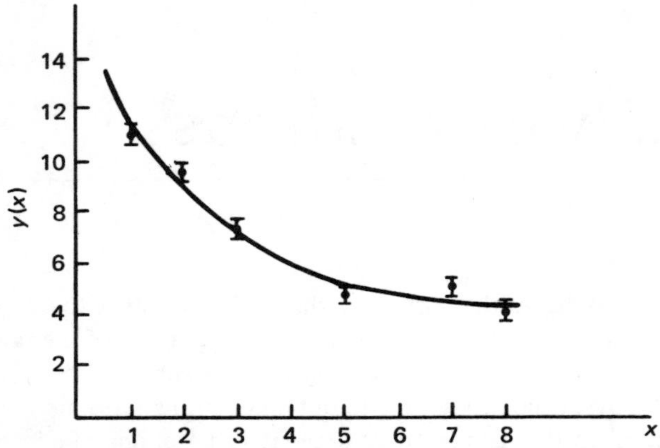

FIG. 11.1 Experimental results consist of the values of the dependent variable $y(x)$ and their errors. The curve was drawn to help the eye.

1 What type of function would represent the data best (e.g., exponential, straight line, parabola, logarithmic)?
2 After the type of function is decided upon, how can one determine the best values of the constants that define the function uniquely?

Since there exist an infinite number of functions, the observer would like to have certain criteria or rules that limit the number of possible functions. While no such formal set of criteria exists, the following suggestions have proved useful.

First, the observer should utilize any a priori knowledge about $y(x)$ and x. Examples are restrictions of x and y within a certain range (e.g., in counting experiments both x and y are positive) or information from theory that suggests a particular function (e.g., counting data follow Poisson statistics).

Second, the observer should try the three simple expressions listed next, before any complicated function is considered.

1 The linear relation (straight line)

$$y(x) = ax + b \tag{11.1}$$

where a and b are constants to be determined based on the data. A linear relationship will be recognized immediately in a linear plot of $y(x)$ versus x.

2 The exponential relationship

$$y(x) = ae^{-bx} \tag{11.2}$$

If the data can be represented by such a function, a plot on semilog paper–i.e., a plot of $\ln y$ versus x–will give a straight line.

3 The power relationship

$$y(x) = ax^b \tag{11.3}$$

If the data can be represented by this expression, a plot on log-log paper—i.e., a plot of $\ln y$ versus $\ln x$—will give a straight line.

Third, the observer should know that a polynomial of degree N can always be fitted exactly to $N+1$ pieces of data (see also Secs. 11.3-11.5).

If no satisfactory fit can be obtained by using any of these suggestions, the analyst should try more complicated functions. Plotting the data on special kinds of graph paper, such as reciprocal or probability proper may be helpful. After the type of function is found, the constants associated with it are determined by a least-squares fit (see Sec. 11.4).

11.3 INTERPOLATION SCHEMES

It was mentioned in Sec. 11.2 that one of the reasons for curve fitting is to be able to evaluate the function $y(x)$ at values of x for which measurements do not exist. An alternative to curve fitting that can be used for the calculation of intermediate $y(x)$ values is the method of interpolation. This section presents one of the basic interpolation techniques—the *Lagrange formula.* Many other formulas exist that the reader can find in the bibliography of this chapter (e.g., see Hildebrand, and Abramowitz's *Handbook of Mathematical Functions*).

Assume that N values of the dependent variable $y(x)$ are known at the N points x_i, $x_a \leqslant x_i \leqslant x_b$ for $i = 1, \ldots, N$. The pairs of data (y_i, x_i) for $i = 1, \ldots, N$, where $y(x_i) = y_i$, may be the results of an experiment or tabulated values. Interpolation means to obtain a value $y(x)$ for $x_a < x < x_b$ based on the data (y_i, x_i), when the point x is not one of the N values for which $y(x)$ is known.

The Lagrange interpolation formula expresses the value $y(x)$ in terms of polynomials (up to degree $N-1$ for N pairs of data). The general equation is

$$y(x) = \sum_{i=0}^{M} P_i(x) y(x_i) \qquad M \leqslant N-1 \tag{11.4}$$

where

$$P_i(x) = \frac{\prod_{j=0, j\neq i}^{M} (x - x_j)}{\prod_{j=0, j\neq i}^{M} (x_i - x_j)} \qquad M \leqslant N-1 \tag{11.5}$$

The error associated with Eq. 11.4 is given by

$$\text{Error}\,[y(x)] = \prod_{j=0}^{M} (x - x_j) \frac{y^{M+1}(\xi)}{(M+1)!} \qquad M \leqslant N-1 \tag{11.6}$$

where $y^{M+1}(\xi)$ is the $(M+1)$ derivative of $y(x)$ evaluated at the point ξ, $x_a < \xi < x_b$. Since $y(x)$ is not known analytically, the derivative in Eq. 11.5 has to be calculated numerically.

Equation 11.4 is the most general. It uses all the available points to calculate any new value of $y(x)$ for $x_a < x < x_b$. In practice, people use only a few points at a time, as the following two examples show.

EXAMPLE 11.1

Derive the Lagrante formula for $M = 1$.

ANSWER

If $M = 1$, Eq. 11.4 takes the form (also using Eq. 11.5):

$$y(x) = \sum_{i=0}^{1} P_i(x)y(x_i) = \frac{x - x_1}{x_0 - x_1} y_0 + \frac{x - x_0}{x_1 - x_0} y_1 \tag{11.7}$$

where $y_i = y(x_i)$. The points x_0 and x_1 could be anywhere between x_a and x_b, but the point x should be: $x_0 \leqslant x \leqslant x_1$.

To calculate $y(x)$ at any x, Eq. 11.7 uses two points, one on either side of x, and for this reason it is called the *Lagrange two-point interpolation formula.* Eq. 11.7 may be written in the form

$$y(x) = y_0 + \frac{x - x_0}{x_1 - x_0}(y_1 - y_0) \tag{11.8}$$

which shows that the two-point formula amounts to a linear interpolation.

The error associated with the two-point formula is obtained from Eq. 11.5:

$$\text{Error}[y(x)] = (x - x_0)(x - x_1)\frac{y^{(2)}(\xi)}{2}$$

where $y^{(2)}(\xi)$ is the second derivative evaluated at $\xi, x_0 < \xi < x_1$.

EXAMPLE 11.2

Derive the Lagrante formula for $M = 2$.

ANSWER

If $M = 2$, Eq. 11.4 takes the form

$$y(x) = \sum_{i=0}^{2} P_i(x)y(x_i) = \frac{(x - x_1)(x - x_2)}{(x_0 - x_1)(x_0 - x_2)} y_0 + \frac{(x - x_0)(x - x_1)}{(x_1 - x_0)(x_1 - x_2)} y_1$$

$$+ \frac{(x - x_0)(x - x_1)}{(x_2 - x_0)(x_2 - x_1)} y_2 \tag{11.9}$$

To calculate $y(x)$ at any point x, Eq. 11.9 uses three points x_0, x_1, x_2 with x, $x_0 \leqslant x \leqslant x_2$, and is called the *Lagrange three-point interpolation formula.* The three-point formula amounts to a parabolic representation of the function $y(x)$ between any three points.

The error associated with the three-point formula is (applying again Eq. 11.6):

$$\text{Error}[y(x)] = (x - x_0)(x - x_1)(x - x_2)\frac{y^{(3)}(\xi)}{3!}$$

where $y^{(3)}(\xi)$ is the third derivative evaluated at $\xi, x_0 < \xi < x_2$.

EXAMPLE 11.3

Calculate the value of the function $f(x)$ at $x = 11.8$ from the table below using the Lagrange two- and three-point interpolation formulas. The data are plotted in Fig. 11.2.

x	$f(x)$
10	30.5
11	33.0
12	35.8
13	36.7
14	37.2

ANSWER

(a) Using the two-point formula (Eq. 11.7), one has $x_0 = 11$, $x_1 = 12$, $y_0 = 33.0$, $y_1 = 35.8$.

$$y(x) = y(11.8) = \frac{11.8 - 12}{11 - 12}(33.0) + \frac{11.8 - 11}{12 - 11}(35.8) = 35.2$$

(b) Using the three-point formula (Eq. 11.9), one has $x_0 = 11$, $x_1 = 12$, $x_2 = 13$, $y_0 = 33$, $y_1 = 35.8$, $y_2 = 36.7$.

$$y(11.8) = \frac{(11.8 - 12)(11.8 - 13)}{(11 - 12)(11 - 13)}(33) + \frac{(11.8 - 11)(11.8 - 13)}{(12 - 11)(12 - 13)}(35.8)$$

$$+ \frac{(11.8 - 11)(11.8 - 12)}{(12 - 11)(13 - 12)}(36.7) = 3.96 + 34.37 - 2.94 = 35.4$$

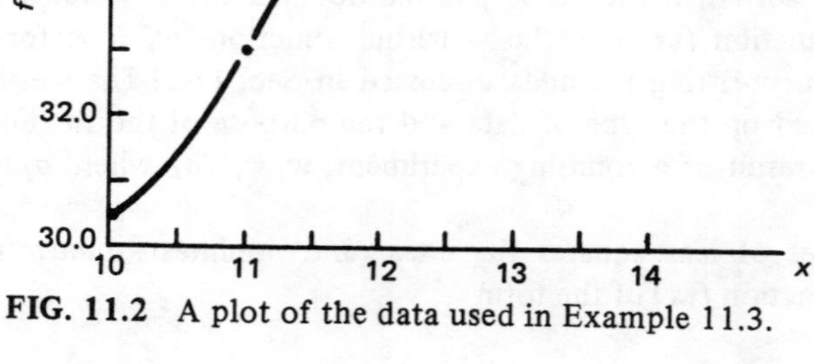

FIG. 11.2 A plot of the data used in Example 11.3.

The error of $y(x)$ associated with the interpolation is, for the first case only,

$$\text{Error } [y(x_I)] = (x - x_0)(x - x_1) \frac{y^{(2)}(\xi)}{2}$$

The second derivative evaluated at $\xi = \frac{1}{2}(x_0 + x_1)$ is

$$y^{(2)}(\xi) = \frac{y_1 - 2y - y_0}{(x_1 - xo)^2}$$

Thus the error is

$$\text{Error } [y(x)] = (11.8 - 11)(11.8 - 12) \frac{35.8 - 2(35.2) + 33}{(1^2)^2} = 0.13$$

or $0.13/35.2 = 0.4$ percent.

11.4 LEAST–SQUARES FITTING

Assume that an observer obtained the experimental data $y_i = y(x_i)|_{i=1;N}$. It is often desirable to find a function that can represent the data, i.e., to find a function that can be fitted to the data. Let such a function be written as

$$f(x, a_1, a_2, \ldots, a_M) \qquad M < N$$

where $a_m|_{m=1,M}$ are parameters to be determined. According to the method of least squares, the best values of the parameters a_M are those which minimize the quantity

$$Q = \sum_{i=1}^{N} w_i[y_i - f(x_i)]^2 \tag{11.10}$$

where $w_i|_{i=1,N}$ are weighting functions. Minimization of Q is achieved by requiring

$$\frac{\partial Q}{\partial a_M} = 0 \qquad m = 1, \ldots, M \tag{11.11}$$

Before Eq. 11.10 is solved, it should be pointed out that the observer decides about the form of the function $f(x)$ and the weighting functions w_i. The form of $f(x)$ is obtained by the curve-fitting methods discussed in Sec. 11.2. The weighting functions are selected based on the type of data and the purpose of the fit. For example, if the data are the result of a counting experiment, $w_i = 1/\sigma_i^2$ where σ_i is the standard deviation of y_i.

There are two types of least-squares fit, linear and nonlinear. *Linear least-squares fit* is based on a function $f(x)$ of the form

$$f(x, a_1, a_2, \ldots, a_M) = \sum_{m=1}^{M} a_m \phi_m(x) \tag{11.12}$$

where the $\phi_m(x)$ are known functions of a.

Nonlinear least-squares fit is based on a function $f(x)$ nonlinear in a_m, such as,

$$f(x, a_1, a_2, \ldots, a_M) = a_1 \cos(a_2 x)$$

The interested reader should consult the bibliography of this chapter (see Bevington) for further information on nonlinear least-squares fit.

For a linear least-squares fit, the parameters $a_m|_{m=1,M}$ are determined from Eq. 11.11, with Eq. 11.12 giving the form of $f(x)$. The result is

$$\frac{\partial Q}{\partial a_k} = \frac{\partial}{\partial a_k} \left\{ \sum_{i=1}^{N} w_i \left[y_i - \sum_{m=1}^{M} a_m \phi_m(x_i) \right]^2 \right\} \qquad k = 1, \ldots, M \tag{11.13}$$

If one defines

$$A_{km} = \sum_{i=1}^{N} w_i \phi_k(x_i) \phi_m(x_i) \qquad k, m = 1, \ldots, M \tag{11.14}$$

and $$B_k = \sum_{i=1}^{N} w_i y_i \phi_k(x_i) \qquad k = 1, \ldots, M \tag{11.15}$$

then Eq. 11.13 takes the form

$$\sum_{m=1}^{M} A_{km} a_m = B_k \qquad k = 1, \ldots, M \tag{11.16}$$

Equation 11.16 forms a system of M linear nonhomogeneous equations for the M unknowns $a_m|_{m=1,M}$ and can be solved by using any of the standard methods (e.g., Kramer's rule). In matrix notation, the solution is

$$a_m = \mathbf{A}^{-1}\mathbf{B} \tag{11.17}$$

where **A** and **B** are matrices with elements given by Eqs. 11.14 and 11.15.

If the function $f(x)$ is a polynomial, then

$$f(x, a_1, a_2, \ldots, a_m) = \sum_{m=1}^{M} a_m x^{m-1} \tag{11.18}$$

Equations 11.14 and 11.15 take the form [since $\phi_k(x) = x^{k-1}$]

$$A_{km} = \sum_k w_i x_i^{k-1} x_i^{m-1} \tag{11.19}$$

$$B_k = \sum_i w_i y_i x_i^{k-1} \tag{11.20}$$

The notation used in Eqs. 11.19 and 11.20 and in the next section is $\Sigma_{i=1}^{N} \to \Sigma_i$.

11.4.1 Least-Squares Fit of a Straight Line

If the function represented by Eq. 11.18 is a straight line, then

$$f(x_i, a_1, a_2) = a_1 + a_2 x_i \tag{11.21}$$

Thus, Eqs. 11.19 and 11.20 become

$$A_{11} = \sum_i w_i \qquad A_{12} = \sum_i w_i x_i = A_{21} \qquad A_{22} = \sum_i w_i x_i^2$$

$$B_1 = \sum_i w_i y_i \qquad B_2 = \sum_i w_i y_i x_i$$

Then, Eqs. 11.16 take the form

$$A_{11} a_1 + A_{12} a_2 = B_1$$

$$A_{21} a_1 + A_{22} a_2 = B_2$$

which are solved to give

$$a_1 = \frac{1}{D}\left[\left(\sum_i w_i y_i\right)\left(\sum_i w_i x_i^2\right) - \left(\sum_i w_i y_i x_i\right)\left(\sum_i w_i x_i\right)\right] \tag{11.22}$$

$$a_2 = \frac{1}{D}\left[\left(\sum_i w_i\right)\left(\sum_i w_i y_i x_i\right) - \left(\sum_i w_i x_i\right)\left(\sum_i w_i y_i\right)\right] \tag{11.23}$$

and $$D = \left(\sum_i w_i\right)\left(\sum_i w_i x_i^2\right) - \left(\sum_i w_i x_i\right)^2 \tag{11.24}$$

The variance of a_1 and a_2 is obtained by using the principle of propagation of error presented in Chap. 2.

$$\sigma_{a_m}^2 = \sum_i \left(\frac{\partial a_m}{\partial y_i}\right)^2 \sigma^2 y_i \qquad m = 1, 2 \tag{11.25}$$

In many cases, the standard deviation of y_i defines the weighting functions and, specifically, analysts use

$$w_i = \frac{1}{\sigma_i^2}$$

Then, Eq. 11.25 gives

$$\sigma_{a_i}^2 = \frac{1}{D} \sum_i \frac{x_i^2}{\sigma_i^2} \tag{11.26}$$

$$\sigma_{a_2}^2 = \frac{1}{D} \sum_i \frac{1}{\sigma_i^2} \tag{11.27}$$

where D is given by Eq. 11.24 with $w_i = 1/\sigma_i^2$. Equations 11.22–11.27 are further simplified if all the σ_i have the same value.

11.4.2 Least-Squares Fit of General Functions

A straight-line least-squares fit is not limited to linear functions of x. It may be used with functions such as the exponential ($y = ae^{bx}$) or the power relationship ($y = ax^b$) after an appropriate transformation of variables. For example, the exponential function can be written as

$$\ln y = \ln a + bx \tag{11.28}$$

which is of the form given by Eq. 11.21 after setting

$$y' = \ln y \qquad a_1' = \ln a \qquad a_2 = b$$

When the variable is transformed, it is necessary to obtain the standard deviation of the new variable. In general, if one sets

$$y' = g(y)$$

then the standard deviation of $y'(x)$ is

$$\sigma_i' = \frac{\partial g(y_i)}{\partial y_i} \sigma_i \tag{11.29}$$

In the example given above, $y = \ln y$ and

$$\sigma_i' = \frac{\partial(\ln y_i)}{\partial y_i} \sigma_i = \frac{\sigma_i}{y_i} \tag{11.30}$$

Therefore, if a transformation is applied to the function, all the σ_i in Eqs. 11.26 and 11.27 should be replaced by the values given by Eq. 11.30.

If the parameters a_m are transformed, the standard deviation of the new constant is again determined by Eq. 11.29. In the example given above, $a' = \ln a$ and

$$\sigma_{a'} = \frac{\sigma_a}{a} \tag{11.31}$$

Table 11.1 presents a number of functions that can be cast into a linear (or polynominal) form by a transformation of variables. It should be emphasized that although the functions shown in Table 11.1 are not linear in x, the least-squares fit is still linear. An example of a function $f(x, a_1, \ldots)$ that represents a nonlinear least-squares fit is given in Sec. 11.4.

$$f(x, a_1, a_2, \ldots) = a_1 \cos(a_2 x)$$

EXAMPLE 11.4

The following table gives neutron-absorption cross-section values and their errors as a function of neutron kinetic energy. Determine the analytic function that fits this data.

T_n (eV)	$\sigma_a(b)$	Error (b)
1	1000	50
4	540	23
10	290	19
20	225	15
50	160	11
80	108	10
100	105	9

ANSWER

If one plots the data on log-log paper, the result is very close to a straight line (see Fig. 11.3). Therefore, the function to use is

$$\ln \sigma_a = a_1 + a_2 \ln T_n$$

Table 11.1 Functions that Can Be Changed into a Form Suitable for a Linear Least-Squares Fit

Function $y(x)$	Transformation			Function used in the least-squares fit
$y = ae^{bx}$	$y' = \ln y$	$a' = \ln a$		$y' = a + bx$
$y = ax^b$	$y' = \ln y$	$a' = \ln a$	$x' = \ln x$	$y' = a + bx'$
$y = a \exp\left[-\frac{(x-a)^2}{2\sigma^2}\right]$	$y' = \ln y$			$y' = a_1 + a_2 x + a_3 x^2$
$y = a_1 + a_2 x^b + a_3 x^{2b}$	$x' = x^b$			$y' = a_1 + a_2 x' + a_3 x'^2$
$y = a_1 x^{b_1} + a_2 x^{b_2}$	$y' = yx^{-b_1}$		$x' = x^{b_2 - b_1}$	$y' = a_1 + a_2 x'$

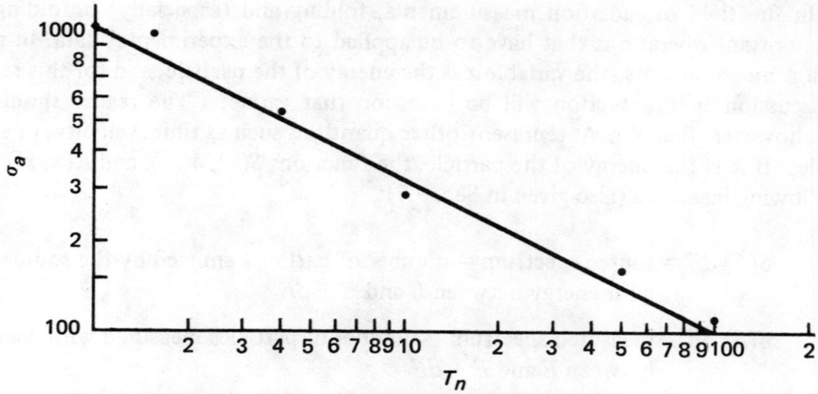

FIG. 11.3 The least-squares fit to the data of Example 11.4.

If one takes $w_i = 1/\sigma_i^2$ and applies Eqs. 11.22-11.27, the result is

$$a_1 = 6.93 \qquad a_2 = 0.506$$

$$\sigma_{a_1} = 0.039 = 0.6\% \qquad \sigma_{a_2} = 0.015 = 3\%$$

11.5 FOLDING AND UNFOLDING

To define the problems of folding and unfolding, consider the functions

$$S(x) \qquad 0 < x < \infty$$

$$M(x) \qquad 0 < x < \infty$$

$$R(x, x') \qquad 0 < x, x' < \infty$$

where the function $R(x, x')$ is normalized to 1:

$$\int_0^\infty R(x, x')\, dx = 1 \tag{11.32}$$

Folding the function $S(x)$ with the function $R(x, x')$ to obtain the function $M(x)$ means to perform the integration

$$M(x) = \int_0^\infty R(x, x')S(x')\, dx' \tag{11.33}$$

Unfolding means to obtain the function $S(x)$, knowing $M(x)$ and $R(x, x')$. Thus, folding is an integration, as shown by Eq. 11.33. Unfolding, on the other hand, entails solving the integral equation, Eq. 11.33–known as the *Fredholm equation*–for the unknown function $S(x)$.

In the field of radiation measurements, folding and (especially) unfolding are very important operations that have to be applied to the experimental data. In most radiation measurements, the variable x is the energy of the particle, and for this reason the discussion in this section will be based on that variable. The reader should be aware, however, that x may represent other quantities such as time, velocity, or space variables. If x is the energy of the particle, the functions $S(x)$, $M(x)$, and $R(x, x')$ have the following meanings (also given in Sec. 9.7):

$S(E)\,dE$ = source spectrum = number of particles emitted by the source with energy between E and $E + dE$

$M(E)\,dE$ = measured spectrum = number of particles measured with energy between E and $E + dE$

$R(E, E')\,dE$ = response of the detector = probability that a particle emitted by the source with energy E' will be recorded as having energy between E and $E + dE$

As explained in Chap. 9, the response function is measured using monoenergetic sources. A monoenergetic source is represented mathematically by the *delta function* (δ-function, which has these properties (Fig. 11.4):

$$\delta(E - E_0) = 0 \qquad E \neq E_0$$

$$\int_{E_1}^{E_2} \delta(E - E_0)\,dE = \begin{cases} 1 & \text{if } E_1 < E_0 < E_2 \\ 0 & \text{otherwise} \end{cases} \tag{11.34}$$

Thus, the δ-function is equal to zero everywhere except at $E = E_0$, which is, of course, what the energy spectrum of a monoenergetic source represents. Because of the property expressed by Eq. 11.34, integrals involving the δ-function are immediately evaluated. For any function $f(E)$ one obtains

$$\int_{E_1}^{E_2} f(E)\delta(E - E_0)\,dE = f(E_0) \qquad E_1 < E_0 < E_2 \tag{11.35}$$

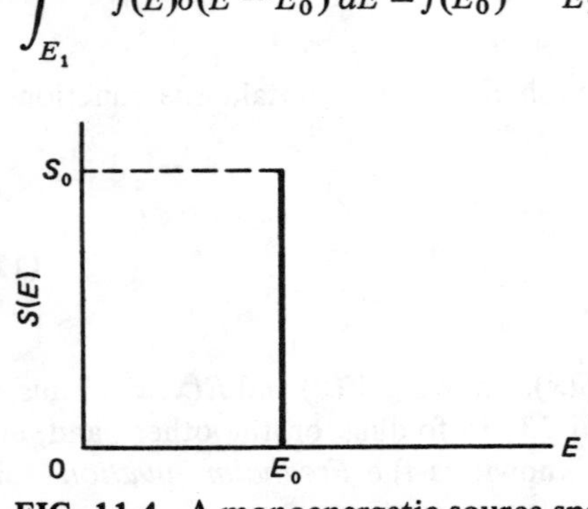

FIG. 11.4 A monoenergetic source spectrum.

because there is no contribution to the integral except at $E = E_0$. For the same reason, if E_0 is outside the limits of integration, then

$$\int_{E_1}^{E_2} f(E)\delta(E - E_0)\,dE = 0 \qquad \begin{cases} E_0 < E_1 \\ E_0 > E_2 \end{cases} \tag{11.36}$$

Assume that a monoenergetic source emitting S_0 particles per second (Fig. 11.4) is used to measure the response function. If one substitutes the expression for this source,

$$S(E) = E_0\,\delta(E - E_0) \tag{11.37}$$

into Eq. 11.33, the result is

$$M(E) = \int_0^\infty R(E_1 E')S_0\,\delta(E' - E_0)\,dE' = S_0 R(E, E_0) \tag{11.38}$$

Equation 11.38 shows that the measured spectrum is indeed equal to the response function in the case of a monoenergetic source.

11.5.1 Examples of Folding

In radiation measurements, *folding* means to obtain the shape of the measured spectrum when the source and the detector response are known. Several examples of folding using a Gaussian distribution as the response function are presented next.

EXAMPLE 11.5

The source spectrum is a step function:

$$S(E) = \begin{cases} S_0 & E \geqslant E_0 \\ 0 & E < E_0 \end{cases}$$

What is the measured spectrum?

ANSWER

$$M(E) = \int_0^\infty \frac{dE'}{\sqrt{2\pi}\sigma} \exp\left[-\frac{(E - E')^2}{2\sigma^2}\right] S(E')$$

$$= S_0 \int_{E_0}^\infty \frac{dE'}{\sqrt{2\pi}\sigma} \exp\left[-\frac{(E - E')^2}{2\sigma^2}\right]$$

$$M(E) = \frac{S_0}{2}\left[1 + \text{erf}\left(\frac{E - E_0}{\sqrt{2}\sigma}\right)\right]$$

where

$$\text{erf}\left(\frac{E-E_0}{\sqrt{2}\sigma}\right) = \text{error function} = \frac{1}{\sqrt{2\pi}} \int_0^{(E-E_0)/\sigma} e^{-t^2/2}\, dt$$

Figure 11.5 shows the three functions involved.

EXAMPLE 11.6

The source spectrum is a square spectrum:

$$S(E) = \begin{cases} S_0 & E_1 \leqslant E \leqslant E_2 \\ 0 & \text{otherwise} \end{cases}$$

What is the measured spectrum?

ANSWER

$$M(E) = \int_0^{\infty} \frac{dE'}{\sqrt{2k}\sigma} \exp\left[-\frac{(E-E')^2}{2\sigma^2}\right] S(E')$$

$$= S_0 \int_{E_1}^{E_2} \frac{dE'}{\sqrt{2\pi}\sigma} \exp\left[-\frac{(E-E')^2}{2\sigma^2}\right]$$

$$= \frac{S_0}{2}\left[\text{erf}\left(\frac{E_2-E}{\sqrt{2}\sigma}\right) - \text{erf}\left(\frac{E_1-E}{\sqrt{2}\sigma}\right)\right]$$

Figure 11.6 shows the functions involved.

EXAMPLE 11.7

The source spectrum is a Gaussian centered at $E = E_0$:

$$S(E) = \frac{S_0}{\sqrt{2\pi}\sigma_S} \exp\left[-\frac{(E-E_0)^2}{2\sigma_S^2}\right]$$

What is the measured spectrum?

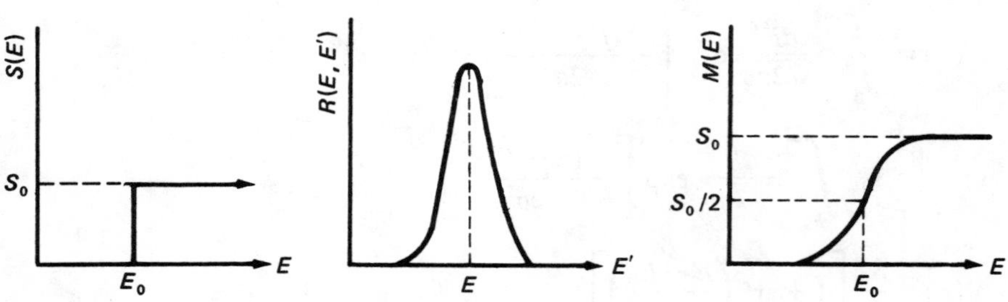

FIG. 11.5 A step function folded with a Gaussian.

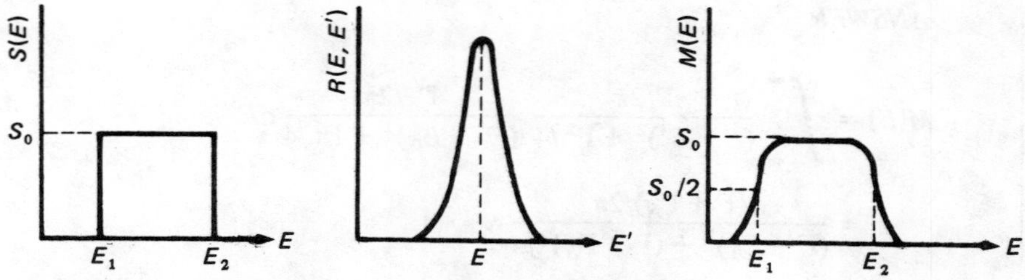

FIG. 11.6 A "square" function folded with a Gaussian.

ANSWER

$$M(E)^{\dagger} = \int_{-\infty}^{\infty} \frac{dE'}{\sqrt{2\pi}\sigma} \exp\left[-\frac{(E-E')^2}{2\sigma^2}\right] \frac{1}{\sqrt{2\pi}\sigma_S} \exp\left[-\frac{(E-E_0)^2}{2\sigma_S^2}\right]$$

$$= \frac{1}{\sqrt{2\pi}(\sigma_S^2+\sigma^2)^{1/2}} \exp\left[-\frac{(E-E_0)^2}{2(\sigma^2+\sigma_S^2)}\right]$$

Figure 11.7 shows the three functions involved. It is worth noting that if a Gaussian is folded with another Gaussian, their standard deviations add in quadrature.

EXAMPLE 11.8
The source spectrum is a Lorentzian centered at E_0:

$$S(E) = \frac{\Gamma_S/2\pi}{(E-E_0)^2 + \Gamma_S^2/2}$$

and the response function is also a Lorentzian with width Γ. What is the measured spectrum?

†The integral of Eq. 11.33 may be extended to $-\infty$ because the Gaussian drops off quickly to a negligible value for $E < 0$.

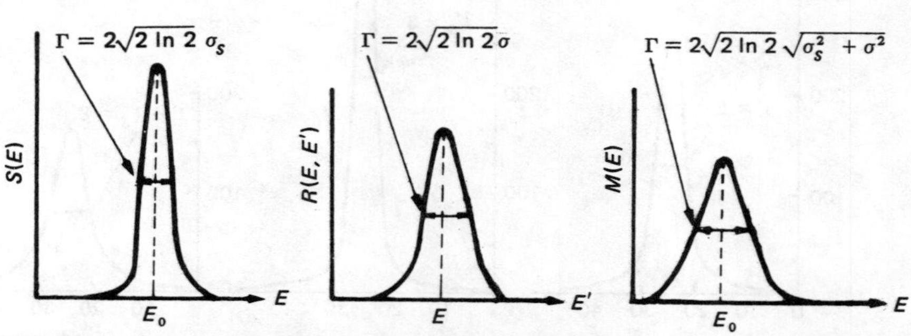

FIG. 11.7 A Gaussian folded with a Gaussian gives a third Gaussian which has a larger width.

ANSWER

$$M(E) = \int_{-\infty}^{\infty} \frac{\Gamma/2\pi}{(E-E')^2 + \Gamma^2/4} \frac{\Gamma_S/2\pi}{(E'-E_0)^2 + \Gamma_S^2/4} dE'$$

$$= \frac{(\Gamma + \Gamma_S)/2\pi}{(E-E_0)^2 + (\Gamma + \Gamma_S)^2/4}$$

Figure 11.8 shows the three functions involved. Notice that by folding a Lorentzian with a Lorentzian, the result is a third Lorentzian with width equal to the sum of the two widths.

11.5.2 The General Method of Unfolding

This section discusses methods of unfolding, assuming that an energy spectrum is measured with a multichannel analyzer or any other device that divides the measured spectrum into energy bins. As stated at the beginning of Sec. 11.5, unfolding means to solve the Fredholm-type integral equation

$$M(E) = \int_0^{\infty} R(E, E')S(E')\,dE' \tag{11.39}$$

for the unknown function $S(E)$. Before possible methods of solution of Eq. 11.39 are discussed, it is important to note that no spectrometer measures $M(E)$. What is measured is the quantity

$$M_i = \int_{E_i}^{E_{i+1}} M(E)\,dE \tag{11.40}$$

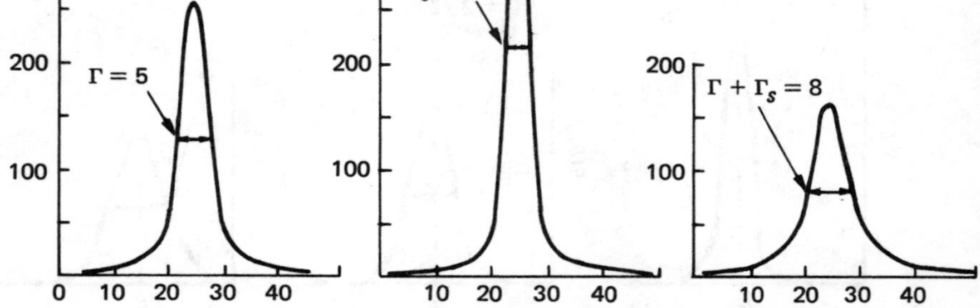

FIG. 11.8 A Lorentzian folded with a Lorentzian gives a third Lorentzian with a width equal to the sum of the widths of the first two distributions.

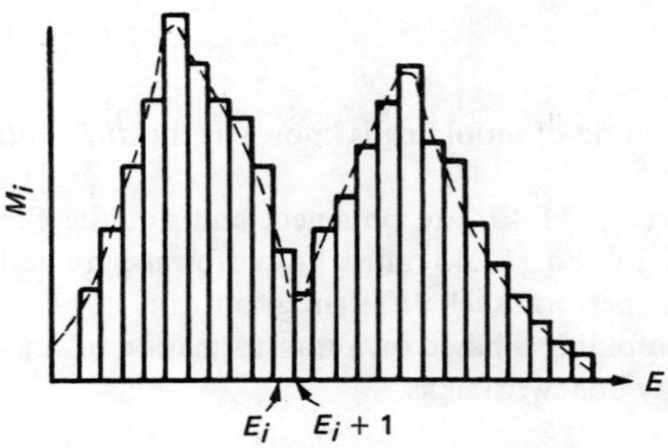

FIG. 11.9 The spectrometer produces a histogram, i.e., the quantities M_i, and not the continuous function $M(E)$ shown by the dashed line.

where $E_{i+1} - E_i = \Delta E_i$ = energy "bin" of the spectrometer. For a multichannel analyzer, ΔE_i represents the width of one of the channels. Therefore, one never measures a continuous function $M(E)$ but obtains instead a histogram consisting of the quantities M_i (see Fig. 11.9). As a first approximation, $M(E_i) \approx M_i/\Delta E_i$.

An analytic solution of Eq. 11.39 is immediately obtained if the detector response is a δ-function. Indeed, if $R(E, E') = \delta(E - E')$, then

$$M(E) = \int_0^\infty \delta(E - E') S(E')\, dE' = S(E)$$

This case is not encountered in practice because there is no detector with such response function; it indicates only that with perfect energy resolution there is no need for unfolding. In general, the more the detector response resembles a δ-function, the more the measured spectrum looks like the source spectrum.

A second type of response that gives an analytic solution, in principle, is a step function† (Fig. 11.10). Let

$$R(E, E') = \begin{cases} \dfrac{C}{E'} & 0 < E \leqslant E' \\ 0 & \text{otherwise} \end{cases} \tag{11.41}$$

where C is a normalization constant. Then Eq. 11.39 takes the form

$$M(E) = \int_E^\infty \frac{C}{E'} S(E')\, dE' \tag{11.42}$$

The lower limit of the integral has been set equal to E because if the response function is that given by Eq. 11.41, no source particles with energy $E' < E$ can contribute to $M(E)$. Upon differentiation,[1-4] Eq. 11.42 gives

†The response of proton-recoil counters resembles a step function (see Chap. 14).

$$S(E) = -\frac{E}{C}\frac{dM(E)}{dE} \tag{11.43}$$

which is the desired solution. This method of unfolding is known as the *differentiation method.*

Since only the quantities M_i (Eq. 11.40) are obtained, and not the function $M(E)$, the differentiation indicated by Eq. 11.43 must be performed numerically. There are several computer codes that perform such differentiation.

The most general method of unfolding is based on a transformation of Eq. 11.39 into a matrix equation. Eq. 11.39 may be rewritten as

$$M(E) = \sum_{j}^{NR} \int_{E_j}^{E_{j+1}} R(E, E')S(E')\, dE' \tag{11.44}$$

where the integral over E' has been written as a sum of integrals over NR energy intervals $\Delta E_j = E_{j+1} - E_j$. Next, Eq. 11.44 is integrated over E to give (see Eq. 11.40)

$$M_i = \int_{E_i}^{E_{i+1}} dE\, M(E) = \sum_{j}^{NR} \int_{E_i}^{E_{i+1}} dE \int_{E_j}^{E_{j+1}} R(E, E')S(E')\, dE' \tag{11.45}$$

Equation 11.45 is still exact. To proceed further, one needs an approximation for the source spectrum $S(E)$. Two approximations and corresponding methods of solving the resulting matrix equation are presented in the next two sections.

11.5.3 An Iteration Method of Unfolding

There are several iteration methods. The method presented here is useful for slowly varying spectra and has been used successfully to unfold beta spectra.[5]

The source spectrum $S(E)$ is approximated over any interval ΔE_j by the expression

$$S(E') = \frac{S_j}{\Delta E_j} \tag{11.46}$$

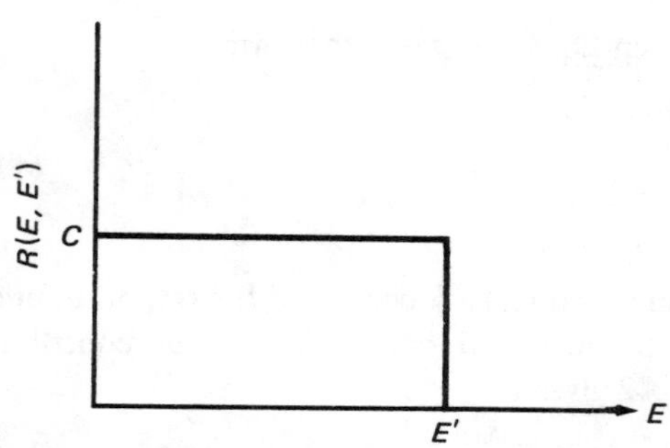

FIG. 11.10 A step-function response.

Using Eq. 11.46 and defining

$$R_{ij} = \frac{1}{\Delta E_j} \int_{E_i} dE \int_{E_j} dE' \, R(E, E') \tag{11.47}$$

then Eq. 11.45 takes the form

$$M_i = \sum_j R_{ij} S_j \qquad i, j = 1, NR \tag{11.48}$$

or, in matrix notation,

$$\mathbf{M} = \mathbf{R} \cdot \mathbf{S} \tag{11.49}$$

A formal solution of Eq. 11.49 is

$$\mathbf{S} = \mathbf{R}^{-1} \mathbf{M} \tag{11.50}$$

where $\mathbf{R}^{-1}$ is the inverse of the matrix with elements given by Eq. 11.47. Although in principle Eq. 11.50 represents a solution to the unfolding problem, in practice the matrix inversion is not always achieved or leads to a solution with a large error.

The iteration method to be discussed here starts with Eq. 11.48 and uses the measured spectrum as the first guess of the iteration procedure.[5]

$$S_j^{(1)} = M_j \qquad j = 1, NR$$

This source spectrum when substituted into Eq. 11.48 gives

$$M_j^{(1)} = \sum_i R_{ij} S_j^{(1)} \qquad j = 1, NR$$

The error of $S_j^{(1)}$ is taken to be

$$S_j - S_j^{(1)} = M_j - M_j^{(1)} \qquad j = 1, NR$$

and the new guess for the second iteration is

$$S_j^{(2)} = S_j^{(1)} + [M_j - M_j^{(1)}] \qquad j = 1, NR$$

Substitution into Eq. 11.48 gives

$$M_j^{(2)} = \Sigma \, R_{ij} S_i^{(2)} \qquad i = 1, NR$$

and so on. The nth iteration uses

$$S_j^{(n)} = S_j^{(n-1)} + [M_j - M_j^{(n-1)}] \qquad j = 1, NR$$

and is the solution to the problem if the difference $|M_j - M_j^{(n)}|$ for $j = 1, \ldots, NR$ is acceptably small. This iteration method converges in less than five iterations and gives good results.

11.5.4 Least-Squares Unfolding

A different approximation for the source spectrum, used with neutrons, assumes that $S(E)$ can be represented as a sum of NS discrete components.[6-8] Therefore, one can write

$$S(E') = \sum_{j=1}^{NS} X_j \delta(E' - E_j) \tag{11.51}$$

Using Eq. 11.51 and defining

$$\mathbf{A}_{ij} = \frac{1}{E_i - E_{i-1}} \int_{E_{i-1}}^{E_i} R(E, E_j)\, dE \tag{11.52}$$

then Eq. 11.45 takes the form

$$M_i = \sum_{j=1}^{NS} A_{ij} X_j \qquad \begin{cases} i = 1, \ldots, NR \\ j = 1, \ldots, NS \end{cases} \tag{11.53}$$

or in matrix notation,

$$\mathbf{M} = \mathbf{AX} \tag{11.54}$$

If $NR = NS$, the formal solution of Eq. 11.54 is, as with Eq. 11.49,

$$\mathbf{X} = \mathbf{A}^{-1}\mathbf{M} \tag{11.55}$$

Because of the difficulties of matrix inversion, a least-squares solution has been attempted with $NR > NS$. If $NR > NS$, no unique solution exists but an acceptable one has been obtained.

The least-squares unfolding starts with Eq. 11.53 and minimizes the quantity

$$Q = \sum_{i=1}^{NR} w_i \left(M_i - \sum_{j=1}^{NS} A_{ij} X_i \right)^2 \tag{11.56}$$

The weighting factors w_i are usually taken to be the inverse of the variance of M_i. The minimization is achieved by setting

$$\frac{\partial Q}{\partial X_k} = 0 \qquad k = 1, \ldots, NS$$

which gives

$$\sum_{i=1}^{NR} w_i A_{ik} \left(M_i - \sum_{j=1}^{NS} A_{ij} X_j \right) = 0 \qquad k = 1, \ldots, NS \tag{11.57}$$

and can be solved for X_j for $j = 1, NS$. Equation 11.57 may be written in matrix form[6]

$$\mathbf{X} = (\mathbf{A}^T \mathbf{W} \mathbf{A})^{-1} \mathbf{A}^T \mathbf{W} \mathbf{M} \tag{11.58}$$

where $\mathbf{A}^T$ = transpose of $\mathbf{A}$.

Computer round-off errors in completing the matrix inversion shown by Eq. 11.58 lead to large oscillations in the solution **X**. The oscillations can be reduced if the least-squares solution is "constrained." Details of least-squares unfolding with constraints are given in references 6 and 7.

11.6 DATA SMOOTHING

The smoothing of raw experimental data is a controversial subject because it represents manipulation of the data without clear theoretical justification. However, smoothing is generally accepted as common practice since experience has shown that it is beneficial in certain cases to the subsequent analysis of the data, for example, in identification of energy peaks in complex gamma energy spectra (Chap. 12) and unfolding of neutron energy spectra (Chap. 14). Data smoothing should be viewed as an attempt to filter out the statistical fluctuations without altering the significant features of the data.

To illustrate how data smoothing is performed, consider again N measurements $y_i|_{i=1,N}$ where $y_i = y(x_i)$. Smoothing, which is applied to the values of y_i, is an averaging process. In the simplest case, one adds a fixed odd number of y_i values, takes the arithmetic average of the sum, and sets the smoothed value of y_i at the center of the group equal to this average. Next, the first point of the group is dropped, the next point is added at the other end of the group, and the process is repeated for all y_i points. In general, the "smoothing" equation takes the form

$$y_i = \frac{1}{M} \sum_{j=-n}^{j=n} C_j y_{i+j} \tag{11.59}$$

where C_j = coefficients that depend on the method of smoothing (see below)
M = normalization constant

n = index showing the number of points used in the smoothing process (The index n means that $2n+1$ points were used for smoothing.)

y_i = smoothed value, replacing the old y_i in the middle of $2n+1$ points

The coefficients C_j are determined by least-squares fitting a polynomial of order m to $2n+1$ data points[9–11] and taking the smoothed value equal to the value of the polynomial in the middle point (Fig. 11.11). To illustrate the method, a few examples are given below. The least-squares fit will be based on Eqs. 11.18–11.20.

THREE–POINT ZEROTH-ORDER SMOOTHING

From Eq. 11.18,

$$f(x) = a_1$$

From Eq. 11.19,

$$A_{11} = \sum_{i=1}^{3} w_i = \sum_i 1 = 3$$

From Eq. 11.20,

$$B_1 = \sum_i^3 w_i y_i = \sum_{i=1}^{3} y_i$$

From Eq. 11.16,

$$a_1 = \frac{B_1}{A_{11}} = \frac{1}{3} \sum_{i=1}^{3} y_i$$

Therefore, if three-point zeroth-order smoothing is applied (Fig. 11.12), the constants of Eq. 11.59 are

$$M = 3 \qquad C_1 = C_2 = C_3 = 1$$

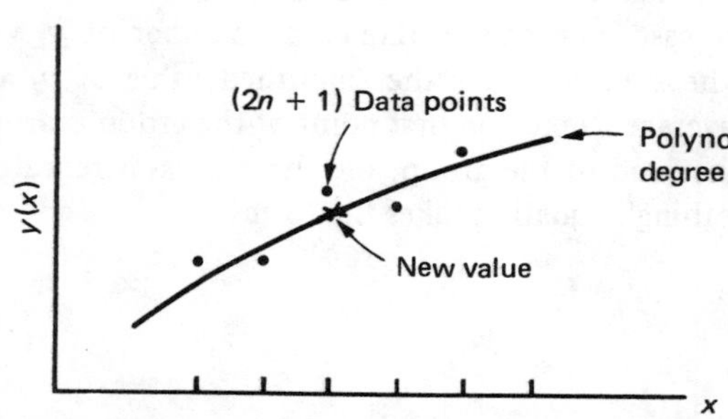

FIG. 11.11 Data smoothing. A polynomial of degree m is fitted to $(2n+1)$ data points, and the smoothed value is equal to the value of the polynomial in the middle point.

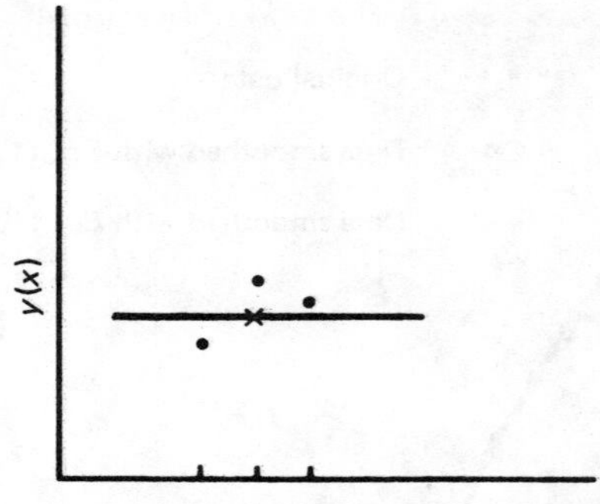

FIG. 11.12 Three-point, zeroth-order smoothing.

and $y_i = \frac{1}{3}(y_{i-1} + y_i + y_{i+1})$ (11.60)

FIVE-POINT ZEROTH-ORDER SMOOTHING

Following the same steps as above, one obtains

$M = 5 \quad C_i = 1 \quad i = 1, 5$

and $y_i = \frac{1}{5}(y_{i-2} + y_{i-1} + y_i + y_{i+1} + y_{i+2})$ (11.61)

THREE-POINT FIRST-ORDER SMOOTHING

From Eq. 11.18, $f(x) = a_i + a_2 x$. Using Eqs. 11.19, 11.20, and 11.16, one can solve for the values of M and C_i. If the x_i points are equally spaced, the result is identical with three-point zeroth-order smoothing (Eq. 11.60). This is true, in general, for equally spaced x_i; that is, the result of smoothing with an even-order polynomial is the same as that with a polynomial of the next higher order. Table 11.2 gives the values of M and C_j for second-order smoothing.

As an example of using the various equations, Fig. 11.13 shows results of three-

Table 11.2 Coefficients for Second-Order Smoothing†

n	5	7	9	11
M	35	21	231	429
C_{-5}				36
C_{-4}			−21	9
C_{-3}		−2	14	44
C_{-2}	−3	3	39	69
C_{-1}	12	6	54	84
C_0	17	7	59	89
C_1	12	6	54	84
C_2	−3	3	39	69
C_3		−2	14	44
C_4			−21	9
C_5				−36

†From reference 9.

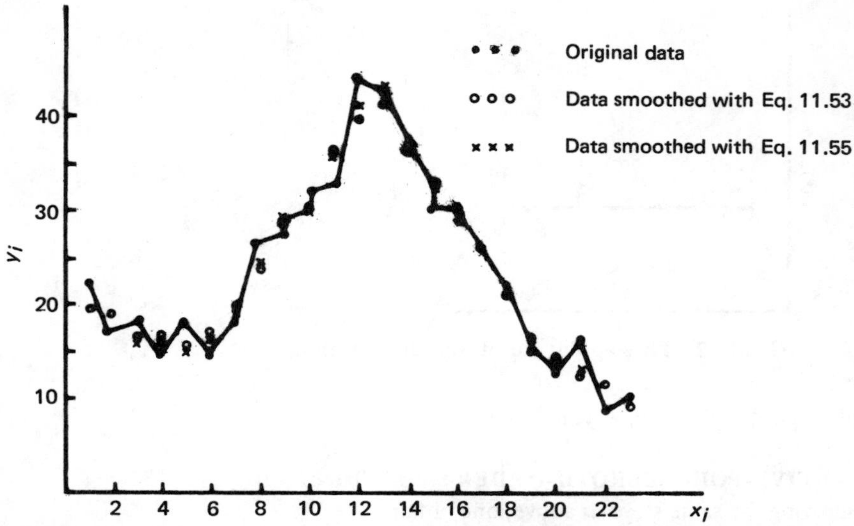

FIG. 11.13 The same data smoothed with two different equations. The lines simply join the original data.

point zeroth-order smoothing and five-point second-order smoothing, i.e., using Eqs. 11.60 and 11.62:

$$y_i = \frac{1}{35}(-3y_{i-2} + 12y_{i-1} + 17y_i + 12y_{i+1} - 3y_{i+2}) \tag{11.62}$$

If the total number of points is N, the number of smoothed points is $N - 2n$ for $(2n + 1)$-point smoothing because the first smoothed point is $i = n + 1$ and the last one is $N - n$. The smoothing process can be repeated if necessary–i.e., one may smooth data which were previously smoothed.[9]

One of the difficulties in data smoothing is the choice of the correct value for n. Unfortunately, there are no strict criteria for the selection of n. The analyst should be guided mainly by experience and by the general effects of smoothing on the results. In particular, if the data represent energy spectra with many peaks, the smoothed spectrum may tend to flatten the peaks and fill the valleys.

The general smoothing equation may be written as

$$Z_i = \sum_j R_{i-j} Y_j \tag{11.63}$$

which has the same form as the folding-unfolding matrix equations (Eq. 11.48). Thus, smoothing may be considered as folding the data (y_i) with the weights (R_{i-j}) as the response function. Taking it one step further, one may perform smoothing by using a continuous function, e.g., a Gaussian. Then

$$Z(x) = \int G(x, x')y(x')\,dx' \tag{11.64}$$

The operation indicated by Eq. 11.64 has been applied to neutron spectroscopic data.[12]

PROBLEMS

11.1 The table below shows radioactive decay data from a certain isotope. Using least-squares fit, determine the half-life of the isotope. What is the error of the half-life as determined by this set of data?

t(min	Counts	t(min)	Counts
0	500	6	164
1	430	7	130
2	310	8	92
3	265	9	89
4	240	10	75
5	186		

11.2 The numbers below represent values of cosine for the corresponding angles.

Angle:	5°	10°	15°	20°	25°	30°
Cosine:	0.99619	0.98481	0.96593	0.93969	0.90631	0.86603

Obtain cosine values, by interpolation, for 22° using Lagrange's three-point interpolation formula. Evaluate the error of your result. Compare the error with its correct value.

11.3 Prove Eqs. 11.22 to 11.24.

11.4 Obtain the least-squares fit equations for a quadratic fit.

11.5 Prove that the result of folding a step function with a Gaussian is

$$M(E) = \frac{S_0}{2}\left[1 + \text{erf}\left(\frac{E - E_0}{\sqrt{2}\sigma}\right)\right]$$

where the source spectrum is

$$S(E) = \begin{cases} S_0 & E \geqslant E_0 \\ 0 & \text{otherwise} \end{cases}$$

11.6 Prove that the result of folding a Gaussian with a Gaussian is

$$M(E) = \frac{1}{\sqrt{2\pi}\sqrt{(\sigma^2 + \sigma_S^2)}} \exp\left[-\frac{(E - E_0)^2}{2(\sigma^2 + \sigma_S^2)}\right]$$

where the source spectrum is centered at E_0 and has a standard deviation σ_S.

11.7 Prove that the result of folding an exponential function e^{-aE} with a Gaussian is

$$M(E) = e^{a^2\sigma^2/2} e^{-aE}$$

11.8 What is the measured spectrum $M(E)$ if the detector response is a step function of the form $R(E, E') = C(E')/E'$ and the source emits two types of particles at energy E_1 and E_2?

11.9 What is the measured spectrum $M(E)$ if the detector response is a step function, as in Prob. 11.8, and the source spectrum is

$$S(E) = \frac{S_0}{E_2 - E_1} \qquad E_1 \leqslant E \leqslant E_2$$

and is zero otherwise.

11.10 The following data represent results of counting an energy peak. How does the FWHM of the peak change if one applies (a) three-point zeroth-order smoothing and (b) five-point second-order smoothing?

Channel	Counts	Channel	Counts
10	12	17	34
11	10	18	26
12	14	19	18
13	14	20	10
14	24	21	12
15	30	22	9
16	40		

BIBLIOGRAPHY

Abramowitz, M., and Stegan, I. A. (eds.), *Handbook of Mathematical Functions*, National Bureau of Standards, Applied Mathematics Series, 55, June 1964.

Bevington, P. R., *Data Reduction and Error Analysis for the Physical Sciences*, McGraw-Hill, New York, 1969.

Hildebrand, F. B., *Introduction to Numerical Analysis*, McGraw-Hill, New York, 1956.

Householder, A. S., *Principles of Numerical Analysis*, McGraw-Hill, New York, 1953.

REFERENCES

1. Bennet, F. F., and Yule, T. J., ANL-7763 (1971).
2. Ciallela, C. M., and Devanney, J. A., *Nucl. Instr. Meth.* 60:269 (1968).
3. Toms, M. E., *Nucl. Instr. Meth.* 92:61 (1971).
4. Johnson, R. H., and Wehring, B. W., ORNL/RSIC-40, p. 33 (1976).
5. Tsoulfanidis, N., Wehring, B. W., and Wyman, M. E., *Nucl. Instr. Meth.* 73:98 (1969).
6. Burrus, W. R., and Verbinski, V. V., *Nucl. Instr. Meth.* 67:181 (1969).
7. Kendrick, H., and Sperling, S. M., GA-9882 (1970).
8. Johnson, R. H., Ingersoll, D. T., Wehring, B. W., and Dorning, J. J., *Nucl. Instr. Meth.* 145:337 (1977).
9. Savitzky, A., and Golay, M. J. E., *Anal. Chem.* 36:1627 (1964).
10. Yule, H. P., *Nucl. Inst. Meth.* 54:61 (1967).
11. Yule, H. P., *Anal. Chem.* 38:103 (1966).
12. Johnson, R. H., Wehring, B. W., and Dorning, J. J., *Nucl. Sci. Eng.* 73:93 (1980).

12

gamma and x-ray spectroscopy

12.1 INTRODUCTION

Gamma and x-rays may be treated either as electromagnetic waves or as particles called photons. An *electromagnetic wave* is characterized by its wavelength λ or frequency ν. A *photon* is a particle having zero charge and zero rest mass, traveling with the speed of light, and having an energy $E = h\nu$, where h = Planck's constant. The wave properties of a photon are used for low-energy measurements only. In all other cases, detection of photons is based on their interactions as particles.

This chapter first examines the mechanisms of detection in photon counters and then discusses the spectroscopic characteristics of the different types of x-ray and γ-ray detectors.

12.2 MODES OF ENERGY DEPOSITION IN THE DETECTOR

Photons are detected by means of the electrons they produce when they interact in the material of which the detector is made. The main interactions are photoelectric effect, Compton scattering, and pair production. The electrons (or positrons) produced by these interactions deposit their energy in the counter and thus generate a voltage pulse that signifies the passage of the photon. The height of the voltage pulse is proportional to the energy deposited in the detector. Since the objective is to measure the energy of the incident photon, the question arises: Is this voltage pulse proportional to the energy of the incident particle? To provide an answer, one must examine how the photon interacts and what happens to its energy.

12.2.1 Energy Deposition by Photons with $E < 1.022$ MeV

A photon with $E < 1.022$ MeV can interact only through the photoelectric or the Compton effect. If a photoelectric interaction takes place, the photon disappears and an electron appears with energy equal to $E - B$, where B is the binding energy of that electron. The range of electrons in a solid, either a scintillator crystal or a semiconductor, is so short that it can be safely assumed that all the electron energy will be deposited in the detector (Fig. 12.1*a*). If the interaction occurs very close to the wall, the electron may deposit only part of its energy in the counter (Fig. 12.1*b*), but the probability of this happening is small. In practice, one assumes that all the photoelectrons deposit all their energy in the detector. This energy is less than the energy of the incident photon by the amount B, the binding energy of the electron. What happens to the energy B?

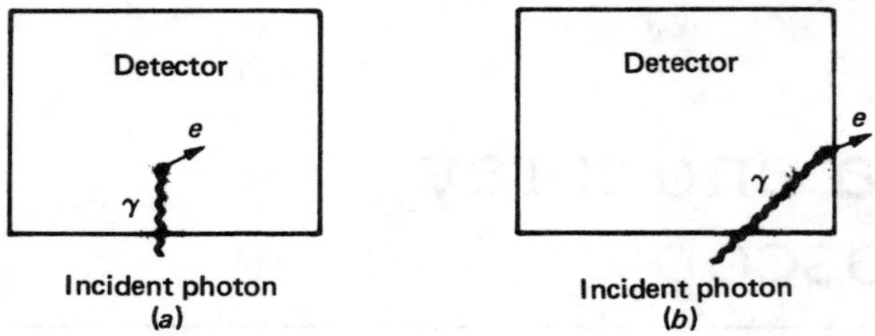

FIG. 12.1 As a result of a photoelectric interaction the photon disappears. In (*a*), all the energy of the electron is deposited in the detector. In (*b*), part of the energy is deposited in the wall.

After a photoelectric effect takes place, an electron from one of the outer atomic shells drops into the empty inner state in about 10^{-8} s. This electronic transition is followed by an x-ray or by an Auger electron (see Chap. 4). The Auger electron will also deposit its energy in the detector. The x-ray with energy in the low keV range (~ 100 keV or less) interacts again photoelectrically and generates another electron.† The net result of these successive interactions is that the part B of the incident photon energy is also deposited in the counter. All these events take place within a time of the order of 10^{-8} s. Since the formation of the voltage pulse takes about 10^{-6} s, both parts of the energy—namely, $E - B$ = energy of photoelectron and B = energy of the x-ray—contribute to the same pulse, the height of which is proportional to $(E - B) + B = E$ = incident photon energy. The conclusion is, therefore, that if the photon interacts via photoelectric effect, the resulting pulse has a height proportional to the incident particle energy.

If Compton scattering takes place, only a fraction of the photon energy is given to an electron. A scattered photon still exists carrying the rest of the energy. The energy of the electron is deposited in the detector. But what happens to the energy of the scattered photon?

The scattered photon may or may not interact again inside the detector. The probability of a second interaction depends on the size of the counter (Fig. 12.2), on

†For thin detectors, or detectors made of high-Z material—e.g., CdTe or HgI_2 —some x-rays may escape, thus forming the so-called "escape peaks" (see Sec. 12.9).

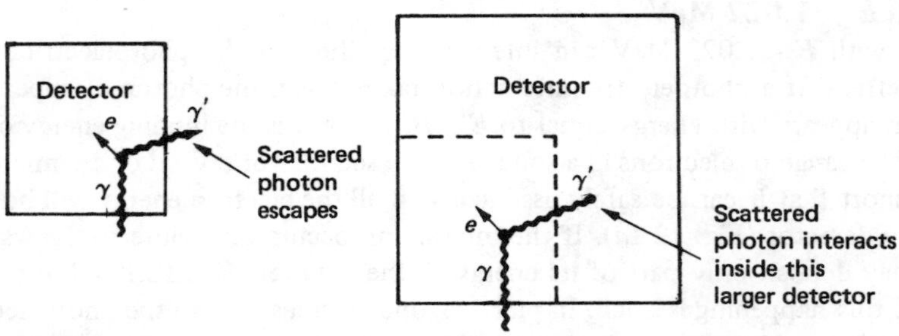

FIG. 12.2 As a result of Compton scattering, part of the photon energy may escape.

the position of the first interaction, on the energy of the scattered photon, and on the material of which the detector is made. Unless the detector is infinite in size, there is always a chance that the scattered photon may escape, in which case a pulse will be formed with height proportional to an energy which is *less* than the energy of the incident photon.

From the study of the Compton effect (Chap. 4), it is known that Compton electrons have an energy range from zero up to a maximum energy T_{max} which is

$$T_{max} = E - \frac{E}{1 + 2E/mc^2} \tag{12.1}$$

where $mc^2 = 0.511$ MeV, the rest mass energy of the electron. Therefore, if the interaction is Compton scattering, pulses are produced from Compton electrons with heights distributed from $V = 0$ V, corresponding to $T_{min} = 0$, up to a maximum height V_{max} corresponding to the maximum energy T_{max}. Figures 12.3 to 12.5 illustrate how a monoenergetic photon spectrum is recorded as a result of photoelectric and Compton interactions.

Figure 12.3 shows the source spectrum. In the case of perfect energy resolution, this monoenergetic source produces in an MCA the measured spectrum shown by Fig. 12.4. Some photons produce pulses that register in channel C_0, corresponding to the source energy E_0, and thus contribute to the main peak of the spectrum, which is called the full-energy peak. The Compton electrons are responsible for the continuous part of the spectrum, extending from zero channel up to channel *CC* and called the *Compton continuum.* The end of the Compton continuum, called the *Compton edge*, corresponds to the energy given by Eq. 12.1. Since no detector exists with perfect energy resolution, the measured spectrum looks like that of Fig. 12.5.

Sometimes the Compton interaction occurs very close to the surface of the detector or in the material of the protective cover surrounding the detector (Fig. 12.6).† Then there is a high probability that the electron escapes and only the energy of the scattered photon is deposited in the detector. The minimum energy E_{min} of the scattered photon is given by

$$E_{min} = \frac{E}{1 + 2E/mc^2} \tag{12.2}$$

†Backscattering may also take place in the source itself, or in the shield surrounding the detector.

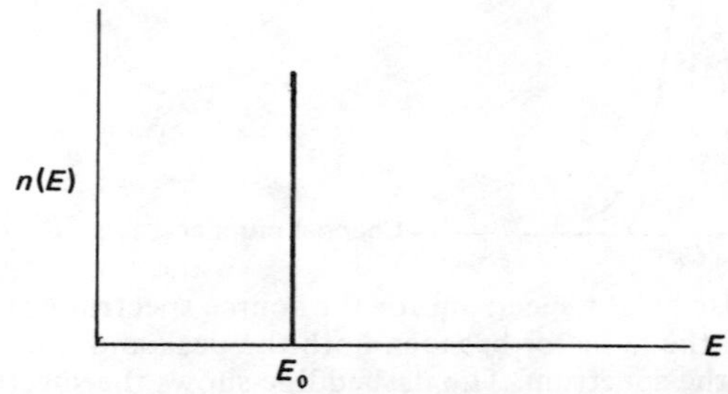

FIG. 12.3 A monoenergetic gamma spectrum (source spectrum).

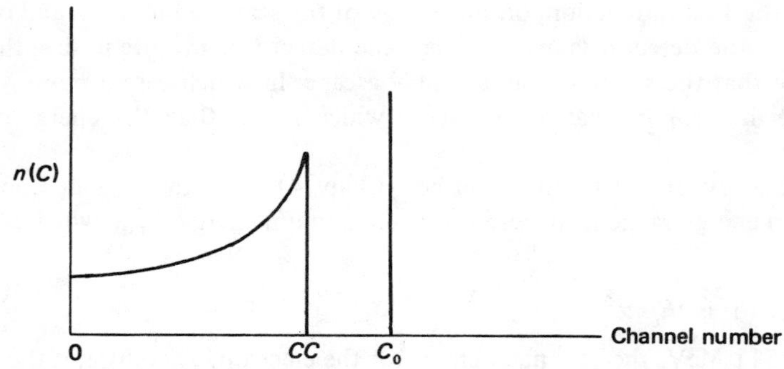

FIG. 12.4 The pulse height spectrum obtained from the source spectrum of Fig. 12.3, in the absence of statistical effects in the detector (perfect energy resolution).

Occasionally a rather broad peak, corresponding to the energy given by Eq. 12.2, is observed in γ-ray spectra. This peak is called the ***backscatter peak*** (Fig. 12.5).

The fraction of counts recorded outside the full-energy peak depends on the energy of the gamma and on the size of the detector. The energy of the photon determines the ratio σ/μ of the Compton scattering coefficient to the total attenuation coefficient. The lower the gamma energy is, the smaller this ratio becomes. Then a greater fraction of particles interacts photoelectrically and is recorded in the full-energy peak, thus reducing the Compton continuum part of the spectrum. As an example, consider gammas with energy 100 keV and 1 MeV, and a Ge(Li) detector. For 100-keV gammas in germanium, the ratio σ/μ is $0.9/3.6 \approx 0.25$ (Fig. 12.27), which indicates that 25 percent of the interactions are Compton and 75 percent photoelectric. The number of pulses in the Compton continuum should be equal to or less than one-third the number recorded under the full-energy peak. At 1 MeV, the

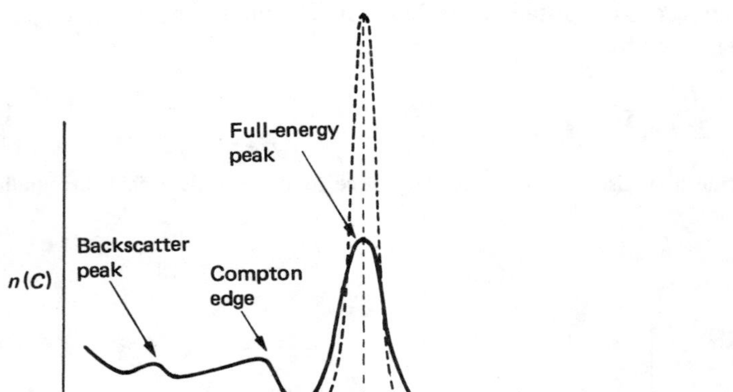

FIG. 12.5 The measured pulse height spectrum for the source spectrum of Fig. 12.3. The statistical effects in the detector broaden both the peak and the Compton continuum part of the spectrum. The dashed line shows the spectrum that would have been recorded in the absence of the Compton continuum.

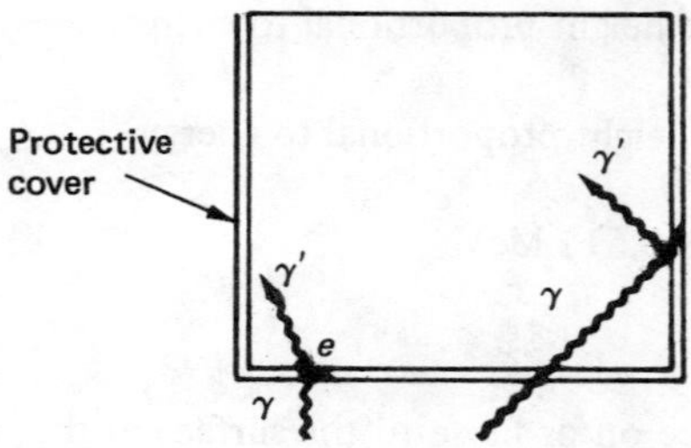

FIG. 12.6 If Compton scattering occurs close to the surface of the detector, the only energy deposited may be that of the scattered photon.

ratio σ/μ is about $0.4/0.42 \approx 0.95$, which means that about 95 percent of the interactions are Compton and only 5 percent photoelectric. Thus, the Compton continuum due to 1-MeV photons is the largest part of the spectrum.

The magnitude of the Compton continuum is also affected by the size of the detector (Fig. 12.2). The larger the detector is, the greater the probability of a second Compton interaction. If the detector size could become infinite, the Compton continuum would disappear.

12.2.2 Energy Deposition by Photons with $E > 1.022$ MeV

If $E > 1.022$ MeV, pair production is possible, in addition to photoelectric effect and Compton scattering. As a result of pair production, the photon disappears and an electron-positron pair appears, at the expense of 1.022 MeV transformed into the pair's rest masses. The total kinetic energy of the electron-positron pair is

$$T_{e^-} + T_{e^+} = T = E - 1.022 \text{ MeV}$$

The kinetic energy of the pair is deposited in the counter (the arguments are the same as for photoelectrons or Compton electrons). Therefore, pulses proportional to the energy $T = E - 1.022$ MeV are certainly produced. But, what happens to the energy of 1.022 MeV?

The positron slows down and reaches the end of its range in a very short time, shorter than the time needed for pulse formation. Sometimes while in flight, but most of the time at the end of its track, it combines with an atomic electron, the two annihilate, and two gammas are emitted, each with energy 0.511 MeV.† There are several possibilities for the fate of these annihilation gammas.

1 The energy of both annihilation gammas is deposited in the detector. Then, a pulse height proportional to energy

$$(E - 1.022) + 1.022 = E$$

is produced.

†There is a small probability that three gammas may be emitted. This event has a negligible effect on spectroscopy measurements.

2 Both annihilation photons escape. A pulse height proportional to energy $E - 1.022$ MeV is formed.
3 One annihilation photon escapes. A pulse height proportional to energy

$$E - 1.022 \text{ MeV} + 0.511 \text{ MeV} = E - 0.511 \text{ MeV}$$

is formed.

If the pair production event takes place on or close to the surface of the detector, it is possible that only one of the annihilation photons enters the counter. In such a case, a pulse height proportional to energy 0.511 MeV is formed.

Peaks corresponding to these energies could be identified, but this does not mean that they are observed in every γ-ray spectrum. The number, energy, and intensity of peaks depends on the size of the detector, the geometry of the source (is it collimated or not?), and the energies of the gammas in the spectrum. If a source emits only one gamma, the measured spectrum will certainly show:

1 The full-energy peak, corresponding to E (this is the highest energy peak)
2 The Compton edge, corresponding to energy

$$E - \frac{E}{1 + 2E/mc^2}$$

Other peaks that may be observed are:

3 Backscatter peak, with energy

$$\frac{E}{1 + 2E/mc^2}$$

4 The *single-escape peak* with energy $(E - 0.511)$ MeV
5 The *double-escape peak* with energy $(E - 1.022)$ MeV

Figure 12.7 presents the spectrum of ^{24}Na. The single- and double-escape peaks due to the 2.754-MeV gamma are clearly shown. The single- and double-escape peaks are very important when complex gamma spectra are recorded. The observer should be extremely careful to avoid identifying them falsely as peaks produced by gammas emitted from the source.

If the source is a positron emitter, a peak at 0.511 MeV is always present. The positron-emitting isotope ^{22}Na is such an example. It emits only one gamma with energy 1.274 MeV, yet its spectrum shows two peaks. The second peak is produced by 0.511-MeV annihilation photons emitted after a positron annihilates (Fig. 12.8).

The Compton continuum, present in gamma energy spectra recorded either by a NaI(Tl) scintillator or by a Ge(Li) detector, is a nuisance that impedes the analysis of complex spectra. It is therefore desirable to eliminate or at least reduce that part of the spectrum relative to the gamma-energy peak. One way to achieve this is to use two detectors and operate them in anticoincidence. Such an arrangement, known as the *Compton-suppression spectrometer*, is shown in Fig. 12.9. A large NaI(Tl) scintil-

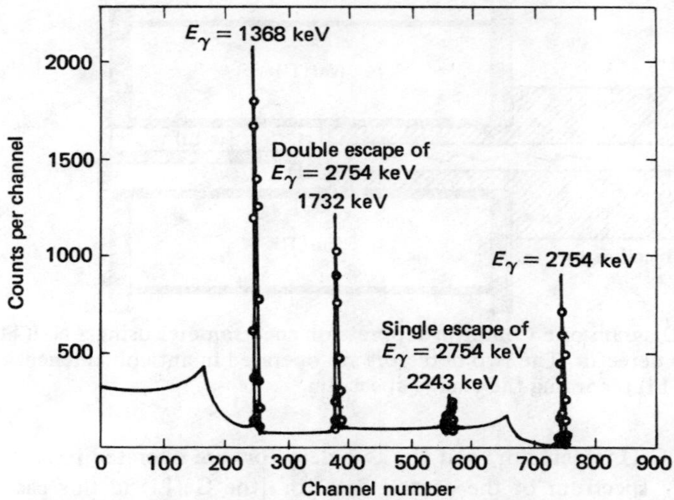

FIG. 12.7 A gamma spectrum showing single- and double-escape peaks (from Chap. 4.4.2 of Bertolini-Coche).

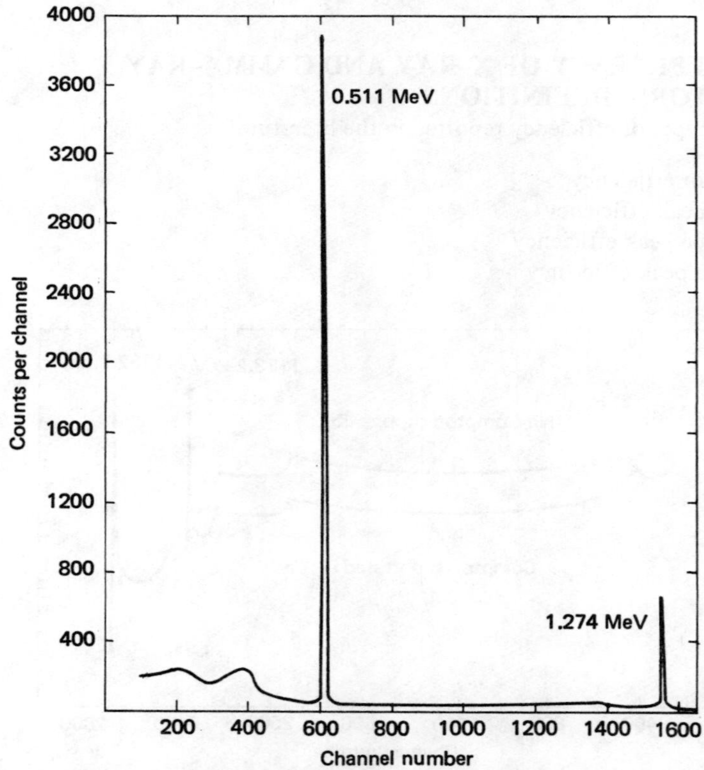

FIG. 12.8 The ^{22}Na spectrum showing the 1.274-MeV peak and the 0.511-MeV peak that is due to annihilation gammas.

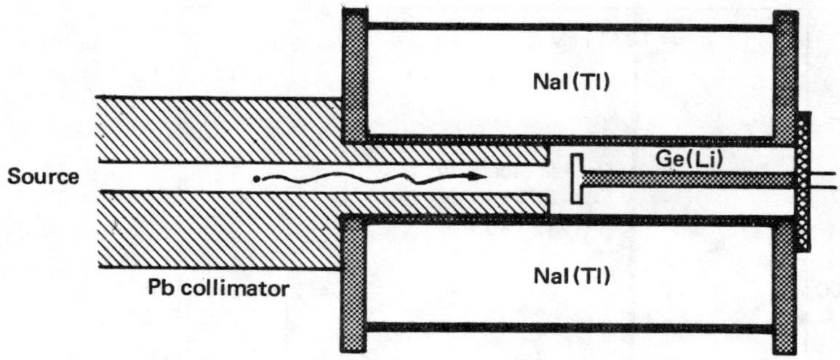

FIG. 12.9 Diagram of a Compton suppression spectrometer using a NaI(Tl) and a Ge(Li) detector. The two detectors are operated in anticoincidence, with the Ge(Li) recording the energy spectrum.

lator surrounds a Ge(Li) detector, and the two detectors are operated in anticoincidence. The energy spectrum of the central detector [the Ge(Li) in this case] will consist of pulses that result from total energy absorption in that detector. Figure 12.10 shows the ^{60}Co spectrum obtained with and without Compton suppression.

12.3 EFFICIENCY OF X-RAY AND GAMMA-RAY DETECTORS: DEFINITIONS

There are four types of efficiency reported in the literature:

1 Total detector efficiency
2 Full-energy peak efficiency
3 Double-escape peak efficiency
4 Single-escape peak efficiency

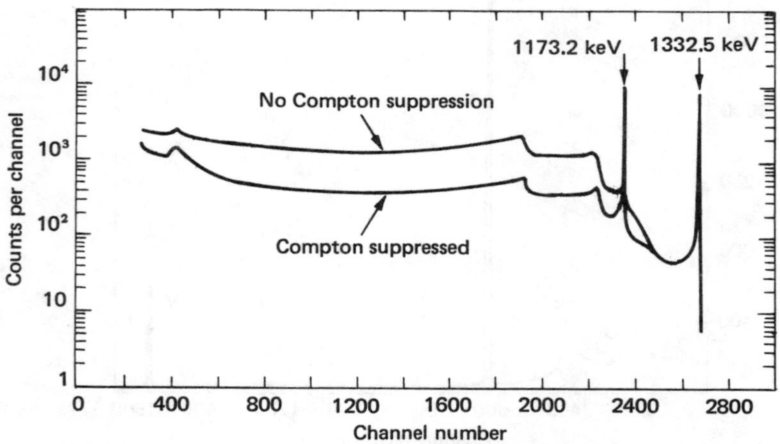

FIG. 12.10 The ^{60}Co spectrum recorded with and without Compton suppression. Notice that the ordinate is in logarithmic scale.

The first two are much more frequently used than the last two. All four efficiencies may be intrinsic, absolute, or relative. The individual definitions are as follows.

Intrinsic total detector efficiency is the probability that a gamma of a given energy which strikes the detector will be recorded. The geometry assumed for the calculation or measurement of this efficiency is shown in Fig. 12.11.

Absolute total detector efficiency is the probability that a gamma emitted from a specific source will be recorded in the detector. The geometry assumed for the absolute efficiency is shown in Fig. 12.12. The intrinsic efficiency (Fig. 12.11) depends on the energy of the gamma, E, and the size of the detector L. The total efficiency (Fig. 12.12) depends on, in addition to E and L, the radius of the detector R and the source-detector distance d. Therefore the total efficiency, as defined here, is the product of intrinsic efficiency times the solid angle fraction (see also Chap. 8).

Full-energy peak efficiency is defined as follows:

$$\begin{pmatrix}\text{Full-energy peak}\\ \text{efficiency}\end{pmatrix} = \begin{pmatrix}\text{total detector}\\ \text{efficiency}\end{pmatrix} \times \frac{\begin{pmatrix}\text{counts in full-}\\ \text{energy peak}\end{pmatrix}}{\begin{pmatrix}\text{total counts in}\\ \text{spectrum}\end{pmatrix}} \tag{12.3}$$

The ratio by which the total detector efficiency is multiplied in Eq. 12.3 is called the *peak-to-total ratio* (P). Figure 12.13 shows how P is measured.

The *double-escape peak efficiency* is important if the energy of the gamma E is greater than about 1.5 MeV, in which case pair production becomes important. The energy of the double-escape peak, equal to $E - 1.022$ MeV, is used for identification of certain isotopes. This kind of efficiency is defined by

$$\begin{pmatrix}\text{Double-escape}\\ \text{peak efficiency}\end{pmatrix} = \begin{pmatrix}\text{total detector}\\ \text{efficiency}\end{pmatrix} \times \frac{\begin{pmatrix}\text{counts in double-}\\ \text{escape peak}\end{pmatrix}}{\begin{pmatrix}\text{total counts in}\\ \text{spectrum}\end{pmatrix}} \tag{12.4}$$

The *single-escape peak efficiency* is important also for $E > 1.5$ MeV, and its definition is analogous to that of the double-escape peak.

$$\begin{pmatrix}\text{Single-escape}\\ \text{peak efficiency}\end{pmatrix} = \begin{pmatrix}\text{total detector}\\ \text{efficiency}\end{pmatrix} \times \frac{\begin{pmatrix}\text{counts in single-}\\ \text{escape peak}\end{pmatrix}}{\begin{pmatrix}\text{total counts in}\\ \text{spectrum}\end{pmatrix}} \tag{12.5}$$

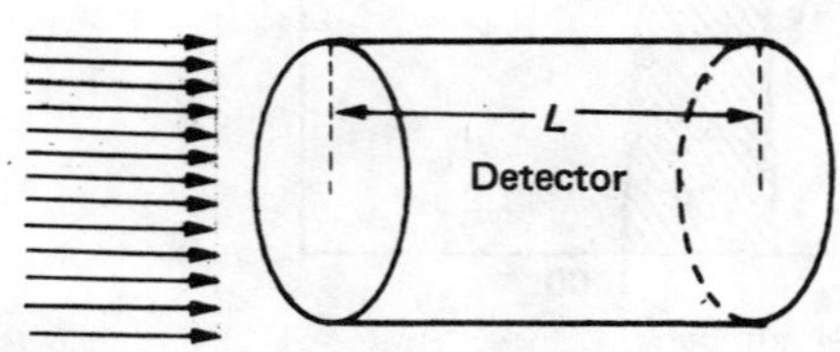

FIG. 12.11 The geometry assumed in the definition of intrinsic efficiency.

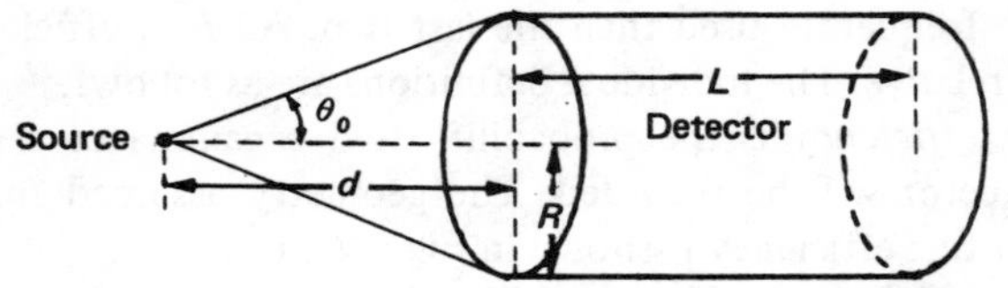

FIG. 12.12 The geometry assumed in the definition of absolute efficiency.

The double- and single-escape peak efficiencies are used with semiconductor detectors only. In the above definitions, if the total detector efficiency is replaced by intrinsic, the corresponding full-energy, single-, and double-escape peak efficiencies are also considered intrinsic.

Relative efficiency may be obtained for all the cases discussed above. In general,

$$(\text{Relative efficiency})_i = \frac{(\text{absolute efficiency})_i}{\text{efficiency of a standard}} \tag{12.6}$$

where the subscript i refers to any one of the efficiencies defined earlier.

12.4 DETECTION OF PHOTONS WITH NaI(Tl) SCINTILLATION COUNTERS

Of all the scintillators existing in the market, the NaI crystal activated with thallium, NaI(Tl), is the most widely used for the detection of γ rays. NaI(Tl) scintillation

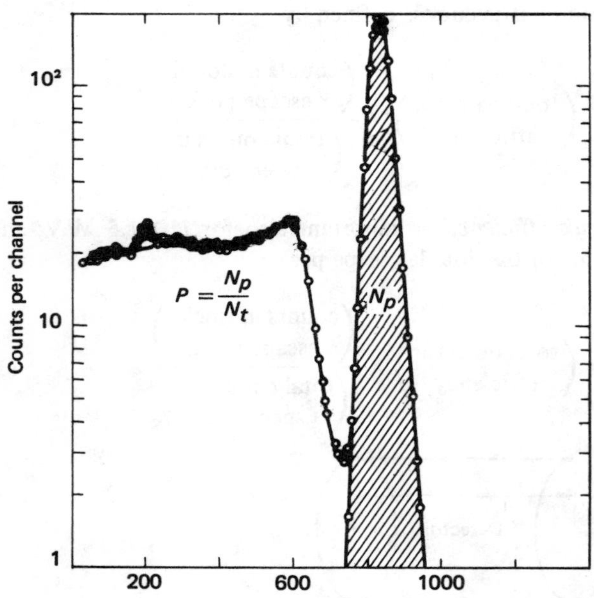

FIG. 12.13 The peak-to-total ratio is equal to the number of counts under the peak (N_p) divided by the total number of counts (N_t).

counters are used when the energy resolution is not the most important factor of the measurement. They have the following advantages over Ge(Li) and Si(Li) detectors:

1 They can be obtained in almost any shape and size. NaI(Tl) crystals with size 0.20 m (8 in) diameter by 0.20 m (8 in) thickness are commercially available.
2 They have rather high efficiency (see Sec. 12.3.1).
3 They cost less than semiconductor detectors.

A disadvantage of all scintillation counters, in addition to their inferior energy resolution relative to Si(Li) and Ge(Li) detectors, is the necessary coupling to a photomultiplier tube.

NaI(Tl) detectors are offered in the market today either as crystals that may be ordered to size or as integral assemblies mounted to an appropriate photomultiplier tube.[1-3] The integral assemblies are hermetically sealed by an aluminum housing. Often, the housing is chrome-plated for easier cleaning. The phototube itself is covered by an antimagnetic μ-metal that reduces gain perturbations caused by electric and magnetic fields surrounding the unit.

The front face of the assembly is usually the "window" through which the photons pass before they enter into the crystal. The window should be as thin as possible to minimize the number of interactions of the incident photons in the materials of the window. Commercially available NaI(Tl) counters used for γ-ray detection have an aluminum window, which may be as thin as 0.5 mm (0.02 in). X-ray scintillation counters usually have a beryllium window, which may be as thin as 0.13 mm (0.005 in). Beryllium is an excellent material because it causes less absorption thanks to its low atomic number ($Z = 4$).

12.4.1 Efficiency of NaI(Tl) Detectors

The intrinsic efficiency of NaI(Tl) detectors (see Fig. 12.11) is essentially equal to $1 - \exp[-\mu(E)L]$, where

$\mu(E)$ = total attenuation coefficient in NaI for photons with energy E

L = length of the crystal

A plot of $\mu(E)$ for NaI as a function of photon energy is shown in Fig. 12.14.

The efficiency increases with crystal size. The user should be aware, however, that when the detector volume increases, the background counting rate increases too. In fact, the background is roughly proportional to the crystal volume, while the efficiency increases with size at a slower than linear rate. Thus, there may be a practical upper limit to a useful detector size for a given experiment.

Calculated total efficiencies of a NaI crystal are given in Fig. 12.15 for several source-detector distances. They have been obtained by integrating Eq. 8.20, which is repeated here (refer to Fig. 12.12 for symbols):

$$\epsilon(E) = \frac{\int_0^{\theta_0} S\{1 - \exp[-\mu(E)r(\theta)]\}\frac{1}{2} \sin\theta \, d\theta}{(S/2)\int_0^{\theta_0} \sin\theta \, d\theta} \tag{8.20}$$

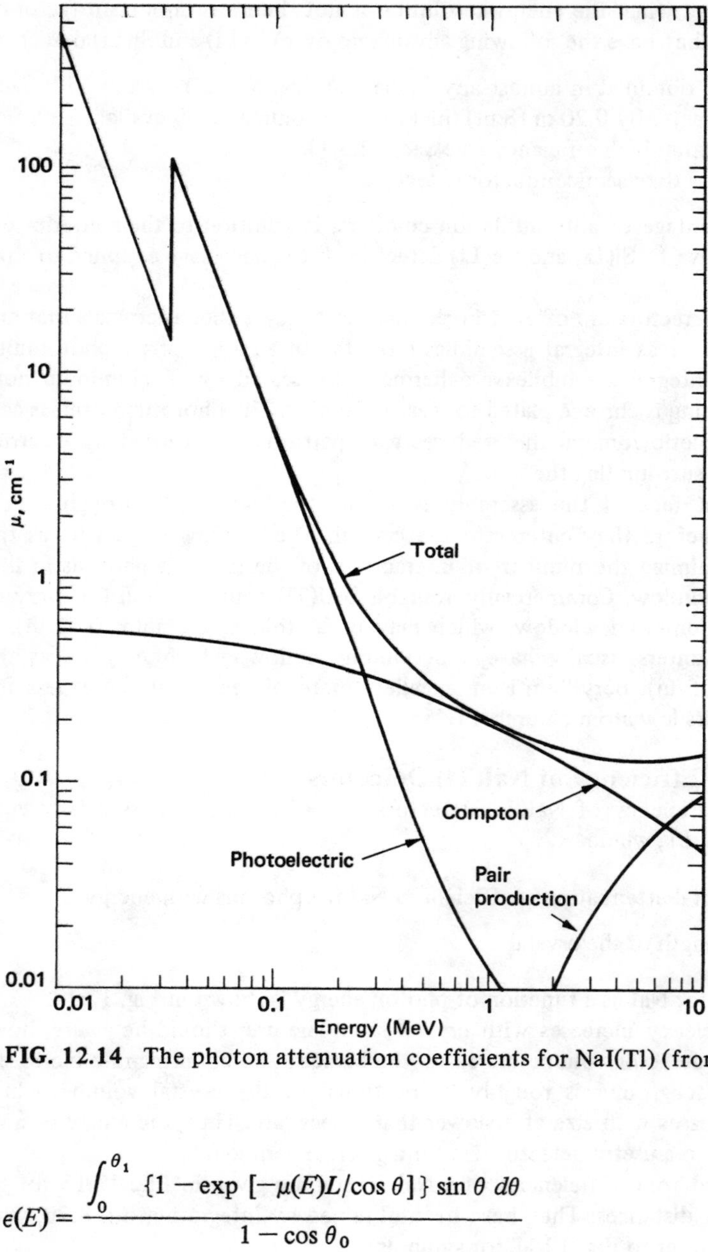

FIG. 12.14 The photon attenuation coefficients for NaI(Tl) (from ref. 3).

or,

$$\epsilon(E) = \frac{\int_0^{\theta_1} \{1 - \exp[-\mu(E)L/\cos\theta]\} \sin\theta \, d\theta}{1 - \cos\theta_0} + \frac{\int_{\theta_1}^{\theta_0} (1 - \exp\{-\mu(E)[(L/\sin\theta) - (d/\cos\theta)]\}) \sin\theta \, d\theta}{1 - \cos\theta_0}$$

where $\theta_1 = \tan^{-1}(R/d + L)$
$\theta_0 = \tan^{-1}(R/d)$

In Fig. 12.15, note that the efficiency decreases with energy up to about 5 MeV. Beyond that point, it starts increasing because of the increase in the pair production probability. Figure 12.16 shows how the peak-to-total ratio (see Fig. 12.13) changes with energy for a source located 0.10 m from detectors of different sizes.

12.4.2 Analysis of Scintillation Detector Energy Spectra

NaI(Tl) scintillators are seldom used as gamma-ray spectrometers because their energy resolution is inferior to that of semiconductor detectors. Despite this fact, a brief discussion of the methods of analysis of NaI(Tl) spectra is instructive because it helps point out differences and similarities between the responses of NaI(Tl) and Ge(Li) detectors.

If a NaI(Tl) scintillator is used to detect a photon spectrum consisting of many gamma energies, the measured spectrum will be the summation of spectra similar to those shown in Fig. 12.5. To identify individual energies from a complex spectrum, one unfolds the measured spectrum (see Chap. 11). Unfolding, in turn, requires the knowledge of the detector response function.

Response functions of NaI(Tl) detectors, obtained by Heath et al.,[4,5] are shown in Fig. 12.17. These authors measured the response for several gamma energies and then used an interpolation scheme to derive the three-dimensional plot of Fig. 12.17. A modified Gaussian of the form

$$y(x) = y_0 \left\{ 1 + \alpha_1 (x - x_0)^4 + \alpha_2 (x - x_0)^{12} \exp\left[-\frac{(x - x_0)^2}{b_0} \right] \right\} \tag{12.7}$$

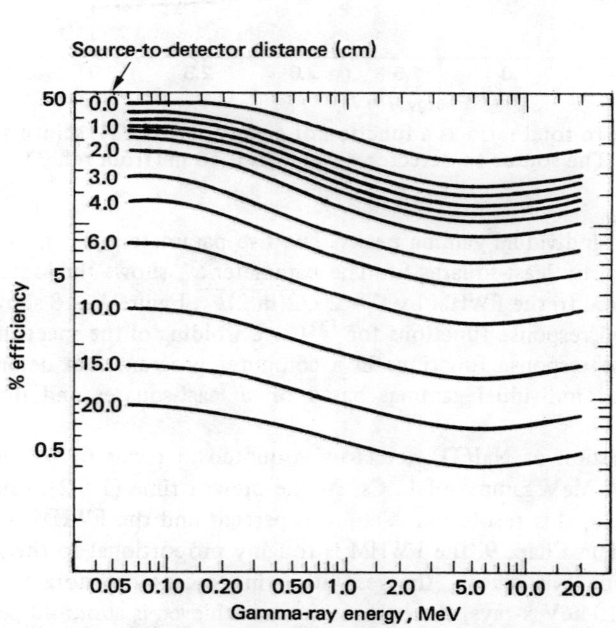

FIG. 12.15 Calculated efficiencies of a 3″ × 3″ (76.2-mm × 76.2-mm) NaI(Tl) scintillator as a function of energy for different source-detector distances (from ref. 3).

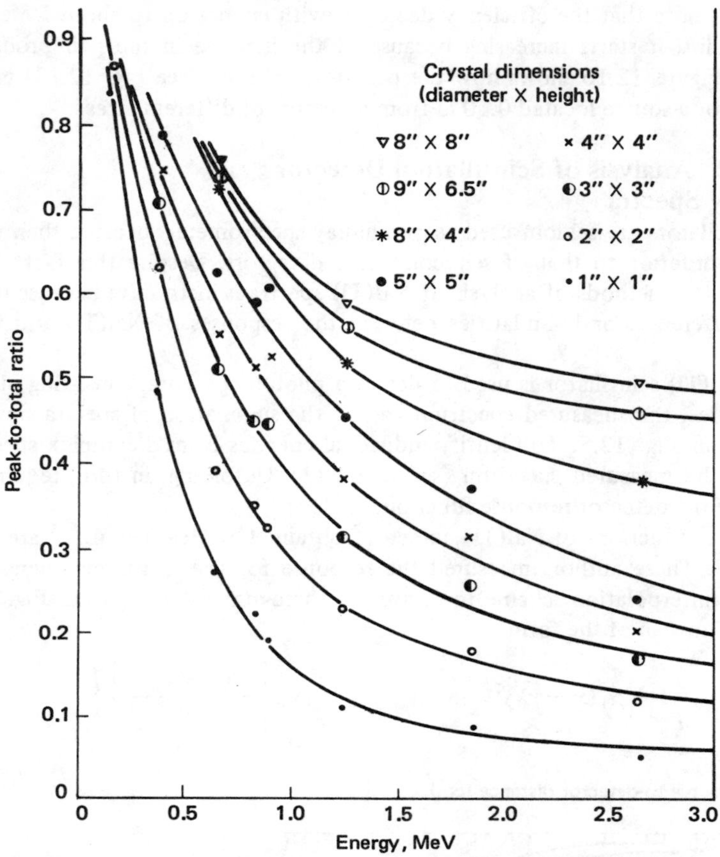

FIG. 12.16 Peak-to-total ratio as a function of energy for NaI(Tl) scintillators of different sizes. The source-to-detector distance is 0.10 m (from ref. 3).

gave a successful fit to individual gamma peaks. The five parameters y_0, x_0, b_0, α_1, and α_2 were determined by least-squares fit. The parameter x_0 shows the location of the peak, and b_0 is related to the FWHM by $\Gamma = 2\sqrt{2(\ln 2)b_0}$. Figure 12.18 shows the measured and calculated response functions for ^{137}Cs. Unfolding of the spectrum was achieved by using these response functions in a computer program that determines energy and intensity of individual gammas based on a least-squares and iteration technique.[5]

The energy resolution of NaI(Tl) detectors is quoted in terms of the percent resolution for the 0.662-MeV gamma of ^{137}Cs. At the present time (1982), using the best electronics available, this resolution is about 6 percent and the FWHM is about 40 keV. As mentioned in Chap. 9, the FWHM is roughly proportional to the square root of the energy. For this reason, the resolution in percent deteriorates as the energy decreases. For 10-keV x-rays, the best resolution achieved is about 40 percent, which makes the FWHM about 4 keV.

12.5 DETECTION OF GAMMAS WITH AN NE 213 ORGANIC SCINTILLATOR

The NE 213 organic scintillator has emerged as one of the leading fast neutron spectrometers (see Chap. 14). As a gamma spectrometer, the NE 213 scintillator has an efficiency lower than that of NaI(Tl) and an energy resolution which is poor compared to that of semiconductor detectors. There are certain applications, however, where high energy resolution is not the most important factor. One such application is detection of gammas in a mixed neutron-gamma field. There, the ability of the NE 213 scintillator to discriminate against neutrons makes it an attractive gamma detector.

Because gammas are detected by the NE 213 scintillator mainly through Compton interactions, the response function of the detector consists of the Compton continuum. The response function has been calculated[6] and measured[7] for several gamma

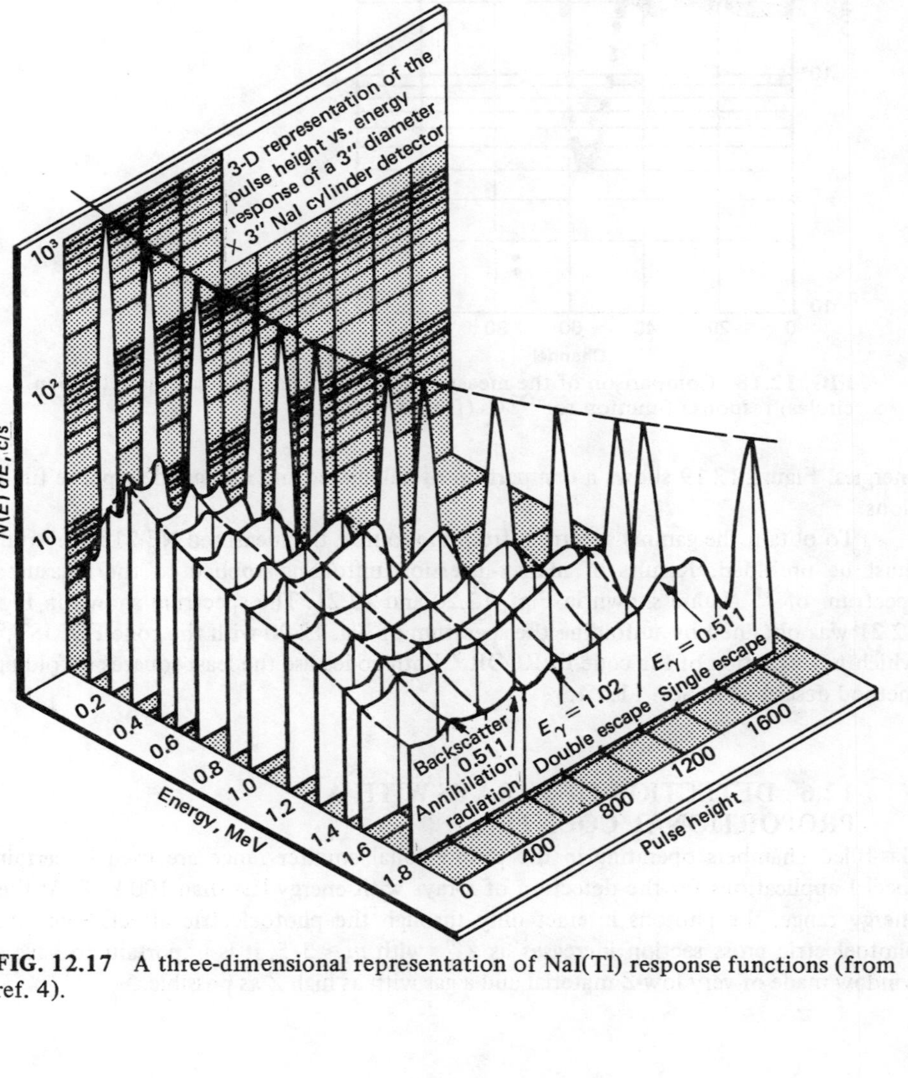

FIG. 12.17 A three-dimensional representation of NaI(Tl) response functions (from ref. 4).

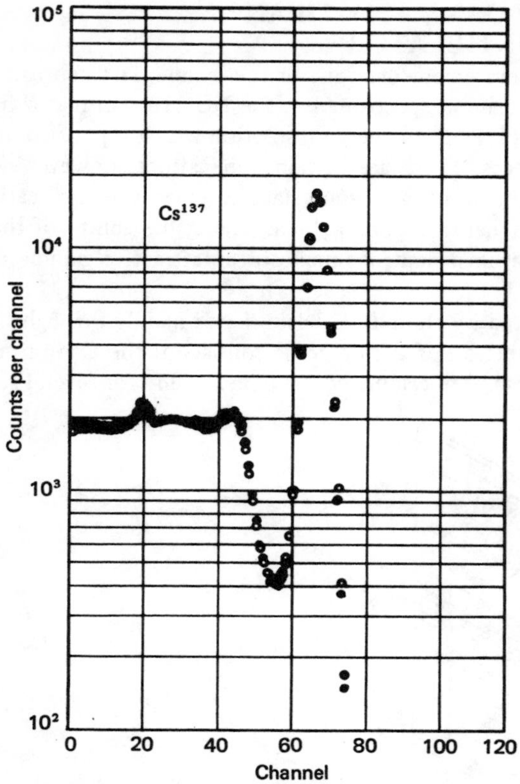

FIG. 12.18 Comparison of the measured (solid circles) and calculated (open circles) response function for ^{137}Cs (from ref. 4).

energies. Figure 12.19 shows a comparison of calculated and measured response functions.

To obtain the gamma spectrum from the source, the measured NE 213 spectrum must be unfolded. Results of matrix-inversion unfolding applied to the measured spectrum of ^{24} Na are shown in Figs. 12.20 and 12.21. The spectrum shown in Fig. 12.21 was obtained by unfolding the spectrum of Fig. 12.20 with the code FORIST,[8] which is a variation of the code FERDOR.[9] Both codes use the least-squares unfolding method described in Sec. 11.5.4.

12.6 DETECTION OF X-RAYS WITH A PROPORTIONAL COUNTER

Gas-filled chambers operating in the proportional counter range are used in certain special applications for the detection of x-rays with energy less than 100 keV. At this energy range, the photons interact only through the photoelectric effect. Since the photoelectric cross section increases as Z^m, with $m \approx 3$-5, it is important to have a window made of very low-Z material and a gas with as high Z as possible.

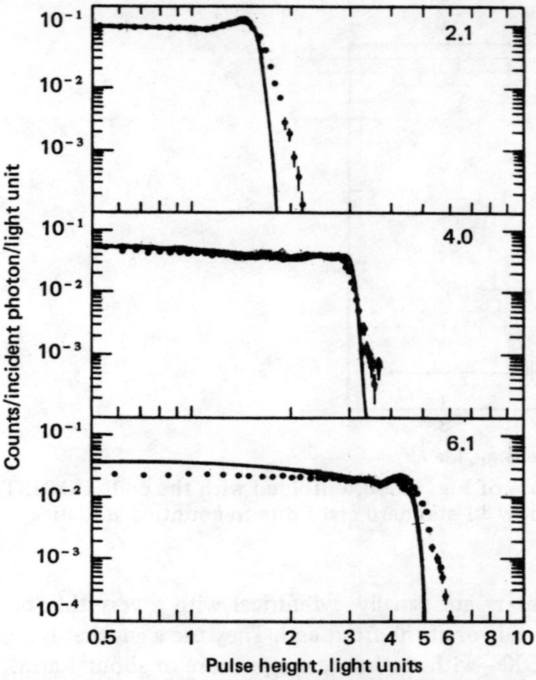

FIG. 12.19 Comparison of measured (points) and calculated (line) NE 213 response functions for 2.1-, 4.0-, and 6.1-MeV gamma rays (from ref. 7).

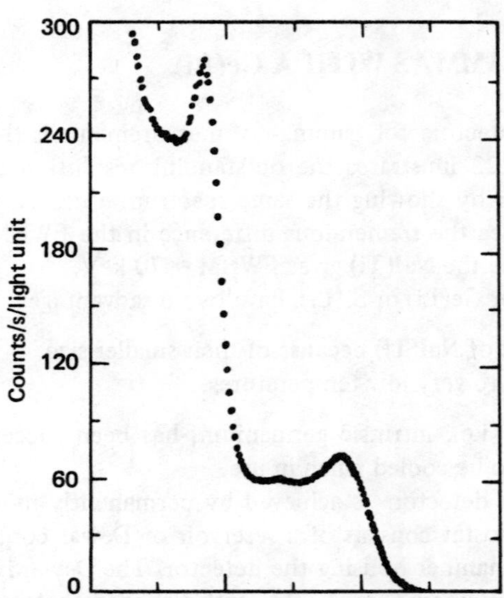

FIG. 12.20 Gamma-ray pulse height spectrum produced by NE 213 for a ^{24}Na source. The two Compton edges are due to the 1.37- and 0.75-MeV gammas (from ref. 7).

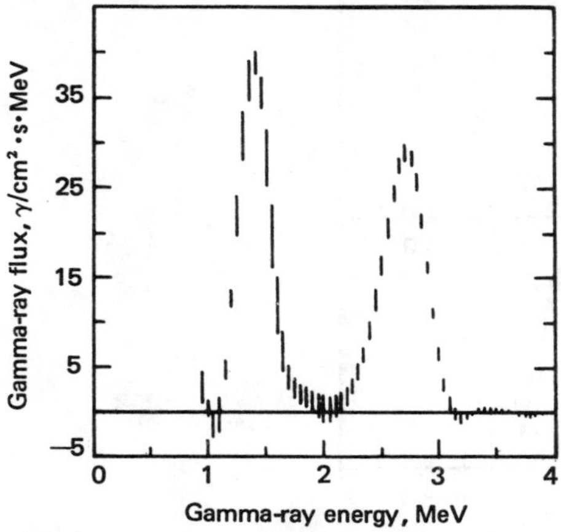

FIG. 12.21 The spectrum of Fig. 12.20, unfolded with the code FORIST. The vertical error bars show ±1 standard error due to counting statistics (from ref. 7).

X-ray proportional counters are usually cylindrical with a very thin beryllium window located either on the side or at the front end. They use a gas that is a mixture of a noble gas—He, Ne, Ar, Kr, Xe—with methane, at a pressure of about 1 atm.

The energy resolution of these counters is such that the FWHM is 1–2 keV at 20 keV. Thus, proportional counters are superior to scintillation counters in this energy range.

12.7 DETECTION OF GAMMAS WITH A Ge(Li) DETECTOR

The main advantage of a Ge(Li) detectors for gamma-ray measurements is their superior energy resolution. Figure 12.22 illustrates the outstanding resolution characteristics of a semiconductor detector by showing the same spectrum obtained with a NaI(Tl) and a Ge(Li) detector. Notice the tremendous difference in the FWHM. The Ge(Li) gives aFWHM = 1.9 keV, while the NaI(Tl) gives FWHM ≈ 70 keV.

Semiconductor detectors, either Ge(Li) or Si(Li), have two disadvantages.

1 Their efficiency is lower than that of NaI(Tl) because of their smaller size.
2 It is necessary to cool them down to very low temperatures.

Recently, high-purity germanium,[10] i.e., intrinsic germanium, has been successfully used as a detector, but it also needs to be cooled when in use.

The cooling of semiconductor detectors is achieved by permanently mounting the detector on a cryostat. The cryostat consists of a reservoir or Dewar containing the cooling medium and a vacuum-chamber housing the detector. The Dewar is made of two or more concentric metal containers (Fig. 12.23) with the space between the two containers being evacuated for thermal insulation. The cooling medium is usually liquid nitrogen (boiling temperature equal to $-196°C = 77$ K). The cooling rod of the "dipstick-type" Dewar is made of copper. In addition to the vertical cryostats

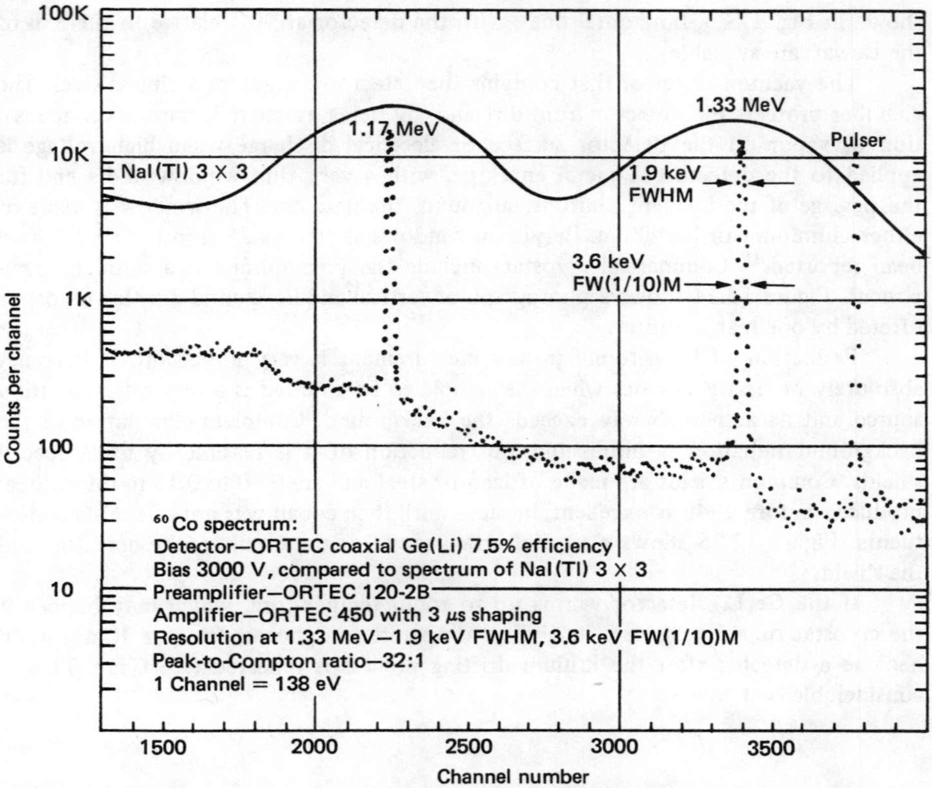

FIG. 12.22 The ^{60}Co spectrum obtained with a NaI(Tl) scintillator and a Ge(Li) detector (from ref. 1).

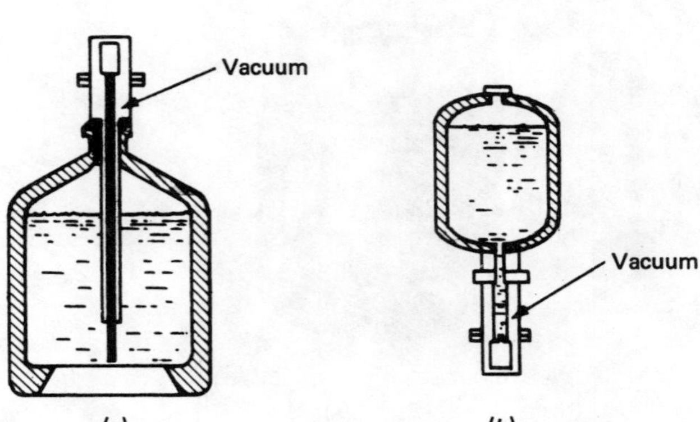

FIG. 12.23 Two types of commercial cryostats: (*a*) dipstick and (*b*) chicken-feeder (from ref. 2).

shown in Fig. 12.23, horizontal ones, with the detector at 90° relative to the axis of the Dewar, are available.

The vacuum chamber that contains the detector is made of stainless steel. The chamber protects the detector from dirt and, by being evacuated, prevents condensation of vapor on the detector surface or electrical discharge when high voltage is applied to the detector. A metal envelope, with a very thin window at its end for the passage of the incident photons, surrounds the detector. The window is made of either aluminum or beryllium. Beryllium windows as thin as 25 μ m (0.001 in) have been reported.[11] Commercial cryostats include the preamplifier as a standard component. Figure 12.24 shows a photograph of vertical and horizontal dipstick cryostats offered by one manufacturer.

Reduction of background in any measurement is very important. It becomes absolutely necessary in cases when the sample to be counted is a very weak radiation source and its activity barely exceeds the background. Complete elimination of the background radiation is impossible, but reduction of it is feasible by using special shields. Common shields are made of lead or steel and are 0.10 to 0.15 m thick. Steel produced before 1940 is excellent, because until then cobalt was not one of its constituents. Figure 12.25 shows a typical arrangement of the cryostat, the detector, and the shield.

If the Ge(Li) detector warms up to room temperature, which may happen if the cryostat runs dry, it loses its ability to operate as a photon detector. It may again become a detector after the lithium drifting process is repeated (see Chap. 7), at a considerable cost.

(a) (b)

FIG. 12.24 A photograph of (*a*) a vertical and (*b*) a horizontal commercially available cryostat (from Canberra).

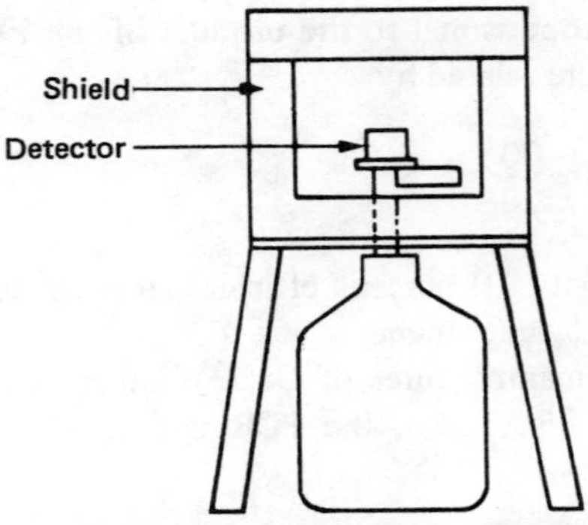

FIG. 12.25 A typical arrangement of the detector, the shield, and the cryostat.

12.7.1 Efficiency of Ge(Li) Detectors

The efficiency of commercial Ge(Li) detectors quoted in the list of specifications is *relative* full-energy peak efficiency referenced in terms of the absolute efficiency of a 76 mm × 76 mm (3 in × 3 in) NaI(Tl) crystal. The measurement (or calculation) is based on the 1.33 MeV peak of ^{60}Co. It is assumed that a ^{60}Co source of known strength is positioned 0.25 m away from the face of the detector. A count is taken for a period of time, and the absolute full-energy peak efficiency of the Ge(Li) detector is determined in accordance with Eq. 12.3 by dividing the total number of counts under the 1.33-MeV peak (shaded area, Fig. 12.26) by the number of photons emitted from the source during the same time period. This absolute efficiency is divided by 1.2×10^{-3}, which is the absolute efficiency of a 3 in × 3 in NaI(Tl) crystal 0.25 m from the source, to give the relative efficiency quoted in the specifications.

Full-energy peak efficiency of Ge(Li) detectors is only about 3–30 percent of that for the NaI(Tl) scintillation counter. Lower efficiency, however, is more than compensated for by the better energy resolution of the semiconductor detector. To make this point better understood, consider a case of 10,000 counts being recorded by a 3 in × 3 in NaI(Tl) detector under the 1.33-MeV peak of ^{60}Co. The FWHM of this peak is about 70 keV. A Ge(Li) detector with 10 percent efficiency will record only 1000 counts under the peak, but the FWHM is only about 2 keV. Since the

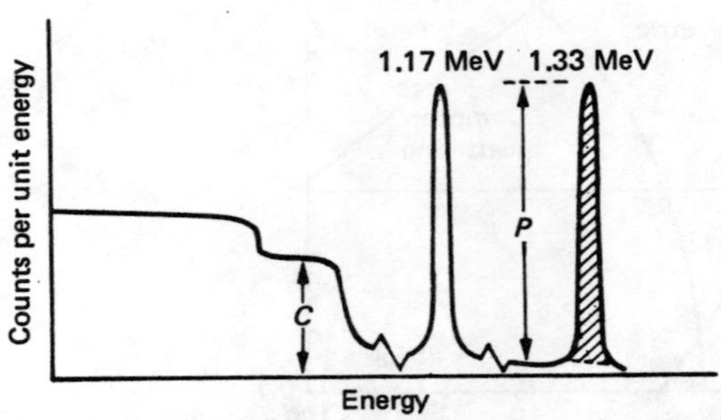

FIG. 12.26 A sketch of the ^{60}Co spectrum, indicating how it is used for efficiency and peak-to-Compton ratio determination.

total number of counts under the peak is proportional to the product of the FWHM times the peak, the heights of the two peaks are related by

$$\frac{\text{Height of Ge(Li) peak}}{\text{Height of NaI(Tl) peak}} \approx \frac{(1000/2)}{(10{,}000/70)} \approx \frac{70}{20} \approx 3.5$$

Thus, even though the Ge(Li) detector is only 10 percent efficient, relative to the NaI(Tl) crystal, it produces a peak which is 3.5 times higher.

Another parameter specified by the manufacturer of Ge(Li) detectors is the *peak-to-Compton ratio* (PCR). Looking at Fig. 12.26, the PCR is defined by the equation

$$\text{PCR} = \frac{\text{height of 1.33-MeV peak}}{\left(\begin{array}{c}\text{average height of Compton} \\ \text{plateau of 1.33-MeV peak}\end{array}\right)}$$

The average of the plateau is taken between 1040 and 1096 keV, in accordance with IEEE Standard No. 325-1971.[12] The PCR is important because it indicates the capability of the detector to identify low-energy peaks in the presence of stronger peaks of higher energy. PCR values of 30:1 are common but higher values have also been reported.

Figure 12.27 shows the total linear attenuation coefficient of gammas in germanium. The intrinsic efficiency depends, as in the case of scintillators, on $\mu(E)$ and

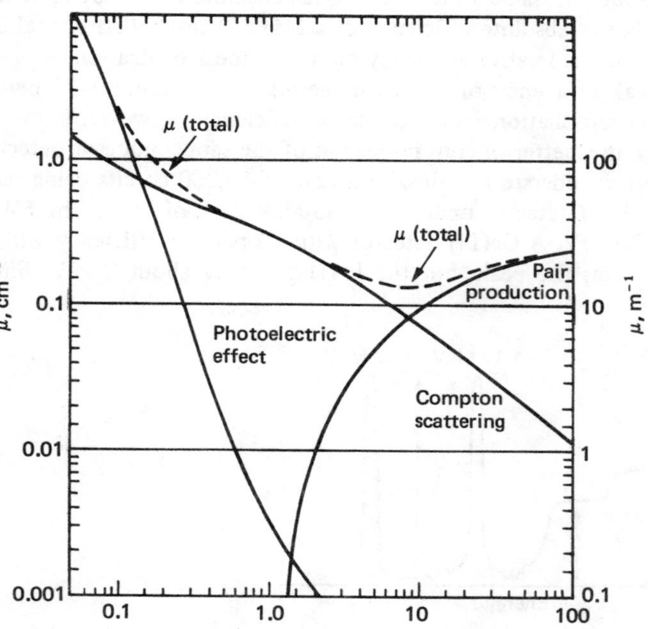

FIG. 12.27 Photon attenuation coefficients for germanium. The dashed line is the approximate total linear attenuation coefficient.

the size of the detector. What is done in practice with Ge(Li) detectors is to measure the full energy peak for a number of energies using calibrated sources or one source emitting many gammas with relative intensities accurately known.[12-18] A table of gamma energies used for calibration is given in App. C.

For the analysis of complex gamma spectra, it is helpful to have an analytic function that represents the efficiency of the detector as a function of energy. Many semiempirical equations have been developed to fit the efficiency of Ge(Li) detectors.[13-16,19-23] Three equations that influenced all the others are:

THE FREEMAN-JENKIN EQUATION[13]

$$\epsilon = 1 - \exp(-\tau x) + \sigma A \exp(-BE) \tag{12.8}$$

where τ = photoelectric coefficient
x = thickness of the detector
σ = Compton coefficient
A, B = constants to be determined from spectra

Equation 12.8 was used to determine the *relative* efficiency of a cylindrical and a trapezoidal detector over the range 500 to 1500 keV, with an accuracy of about 1 percent.

THE MOWATT EQUATION[20]

$$\epsilon = a_1 T \exp(-\mu_{Ge} a_2) \left[\frac{\tau + \sigma a_3 \exp(-a_4 E)}{\tau + \sigma}\right] 1 - \exp(-\mu_{Ge} a_5) \tag{12.9}$$

where $T = \Pi_i \exp(-\mu_i x_i)$ = product of attenuation factors outside the intrinsic region
a_1 = normalization factor
a_2 = the germanium front dead layer
a_3, a_4 = constants to be determined from spectra
a_5 = effective detector depth

Equation 12.9 is an improvement of Eq. 12.8 because it takes into account absorption in the window (through the factor T) and in the dead layer of the detector [through the factor $\exp(-\mu_{Ge} a_2)$]. Mowatt's equation, developed for planar detectors, gives the efficiency with an accuracy of 1.5 percent over the range 100 to 1400 keV.

McNELLES-CAMPBELL EQUATION[15]

$$\epsilon = \left(\frac{a_1}{E}\right)^{a_2} + a_3 \exp(-a_4 E) + a_5 \exp(-a_6 E) + a_7 \exp(-a_8 E) \tag{12.10}$$

where a_1 through a_8 are constants to be determined from measured gamma spectra. Equation 12.10, developed for coaxial detectors, predicts the efficiency with an accuracy of 0.2 percent over the energy range 160 to 1333 keV. Further testing of Eq. 12.10 showed[23] that the last term involving the constants a_7 and a_8 has a negligible effect to the result.

Attempts have been made to calculate the efficiency, especially using the Monte

Carlo method.[24,25] The calculation is not as simple as for NaI(Tl) detectors with regular geometric shapes that can be measured very accurately. Neither the calculation nor the semiempirical equations can replace accurate measurements of the efficiency using calibrated sources. A typical efficiency curve is shown in Fig. 12.28. The decrease in efficiency as energy increases follows, roughly, the change in $\mu(E)$ (Fig. 12.27). The decrease on the low-energy side is due to increased absorption in the window of the cryostat and the dead layer in front of the sensitive detector volume.

12.7.2 Energy Resolution of Ge(Li) Detectors

The energy resolution of a Ge(Li) detector is given in terms of the FWHM (Γ).† The width Γ consists of the following two components:

Γ_d = width due to detector effects

Γ_e = width due to effects of electronics

Since these two components are uncorrelated, they are added in quadrature to give the total width,

$$\Gamma = \sqrt{\Gamma_d^2 + \Gamma_e^2} \tag{12.11}$$

As shown in Chap. 9, the width Γ_d is energy-dependent and is given by

$$\Gamma_d = 2\sqrt{(2 \ln 2)FEw} \tag{12.12}$$

where F is the Fano factor and w is the average energy needed to produce an electron-hole pair. For germanium, at the operational temperature of 77 K, $w = 2.97$ eV. Thus,

$$\Gamma_d \text{ (keV)} = 0.1283\sqrt{FE \text{ (keV)}} \tag{12.13}$$

†Manufacturers of Ge(Li) detectors usually quote not only the FWHM but also the full width at one-tenth of the maximum (FWTM).

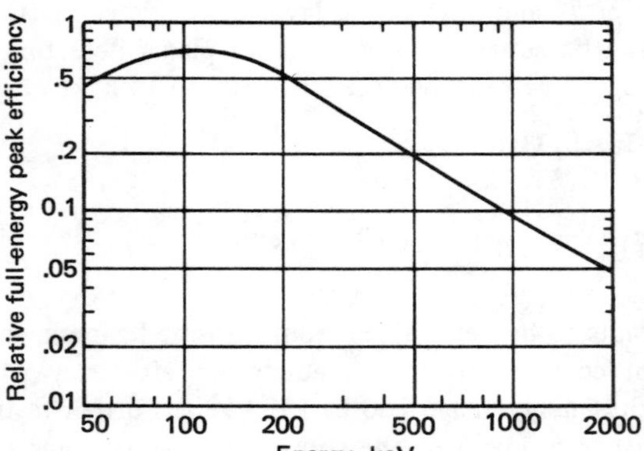

FIG. 12.28 Efficiency of a Ge(Li) detector as a function of energy (from ref. 12).

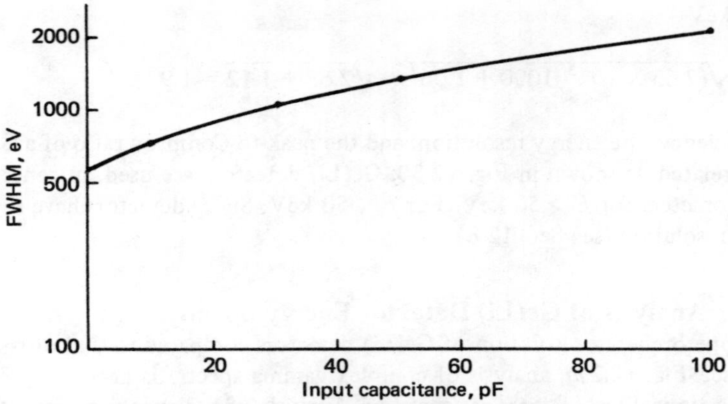

FIG. 12.29 The dependence of Γ_e on input capacitance for a charge-sensitive preamplifier (from ref. 2).

The width Γ_e increases when the detector capacitance decreases. The detector capacitance, in turn, generally increases with detector size and may change with detector bias. Good Ge(Li) detectors have a flat capacitance-bias relationship, over most of the range of bias voltage applied.

The capacitance of the detector has an effect on the energy resolution because it influences the performance of the charge-sensitive preamplifier that accepts the detector signal. The contribution of the preamplifier to the value of Γ_e increases with the input capacitance. One of the manufacturers, Canberra, reports a 0.570-eV Γ_e with zero input capacitance and a slow increase with higher values as shown in Fig. 12.29. Clearly the resolution improves if the capacitance is kept low. The other component of the input capacitance comes from items like electronics and cables. Reduction of the length of input cable and of connectors' capacitance is helpful. For the best resolution with a given system, the preamplifier should be located as close to the detector as possible.

Large Ge(Li) detectors commercially available today have capacitance as high as 30 pF, which results in a value of $\Gamma_e = 1.06$ keV (Fig. 12.29). Combining the two contributions Γ_d and Γ_e in accordance with Eq. 12.11, one gets

$$\Gamma\,(\text{keV}) = \sqrt{(0.1283)^2 FE\,(\text{keV}) + 1.06^2} \tag{12.14}$$

A typical value of the Fano factor for $1 < E < 10$ MeV is 0.16. Substitution in Eq. 12.14 gives

$$\Gamma\,(\text{keV}) = \sqrt{(2.63 \times 10^{-3})E\,(\text{keV}) + 1.06^2} \tag{12.15}$$

Equation 12.15 shows that for low energies, the resolution is determined by electronic noise. For higher energies, the energy contribution predominates. Consider two cases:

1 $E = 100$ keV

$$\Gamma = \sqrt{(2.63 \times 10^{-3})100 + 1.06^2} = \sqrt{0.263 + 1.12} = 1.2 \text{ keV}$$

2 $E = 1000$ keV

$$\Gamma = \sqrt{(2.63 \times 10^{-3})1000 + 1.06^2} = \sqrt{2.63 + 1.12} = 1.9 \text{ keV}$$

The efficiency, the energy resolution, and the peak-to-Compton ratio of a Ge(Li) detector are related as shown in Fig. 12.30. Ge(Li) detectors are used, in general, as gamma spectrometers for $E > 50$ keV. For $E < 50$ keV, Si(Li) detectors have slightly better energy resolution (see Sec. 12.8).

12.7.3 Analysis of Ge(Li) Detector Energy Spectra

Despite the superb energy resolution of Ge(Li) detectors compared to that of NaI(Tl) scintillators (see Fig. 12.22), analysis of complex gamma spectra is necessary. Figure 12.31 shows a typical Ge(Li) energy spectrum. Analysis of the spectrum entails first the assignment of energy to the peaks of the spectrum and second the determination of the number of counts (i.e., the area) for each peak.

The energy assignment to the peaks of the spectrum is accomplished by calibrating the detector with a source that emits gammas of known energy and intensity. As explained in Sec. 9.10, calibration means to determine the constants of the equation

$$E = a_1 + a_2 C + a_3 C^2 \tag{12.16}$$

where C is the channel number (for most systems a_3 is very small, or zero). When the energy assignment is performed, the observer should be aware of the following general features of a gamma spectrum recorded by a Ge(Li) detector:

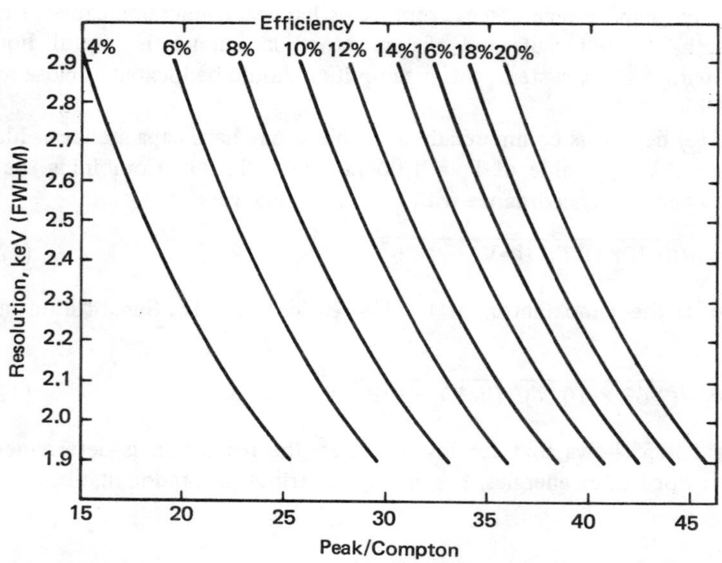

FIG. 12.30 Relationship between the Ge(Li) detector resolution, the peak-to-Compton ratio, and the relative efficiency at 1.33 MeV (from ref. 2).

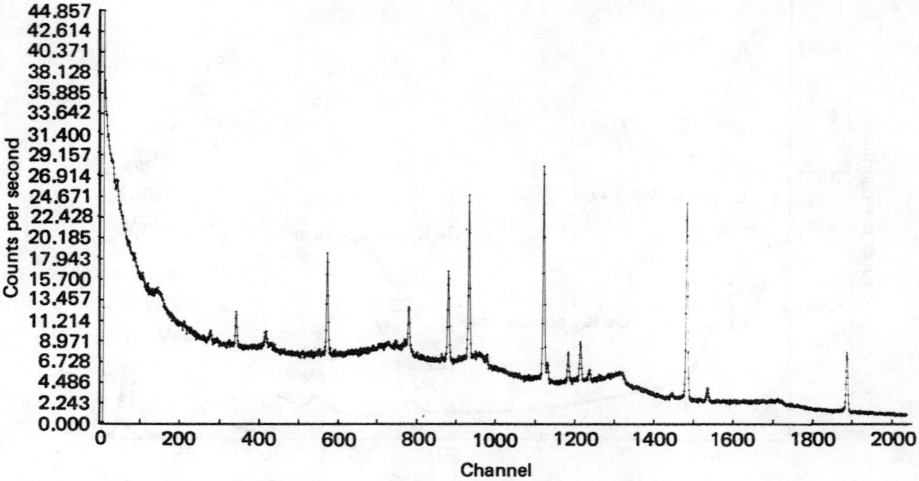

FIG. 12.31 A gamma-ray energy spectrum recorded with a Ge(Li) detector. (This the spectrum from a human hair sample irradiated in a reactor.)

1 For $E_\gamma < 2$ MeV, the full-energy peak is intense and almost Gaussian in shape.
2 At higher energies ($E_\gamma > 2$ MeV) the double-escape peak ($E_{DE} = E_\gamma - 1.022$ MeV) becomes prominent. The single-escape peak is present too (see Fig. 12.7).
3 In spectra taken with thin Ge(Li) detectors, one may see the germanium "escape" peaks. The escape peaks (EP) have energy equal to

$$E_{EP} = E_\gamma - E_k$$

where E_k is the K x-ray energy of germanium. The escape peaks are due to the loss of the energy carried away by the escaping K x-ray of germanium. This energy is equal to 9.9 keV for the K_α and 11.0 keV for the K_β x-ray of germanium. Figure 12.32 shows the ^{139}Ce x-ray spectrum with the escape peaks marked.
4 When the front surface of the detector is covered by a metal, characteristic x-rays of that metal are emitted if the incident radiation consists of photons with energy greater than the K x-ray energy of that metal. Gold, which is frequently used, emits a K_α x-ray with energy 68 keV and five L x-rays between 9 and 13 keV.

The determination of the area under a peak–i.e., the absolute intensity of a particular gamma energy–is not as straightforward as the assignment of energy, because the area under a peak includes contributions from other gammas. The methods that have been developed for the determination of the area can be classified into two groups: methods that treat the data (i.e., counts per channel) directly, and methods that fit a known function to the data.

Methods that treat the data directly give the area under the peak by adding the counts from all the channels in the region of the peak and subtracting a "base background." The methods differ in the way they define the "base" and the number of channels that define the peak (a review of the methods is given in references 28 and

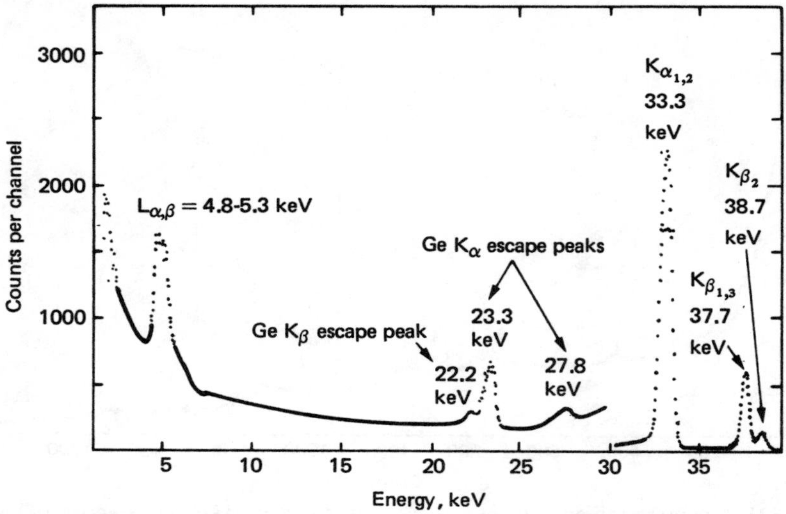

FIG. 12.32 The x-ray spectrum resulting from the decay of ^{139}Ce. The Ge escape peaks are clearly seen (from ref. 27).

29). Figures 12.33 and 12.34 show graphically three of the methods. The first two (Fig. 12.33) use a straight line to separate the peak from the base,[28,30] and calculate the net area under the peak (NPA) using the following equations (refer to Fig. 12.33 for explanation of symbols):

$$\text{NPA} = \sum_{i=L}^{R} a_i - (a_L + a_R)\frac{R - L + 1}{2} \tag{12.17}$$

where a_i = numbers of counts in channel i
L = channel number at left limit of photopeak
R = channel number at right limit of photopeak

$$\text{NPA} = \sum_{i=-n}^{n} a_i - \left(n + \frac{1}{2}\right)(b_n + b_{-n}) \tag{12.18}$$

where n = number of channels on either side of peak
b_i = background in channel i as determined from a straight line drawn between channels L and R

The third one, due to Quittner,[31] fits a polynomial, using the least-squares technique, to the data from $2k_L + 1$ and $2k_R + 1$ channels (Fig. 12.34) on either side of the peak. The base line is constructed in such a way that at X_L and X_R (shown in Fig. 12.34) it has the same magnitude (p_L, p_R) and shape (q_L, q_R) as the fitted polynomial. The net area under the peak is now given by the equation

$$\text{NPA} = \sum_{i=-L}^{R} (a_{x_p+i} - b_{x_p+i}) \tag{12.19}$$

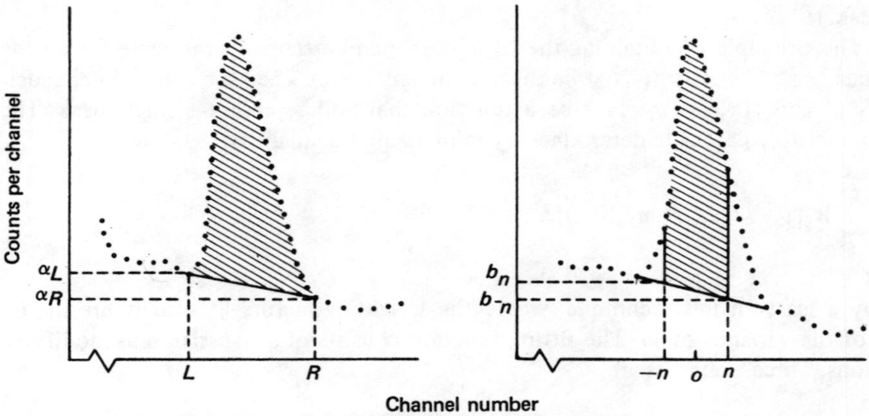

FIG. 12.33 Determination of the net area under the peak using a straight line to define the base.

where the value of b is, in terms of a third degree polynomial, equal to

$$b(x) = p_L + q_L(x - x_L) + \left[\frac{-q_R - 2q_L}{l_L + l_R} + \frac{3(p_R - p_L)}{(l_L + l_R)^2}\right](x - x_L)^2$$
$$+ \left[\frac{q_L + q_R}{(l_L + l_R)^2} + \frac{2(p_L - p_R)}{(l_L + l_R)^3}\right](x - x_L)^3 \tag{12.20}$$

Quittner's method is quite accurate if the peaks are separated by about 20 channels.

Today, the most widely used methods are those which fit an analytic function to each peak. After the fit is accomplished, the analytic function is used for the

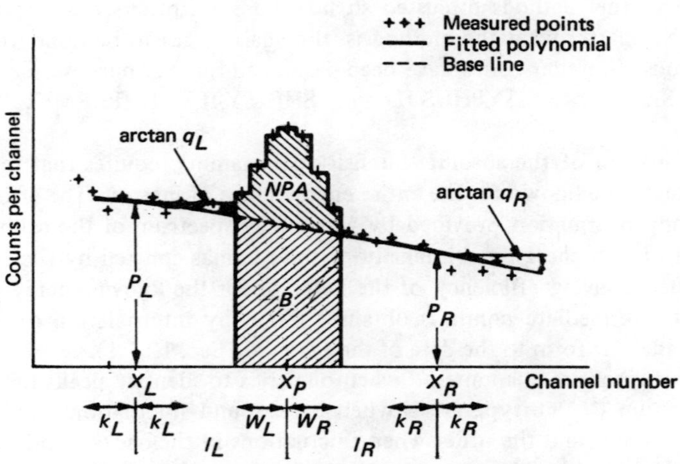

FIG. 12.34 Determination of the area under the peak using the method of Quittner (from ref. 31).

calculation of quantities of interest such as the area and the centroid (position) of the peak.

The principle of obtaining the fit is simple and essentially the same for all the methods.[4,5,32-40] Let $y_i|_{i=1,N}$ be the experimental point–i.e., the counts in channels $x_i|_{i=1,N}$–and $f(x, a_1, a_2, \ldots)$ be a function that will represent a single peak. The parameters $a_1, a_2, \ldots$ are determined by minimizing the quantity

$$\sum_i W_i[y_i - f(x_i, a_1, a_2, \ldots)]^2$$

i.e., by a least-squares technique, where the weighting factors W_i usually are the inverse of the variance of y_i. The fitting function consists of a Gaussian plus modifying functions. Three examples are

1 $$f = G(x) + B + Cx = y_0 \exp\left[-\frac{(x - x_0)^2}{2\sigma^2}\right] + B + Cx$$

Here, a Gaussian describes the peak, and the linear function $B + Cx$ describes the background.[32]

2 $$f = G(x)[1 + a_1(x - x_0)^{m_1} + a_2(x - x_0)^{m_2}] \qquad \text{(from ref. 33)}$$

3 $$f = G(x) + B + S + D \qquad \text{(from ref. 40)}$$

where B = linear background
S = step function $= A\{1 - \text{erf}\,[(x_0 - x)/\sigma\sqrt{2}]\}$
D = tail function having an exponential form on the left of the peak and a Gaussian on the right

In addition to the methods discussed so far, there are others that use mixed techniques.[41,42] No matter what the method is, the analysis has to be done by computer and numerous computer codes have been developed for that purpose. Examples are the codes SAMPO,[34] SISYPHUS-II and SHIFTY-II,[5] HYPERMET,[42] and GAUSS.[43]

The determination of the absolute intensity of a gamma requires that the efficiency of the detector be known for the entire energy range of interest. The efficiency is determined from information provided by the energy spectrum of the calibration source. Using the known energy and intensity of the gammas emitted by the source, a table is constructed giving efficiency of the detector for the known energy peaks. The efficiency at intermediate points is obtained either by interpolation or, better yet, by fitting an analytic form to the data of the table (see Sec. 12.7.1).

Two types of error are encountered when one tries to identify peaks in a complex gamma spectrum.[44] One type arises when background fluctuations are falsely identified as photopeaks, and the other when fluctuations in the background obscure true photopeaks. Reference 44 discusses criteria, in the form of confidence limits, that can be used in order that both types of errors may be avoided.

12.7.4 Timing Characteristics of the Pulse

For certain measurements, like coincidence-anticoincidence counting or experiments involving accelerators, the time resolution of the signal is also important, in addition to energy resolution. For timing purposes, it is essential to have pulses with constant risetime.

No detector produces pulses with exactly the same risetime. This variation is due to the fact that electrons are produced at different points inside the detector volume, and thus traverse different distances before they reach the point of their collection. As a result, the time elapsing between production of the charge and its collection is not the same for all the carriers.

Consider a true coaxial detector, shown in Fig. 12.35 (see also Fig. 7.26). Since the electric field is radial, electrons and holes will follow a trajectory perpendicular to the axis of the detector. The maximum time required for collection of the charge corresponds to electron-holes being produced either at A or C. That time t is equal to $t \approx (AC)/v$, where AC is the detector thickness and v is the velocity of electrons or holes. For a detector bias of about 2000 V and the size shown in Fig. 12.35*a*, $v \approx 0.1$ mm/ns $= 10^5$ m/s, which gives a maximum collection time of 120 ns. The best risetime corresponds to electron-holes generated at point B (Fig. 12.35*a*) and is equal to about 60 ns.

The pulse risetime is essentially equal to the collection time. For the detector shown in Fig. 12.35, the risetime will vary between 60 and 120 ns. For other detectors geometries like wrap-around coax or trapezoidal coax (see Fig. 7.26), the variation in risetime is greater because the electrons and holes, following the electric field lines, may travel distances larger than the thickness of the detector core (Fig. 12.35*b*). The variation in risetime for the detector of Fig. 12.35 will be between 60 and 200 ns. The distribution of pulse risetimes for commercial detectors is a bell-type curve, not exactly Gaussian, with a FWHM of less than 5 ns (see Fig. 10.24).

12.8 HYPERPURE GERMANIUM (HPGe) DETECTORS AS GAMMA SPECTROMETERS

As described in Sec. 7.5.6, hyperpure germanium (HPGe) detectors are made of high-purity germanium (impurity concentration 10^{16} atoms/m^3 or less). HPGe detec-

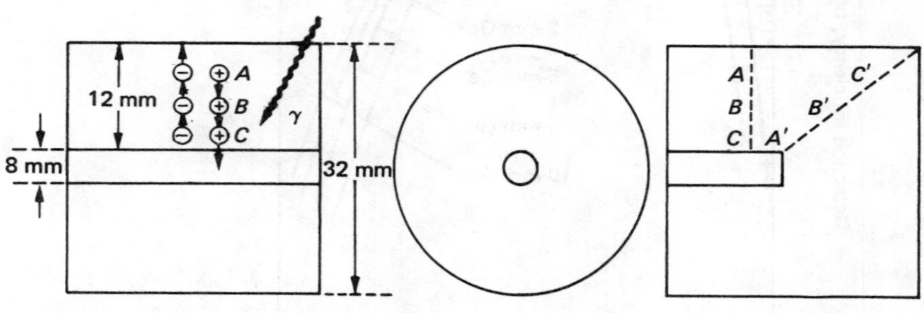

FIG. 12.35 (*a*) In a true coaxial detector, electrons and holes travel along the direction ABC. (*b*) In wrap-around coaxial detectors, the carriers may travel along ABC but also along the longer path $A'B'C'$.

tors have excellent resolution characteristics over the energy range 3 keV $\lesssim E \lesssim$ 1 MeV. A recent breakthrough reported by one of the manufacturers[1] makes it possible to get rid of the front metal contact of the detector, which in turn eliminates the interfering x-rays produced by the metal contact itself when the incident radiation consists of photons with energy greater than the K absorption edge of the metal.

HPGe detectors need to be cooled only when in use. They may be cycled to room temperature without their performance being damaged. This feature is a great advantage, especially when there is a need to ship the detector.

The energy resolution of HPGe detectors is usually reported for the 5.9-keV line from ^{55}Fe (K_α x-ray of ^{55}Mn) and the 122-keV line from ^{57}Co (γ ray of ^{57}Fe). The resolution depends on the size of the detector. For a detector with a 16-mm diameter and a 10-mm depth, the FWHM is reported to be 220–290 eV for the 5.9-keV line and 510–590 eV for the 122-keV line.[2] The efficiency of HPGe detectors is excellent for the energy 3 keV to 100 keV, as Fig. 12.36 shows. Their cryostat is equipped with a very thin beryllium window, which affects the efficiency for $E < 8$ keV.

12.9 CdTe AND HgI_2 DETECTORS AS GAMMA SPECTROMETERS

A great advantage for CdTe and HgI_2 detectors,[45–51] compared to Ge(Li) and Si(Li) detectors, is that they can operate at room temperature (see also Sec. 7.5.7). At this time (1982), they can be obtained in relatively small volumes, but they still have an efficiency of about 75 percent at 100 keV because of the high atomic number of the elements involved. The energy resolution of CdTe detectors is 18 percent at 6 keV and 1.3 percent at 662 keV. The corresponding numbers for HgI_2 are 8 percent and 0.7 percent.[51]

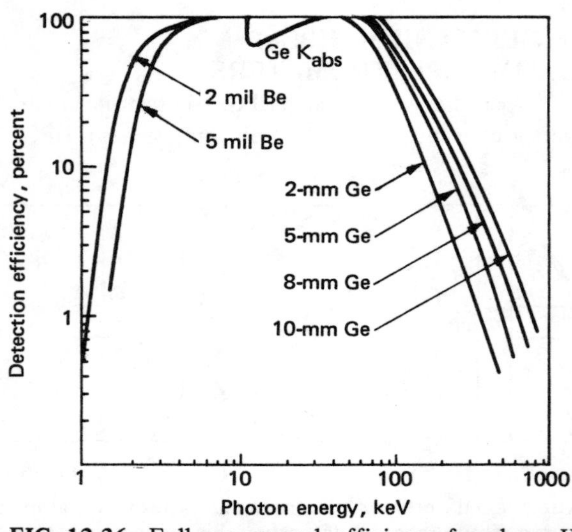

FIG. 12.36 Full-energy peak efficiency for planar HPGe detectors (from ref. 2).

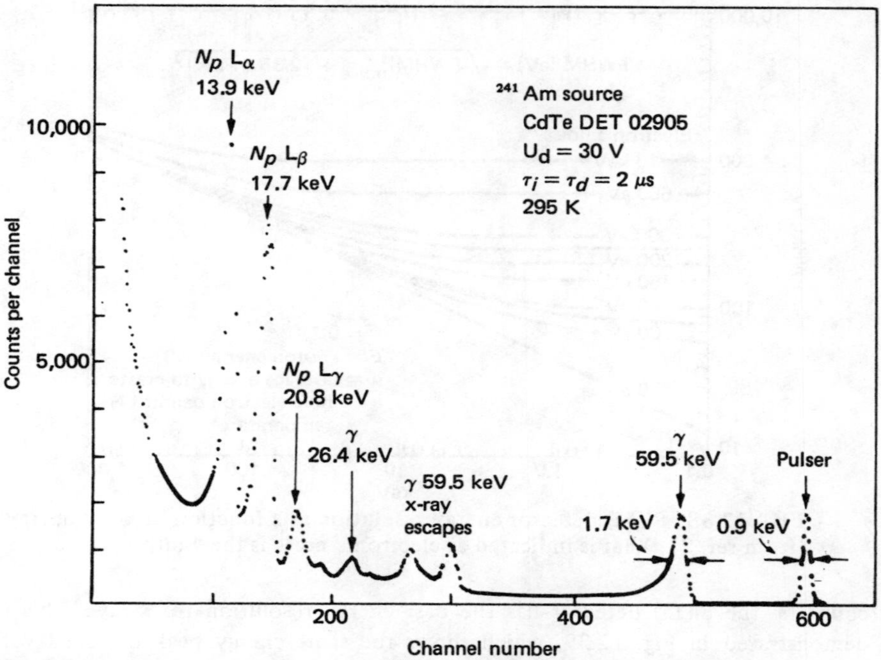

FIG. 12.37 The 59.5-keV x-ray of ^{241}Am detected with a CdTe detector (from ref. 48).

The HgI_2 detectors are very useful for the measurement for x-rays with energy less than 10 keV. In that energy range, resolutions of 245 eV at 1.25 keV[52] and 295 eV at 5.9 keV[53] have been reported.

As with small Ge(Li) detectors, x-ray escape peaks are present in spectra taken with these CdTe and HgI_2. Figure 12.37 shows the spectrum of the 59.5-keV gamma taken with a CdTe detector. The escape peaks due to the K_α x-rays of cadmium (23 keV) and tellurium (27 keV) are clearly seen.

12.10 DETECTION OF X-RAYS WITH A Si(Li) DETECTOR

Si(Li) detectors are generally used as x-ray spectrometers for $E < 50$ keV. They need cooling; therefore, they require a cryostat. Their energy resolution and efficiency are better than those of a Ge(Li) detector for $E < 50$ keV. Figure 12.38 shows how the resolution changes with energy. The FWHM is again given by Eq. 12.11. Using a value of $w = 3.7$ eV per electron-hole pair for silicon at 77 K and Fano factor equal to 0.1, the width Γ becomes

$$\Gamma\,(\text{eV}) = \sqrt{2.05E\,(\text{eV}) + \Gamma_e^2} \tag{12.21}$$

The width Γ_e is indicated as "electronic noise" in Fig. 12.38. Of the three types of x-ray detectors mentioned–scintillation, proportional, and semiconductor

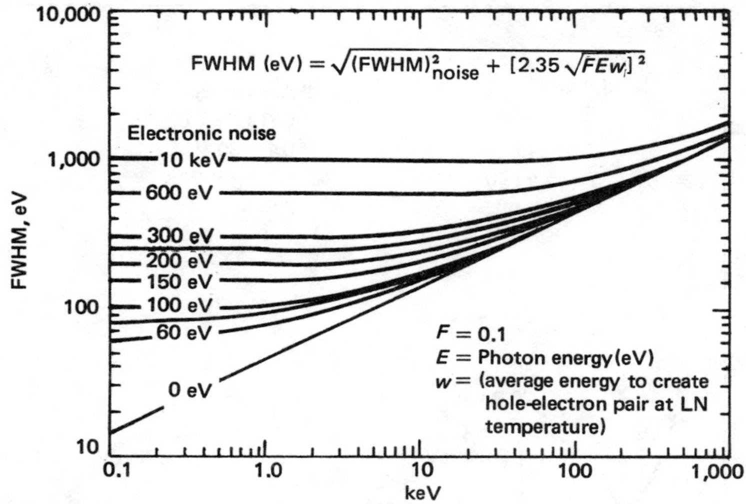

FIG. 12.38 Si(Li) detector energy resolution as a function of x-ray energy (from ref. 2). What is indicated as electronic noise is the width Γ_e.

counters—the Si(Li) detector has the best energy resolution for x-rays. This fact is demonstrated in Fig. 12.39, which shows the same energy peak obtained with the three different detectors. Notice that only the Si(Li) detector can resolve K_α and K_β lines, an ability absolutely necessary for the study of fluorescent x-rays for most elements above oxygen. The iron fluorescence spectrum obtained with a Si(Li) detector is shown in Fig. 12.40.

The efficiency of Si(Li) detectors depends on x-ray energy, as shown in Fig. 12.41. The thin beryllium window of the detector affects the efficiency for $E < 3$ keV. At the higher energy range ($E > 15$ keV), the efficiency falls off again because the probability of x-ray interaction in Si decreases as the photon energy increases. The total linear attenuation coefficient for photons moving in silicon is shown in Fig. 12.42.

12.11 DETECTION OF X-RAYS WITH A CRYSTAL SPECTROMETER

The measurement of x-ray energy by wavelength-dispersive crystal spectrometers is based on the phenomenon of differaction of electromagnetic waves from a crystal (see Compton and Allison, and Cullity). Consider an x-ray of wavelength λ incident at an angle θ on a crystal with interplanar spacing d (Fig. 12.43). The incident x-rays will be scattered by the atoms of the crystal in all directions, but as a result of the periodicity of the atom positions there are certain directions along which constructive interference of the scattered photons takes place. The direction of constructive interference of the diffracted beam is given by the Bragg condition

$$n\lambda = 2d \sin \theta \tag{12.22}$$

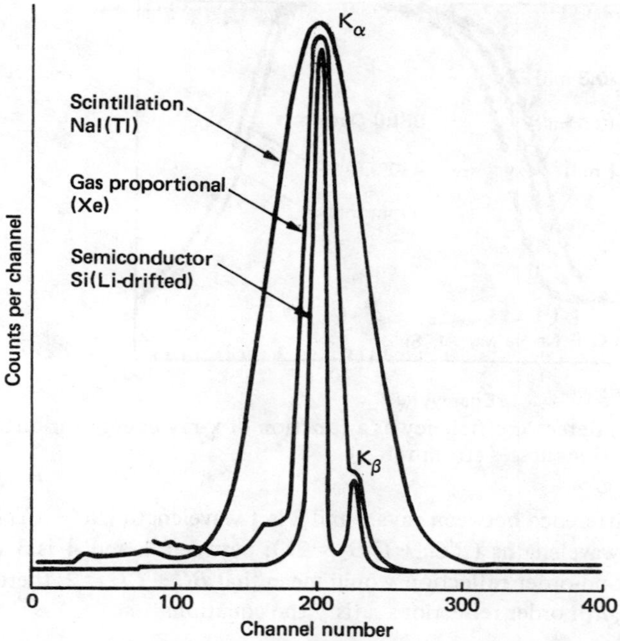

FIG. 12.39 Demonstration of the superior energy resolution of Si(Li) detectors by showing the same peak recorded with a NaI(Tl) scintillator and a gas-filled proportional counter (from ref. 11).

where n is the order of reflection (= 1, 2, 3, . . .). Equation 12.22 states that for constructive interference, the path difference between any two rays (photons) must be an integral number of wavelengths. In Fig. 12.43, the path difference for rays 1 and 2 is $BC + CD = 2d \sin\theta = n\lambda$.

The order of reflection n may take any value (for a given θ and d) provided $\sin\theta < 1$, and is physically equal to the number of wavelgnths by which the paths of x-rays scattered by two adjacent planes are different. Figure 12.43 shows first-order

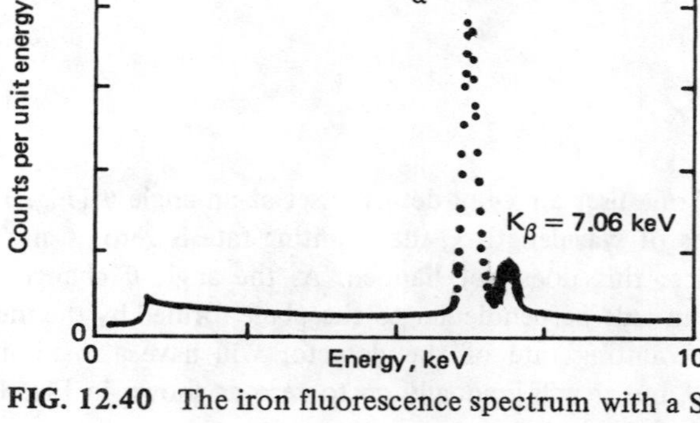

FIG. 12.40 The iron fluorescence spectrum with a Si(Li) detector (from ref. 1).

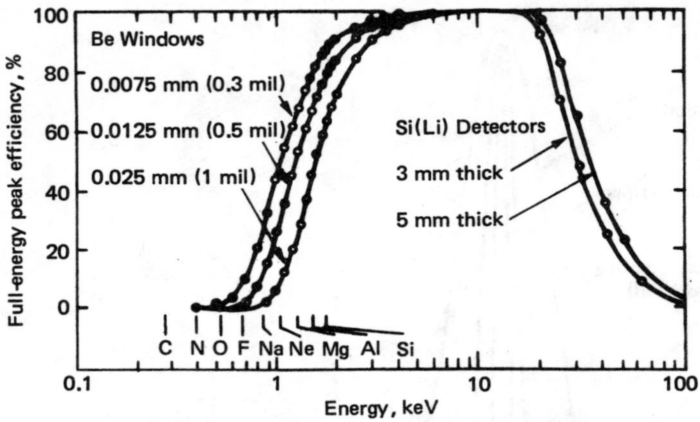

FIG. 12.41 Si(Li) detector efficiency as a function of x-ray energy for different beryllium window thicknesses (from ref. 2).

reflection if the path difference between rays 1 and 2 is 1 wavelength ($BC + CD = \lambda$); between 1 and 3 is 2 wavelengths ($B'C' + C'D' = 2\lambda$); between 1 and 4 is 3 wavelengths, and so on. Second-order reflection would mean that $BC + CD = 2$; therefore, first, second, third, . . . , *n*th order reflections satisfy the equations

$$\lambda = 2d \sin \theta_1$$

$$2\lambda = 2d \sin \theta_2$$

$$3\lambda = 2d \sin \theta_3$$

$$\vdots$$

$$n\lambda = 2d \sin \theta_n$$

Reflections of order higher than 1 act as "contamination" in x-ray measurements. Fortunately, the first-order reflection is always much more intense than higher-order ones (second-order reflection is about 20 percent the intensity of the first).

The diffracted photons are detected by a counter set to accept the radiation emerging at an angle θ (Fig. 12.43), i.e., set to accept photons of wavelength λ, or energy

$$E = \frac{123.99}{\lambda \text{ (nm)}} \text{ keV} \tag{12.23}$$

According to Eq. 12.22, if one uses an x-ray detector set at an angle θ (Fig. 12.44) to detect diffracted photons of wavelength λ, the counting rate is zero at any angle different than θ. In practice, this does not happen. As the angle θ changes from rotating the crystal around an axis perpendicular to the plane formed by the incident and diffracted beams, the counting rate of the detector will have a maximum at $\theta = \theta_0$, where $\lambda = 2d \sin \theta_0$, but it will gradually go to zero as shown in Fig. 12.45.

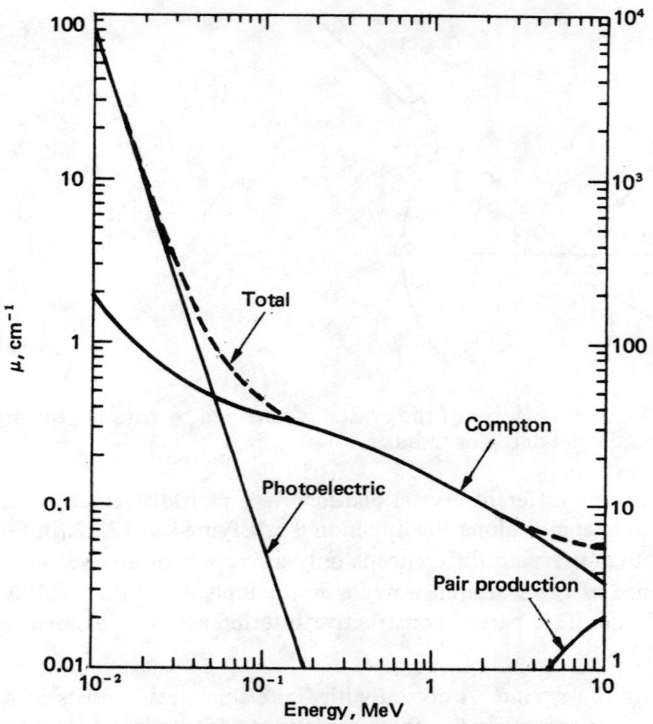

FIG. 12.42 Photon attenuation coefficient for silicon.

The curve of Fig. 12.45, known as the *rocking curve* of the crystal, is the result of incomplete destructive interference of the diffracted x-rays. The rocking curve is the equivalent of the Gaussian response function of a detector to a monoenergetic photon source. However, its origin is not statistical, as the following discussion shows.

As stated earlier, the Bragg condition (Eq. 12.22) indicates that the waves

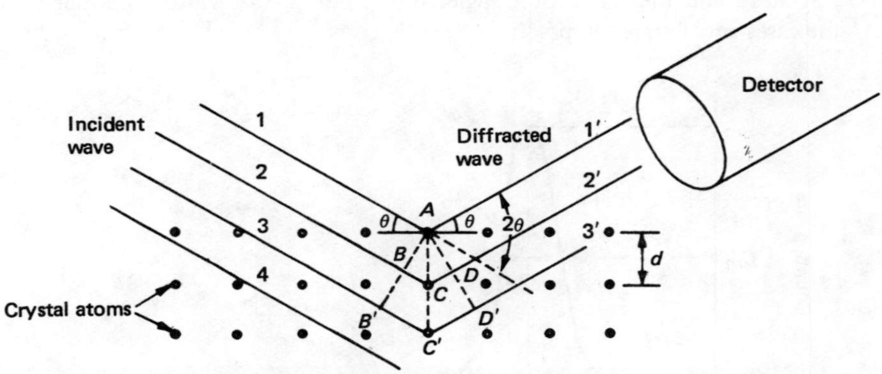

FIG. 12.43 Diffraction of electromagnetic waves by a crystal. The detector counts photons being diffracted at an angle θ, i.e., photons having wavelength λ satisfying Eq. 12.18.

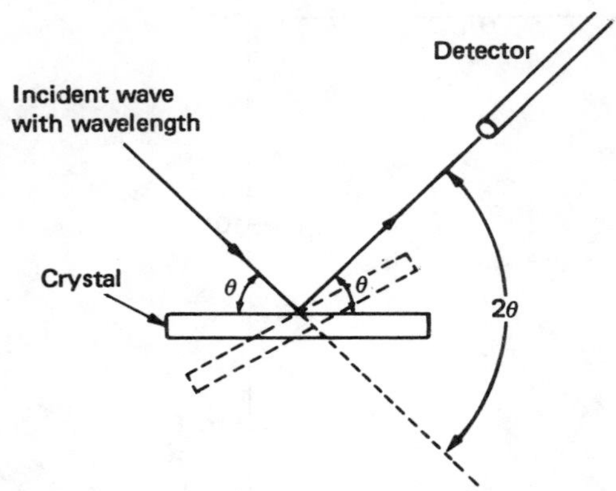

FIG. 12.44 The rocking curve of the crystal is obtained by rotating the crystal and keeping source and detector fixed.

(photons) scattered by the different crystal planes have a path difference equal to an integral number of wavelengths along the direction θ satisfying Eq. 12.22. But what if the angle θ is such that the path difference is only a fraction of a wavelength? The destructive interference of the scattered waves is not complete and the result is radiation of lower amplitude. This partial constructive interference may happen because of three reasons:

1 *Finite thickness of the crystal.* A crystal with finite thickness consists of a finite number of planes. For any angle θ_0, there are planes with no matching partner to create the correct phase difference in scattered radiation for complete destructive interference. The width Γ_t of the rocking curve, due to the finite thickness of the crystal, is given by (see Cullity)

$$\Gamma_t = \frac{0.9\lambda}{t \cos \theta_0} \tag{12.24}$$

where t is the crystal thickness. Notice that Γ_t depends on the wavelength, the thickness, and the diffraction angle. If $t \gtrsim 1$ mm, the width Γ_t is negligible for the cases encountered in practice.

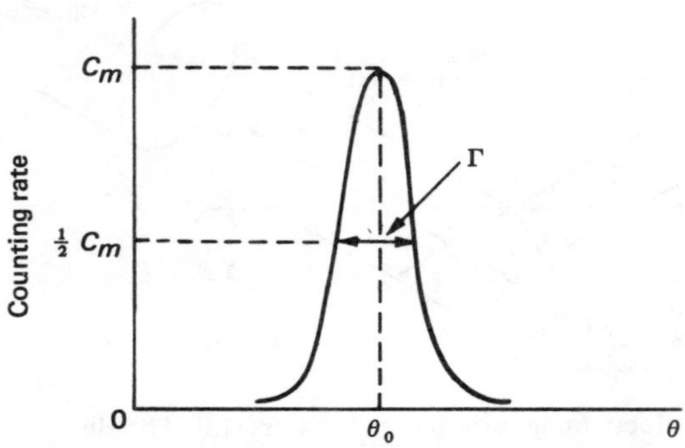

FIG. 12.45 The rocking curve of a crystal.

2 *Nonparallel incident rays.* It is impossible to produce a beam consisting of exactly parallel rays. Thus, at any angle θ_0 there are x-rays hitting the crystal at an angle slightly different from θ_0. The magnitude of this effect depends on the collimation of the beam.
3 *Mosaic structure of the crystal.* Real crystals do not consist of a single uninterrupted lattice. Instead, they are made up of small regions, each region being a perfect lattice but having a slightly different orientation with respect to the next one (Fig. 12.46).

One of the most difficult and important tasks in the use of a crystal spectrometer is correct alignment of the instrument and accurate measurement of the diffraction angle. At any position, the diffraction angle θ is taken as the angle corresponding to the maximum of the rocking curve. In most cases, what is measured is 2θ (Fig. 12.44) instead of θ. Of course, for an accurate measurement, the experimenter must know the crystal parameters and the crystal planes giving the reflection, in addition to the angle θ.

12.11.1 Types of Crystal Spectrometers

Crystal spectrometers can be used in two ways (Fig. 12.47). The first is to use the spectrometer as a means to obtain monochromatic radiation, which then strikes a sample (Fig. 12.47*a*). The second is to use it for the measurement of the energy of the radiation emerging from the sample (Fig. 12.47*b*). In most cases the sample emits fluorescent radiation, and for this reason the instrument is called a *fluoroescent spectrometer.* The diffracted radiation reaching the counter (Fig. 12.47) is strongly attenuated, first at the crystal and a second time at the sample. A very small fraction of the incident radiation reaches the counter. It is therefore important to design a spectrometer that provides maximum intensity. Considerable improvement in intensity is achieved if the crystal, the x-ray source, and the detector are located on the same circle, and the crystal is bent to twice the radius of that circle (Fig. 12.48).[54-56] In addition to the reflection spectrometer shown in Fig. 12.48, called the *Johansson type*, others based on transmission through the crystal have also been constructed.

12.11.2 Energy Resolution of Crystal Spectrometers

The resolution of crystal spectrometers is traditionally expressed as the *resolving power*, which is the inverse of the energy resolution:

$$\text{Resolving power} = \frac{E}{\Delta E} = \frac{\lambda}{\Delta\lambda} = \frac{\tan\theta}{\Delta\theta} = \frac{nk}{2d(\Delta\theta)E}\left(1 - \frac{n^2k^2}{4d^2E^2}\right)^{-1/2} \qquad (12.25)$$

where $k = 123.99$ keV·nm, the constant of Eq. 12.23. The uncertainty of the angle $\Delta\theta$ depends on crystal properties (rocking curve), the size of the source, and the size of the detector. The experimentally obtained rocking curve (Fig. 12.45) includes all the contributions to $\Delta\theta$; thus its width Γ is

$$\Gamma = \sqrt{\Gamma_t^2 + \Gamma_g^2} \qquad (12.26)$$

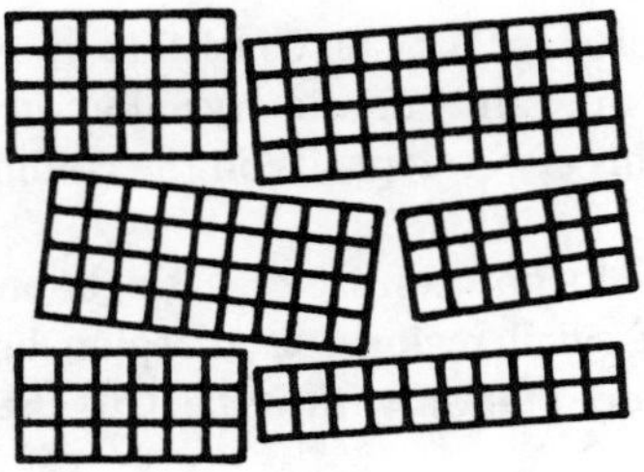

FIG. 12.46 The mosaic structure of real crystals. Each block represents a perfect crystal.

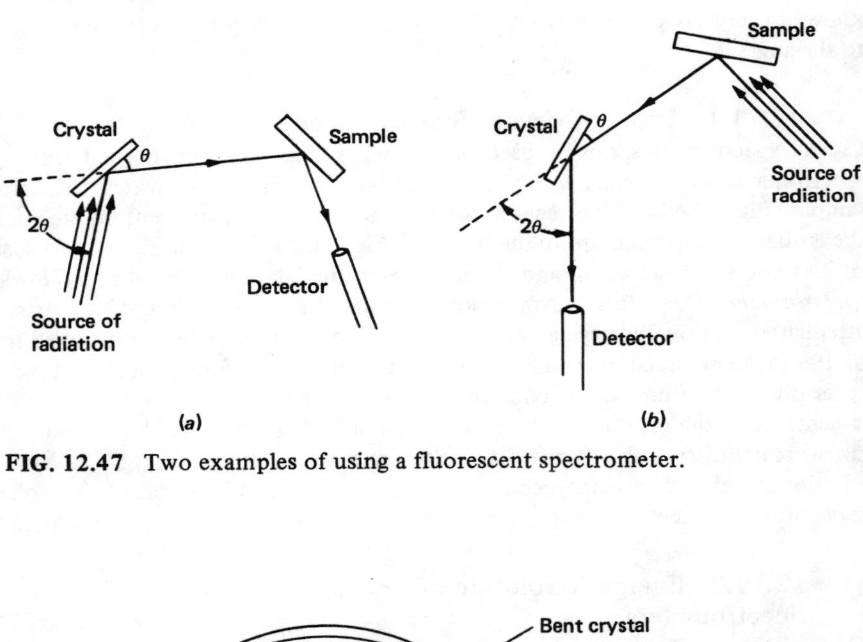

FIG. 12.47 Two examples of using a fluorescent spectrometer.

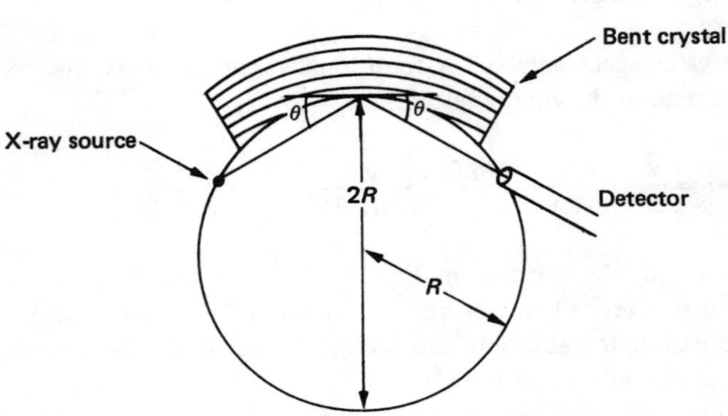

FIG. 12.48 The geometry of a bent-crystal spectrometer.

where Γ_g = width due to the geometry (collimator, source, detector)
Γ_t = width due to crystal imperfections and size, given by Eq. 12.24

For practical cases of interest, $\Gamma_t \ll \Gamma_g$ and $\Gamma \approx 0.5°$. The resolving power and the resolution of a crystal spectrometer as a function of energy are shown in Fig. 12.49. It should be pointed out that the x-ray detector used with the crystal spectrometer (Fig. 12.45) need not have an extremely good energy resolution, because it is the resolution of the analyzing crystal that determines the spectroscopic capabilities of the system, not that of the detector. The x-ray counter may be a proportional counter or a Si(Li) detector.

EXAMPLE 12.1

What is the energy resolution of an x-ray crystal spectrometer for 6-keV x-rays using a crystal with $d = 0.15$ nm and $\Delta\theta = 0.3°$?

ANSWER

Using Eq. 12.25, the resolution is ($n = 1$)

$$\frac{\Delta E}{E} = \frac{2d(\Delta\theta)E}{k}\sqrt{1 - \frac{k^2}{4d^2E^2}} = \frac{2(0.15)(0.0052)(6)}{123.99}\sqrt{1 - \frac{123.99^2}{4(0.15)^2(6)^2}}$$
$$= 0.005 = 0.5\%$$

One can improve the resolution by decreasing $\Delta\theta$ or using a crystal with a smaller interplanar distance.

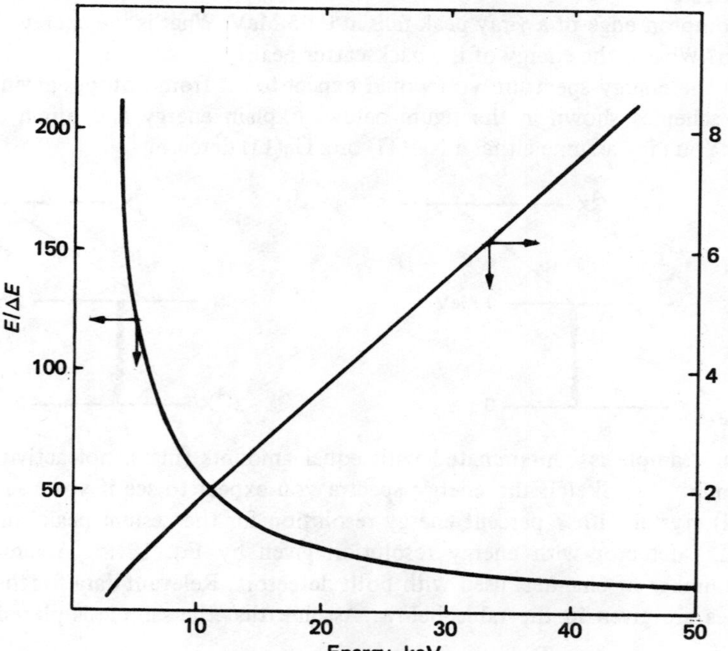

FIG. 12.49 Resolving power and energy resolution of a crystal spectrometer as a function of x-ray energy (data used: $n = 1$, $d = 0.2$ nm, $\Delta\theta = \frac{1}{3}°$).

TABLE 12.1 Advantages and Disadvantages of a Crystal Spectrometer and a Si(Li) Detector

	Crystal spectrometer	Si(Li) detector
Energy resolution (6.4 keV)	~10 eV	~150 eV
Data collection	One energy at a time	All energies at once
Efficiency	Low	High
Position, relative to sample	Certain distance away from sample	Very close to sample

The energy range over which crystal spectrometers can be used is determined, in principle, from the Bragg condition (Eq. 12.22) and the requirement that $0 < \sin\theta < 1$. Using Eq. 12.22 and assuming $\theta_{min} \approx 0.5°$, $d = 0.2$ nm,† and $n = 1$, the limits are

$$3 \text{ keV} < E < 350 \text{ keV}$$

In practice, the upper limit is about 50 keV because Si(Li) detectors have better resolution beyond that energy. Advantages and disadvantages of crystal spectrometers versus Si(Li) detectors are summarized in Table 12.1.

PROBLEMS

12.1 The Compton edge of a γ-ray peak falls at 0.95 MeV. What is the energy of the photon? What is the energy of the backscatter peak?

12.2 Sketch the energy spectrum you would expect to get from isotopes having the decay schemes shown in the figure below. Explain energy and origin of all peaks. You may assume either a NaI(Tl) or a Ge(Li) detector

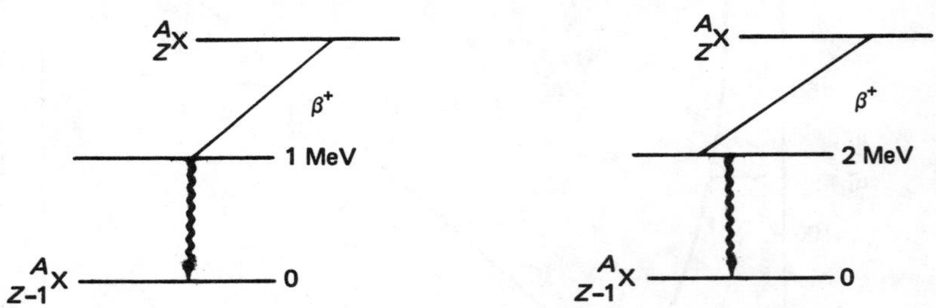

12.3 A liquid sample is contaminated with equal amounts (mass, not activity) of ^{131}I and ^{137}Cs. Sketch the energy spectra you expect to see if you use (a) a NaI(Tl) crystal with 7 percent energy resolution for the cesium peak, and (b) a Ge(Li) detector with energy resolution given by Eq. 12.15. Assume the same number of channels used with both detectors. Relevant data for the two isotopes are given in the table below. Assume that the sample is placed at a

†Crystals frequently used are mica, $(33\bar{1})$ planes, $d = 0.15$ nm; LiF, (200) planes, 0.201 nm; NaCl, (200) planes, 0.082 nm.

distance of 0.25 m from the detectors. The NaI(Tl) is a 3 in × 3 in crystal. The efficiency of the Ge(Li) detector is given by Fig. 12.28.

^{131}I		^{137}Cs	
E (MeV)	Intensity (%)	*E* (MeV)	Intensity (%)
0.284	5	0.662	85
0.364	82		
0.637	6.8		
0.723	1.6		
Half-life = 8.05 d		Half-life = 30 y	

12.4 An isotope emits two gammas with energies 0.8 and 1.2 MeV and intensities 30 and 100 percent, respectively. Assume that a Ge(Li) detector 5 mm thick is used for the measurement of this spectrum. Also assume that all the photons are normally incident upon the detector. Calculate the ratio of counts under the 0.8 MeV spectrum to counts under the 1.2 MeV spectrum.

12.5 What is the maximum width above which the two peaks of iron shown in Fig. 12.40 cannot be resolved?

12.6 Will the peaks of Fig. 12.40 be resolved with a gas-filled proportional counter, assuming the best possible resolution for that type of counter?

12.7 What is the efficiency of a 50-mm-long proportional counter filled with a mixture of xenon (20 percent) and methane at a pressure of 1 atm for a parallel beam of 10-keV x-rays (geometry of Fig. 12.11)?

12.8 Verify the efficiency values for 1-keV x-rays, shown in Fig. 12.41 for a Si(Li) detector.

12.9 Other things being equal, what is the ratio of intrinsic efficiencies of Si(Li) and Ge(Li) detectors 3 mm thick for 50-keV x-rays?

12.10 Prove that the energy of a photon is related to its wavelength by

$$E = \frac{123.99}{\lambda\,(\text{nm})}\ (\text{keV})$$

12.11 A crystal spectrometer will be used for analysis of fluorescent x-rays. If the required resolution is 0.1 percent and the angular aperture cannot be less than 0.5°, what is the lattice interplanar distance needed? Assume first-order reflection. The energy of the x-rays is between 4 and 6 keV.

BIBLIOGRAPHY

Compton, A. H., and Allison, S. K., *X Rays in Theory and Experiment*, Van Nostrand, New York, 1943.

Cullity, B. D., *Elements of X-Ray Diffraction*, Addison-Wesley, Reading, Mass., 1956.

Evans, R. D., *The Atomic Nucleus*, McGraw-Hill, New York, 1972.

Knoll, G. F., *Radiation Detection and Measurement*, Wiley, New York, 1979.

Price, W. J., *Nuclear Radiation Detection*, McGraw-Hill, New York, 1964.

A Handbook of Radioactivity Measurements Procedures, NCRP Report No. 58 (1978).

REFERENCES

1 Ortec catalogue and brochures (1976 and 1979).
2 Canberra catalogue and brochures (1979 and 1980).
3 Harshaw Chemical Company Brochure (1978).
4 Heath, R. L., Helmer, R. G., Schmittroth, L. A., and Cazier, G. A., *Nucl. Instr. Meth.* **47**:281 (1967).
5 Helmer, R. G., Heath, R. L., Schmittroth, L. A., Jayne, G. A., and Wagner, L. M., *Nucl. Instr. Meth.* 47:305 (1967).
6 Lurie, N. A., Harris, L., Jr., and Young, T. C., *Nucl. Instr. Meth.* **129**:543 (1975).
7 Ingersoll, D. T., and Wehring, B. W., *Nucl. Instr. Meth.* **147**:551 (1977).
8 "FORIST Spectra Unfolding Code," RSIC Computer Code Collection PSR-92, ORNL (1975).
9 "COOLC and FERDOR Spectra Unfolding Codes," RSIC Computer Code Collection, RSR-17, ORNL.
10 Baedecker, P. A., Rowe, J. J., and Steinnes, E., *Journal Radiol. Chem.* **40**:115 (1977).
11 Muggleton, A. H. F., *Nucl. Instr. Meth.* **101**:113 (1972).
12 "Calibration and Usage of Germanium Detectors for Measurement of Gamma-Ray Emission of Radionuclides," ANSI N 42.14-1978 (1978).
13 Freeman, J. M., and Jenkin, J. G., *Nucl. Instr. Meth.* **43**:269 (1966).
14 Campbell, J. L., and McNelles, L. A., *Nucl. Instr. Meth.* **101**:153 (1972).
15 McNelles, L. A., and Campbell, J. L., *Nucl. Instr. Meth.* **109**:241 (1973).
16 Abreu, M. C., Maio, A. A., and D'Oliveira, M. J. T., *Nucl. Instr. Meth.* **123**:295 (1975).
17 Berenyi, D., Meszaros, E., Seif el Nasr, S. A. H., and Bucso, J., *Nucl. Instr. Meth.* **124**:505 (1975).
18 Waibel, E., and Grosswendt, B., *Nucl. Instr. Meth.* **131**:133 (1975).
19 Tokcan, G., and Cothern, C. R., *Nucl. Instr. Meth.* **64**:219 (1968).
20 Mowatt, R. S., *Nucl. Instr. Meth.* **70**:237 (1969).
21 Papadellis, T., and Hontzeas, S., *Nucl. Instr. Meth.* **73**:210 (1969).
22 Hajnal, F., and Klusek, C., *Nucl. Instr. Meth.* **122**:559 (1974).
23 Singh, R., *Nucl. Instr. Meth.* **136**:543 (1976).
24 Wainio, K. M., and Knoll, G. F., *Nucl. Instr. Meth.* **44**:213 (1966).
25 de Castro Furia, N. V., and Levesque, R. J. A., *Nucl. Instr. Meth.* **46**:325 (1967).
26 Mann, H. M., Bilger, H. R., and Sherman, I. S., *IEE, NS-13* **3**:352 (1966).
27 Palms, J. M., Venugopala Rao, P., and Wood, R. D., *Nucl. Instr. Meth.* **64**:310 (1968).
28 Baedecker, P. A., *Anal. Chem.* **43**:405 (1971).
29 Kokta, L., *Nucl. Instr. Meth.* **112**:245 (1973).
30 Covell, D. F., *Anal. Chem.* **31**:1785 (1959).
31 Quittner, P., *Nucl. Instr. Meth.* **76**:115 (1969).
32 Mariscotti, M. A., *Nucl. Instr. Meth.* **50**:309 (1967).
33 Helmer, R. G., Heath, R. L., Putnam, M., and Gipson, D. H., *Nucl. Instr. Meth.* **57**:46 (1967).
34 Routti, J. T., and Prussin, S. G., *Nucl. Instr. Meth.* **72**:125 (1969).
35 Robinson, D. C., *Nucl. Instr. Meth.* **78**:120 (1970).
36 Kern, J., *Nucl. Instr. Meth.* **79**:233 (1970).
37 McNelles, L. A., and Campbell, J. L., *Nucl. Instr. Meth.* **127**:73 (1975).
38 Jorch, H. H., and Campbell, J. L., *Nucl. Instr. Meth.* **143**:551 (1977).
39 DeLotto, I., and Ghirardi, A., *Nucl. Instr. Meth.* **143**:617 (1977).
40 Campbell, J. L., and Jorch, H. H., *Nucl. Instr. Meth.* **159**:163 (1979).
41 Connelly, A. L., and Black, W. W., *Nucl. Instr. Meth.* **82**:141 (1970).
42 Phillips, G. W., and Marlow, K. W., *Nucl. Instr. Meth.* **137**:525 (1976).
43 Cline, J. E., Putnam, M. H., and Helmer, R. G., "GAUSS-VI, A Computer Program for the Automatic Analysis of Gamma-Ray Spectra from Ge(Li) Spectrometers," ANCR-1113 (1973).
44 Head, J. H., *Nucl. Instr. Meth.* **98**:419 (1972).
45 Siffert, P., Gonidec, J. P., and Cornet, A., *Nucl. Instr. Meth.* **115**:13 (1974).
46 Jones, L. T., and Woollam, P. B., *Nucl. Instr. Meth.* **124**:591 (1975).
47 Siffert, P., *Nucl. Instr. Meth.* **150**:1 (1978).
48 Dabrowski, A. J., Iwanczyk, J., Szymczyk, W. M., Kokoschinegg, P., and Stelzhammer, J., *Nucl. Instr. Meth.* **150**:25 (1978).

49 Schieber, M., Beinglass, I., Dishon, G., Holzer, A., and Yaron, G., *Nucl. Instr. Meth.* **150**:71 (1978).
50 Shaler, J., *Nucl. Instr. Meth.* **150**:79 (1978).
51 Whited, R. C., and Schieber, M. M., *Nucl. Instr. Meth.* **162**:113 (1979).
52 Dabrowski, A. J., Iwanczyk, J. S., Barton, J. B., Huth, G. C., Whited, R., Ortale, C., Economou, T. E., and Turkevich, A. L., *IEEE Trans. Nucl. Sci.* **NS-28**(1):536 (1981).
53 Iwanczyk, J. S., Dabrowski, A. J., Huth, G. C., Del Duca, A., and Schnepple, W., *IEEE Trans. Nucl. Sci.* **NS-28**(1):579 (1981).
54 Johansson, T., *Z. Physik* **82**:507 (1933).
55 Cauchois, Y., *Ann. Phys.* **1**:215 (1934).
56 DuMond, J. W. M., *Rev. Sci. Instr.* **18**:626 (1947).

13

■ charged-particle spectroscopy

13.1 INTRODUCTION

A charged particle going through any material will have interactions affecting its detection in two ways. First, the energy spectrum is distorted because of the energy loss caused by the interactions in any mass interposed between source and detector. Second, a particle entering the active detector volume will interact there at least once and will be detected, i.e., the efficiency is practically 100 percent.

Since any energy loss outside the detector is undesirable, the task of the experimenter is to design a spectrometer with zero mass between the source and the detector. Such an ideal system cannot be built, and the only practical alternative is a spectrometer that results in such a small energy loss outside the detector that reliable corrections can be applied to the measured spectrum.

In certain measurements, the particles do not stop in the detector, but they go through it and emerge with only a fraction of their energy deposited in the detector. Then a correction to the spectrum of the exiting particles will have to be applied because of *energy straggling*, a term used to describe the statistical fluctuations of energy loss. Energy straggling should not be confused with the statistical effects that result in the finite energy resolution of the detector.

For heavy ions, a phenomenon called the *pulse height defect* (PHD) seems to have an important effect on energy calibration. As a result of the PHD, the relationship between pulse height and ion energy is mass dependent. In semiconductor detectors, experiments have shown that the PHD depends on the orientation of the incident ion beam relative to the crystal planes of the detector. This phenomenon is called channeling.

To avoid unnecessary energy loss, the source of the charged particles should be prepared with special care. The heavier the ion, the more important the source thickness becomes and the more difficult the source preparation is.

This chapter discusses the subjects of energy loss and straggling, pulse height defect, energy calibration methods, and source preparation, from the point of view of their effect to spectroscopy. All the effects are not equally important for all types of particles. Based on similarity in energy loss behavior, the charged particles are divided into three groups, as in Chap. 4:

1 Electrons and positrons
2 Alphas, protons, deuterons, tritons
3 Heavy ions ($Z > 2, A > 4$)

Energy straggling, which is a phenomenon common to all particles is discussed first. Then the other effects are analyzed separately for each particle group.

13.2 ENERGY STRAGGLING

If a monoenergetic beam of charged particles traverses a material of thickness Δx, where Δx is less than the range of the particles in that medium, the beam will emerge from the material with a distribution of energies. The broadening of the beam is due to the statistical fluctuations of the energy loss processes. Simply stated, the incident particle participates in a great number of collisions as it travels the distance Δx, and loses a certain fraction of its energy in every collision. However, neither the number of collisions nor the energy lost per collision is constant, resulting in a distribution of energies called *energy straggling*.

Energy straggling plays no role in the measurement of the total energy of the charged particle. It does play a significant role, however, in transmission-type experiments where the particle emerges from a detector after depositing only a fraction of its energy in it.

Consider a monoenergetic beam of particles with kinetic energy T_0 (Fig. 13.1) going through a thickness Δx that is a fraction of particle range. The average energy $\bar{T}$ of the emerging particles is

$$\bar{T} = T_0 - \int_0^{\Delta x} \left(\frac{dE}{dx}\right) dx \tag{13.1}$$

where dE/dx is the stopping power of the medium for the incident particle (see Chap. 4). In most cases, $\bar{T} < T_p$, where T_p is the most probable energy of the particles after going through the thickness Δx.

The shape of the energy distribution shown in Fig. 13.1 is determined by the parameter k,

$$k = \frac{\xi}{\Delta E_{max}} \tag{13.2}$$

where ξ is roughly equal to the mean energy loss of the particle traversing the thickness Δx, and ΔE_{max} is the maximum energy transfer to an atomic electron in one collision. The expressions for ξ and ΔE_{max} are

$$\xi = 2\pi r_0^2 mc^2 \frac{Z_1^2 N Z_2}{\beta^2} \Delta x \tag{13.3}$$

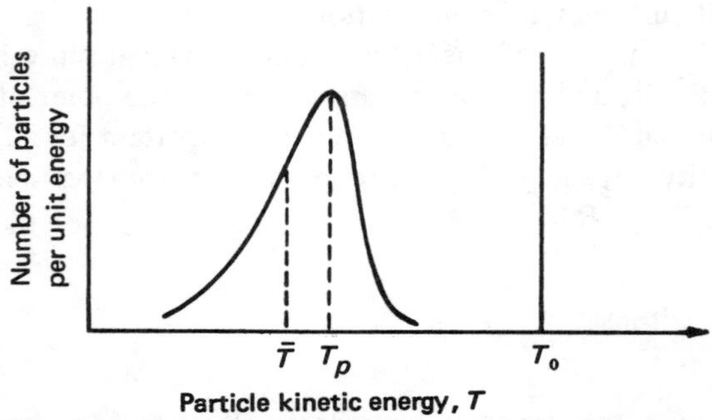

FIG. 13.1 A monoenergetic beam will exhibit an energy distribution after going through a material of thickness $\Delta x < R$.

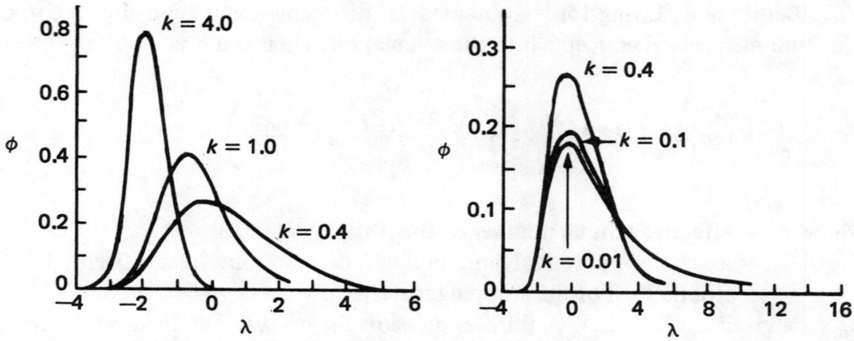

FIG. 13.2 The Vavilov distribution shown for various values of the parameter k. The quantity ϕ is a measure of the probability that a particle will lose energy between T and $T + dT$ in traversing thickness x. The parameter $\lambda = (T - T)/\xi - 0.423 - \ln x$ (from ref. 11).

$$\Delta E_{\max} = 2mc^2 \frac{\beta^2}{1-\beta^2}\left[1 + \left(\frac{m}{M_1}\right)^2 + 2\frac{m}{M_1}\frac{1}{\sqrt{1-\beta^2}}\right] \tag{13.4}$$

All the symbols in Eqs. 13.3 and 13.4 have been defined in Sec. 4.3, except Z_1, the charge of the incident particle, and Z_2, the atomic number of the stopping material. For nonrelativistic particles ($\beta \ll 1$), which are much heavier than electrons, Eq. 13.4 takes the form

$$E_{\max} = \frac{4mM_1}{(m + M_1)^2} T \tag{13.4a}$$

For small values of k ($k \leqslant 0.01$), a small number of collisions takes place in the stopping medium and the resulting distribution is asymmetric with a low energy tail. Landau[1] first investigated this region and obtained a universal asymmetric curve. The case of intermediate k values ($0.01 < k < 10$) was first investigated by Symon[2] and later by Vavilov.[3] The Vavilov distribution was checked and was found to agree with experiment.[4] For small k, the Vavilov distribution takes the shape of the Landau result, while for large k, when the number of collisions is large, it becomes a Gaussian. Figure 13.2 shows how the distribution changes as a function of k. Many other authors have studied special cases of the energy straggling problem.[5–11]

The variance of the energy straggling distribution was first calculated† by Bohr[12] using a classical model. Bohr's result is

$$\sigma_E^2 = \overline{(\Delta T)^2} - (\overline{\Delta T})^2 = 4\pi r_0^2 (mc^2)^2 Z_1^2 Z_2 N \,\Delta x \tag{13.5}$$

where ΔT = energy loss in a specific case.

$\overline{\Delta T}$ = average energy loss (given by Eq. 13.1) = $T_0 - \overline{T}$

The width Γ of the distribution is equal to $(2\sqrt{2 \ln 2})(\sigma_E)$.

†The calculation is presented in Evans, *The Atomic Nucleus*, and Sergré, *Nuclei and Particles* (see bibliography of this chapter).

Bethe and Livingston[13] obtained a different expression by incorporating quantum-mechanical concepts into the calculation. Their result is

$$\sigma_E^2 = 4\pi r_0^2 (mc^2)^2 Z_1^2 N \left(Z_2' + \sum_i \frac{8}{3} \frac{I_i Z_i}{E_{\max}} \ln \frac{\Delta E_{\max}}{I_i} \right) \Delta x \tag{13.6}$$

where Z_2' = effective atomic number of the stopping material

I_i, Z_i = ionization potential and number of electrons, respectively, in the ith atomic shell of the stopping material

$E_{\max}$ is given by Eq. 13.4. A third expression for σ_E^2 was obtained by Titeica.[14] It is worth noting that Bohr's result (Eq. 13.5) is independent of the particle energy, while the Bethe-Livingston (Eq. 13.6) and the Titeica result have a small energy dependence.

The expressions for σ_E^2 mentioned above were all obtained by taking into account electronic collisions only. Nuclear collisions (see Chap. 4) are rare, but they cause large energy losses. As a result, they do not contribute significantly to the average energy loss but they do influence the energy distribution by giving it a low-energy tail. (The energy *loss* distribution will have a high-energy tail.)

The width of the energy distribution after the beam traverses a thickness Δx consists of a partial width Γ_s due to straggling and a second one Γ_d due to the resolution and noise of the detection system. The total width Γ is obtained by adding the two partial widths in quadrature:

$$\Gamma = \sqrt{\Gamma_s^2 + \Gamma_d^2} \tag{13.7}$$

The energy straggling is measured with an experimental set-up shown schematically in Fig. 13.3. A source, a detector, and a movable absorber are housed in an evacuated chamber, to avoid any energy loss as the particles travel from the source to the detector. The width Γ_d is measured first by recording the particle energy spectrum with the absorber removed. Then the absorber is put into place and the measured spectrum gives the width Γ. The straggling width is, using Eq. 13.7,

$$\Gamma_s = \sqrt{\Gamma^2 - \Gamma_d^2} \tag{13.8}$$

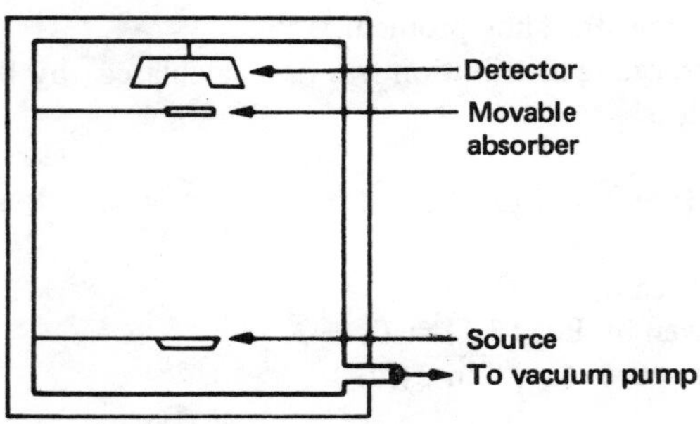

FIG. 13.3 The experimental set-up used in the study of energy straggling.

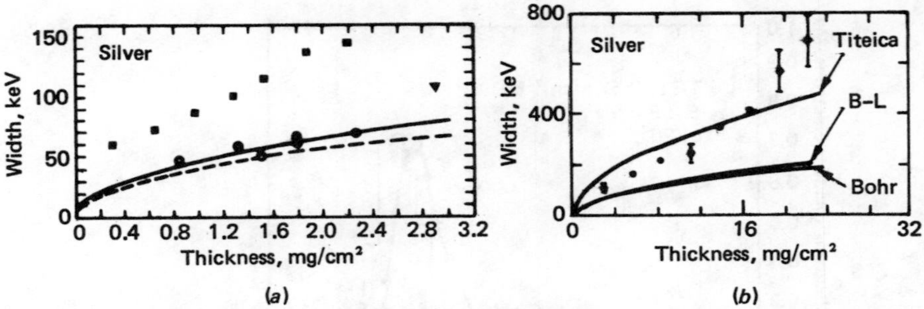

FIG. 13.4 Alpha-particle energy straggling for (*a*) thin, and (*b*) thick foils of silver. [(*a*) Data (○) from ref. 16; (■) measurements of Sykes and Harris (ref. 59); (— — —) calculation from ref. 12; (———) calculation from ref. 13; (▼) measurements from ref. 15; (*b*) Experimental points from ref. 15; the lines are results of calculations.]

By using absorbers of different thicknesses, the width Γ_s may be studied as a function of Δx. Measurements of this type have been performed by many people, especially with alpha particles.[15,16] For small thicknesses, the experimental results agree with theory, while for large thicknesses the theory underestimates the width. Figure 13.4 shows results for thin and thick silver foils. It should be noted that according to the theory (Eqs. 13.5 and 13.6), the width Γ_s is proportional to $\sqrt{\Delta x}$, assuming that Z does not change as the particle traverses the thickness Δx.

Energy straggling is more pronounced for electrons than for heavier particles for three reasons. First, electrons are deflected to large angles and may lose up to half of their energy in one collision. Second, large angle scattering increases their pathlength. Third, electrons radiate part of their energy as bremsstrahlung. All three effects tend to increase the fluctuations of the energy loss. Results of electron transmission and straggling measurements have been reported by many observers. A typical spectrum of straggled electrons is shown in Fig. 13.5, which compares the experimental result[17] with a Monte Carlo calculation.[18]

Range straggling is a phenomenon related to energy straggling by the equation

$$\sigma_R^2 = \left(\frac{dE}{dx}\right)^{-2} \sigma_E^2 \tag{13.9}$$

where σ_R^2 is the range variance. Range straggling refers to the pathlength distribution of monoenergetic particles traversing the same absorber thickness (for more details see Sec. 22.5 of Evans). For spectroscopy measurements, only energy straggling is important.

13.3 ELECTRON SPECTROSCOPY

Under the common title of electron spectroscopy, this section discusses the most important problems of electron, positron, and beta-ray energy measurements: backscattering, energy resolution and detector response function, energy calibration, and source preparation.

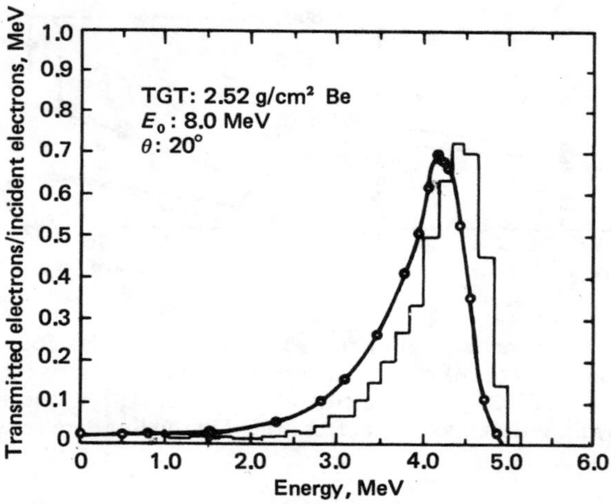

FIG. 13.5 Energy spectrum of 8-MeV electrons transmitted through 2.52 g/cm^2 of Be and observed at 20° from the direction of the incident beam. The histogram is the result of Monte Carlo calculation (ref. 18). The experimental points are from ref. 17.

13.3.1 Electron Backscattering

Electrons moving through a detector behave differently from heavier charged particles in two respects. First, as a result of successive collisions with atomic electrons, the incident electrons may be deflected by more than 90°, i.e., they may be backscattered. Second, electrons slowing down lose part of their energy as bremsstrahlung.† In general, the effect of bremsstrahlung production on spectrum distortion is small. Backscattering in the detector, on the other hand, is important, and corrections to the measured spectrum because of it have to be applied. The effect of backscattering on electron energy spectra is discussed in this section.

Consider a monoenergetic electron beam of energy T_0 impinging normally upon a detector of thickness x where $x > R(T_0)$ (Fig. 13.6), and $R(T_0)$ is the range of

†Every charged particle slowing down radiates part of its energy. For particles other than electrons, however, and for the energies considered here, the bremsstrahlung can be ignored (see Chap. 4).

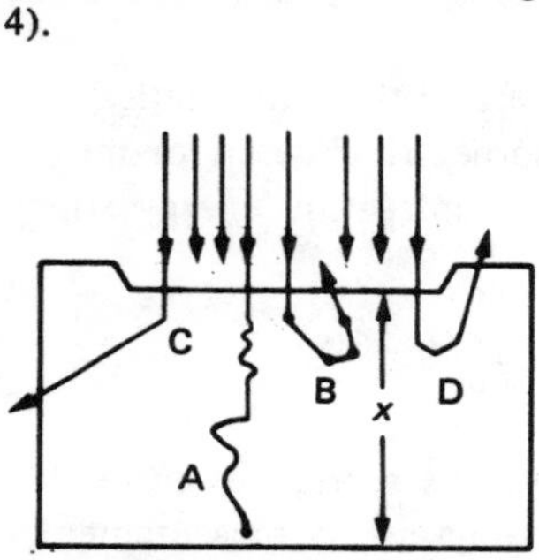

FIG. 13.6 Some of the electrons incident upon the detector are backscattered and deposit only part of their energy in it.

electrons of energy T_0 in the material of which the detector is made. Most of the incident electrons will deposit all their energy in the detector (electron A, Fig. 13.6) and thus generate a pulse proportional to T_0. But some electrons (like B or C or D, Fig. 13.6) are scattered out of the detector before they deposit all their energy in it. Such particles will give rise to a pulse smaller than that corresponding to energy T_0. As a result of electron backscattering, the energy spectrum of a monoenergetic source will have a full-energy peak and a low-energy tail, as shown in Fig. 13.7.[19–23] The fraction of electrons recorded in the tail

1 Increases with the atomic number of the detector material
2 Changes slowly with the energy T_0
3 Increases as the incident angle of the beam deviates from the normal

An electron energy spectrum measured with a plastic scintillator is shown in Fig. 13.8. It is represented extremely well by the following analytic function, which was developed by Tsoulfanidis, Wehring, and Wyman[20] and is shown in Fig. 13.9.

$$R(E,E') = \frac{1}{2}\,\frac{b}{E'}\,\mathrm{erfc}\left(\frac{E-E'}{\sigma\sqrt{2}}\right) + \frac{1-b}{\sigma\sqrt{2\pi}}\exp\left[-\frac{1}{2}\,\frac{(E-E')^2}{\sigma^2}\right] \tag{13.10}$$

where

$$\mathrm{erfc}\,(x) = 1 - \frac{2}{\sqrt{\pi}}\int_0^x \exp\,(-t^2)\,dt$$

and σ = standard deviation of the Gaussian

b = fraction of electrons in the tail

The backscattering fraction b is given by

$$b = \left[1 + \sqrt{2\pi}\left(\frac{g}{\tau} - \frac{1}{2}\right)\frac{\sigma}{E'}\right]^{-1} \tag{13.11}$$

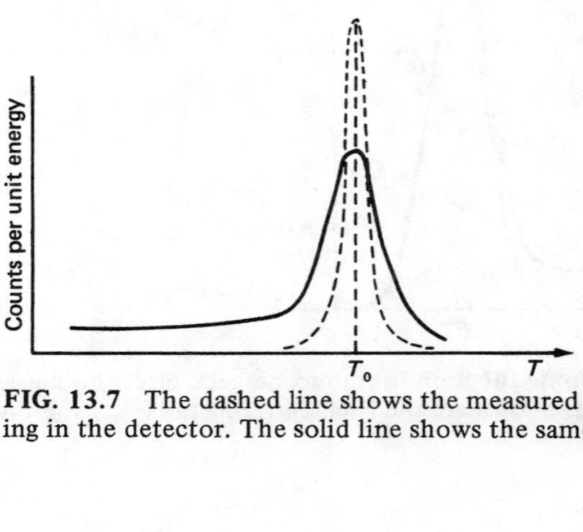

FIG. 13.7 The dashed line shows the measured spectrum without backscattering in the detector. The solid line shows the same spectrum with backscattering.

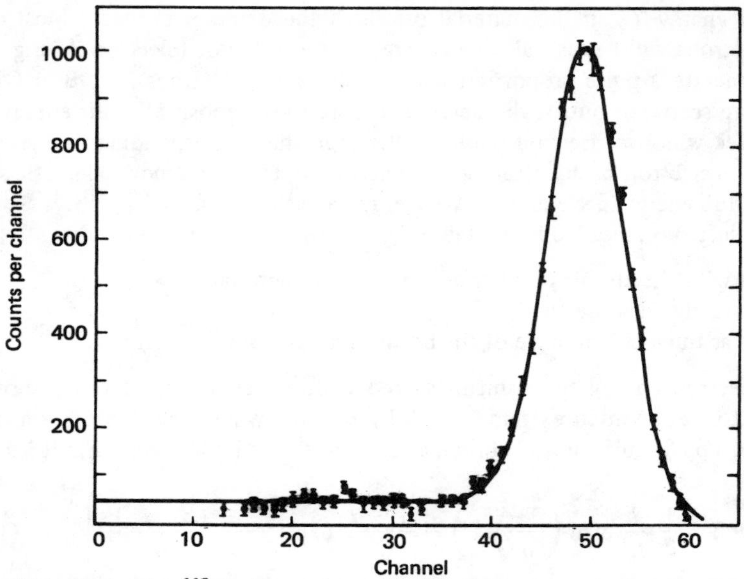

FIG. 13.8 The ^{113}Sn internal conversion electron spectrum obtained with a plastic scintillator. The solid line was obtained using Eq. 13.10 (from ref. 20).

where (Fig. 13.9) g is the height of the peak and τ is height of the backscattering tail.

Similar results have been obtained with a Si(Li) detector.[20,23] More references on the subject are given in Bertolini and Coshe (see their Sec. 4.3.3). Semiempirical formulas giving the value of b as a function of Z and T have been developed by many authors,[24,25] but such equations are of limited general value because the response function and the backscattering depend on the geometry of the system; for this reason, response function and backscattering should be measured for the actual experimental setup of the individual observer.

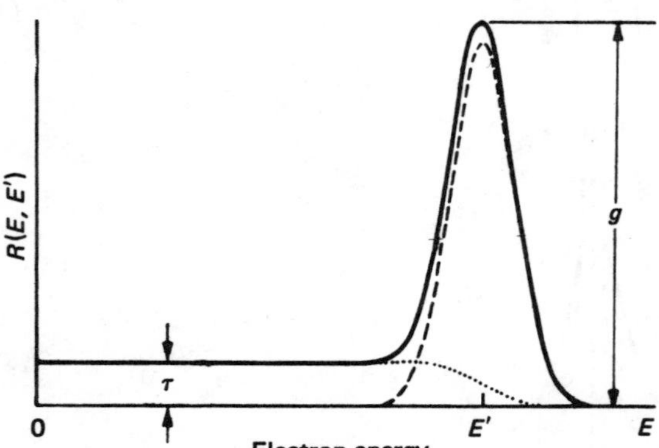

FIG. 13.9 Analytical response function for monoenergetic electrons measured by a thin plastic scintillator. Shown are the backscattering tail (· · ·) and the Gaussian (— — —).

13.3.2 Energy Resolution and Response Function of Electron Detectors

The best energy resolution for electrons is obtained using silicon semiconductor detectors, with the possible exception of magnetic spectrometers. Silicon detectors may be surface-barrier or Si(Li) detectors. The surface-barrier detectors operate at room temperature, while the Si(Li) detectors give best results when cooled to liquid nitrogen temperatures. The energy resolution of semiconductor detectors is determined by the electronic noise alone. It deteriorates as the area and the sensitive depth of the detector increase. For commercial detectors available now (1980), the FWHM ranges from about 7 to 30 keV.

The energy resolution of scintillators, plastic scintillators in particular, is much worse. It is of the order of 8-10 percent at 1 MeV, which gives a FWHM of 80-100 keV. For scintillators the FWHM is roughly proportional to $\sqrt{E}$. Plastic scintillators have two advantages over semiconductor detectors: the backscattering fraction is less for scintillators because of their lower atomic number, and the timing characteristics are extremely useful for certain types of measurements. The pulse risetime is about 0.1 ns for a plastic scintillator, while for a silicon detector it is between 1 and 10 ns.

The response function of electron detectors is of the form shown in Fig. 13.9. Because of the low-energy tail, if one measures a continuous spectrum (e.g., one beta spectrum or a mixture of beta spectra), the measured spectrum will be higher than the source spectrum at the low-energy end and lower at the high-energy end, as shown in Fig. 13.10. Therefore, spectrum unfolding is necessary to shift back to their proper

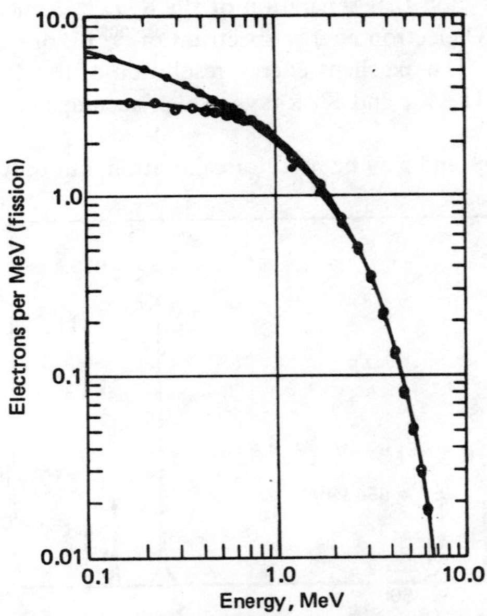

FIG. 13.10 Measured (●) and unfolded (○) beta spectra from ^{235}U fission fragments. Unfolding removes the effect of backscattering in the detector; thus, it reduces the spectrum at the low-energy end (from ref. 26).

energy all the betas that were incorrectly recorded at lower energies because of backscattering. The iteration method of unfolding, described in Sec. 11.5.3, is suitable for beta spectra. The spectrum shown in Fig. 13.10 was unfolded using that method.

13.3.3 Energy Calibration of Electron Spectrometers

The energy calibration of any spectrometer requires the use of sources of known energy and preferably of monoenergetic sources. Monoenergetic electron sources are provided by accelerators and by radioisotopes emitting internal conversion (IC) electrons (see Chap. 3).

The advantage of the accelerators is their ability to provide a monoenergetic beam with any desired energy from zero up to the upper limit of the machine. The disadvantages are their expensive operation and the fact that the spectrometer has to be moved to the accelerator beam.

Internal-conversion emitters are relatively inexpensive to obtain and very easy to handle. They have the disadvantage that they emit not only IC electrons but also gammas. Thus, when a spectrum is recorded, the result includes both IC electrons and Compton electrons created by gammas which interact in the detector. One may eliminate the Compton electrons by utilizing the x-rays that are also given off by the IC source. The x-rays are emitted in coincidence with the IC electrons, while the gammas, and therefore the Compton electrons too, are not. Thus, if the IC electrons are counted in coincidence with the x-rays, the Compton electrons will not be recorded.

Internal-conversion sources emit K, L, . . . , electrons. The energy resolution of silicon semiconductor detectors is so good that separation of the K, L, . . . , electrons is possible. Figure 13.11 shows the IC electron energy spectrum of ^{207}Bi, one of the most widely used calibration sources. The excellent energy resolution of the detector distinguishes K, L, and M electrons. The K_α and K_β x-rays, which accompany the IC process, are also known.

Pure beta-emitting isotopes exist and may be used for calibration, but only after

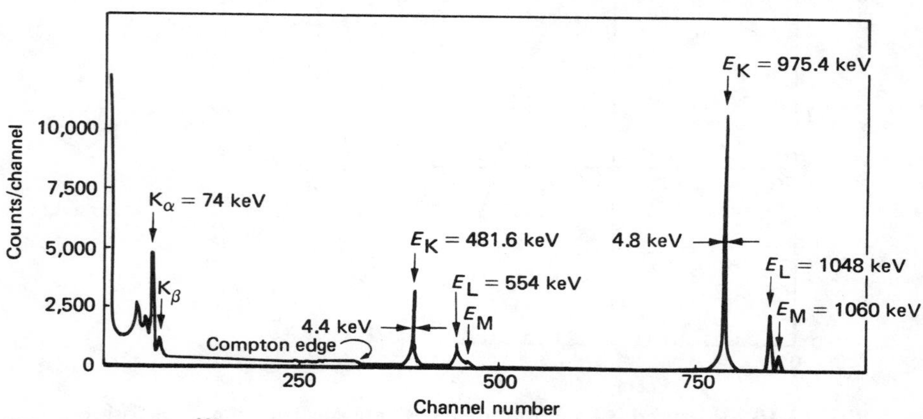

FIG. 13.11 The ^{207}Bi internal conversion electron spectrum obtained with a silicon semiconductor detector (from ref. 27).

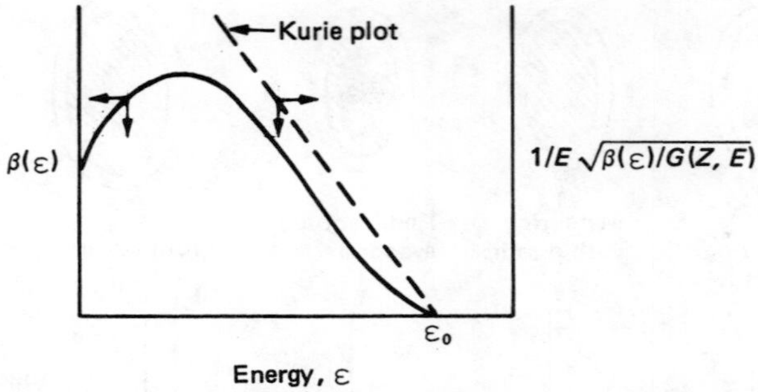

FIG. 13.12 A typical beta energy spectrum (solid line) and its Kurie plot (dashed line).

the energy spectrum is cast into a form called the *Kurie plot.* The beta spectrum is continuous and extends from zero energy up to a maximum end point kinetic energy (see Fig. 13.12). Because of the shape of the spectrum, it is impossible to accurately determine the end point energy. However, from the theory of beta decay, it is known that the beta spectrum may be written as[28]

$$\frac{1}{E}\sqrt{\frac{\beta(\epsilon)}{G(Z,E)}} = k(\epsilon_0 - \epsilon) \tag{13.12}$$

where ϵ = beta kinetic energy in units of mc^2

$E = T + mc^2 = \epsilon + 1$ = total energy

$\beta(\epsilon)\, d\epsilon$ = number of betas with energy between ϵ and $\epsilon + d\epsilon$

$G(Z, E)$ = modified fermi function of β decay

k = constant independent of energy (for allowed transitions)

If the left-hand side of Eq. 13.12 is plotted against ϵ, the result is a straight line that crosses the energy axis at $\epsilon = \epsilon_0$. The Kurie plot is a straight line for allowed beta transitions. A "forbidden" beta decay will show an upward curvature at the end.[28]

13.3.4 Source Preparation

Preparation of electron or beta sources is accomplished by observing these two rules:

1 The source backing material should be as thin as possible, to avoid backscattering from the source.
2 The source cover should be as thin as possible, to avoid any unnecessary energy loss there.

One way to make a source is by following these steps:

1 A liquid solution of the radioisotope is obtained.
2 A thin piece of Mylar or metallic foil (a few microns thick) is mounted on a metal ring (see Fig. 13.13). This is the source support.
3 Using a pipette, a small amount of the radioisotope is placed on the source support.
4 The source support is then placed under a heat lamp until the liquid evaporates. The heat lamp should be located inside a properly monitored fume hood.

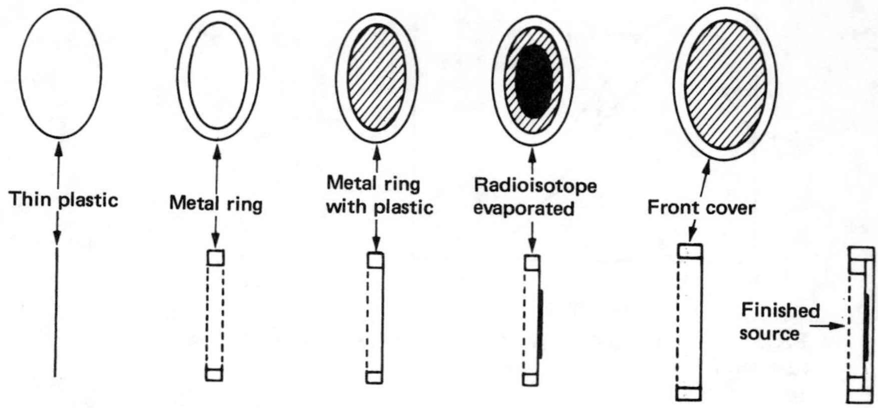

FIG. 13.13 Preparation of an electron source.

5 Steps 3 and 4 are repeated as many times as necessary to achieve the required source strength.
6 The source is sealed by placing a ring with an extremely thin cover (again a few microns thick) on its front face, as shown in Fig. 13.13.

There are many commercial outfits that sell β sources with specified characteristics of intensity, source backing, and source cover. Detailed description of source preparation, for all types of charged particles, is given in a review article by Eijk, Oldenhof, and Zehner,[29] and also in references 30–32.

13.4 ALPHA, PROTON, DEUTERON, AND TRITON SPECTROSCOPY

Protons, deuterons, tritons, and alpha particles behave similarly as far as energy loss and straggling are concerned. As they travel in a medium they are deflected very little from their direction of incidence, as a result of which backscattering is insignificant and their range is almost equal to their pathlength.

To avoid significant energy loss, the particles must go through as small a mass as possible when they move from the source to the detector. This is accomplished by making the source cover and the detector window as thin as possible. The entrance window of such detectors consists of a metallic layer, usually gold, with a thickness of 4×10^{-4} kg/m^2 (40 μg/cm^2) or less. The measurements are performed in an evacuated chamber to avoid energy loss in air.

The discussion in the rest of this section uses examples from alphas, but the points made are valid for the other particles of this group. Alphas have been studied and used much more extensively than the others, providing a basis for discussion.

13.4.1 Energy Resolution and Response Function of Alpha Detectors

The best energy resolution is obtained with silicon surface-barrier detectors. Most detector manufacturers quote the resolution obtained for the 5.477-MeV alphas of ^{241}Am. A typical spectrum obtained with a detector having 25 mm^2 active area and 100 μm sensitive depth is shown in Fig. 13.14. The resolution deteriorates somewhat as the detector size increases. Since the response function of alpha detectors is a narrow Gaussian, there is no need to unfold a measured monoenergetic alpha spectrum.

13.4.2 Energy Calibration

All isotopes with $Z > 82$ emit alphas in the energy range 4 MeV $< T <$ 8 MeV, each isotope giving off more than one group of alphas. A particular isotope is selected to be used for calibration based on the energy of the alphas, the presence of other interfering radiations, and its half-life. For example, the isotope $^{241}_{95}$Am is very popular because it has a 458-year half-life and its only other radiation emitted is ^{237}Np x-rays. Other isotopes frequently used are ^{210}Po, ^{276}Ra, and ^{252}Cf. Alpha sources with $T_\alpha > 8$ MeV, as well as sources of protons, deuterons, and tritons of any energy, can be provided by accelerators only.

13.4.3 Source Preparation

The main precaution taken in the preparation of an alpha source using an alpha-emitting radioisotope is to cover its front face with the thinnest possible layer of material. Commercial sources are made by sandwiching the radioisotope between two thin foils. Figure 13.15 shows a ^{241}Am source made by the Amersham Corporation.

Alpha-emitting isotopes are considered extremely hazardous when ingested, and in particular when they enter the lungs. To avoid accidental exposure, the user should always be certain that the source cover has not been damaged. The user should also be aware that all alpha sources emit a small number of neutrons, produced either through (α, n) reactions with the source-supporting material or from spontaneous fission of the radioisotope itself.

13.5 HEAVY-ION ($Z > 2$) SPECTROSCOPY

Heavy-ion spectroscopy is different from that of lighter charged particles because of the pulse height defect (PHD), which makes the energy calibration equation mass-dependent.

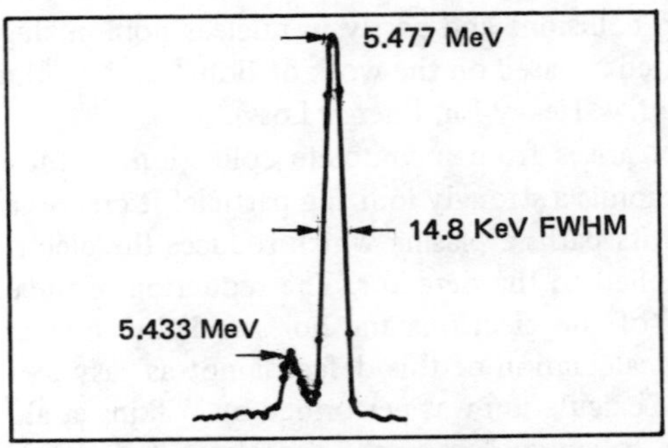

FIG. 13.14 The ^{241}Am alpha spectrum obtained with a silicon surface-barrier detector (from Canberra).

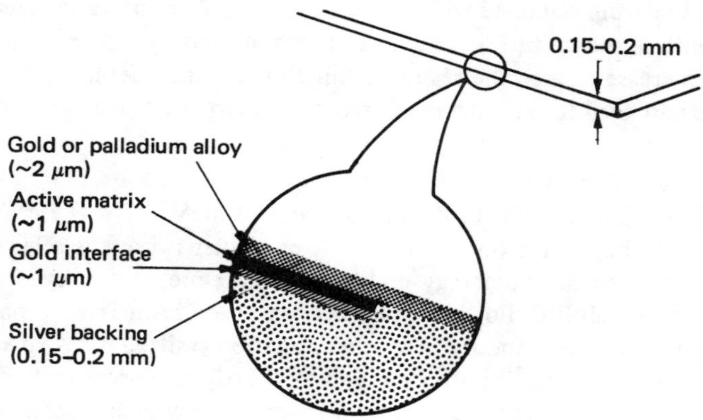

FIG. 13.15 A ^{241}Am source (from Amersham Co.).

13.5.1 The Pulse Height Defect

The measurement of particle energy with any type of detector is based on the assumption that the charge collected at the output of the detector is proportional to the energy of the incident particle. The assumption is valid if all the particle energy is lost in ionizing collisions and all the charge produced is collected, i.e., no recombination takes place.

It has been known for many years that heavy ions moving in gases lose part of their energy in nonionizing collisions with nuclei. The same phenomenon, PHD, has been observed in semiconductor detectors. The PHD, which is negligible for alphas and lighter particles, is defined as the difference between the energy of a heavy ion and that of a light ion (usually an alpha particle) that generates the same pulse height in the detector. Experimental results showing this phenomenon are presented in Fig. 13.16.[33] Notice that for alphas and carbon ions, the relationship between energy and pulse height is linear. Sulfur and silver ions show a small PHD. Heavier ions (Au, U) show a significant PHD.

The PHD is the result of three contributing defects:[33–38]

1. The *nuclear defect* is due to nuclear collisions. As a result of such collisions, the moving ion imparts energy upon other nuclei. The recoiling nuclei lose their energy partly in electronic ionizing collisions and partly in nuclear nonionizing ones. The nuclear defect has been calculated[36] based on the work of Bohr[1] and of Lindhard, Scharff, and Schiott[39] (see also Chap. 4, Heavy-Ion Energy Loss).

2. The *recombination defect* arises from incomplete collection of the charge produced in the detector. A heavy ion is a strongly ionizing particle. It creates a dense plasma of electron-hole pairs along its path, a plasma which reduces the electric field established by the external bias applied to the detector. The reduction of the electric field intensity hinders the drifting of the electrons and holes and thus increases the probability of recombination. The calculation of this defect is not as easy as that of the nuclear one, but an approximate calculation was performed by Wilkins et al.[33]

3. The *window defect* is due to energy loss in the dead layer (window) of the front surface of the detector. It can be obtained from the thickness of the window

and the stopping power of the ion. The thickness of the window can be measured by determining the change in pulse height as a function of the incident angle.

The PHD for iodine and for argon ions has been measured by Moak, Dabbs, and Walker,[40] using the *channeling effect* in silicon. Pulse-height distributions were measured by first aligning the direction of incident ions with the [110] crystal axis of the silicon surface-barrier detector and then by letting the ions impinge at an angle with respect to the same axis. In the first case, the ions moved along the channel between two planes (*channeled* ions); in the second, they did not (*unchanneled* ions). The channeled ions showed an energy resolution about three times better than that of unchanneled ones, and essentially no PHD (Fig. 13.17). This result can be explained by assuming that the channeled ions travelling between atomic planes lose all their energy in ionizing collisions, all the way to the end of their track. Similar results have been obtained with ^{235}U fission fragments[34] and ^{252}Cf fission fragments.[41]

The lack of nuclear collisions for channeled ions is not the only phenomenon that affects the pulse height. It is known that the electron density is much reduced along the channel. As a result, the electronic stopping power is lower and, consequently, so is the charge density produced by the heavy ion. Thus, not only the nuclear but also the reccombination defect is reduced for the channeled ions.

The PHD increases slowly with ion energy, as shown in Fig. 13.18.

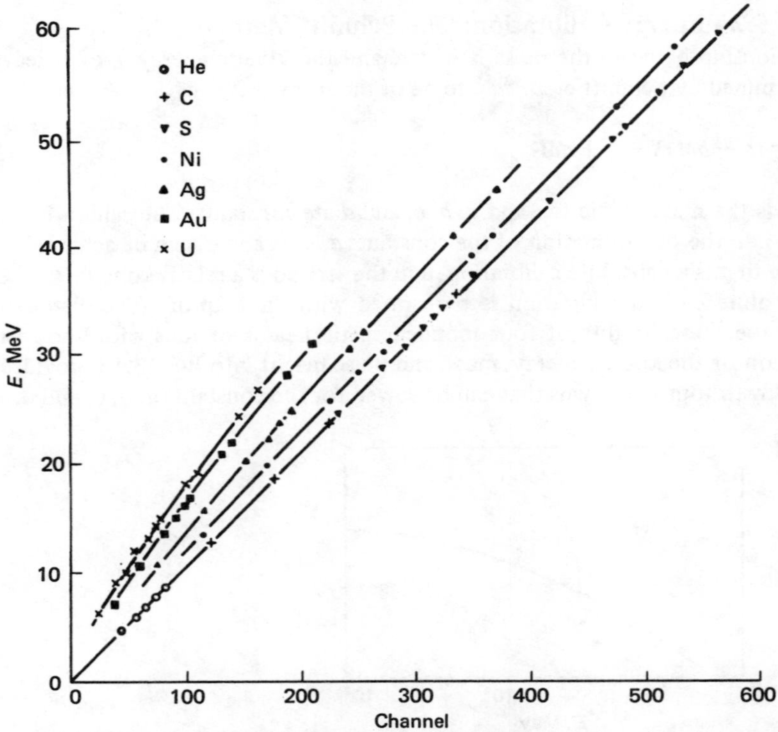

FIG. 13.16 The relationship between energy and pulse height (channel) for ions with different masses (from ref. 33).

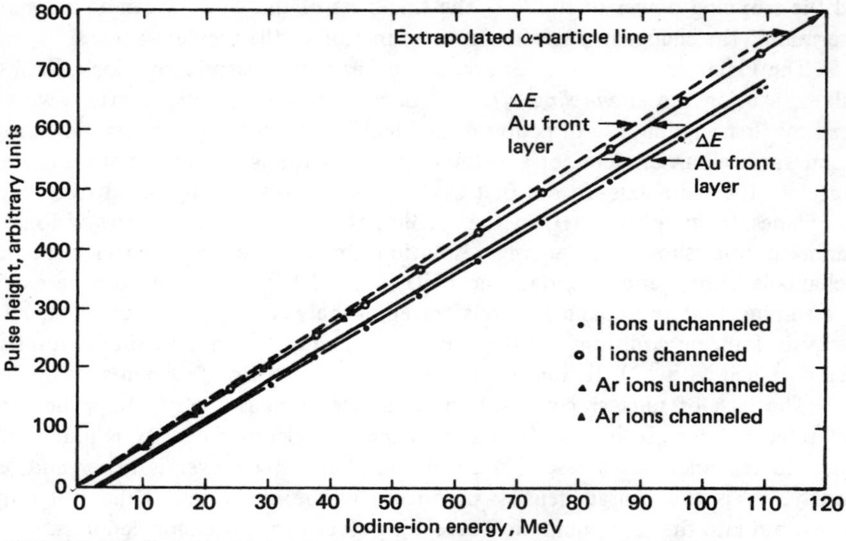

FIG. 13.17 Pulse-height response of a Si surface-barrier detector for "channeled" and "unchanneled" ions (from ref. 38).

13.5.2 Energy Calibration: The Schmitt Method

The relationship between the pulse height X and the kinetic energy T of a heavy ion was determined by Schmitt et al.[42,43] to be of the form

$$T = (a + bM)X + c + dM \tag{13.13}$$

where M is the mass of the ion and a, b, c, and d are constants. The calibration of the detector, i.e., the determination of the constants a, b, c, and d, can be achieved in two ways. The first is an absolute calibration, and the second is a relative one.

Absolute energy calibration is performed with the help of an accelerator. One measures the pulse heights of four monoenergetic beams of ions with known mass. Substitution of the known energy, mass, and pulse height into Eq. 13.13 provides four equations with four unknowns that can be solved for the constants a, b, c, and d.

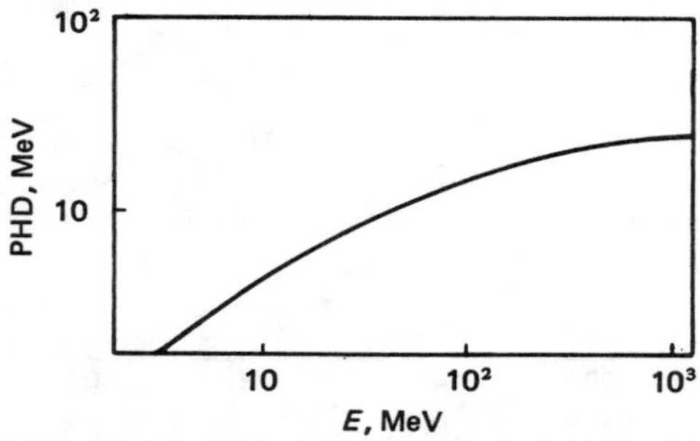

FIG. 13.18 The dependence of the pulse height defect on ion energy (from ref. 33). The numbers correspond, approximately, to uranium ions.

For fission-fragment measurements, a relative calibration is used. The calibration constants of Eq. 13.13 are determined in terms of two pulse heights H and L of a fission-fragment spectrum (Fig. 13.19), where H and L represent pulse height corresponding to the mid-point at three-quarters maximum of the heavy or light fragment peak, respectively. The equations for the constants are

$$a = \frac{a_1}{L - H} \tag{13.14}$$

$$b = \frac{a_2}{L - H} \tag{13.15}$$

$$c = a_3 - aL \tag{13.16}$$

$$d = a_4 - bL \tag{13.17}$$

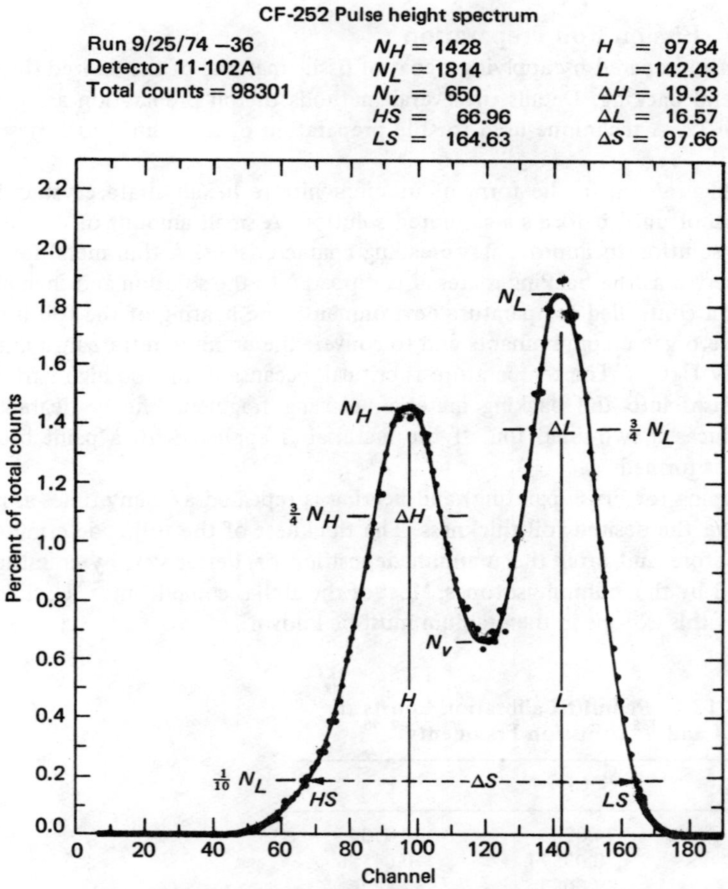

FIG. 13.19 The ^{252}Cf fission-fragment spectrum used for the determination of the detector calibration constants (from ref. 44).

and the constants a_1, a_2, a_3, a_4 for ^{252}Cf and ^{235}U fission fragments are given in Table 13.1.

The constants a_1, a_2, a_3, and a_4 do not depend on the detector. The quality of the detector with respect to energy resolution is determined from a set of criteria developed by Schmitt and Pleasonton[43] and shown in Table 13.2. Figure 13.19 explains the symbols used (see also reference 45).

13.5.3 Calibration Sources

Monoenergetic heavy ions necessary for energy calibration can be provided only by accelerators. Fission fragments, which are heavy ions, cover a wide spectrum of energies (Fig. 13.19). The isotope ^{252}Cf is a very convenient source of fission fragments produced by the spontaneous fission of that isotope. Uranium, plutonium, or thorium fission fragments can only be produced after fission is induced by neutrons; therefore, a reactor or some other intense neutron source is needed.

13.5.4 Fission Foil Preparation

Fission foils are prepared by applying a coat of fissile material of the desired thickness on a thin metal backing. Details of several methods of foil preparation are given in references 29-32. A technique used for the preparation of uranium foils is described here.

Enriched uranium in the form of uranium nitrate hexahydrate crystals is dissolved in ethanol until it forms a saturated solution. A small amount of collodion is added to the solution to improve its spreading characteristics. A thin metal foil—e.g., nickel—that serves as the backing material is dipped into the solution and then heated in an oven in a controlled temperature environment. The heating of the foil is necessary to remove organic contaminants and to convert the uranium nitrate into uranium oxides (mostly U_3O_8). The temperature is critical, because if it is too high part of the uranium diffuses into the backing material, causing fragment energy degradation. Dipping produces a two-sided foil. If the material is applied with a paint brush, a one-sided foil is formed.

The dipping (or brush-painting) and heating is repeated as many times as necessary to achieve the desired foil thickness. The thickness of the foil is determined by weighing it before and after the uranium deposition or, better yet, by counting the alphas emitted by the uranium isotopes. Most of the alphas come from ^{234}U; therefore the fraction of this isotope in the uranium must be known.

TABLE 13.1 Schmitt Calibration Constants for ^{252}Cf and ^{235}U Fission Fragments[42,46]

Constant	^{252}Cf	^{235}U
a_1	24.0203	30.9734
a_2	0.03574	0.04596
a_3	89.6083	87.8626
a_4	0.1370	0.1345

TABLE 13.2 Acceptable Parameters for a ^{252}Cf Fission-Fragment Spectrum

Spectrum parameters	Expected values	Spectrum of Fig. 13.19
NL/NV	2.9	2.79
NH/NV	2.2	2.20
NL/NH	1.3	1.27
$\Delta L/(L-H)$	0.36	0.37
$\Delta H/(L-H)$	0.44	0.43
$(H-HS)/(L-H)$	0.69	0.69
$(LS-L)/(L-H)$	0.48	0.50
$\Delta S/(L-H)$	2.17	2.19

13.6 THE TIME-OF-FLIGHT SPECTROMETER

The time-of-flight (TOF) method, which is also used for the measurement of neutron energy (see Sec. 14.8), has been applied successfully for the determination of the mass of fission fragments and other heavy ions.

The principle of TOF is simple. A beam of ions is directed along a flight path of length L (Fig. 13.20). The time t it takes the ions to travel the distance L determines their speed $V = L/t$. This information, combined with the measurement of the energy of the particle, gives the mass (nonrelativistically):

$$M = \frac{2Et^2}{L^2} \tag{13.18}$$

The errors in determining the mass come from uncertainty in energy, ΔE, in time, Δt, and in length of the flight path, ΔL. The mass resolution is then given by

$$\frac{\Delta M}{M} = \sqrt{\left(\frac{\Delta E}{E}\right)^2 + \left(\frac{2\,\Delta t}{t}\right)^2 + \left(\frac{2\,\Delta L}{L}\right)^2} \tag{13.19}$$

Usually, the system is designed in such a way that $\Delta L/L$ is negligible compared to the other two terms of Eq. 13.19. Assuming that this is the case, consider the sources of uncertainty in energy and time.

The uncertainty $\Delta E/E$ is the resolution of the detector measuring the energy of the ion. At this time (1982), the best energy resolution that can be achieved with silicon surface-barrier detectors is 1.5 to 2 percent. The resolution can be improved

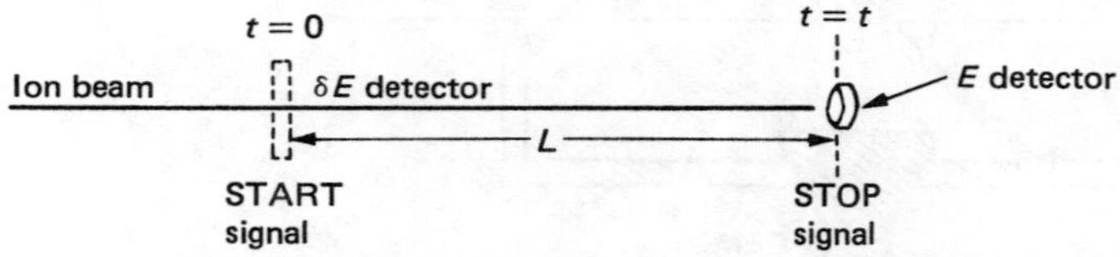

FIG. 13.20 The principle of time-of-flight for the determination of the mass of heavy ions.

with magnetic or electrostatic analyzers (DiIorio and Wehring[47] achieved 0.3 percent energy resolution using an electrostatic analyzer).

The time t it takes the particle to travel the distance L is the difference between a START and a STOP signal (Fig. 13.20). The STOP signal is generated by the detector, which measures the energy of the ion. This detector is usually a surface-barrier detector. The START signal is generated by a transmission counter, also called the δE detector. The ion loses a tiny fraction of its energy going through the START detector.

Several types of δE detectors have been used.[48] Examples are totally depleted surface-barrier detectors,[49,50] thin ($\sim 10^{-3}$ kg/m^2 = 100 μg/cm^2) plastic scintillators,[51] ionization chambers,[52] and secondary-electron emission detectors.[47,53–56] Secondary-electron emission detectors fall into two categories. In the first, the ions traverse a thin foil (e.g., carbon foil 10^{-4} kg/m^2 = 10 μg/cm^2 thick) and generate secondary electrons that are accelerated and focused to strike a scintillator coupled to a photomultiplier tube. In the second category belong the *channel electron multipliers* (CEM) and the *microchannel plates* (MCP).

A CEM is essentially a thin glass tube ($\sim$1 mm diameter) shaped into a spiral, with its inside surface coated with a semiconducting material that is also a good secondary electron emitter. An accelerating field is created in the tube by applying a high voltage along its length. Electrons multiply as they proceed down the tube. Figure 13.21 shows one possible arrangement for the use of a CEM.

An MCP is a glass disk perforated with a large number of small diameter (10–100 μm) holes or channels. Each channel is a glass tube coated with a resistive secondary electron emitting material. If a voltage is applied, each channel acts as an electron multiplier.

The state-of-the-art of δE detector systems is such that $\Delta t < 100$ ps has been achieved and the flight path L can become long enough so that the time $t \approx$ 100–300 ns. Thus, the time resolution of TOF measurements is

$$\frac{\Delta t}{t} \approx \frac{100 \times 10^{-3} \text{ ns}}{200 \text{ ns}} \approx 10^{-2} \approx 0.05\%$$

and the mass resolution (Eq. 13.19) is essentially limited by the energy resolution. The mass measurement is actually the measurement of the mass number A ($\Delta M/M = \Delta A/A$).

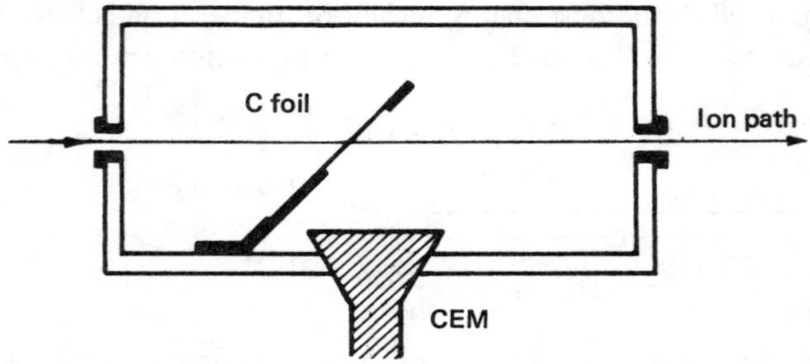

FIG. 13.21 The use of a CEM as a δE detector.

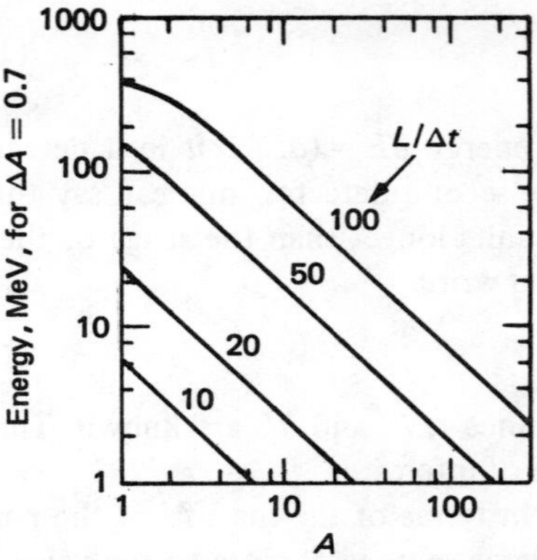

FIG. 13.22 Energy of ion at which adjacent isotopes can be resolved, as a function of A, for different values of $L/\Delta t$ (from ref. 57).

Since A is an integer, the lowest limit for mass resolution is $\Delta A < 1$. Assuming $\Delta A = 0.7$, Fig. 13.22 gives the ion energy necessary for such resolution as a function of A and $L/\Delta t$.[57] For heavy ions, mass resolution as low as $\Delta A = 0.2$ has been reported.[50]

If the mass of the ion is known, the TOF technique can be used to determine the energy of the ion with a resolution much better than with any detector in use today. Indeed, if $\Delta M = 0$, Eq. 13.19 gives

$$\frac{\Delta E}{E} = 2\frac{\Delta t}{t} \approx \frac{200 \times 10^{-3}\ \text{ns}}{200\ \text{ns}} \approx 0.1\%$$

13.7 DETECTOR TELESCOPES ($E\ dE/dx$ DETECTORS)

The TOF method discussed in the previous section measures the energy and the mass of the ion. This section presents a method that identifies the atomic number Z and the mass number A of the particle.[58]

Identification of A and Z is possible by making use of a *detector telescope* consisting of a very thin detector measuring dE/dx and a thick detector that stops the particle. The geometric arrangement is similar to that shown in Fig. 13.20. The particle traverses the thin detector after depositing there an energy equal to $(dE/dx)t$ (where t is the detector thickness), and stops in the "E detector." The total energy of the particle is obtained from the sum of the two detector signals. The product $E\ dE/dx$ can be written, using Eq. 4.2 or 4.32, as

$$E\frac{dE}{dx} = k_1 M Z_{\text{ef}}^2 \ln\left(k_2 \frac{E}{M}\right) \tag{13.20}$$

where Z_{ef} is the effective charge of the ion. Since the logarithmic term changes very slowly with energy, Eq. 13.20 gives a value for MZ_{ef}^2.

Another method, giving better results, is based on the fact that the range of heavy ions is given, over a limited energy range, by an equation of the form

$$R = C \frac{E^b}{MZ_{\text{ef}}^2} \tag{13.21}$$

where b is a constant. If a particle deposits energy $\delta E = (dE/dx)t$ in a detector of thickness t and then deposits energy E in the second detector, one can say that the range of the particle with energy $E + \delta E$ is t units longer than the range of the same particle with energy E. Using Eq. 13.21, one can write

$$t \approx MZ_{\text{ef}}^2 [(E + \delta E)^b - E^b] \tag{13.22}$$

Thus, Eq. 13.22 provides the value of MZ_{ef}^2 since t, E, and δE are known. The constant b is also assumed to be known for the ion of interest.

Equations 13.20 to 13.22 were written in terms of the mass M of the particle. For nonrelativistic particles, M has a nonintegral value very close to the value of A, which in turn is given by an integer. This is fortunate because Z also assumes integral values only, and the product MZ^2 assumes unique values for many particles. For example, for protons, deuterons, tritons, and alphas, the value of MZ^2 is 1, 2, 3, and 16, respectively.

PROBLEMS

13.1 Prove that the maximum energy transfer to an electron as a result of a collision with a particle of mass M and speed $v = c$ is equal to

$$\Delta E_{\text{max}} = 2mc^2 \frac{\beta^2}{1-\beta^2} \left[1 + \left(\frac{m}{M}\right)^2 + 2\frac{m}{M}\frac{1}{\sqrt{1-\beta^2}}\right]$$

13.2 Show that ΔE_{max} of Prob. 13.1 takes the form

$$\Delta E_{\text{max}} = \frac{4mM}{(m+M)^2} T$$

if $T = \frac{1}{2}Mv^2$ (nonrelativistic particle) and $m \ll M$.

13.3 Calculate the energy loss of a 6-MeV alpha particle going through an aluminum foil with thickness equal to one-fourth of the range of the alpha (remember dE/dx is not constant as the particle slows down).

13.4 A monoenergetic beam of 10-MeV alpha particles goes through a nickel foil with a thickness equal to 1/30 of the range of the alphas. What is the width of the emerging beam (in keV)? If the emerging particles are counted by a silicon detector with a resolution of 15 keV, what is the total width of the measured alpha beam (measurement performed in vacuum)?

13.5 What is the thickness of a nickel foil that will cause energy loss equal to 1/50 of the energy of a 10-MeV proton traversing it?

13.6 Show that the backscattering fraction of electrons is given by

$$b = \left[1 + \sqrt{2\pi}\left(\frac{g}{\tau} - 0.5\right)\frac{\sigma}{E'}\right]^{-1}$$

where g, τ, σ, and E' are defined in Fig. 13.9.

13.7 If the stopping power for a heavy ion can be represented by the equation

$$\frac{dE}{dx} = k\sqrt{E} + b$$

where k and b are constants, what is the expression for the range as a function of energy for such ions? What is the energy of the ion as a function of distance traveled?

13.8 If the expression for dE/dx given for Prob. 13.7 is valid, what fraction of the initial kinetic energy of the ion is lost in the first half of its range?

13.9 A thin fission foil was prepared using natural uranium. Calculate the ratio of alpha activities due to the isotopes ^{234}U, ^{235}U, and ^{238}U. The abundance a and half-lives of the three isotopes are:

^{234}U: $a = 0.0057\%$ $T_{1/2} = 2.47 \times 10^5$ y

^{235}U: $a = 0.711\%$ $T_{1/2} = 7.10 \times 10^8$ y

^{238}U: $a = 99.283\%$ $T_{1/2} = 4.51 \times 10^9$ y

Assume that every decay gives off an alpha; i.e., neglect decay by spontaneous fission.

13.10 Verify the curves of Fig. 13.22.

BIBLIOGRAPHY

Bertolini, G., and Coche, A. (eds.), *Semiconductor Detectors*, Wiley Interscience, New York, 1968.
Evans, R. D., *The Atomic Nucleus*, McGraw-Hill, New York, 1972.
Segré, E., *Nuclei and Particles*, W.A. Benjamin, New York, 1965.
Siegbahn, K. (ed.), *Alpha, Beta and Gamma-Ray Spectroscopy*, North Holland Publishing Co., Amsterdam, 1965, vols. 1 and 2.

REFERENCES

1 Landau, W., *J. Phys. USSR* 8:201 (1944).
2 Symon, K., thesis, Harvard 1958, "Experts in High Energy Particles," Prentice-Hall, Englewood Cliffs, N.J., 1952.
3 Vavilov, P. V., *JETP* 5:749 (1957).
4 Maccabee, H. D., Raju, M. R., and Tobias, L. A., *IEEE Nucl. Sci.* 13(3):176 (1966).
5 Blunk, O., and Leisegang, S., *Z. Physik* **128**:500 (1950).
6 Tschalar, C., *Nucl. Instr. Meth.* **61**:141 (1968).
7 Tschalar, C., *Nucl. Instr. Meth.* **64**:237 (1968).
8 Clarke, N. M., *Nucl. Inst. Meth.* **96**:497 (1971).
9 Ispirian, K. A., Murgarian, A. T., and Zuerev, A. M., *Nucl. Instr. Meth.* **117**:125 (1974).
10 Ermilova, V. C., Kotenko, L. P., and Merzon, G. I. *Nucl. Instr. Meth.* **145**:444 (1977).
11 Skyrme, D. J., *Nucl. Instr. Meth.* **57**:61 (1967).
12 Bohr, N., *Phil. Mag.* **30**:581 (1915).
13 Livingston, M. S., and Bethe, H., *Rev. Mod. Phys.* **9**:245 (1937).
14 Titeica, S., *Bull. Soc. Roumaine Phys.* **38**:81 (1939).
15 Comfort, J. R., Decker, J. F., Lynk, E. T., Scully, M. D., and Quinton, A. R., *Phys. Rev.* **150**: 249 (1966).

16 Strittmatter, R. B., and Wehring, B. W., *Nucl. Instr. Meth.* **135**:173 (1976).
17 Lonergan, J. A., Jupiter, C. P., and Merkel, G., *J. Appl. Phys.* **41**:678 (1970).
18 Berger, M. J., and Seltzer, S. M., "ETRAN Monte Carlo Code System for Electron and Photon Transport Through Extended Media," NBS-9836 and NBS-9837 (1968).
19 Rogers, P. C., and Gordon, G. E., *Nucl. Instr. Meth.* **37**:259 (1965).
20 Tsoulfanidis, N., Wehring, B. W., and Wyman, M. E., *Nucl. Instr. Meth.* **73**:98 (1969).
21 Wohn, F. K., Clifford, J. R., Carlson, G. H., and Talbert, W., Jr., *Nucl. Instr. Meth.* **101**:343 (1972).
22 Schupferling, H. M., *Nucl. Instr. Meth.* **123**:67 (1975).
23 Dakuba, T., and Gilboy, W. B., *Nucl. Instr. Meth.* **150**:479 (1978).
24 Tabata, T., Ito, R., and Okabe, S., *Nucl. Instr. Meth.* **94**:509 (1971).
25 Kuzminikh, V. A., and Vorobiev, S. A., *Nucl. Instr. Meth.* **129**:561 (1975).
26 Tsoulfanidis, N., Wehring, B. W., and Wyman, M. E., *Nucl. Sci. Eng.* **43**:42 (1971).
27 Meyer, O., and Langmann, H. J., *Nucl. Instr. Meth.* **39**:119 (1966).
28 Segré, E., *Nuclei and Particles*, Benjamin, New York, 1965, p. 348.
29 Van de Eijk, W., Oldenhof, W., and Zehner, W., *Nucl. Instr. Meth.* **112**:343 (1973).
30 Lowenthal, G. C., and Wyllie, H. A., *Nucl. Instr. Meth.* **112**:353 (1973).
31 Dobrilovic, L., and Simovic, M., *Nucl. Instr. Meth.* **112**:359 (1973).
32 Adair, H. L., and Kuehn, P. R., *Nucl. Instr. Meth.* **114**:327 (1974).
33 Wilkins, B. D., Fluss, M. J., Kaufman, S. B., Cross, C. E., and Steinberg, E. P., *Nucl. Instr. Meth.* **92**:381 (1971).
34 Sullivan, W. J., and Wehring, B. W., *Nucl. Instr. Meth.* **116**:29 (1974).
35 Caywood, J. M., Mead, C. A., and Mayer, J. W., *Nucl. Instr. Meth.* **79**:329 (1970).
36 Haines, E. L., and Whitehead, A. B., *Rev. Sci. Instr.* **37**:190 (1966).
37 Finch, E. C., and Rodgers, A. L., *Nucl. Instr. Meth.* **113**:29 (1973).
38 Finch, E. C., *Nucl. Instr. Meth.* **113**:41 (1973).
39 Lindhard, J., Scharff, M., and Schiott, H. E., *Fys. Medd.* **33**:1 (1963).
40 Moak, C. D., Dabbs, J. W. T., and Walker, W. W., *Rev. Sci. Inst.* **1131**:37 (1966).
41 Britt, H. C., and Wegner, H. E., *Rev. Sci. Instr.* **34**:274 (1963).
42 Schmitt, H. W., Kiker, W. E., and Williams, C. W., *Phys. Rev.* **137**:B837 (1965).
43 Schmitt, H. W., and Pleasonton, F., *Nucl. Instr. Meth.* **40**:204 (1966).
44 Bucher, R. G., Ph.D. thesis, University of Illinois, 1975.
45 Long, A. B., Wehring, B. W., and Wyman, M. E., *Rev. Sci. Inst.* **1566**:39 (1968).
46 Schmitt, H. W., Gibson, W. M., Neiler, J. H., Walter, F. J., and Thomas, T. D., *Proceedings of IAEA Conf. on the Physics and Chemistry of Fission,* **1**:531 (1965).
47 DiIorio, G. J., and Wehring, B. W., *Nucl. Instr. Meth.* **147**:487 (1977).
48 Betts, R. R., *Nucl. Instr. Meth.* **162**:531 (1979).
49 Pleyer, H., Kohlmeyer, B., Schneider, W. F. W., and Bock, R., *Nucl. Instr. Meth.* **96**:363 (1971).
50 Zediman, B., Henning, W., and Kovar, D. G., *Nucl. Instr. Meth.* **118**:361 (1974).
51 Gelbke, C. K., Hildenbrabd, K. D., and Bock, R., *Nucl. Instr. Meth.* **95**:397 (1971).
52 Barrette, J., Braun-Munzinger, P., and Gelbke, C. K., *Nucl. Instr. Meth.* **126**:181 (1975).
53 Dietz, E., Bass, R., Reiter, A., Friedland, V., and Hubert, B., *Nucl. Instr. Meth.* **97**:581 (1971).
54 Schneider, W. F. W., Kohlmeyer, B., and Bock, R., *Nucl. Instr. Meth.* **123**:93 (1975).
55 Pfeffer, W., Kohlmeyer, B., and Schneider, W. F. W., *Nucl. Instr. Meth.* **107**:121 (1973).
56 Wiza, J. L., *Nucl. Instr. Meth.* **162**:587 (1979).
57 Butler, G. W., Poskanzer, A. M., and Landis, D. A., *Nucl. Instr. Meth.* **89**:189 (1970).
58 Goulding, F. S., and Harvey, B. G., *Ann. Rev. Nucl. Sci.* **25**:167 (1975).
59 Sykes, D. A., and Harris, S. J., *Nucl. Instr. Meth.* **101**:423 (1972).

14

neutron detection and spectroscopy

14.1 INTRODUCTION

Since neutrons do not directly ionize atoms, they are detected "indirectly" upon producing a charged particle or a photon, which is then recorded with the help of an appropriate detector. The charged particle or the photon is the result of a neutron interaction with a nucleus. If the mechanism of the interaction is known, information about the neutron can be extracted by studying the products of the reaction. Many types of interactions are used, divided into absorptive and scattering reactions.

Absorptive reactions are (n, α), (n, p), (n, γ), or $(n$, fission). In the case of an (n, γ) reaction, the neutron may be detected through the interactions of the gamma emitted at the time of the capture, or it may be detected through the radiation emitted by the radioisotope produced after the neutron is captured. The radioisotope may emit β^- or β^+ or γ or a combination of them. By counting the activity of the isotope, information is obtained about the neutron flux that produced it. This is called the activation method. If the reaction is fission, two fission fragments are emitted; being heavy charged particles, these are detected easily.

The main scattering reaction used is neutron-proton collision, called the *proton-recoil method.* The knocked-out proton is the particle recorded.

With the exception of the proton-recoil method, which functions for fast neutrons only ($E_n > 1.0$ keV), all the other interactions can be used with neutrons of any energy. However, at every neutron energy, one method may be better than another. The best method will be selected based on the neutron energy, the purpose of the experiment (is number or energy of neutrons measured, or both?), and the physical constraints of the measurement (e.g., inside a reactor core or outside).

This chapter discusses in detail all the neutron detection methods mentioned above, as well as the Bragg crystal spectrometer and the time-of-flight method. Other specialized neutron detectors, such as self-powered neutron detectors (SPND), thermoluminescent dosimeters (TLD), and fission track recorders, are described in Chap. 17.

14.2 NEUTRON DETECTION BY $(n$, CHARGED PARTICLE) REACTION

There are many nuclear reactions of the type (n, charged particle) used for neutron detection. In general, endothermic reactions are used for fast neutrons, and exothermic ones for thermal neutrons. The endothermic reactions will be discussed in Sec. 14.6.

The mose useful exothermic reactions are listed in Table 14.1, along with their Q values and the value of the cross section for thermal neutrons.

TABLE 14.1 Exothermic Reactions Used for Neutron Detection

Reaction	Charged particles produced	Q Value (MeV)	σ (b) for $E_n = 0.025$ eV
${}^{10}_{5}B(n, \alpha){}^{7}_{3}Li$	α, ${}^{7}Li$	2.78	3840
${}^{6}_{3}Li(n, \alpha){}^{3}_{1}H$	α, ${}^{3}H$	4.78	937
${}^{3}_{2}He(n, p){}^{3}_{1}H$	p, ${}^{3}H$	0.765	5400

The charged particles from any one of the reactions of Table 14.1 share an amount of kinetic energy equal to $Q + E_n$, where E_n is the neutron kinetic energy. The large Q values make detection of the products very easy, regardless of the value of E_n, but at the same time make measurement of the energy of slow neutrons impossible. The neutron energy would be measured from the pulse height, which is proportional to $Q + E_n$. But, if the pulse corresponds to energy of the order of MeV (because of the Q value), a small change in E_n will produce a variation in the pulse that is undetectable. For example, the fractional change of an 1-MeV pulse due to 1 keV change in neutron kinetic energy is $(1.001 - 1.000)/1.000 = 0.1$ percent, which is less than the best energy resolution of alpha-particle detectors. Therefore, the measurement of neutron energy is possible only when E_n amounts to a considerable fraction of the Q value.

14.2.1 The BF_3 Counter

The (n, α) reaction with ${}^{10}_{5}B$ is probably the most useful reaction for the detection of thermal neutrons, because

1 The reaction cross section is large.
2 The energy dependence of the cross section is of the $1/v$ type.
3 ${}^{10}_{5}B$ is a constituent of the compound BF_3, which may be used as the gas of a proportional gas counter.

The BF_3 counter is a proportional counter filled with BF_3 gas, usually enriched to more than 90 percent in ${}^{10}B$ (about 20 percent of natural boron is ${}^{10}B$; the rest is ${}^{11}B$). The BF_3 counter detects the alpha and the lithium particles produced by the reaction

$${}^{10}_{5}B + {}^{1}_{0}n \rightarrow {}^{4}_{2}He + {}^{7}_{3}Li + 2.78 \text{ MeV}$$

With thermal neutrons, the ${}^{7}Li$ nucleus is left in an excited state about 96 percent of the time. In that case, the Q value of the reaction is 2.30 MeV and the ${}^{7}Li$ nucleus goes to the ground state by emitting a gamma with energy equal to $2.78 - 2.30 = 0.480$ MeV. This photon is also sometimes used for the detection of the neutron.

The relationship between counting rate and neutron flux is derived as follows. Let

$n(E)\,dE$ = number of neutrons/m^3 with kinetic energy between E and $E + dE$

$\phi(E)\,dE = v(E)n(E)\,dE$ = neutron flux consisting of neutrons with kinetic† energy between E and $E + dE$

$v(E)$ = neutron speed for energy E (m/s)

E_m = upper limit of neutron energy considered

N = number of ^{10}B atoms per unit volume

V = volume of the counter

$\sigma(E) = \sigma(v)$ = cross section of the (n, α) reaction for neutron energy E

Assuming that the neutron flux is uniform over the detector volume, the reaction rate R is given by

$$R \text{ (reactions/s)} = VN \int_0^{E_m} \sigma(E)\phi(E)\,dE \tag{14.1}$$

The ^{10}B cross section has a $1/v$ dependence‡ over a wide range of neutron energies; i.e., it can be written as

$$\sigma(E) = \sigma_0 \frac{v}{v(E)} = \sigma_0 \sqrt{\frac{E_0}{E}} \tag{14.2}$$

where σ_0 is the cross section at some known speed $v_0 = \sqrt{2E_0/M}$, and M is the neutron mass. If Eq. 14.2 is substituted into Eq. 14.1, the reaction takes the form

$$R = NV\sigma_0 v_0 n \text{ (reactions/s)} \tag{14.3}$$

where n is the total number of neutrons per unit volume, or

$$n = \int_0^{E_m} n(E)\,dE \tag{14.4}$$

Equation 14.3 shows that the reaction is proportional to the total neutron density. BF_3 counters are most frequently used for the detection of thermal neutrons, for which one can calculate an average neutron speed $\bar{v}$ given by

$$\bar{v} = \frac{\int_0^{E_m} v(E)n(E)\,dE}{\int_0^{E_m} n(E)\,dE} \tag{14.5}$$

†In Chap. 3, the symbol T was used to denote kinetic energy. That was necessary because the discussion involved kinetic and total energy. In this chapter, E is used for the kinetic energy of the neutron.

‡The ^{11}B cross section is quite different. Equation 14.2 represents only ^{10}B cross section.

and a total flux ϕ, given by

$$\phi = \int_0^{E_m} \phi(E)\, dE = \int_0^{E_m} v(E) n(E)\, dE \tag{14.6}$$

Under these conditions, Eq. 14.4 takes the form

$$R = NV\sigma_0 \frac{v_0}{\bar{v}} \phi \tag{14.7}$$

Thus, if $\bar{v}$ is known, the reaction rate is proportional to the total neutron flux ϕ. For example, for the commonly encountered Maxwell–Boltzmann distribution of thermal neutrons, $\bar{v} = 2v_p/\sqrt{\pi}$, where v_p is the most probable neutron speed.

The derivation of Eqs. 14.3 and 14.7 was based on the assumption that the neutron flux is uniform over the volume of the counter. A measure of the flux uniformity is the value of the factor $\exp(-\Sigma_t l)$, where

Σ_t = total macroscopic cross section for the gas of the counter, averaged over all the neutron energies present

l = a characteristic dimension of the counter (usually the diameter or the length of a cylindrical counter)

If $\exp(-\Sigma_t l) \approx 1$, the flux may be taken as uniform over the reactor volume.

EXAMPLE 14.1

Consider a BF_3 counter with a diameter 0.05 m ($\simeq$2 in) and length 0.30 m ($\simeq$ 12 in) filled with BF_3 gas, 96 percent enriched to ^{10}B, at a pressure of 1 atm and used for the detection of 0.0253 eV neutrons. Should the user take into account flux depression in the counter?

ANSWER

The factor $\exp(-\Sigma_t l)$ should be calculated. For the worst case, consider l = length of the counter. The total macroscopic cross section is

$$\Sigma_t = N(BF_3)[0.96\sigma(^{10}B) + 0.04\sigma(^{11}B) + 3\sigma(F)]$$

$$\approx \frac{0.6023}{22{,}400}(0.96)(3840) = 9.91\ \text{m}^{-1} = 0.0991\ \text{cm}^{-1}$$

For the calculation of Σ_t, the cross sections of ^{11}B and F were neglected because they are much smaller than that of ^{10}B. The "depression factor" is

$$\exp(-\Sigma_t l) = \exp[(-9.91)(0.30)] = \exp(-2.97) = 0.05$$

If l is the diameter of the counter, then

$$\exp(-\Sigma_t l) = \exp[(-9.91)(0.05)] = \exp(-0.496) = 0.61$$

One concludes that flux depression is considerable in this case, and the flux cannot be taken as uniform over the counter volume. If the BF_3 counter is used for the detection of a polyenergetic neutron spectrum, instead of a monoenergetic neutron source, average cross sections should be used for the calculation (Sec. 4.9.4 explains how average cross sections are obtained).

As a first approximation, the efficiency of a BF_3 counter is equal to

$$\epsilon = \frac{\Sigma_a}{\Sigma_t} [1 - \exp(-\Sigma_t l)] [\exp(-\Sigma_t^w t_w)] \tag{14.8}$$

where Σ_t and Σ_a are total and absorptive neutron macroscopic cross sections, respectively, for BF_3, and l is the dimension of the counter parallel to the direction of the neutron beam. Σ_t^w and t_w are the total macroscopic cross section and thickness, respectively, for the material of which the wall or the end window of the counter are made. Equation 14.8 was derived under the assumptions that

1 All neutrons travel the same distance inside the counter (parallel beam).
2 Every neutron interaction in the wall or the end window of the counter removes the neutron from the beam.

EXAMPLE 14.2

What is the efficiency of a BF_3 counter enriched to 96 percent in ^{10}B, 0.04 m (1.57 in) in diameter, 0.30 m (~12 in) long, for a parallel beam of 1-eV neutrons? The BF_3 pressure in the counter is 53,329 Pa (40 cmHg). Consider two cases:

(a) The beam is directed parallel to the axis of the counter.
(b) The beam is perpendicular to the axis of the counter.

Assume that the wall and the window of the counter are made of aluminum and are 2 mm thick. Take the total neutron cross section for Al at 1 eV to be 1.5 b.

ANSWER

Equation 14.8 will be used. At 1 eV for BF_3 enriched to 96 percent to ^{10}B, $\Sigma_a \approx \Sigma_t \approx \Sigma_a(^{10}B)$. To find the microscopic cross section at 1 eV, use Eq. 14.2 and the value of $\sigma_a = 3840$ b at $E_n = 0.025$ eV.

$$\sigma_a(1\text{ eV}) = 3840 \frac{v_0(0.025\text{ eV})}{v(1\text{ eV})} = 3840\sqrt{\frac{0.025}{1}} = 607\text{ b}$$

$$\Sigma_a = 0.96\left(\frac{53{,}329}{101{,}325}\right)\left(\frac{0.6023}{22{,}400}\right)(607) = 0.82\text{ m}^{-1} = 0.0082\text{ cm}^{-1}$$

For aluminum:

$$\Sigma_t^w t_w = 1.5(2.7)\left(\frac{0.6023}{27}\right)(0.2) = 0.018$$

(a) If the beam is parallel to the counter axis, $l = 0.30$ m and

$$\epsilon = [1 - \exp(-\Sigma_a l)] [\exp(-\Sigma_t^w t_w)] = (0.22)(0.982) = 0.216 = 21.6\%$$

(b) If the beam is perpendicular to the axis, all the neutrons do not travel the same distance inside the cylindrical counter. Assuming that the incident neutrons form a narrow beam that hits the counter at the center, l = diameter = 0.04 m. Then,

$$\epsilon = \{1 - \exp[(-0.82)(0.04)]\}(0.982) = (0.032)(0.982) = 0.032 = 3.2\%$$

The specifications of commercial BF_3 counters consist of sensitivity, dimensions, composition of the filling gas, operating voltage, and maximum operating temperature.

The *sensitivity* S is defined as the ratio

$$S = \frac{\text{true net counting rate}}{\text{neutron flux}} = \frac{r}{\phi} \tag{14.9}$$

and is given in terms of counts/s per neutrons/m$^2\cdot$s. The parameters affecting the sensitivity can be seen by noting that

$$r = \epsilon_p R = g - b \tag{14.10}$$

where ϵ_p = efficiency of the counter for detection of the charged particles produced
b = background counting rate
g = gross counting rate
R = reaction rate given by Eq. 14.7

Since the charged particles are generated inside the volume of the detector, the efficiency ϵ_p is practically equal to 1. Also, the background rate may be made negligible because the pulses produced by the charged particles are well above the electronic noise. Thus, a proper discriminator level may be set to eliminate almost all the background. Under these conditions, the equation for the sensitivity becomes

$$S = \frac{R}{\phi} = NV\sigma_0 \frac{v_0}{\bar{v}} \tag{14.11}$$

Equation 14.11 indicates that for a certain neutron spectrum, the sensitivity is proportional to boron density (i.e., pressure of the BF_3 gas) and volume of the counter.

The number of boron atoms decreases with exposure, and so does the sensitivity. The decrease is expressed by a factor having the form

$$\exp(-\sigma_a \phi t)$$

where ϕt in neutrons/m^2 is the fluence to which the counter was exposed. Since the average value of σ_a for thermal neutrons reacting with ^{10}B is of the order of 10^{-25} m^2 (1000 b), the fluence necessary to cause an appreciable change in sensitivity is of the order of 10^{25} neutrons/m^2.

Typical specifications of commercial counters are the following.

Sensitivity: <1 to 0.005 counts/s per neutrons/m$^2\cdot$s (50 counts/s per neutrons/cm$^2\cdot$s).
Dimensions: Almost any dimensions.
Pressure of BF_3: From a little less than 1 to about 2 atm (202 kPa). An increase in pressure requires an increase in the operating voltage.

Operating voltage: BF_3 counters show an almost flat plateau (see Chap. 5) extending over 1000 V or more. Typical operating voltages range from 1000 to 3000 V.
Temperature: Maximum operating temperature is about 100°C.

14.2.2 Boron-Lined Counters

Boron-lined counters are gas-filled proportional counters that employ the same reaction as the BF_3 counter, except that the ^{10}B is coated on the walls of the counter. Since the (n, α) reactions take place in a thin layer close to the wall (Fig. 14.1), only one of the two particles has a chance of entering the sensitive volume of the counter and producing a pulse. The other stops in the wall. The sensitivity increases with the thickness of the ^{10}B coating. That thickness, however, cannot exceed the range of the alphas produced. The advantage of boron-lined over BF_3 counters is the flexibility of using a gas more appropriate than BF_3. Then the operating voltage is less and the counter is less sensitive to gamma rays. For this reason, boron-lined counters may be used to detect neutrons in intense gamma fields.

Both BF_3 and boron-lined counters are mainly used for the measurement of the number (not energy) of thermal neutrons.

14.2.3 ^{6}Li Counters

Neutron detection by ^{6}Li is based on the reaction

$$^6_3\text{Li} + {}^1_0n \rightarrow {}^4_2\text{He} + {}^3_1\text{H} + 4.78 \text{ MeV}$$

The cross section for this reaction is of the $1/v$ type up to 10 keV, with a value of 937 b at 0.025 eV. Lithium is used either as LiF or as the component of a scintillator.

A neutron spectrometer based on LiF[1] consists of a thin slice of ^{6}LiF (~1.50 kg/m^2 = 150 g/cm^2) sandwiched between two surface-barrier silicon detectors. When neutrons strike the LiF, charged-particle pairs (^{4}He–^{3}H) are produced and are detected simultaneously by the two detectors. The pulses from the detectors are amplified and then summed to produce a single pulse, which is proportional to the energy of the neutron plus the Q value of the reaction. The energy resolution of such spectrometers is about 400 keV for thermal neutrons. For fast neutrons, the resolution varies with neutron energy.

There are many inorganic scintillators based on lithium. ^{6}LiI(Eu) has been used for neutron energy measurements from 1 to 14 MeV with 10 percent energy resolution.[2] It has good efficiency for low-energy neutrons, but activation of iodine creates some problems. The most widely used lithium scintillator was developed by Ginther

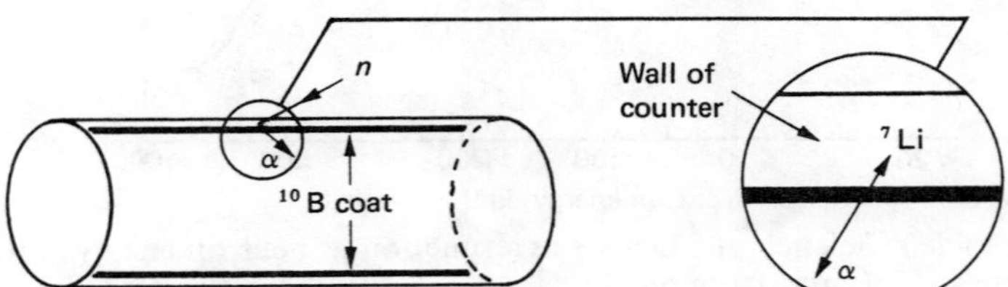

FIG. 14.1 A ^{10}B-lined counter.

and Schulman[3] and Voitovetskii et al.[4] It is a cerium-activated scintillating glass containing Li_2O. The proportion of the cerium activator affects the efficiency of luminescence. A series of measurements of many properties of commercially available glasses has been reported recently by Spowart.[5,6] Today one can buy these glasses in a large variety of thickness (0.5 to 25 mm), sizes (up to 125 mm in diameter), Li content (up to 11 percent), and ^{6}Li enrichment (up to 95 percent). The efficiency of ^{6}Li glass as a function of neutron energy is shown in Fig. 14.2.

To increase the efficiency, ^{6}Li glass scintillators with thickness about 13 mm and diameter 110 mm have been optically coupled to one or more photomultiplier tubes through light pipes.[8,9] To avoid moderation of the incident neutrons, the light pipe should not contain hydrogenous material. One problem with such thick scintillators is considerable scattering of the incident neutrons. The scattered neutrons add an exponential tail to the primary neutron signal, a tail that should be included in the time resolution function of the instrument in time-of-flight measurements.

A different type of arrangement, using a ^{6}LiI(Eu) scintillator to detect the neutrons after they are moderated, is the *Bonner ball*.[10,11] Using the Bonner ball, neutrons are detected by a ^{6}LiI(Eu) scintillator placed at the center of polyethylene spheres with sizes ranging from 51 to 305 mm (2 to 12 in) in diameter.

Using the Bonner ball, the neutron energy is determined based on the difference in efficiency for the spheres of different sizes. For neutrons of a certain energy, the size of the sphere affects the degree of moderation, which in turn affects the detection efficiency of the lithium scintillator. Using monoenergetic neutron sources of known strength with polyethylene spheres of different sizes, a calibration table is obtained consisting of the counting rate as a function of neutron energy and ball diameter. This calibration table is then used for the determination of the number and energy of an unknown neutron spectrum. The Bonner ball is particularly useful for the measurement of continuous energy spectra.

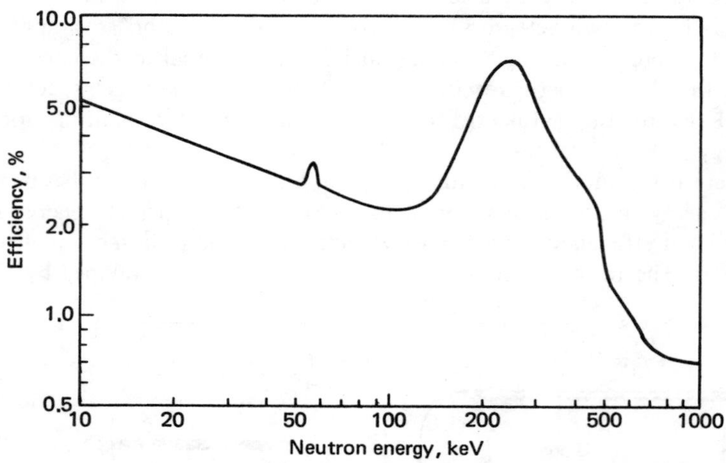

FIG. 14.2 The efficiency of ^{6}Li glass as a function of neutron energy, with glass thickness 12.7 mm (from ref. 7).

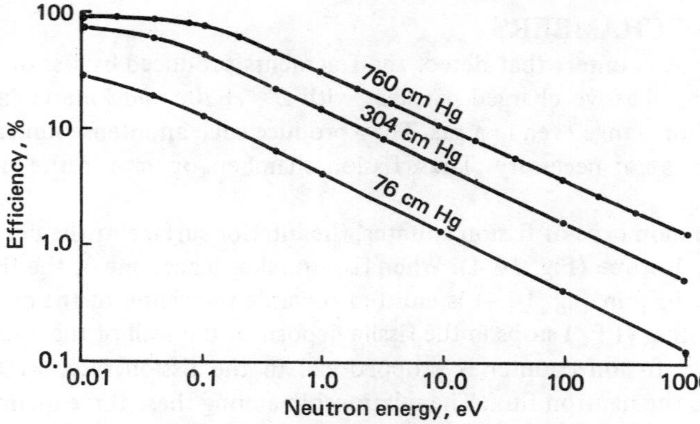

FIG. 14.3 Efficiency of ^{3}He-filled proportional counter as a function of neutron energy.

14.2.4 ^{3}He Counters

Neutron detection by ^{3}He is based on the reaction

$$^3_2\text{He} + ^1_0n \rightarrow ^1_1\text{H} + ^3_1\text{H} + 765\ \text{keV}$$

The cross section for this reaction is quite high for thermal neutrons (5400 b at 0.025 eV) and varies as $1/v$ from 0.001 eV to 0.04 eV.

One type of helium neutron spectrometer consists of two surface-barrier silicon detectors facing each other, with the space between them filled with helium at a pressure of a few atmospheres (this is similar to the ^{6}LiF spectrometer described in Sec. 14.2.3). The energy resolution is about 100 keV for thermal neutrons. For fast neutrons, the resolution varies with neutron energy.

Proportional counters filled with ^{3}He are widely used, especially in time-of-flight measurements (see Sec. 14.8). The efficiency of the counter can be increased by increasing the pressure. Figure 14.3 shows the efficiency of a ^{3}He counter as a function of neutron energy for three different pressures. One of the problems of ^{3}He counters is the *wall effect.* If the reaction takes place closely to the wall of the counter, there is a high probability that only a fraction of the charged-particle energy will be deposited in the counter. As a result, smaller size pulses are produced which do not come under the main peak. There are two ways to overcome this effect. One is to use a large-diameter counter, in which case the fraction of reactions occurring near the wall is smaller. The other is to increase the stopping power of the gas. Increase in stopping power is achieved either by increasing the pressure of the gas or by adding a small fraction of another gas, such as krypton, with a higher stopping power. Increase in stopping power is accompanied, however, by an increase in gamma sensitivity and a decrease in pulse risetime. Fast risetime is important for time-of-flight measurements.

The pressure of the gas and the operating voltage are higher in ^{3}He than in BF_3 counters. The pressure of the ^{3}He is usually between 404 and 1010 kPa (4–10 atm), and the operating voltage is 3000–5000 V.

14.3 FISSION CHAMBERS

Fission chambers are gas counters that detect the fragments produced by fission. The fission fragments, being massive charged particles with $Z \approx +20e$ and kinetic energy 60-100 MeV, have a short range even in a gas. They produce such an intense ionization that gas multiplication is not necessary. Thus, fission chambers operate in the ionization region.

In the most common type of fission counter, the interior surface of the detector is coated with a fissile isotope (Fig. 14.4). When fission takes place, one of the fission fragments (denoted as FF_1 in Fig. 14.4) is emitted towards the center of the counter and is detected. The other (FF_2) stops in the fissile deposit or the wall of the counter. The counting rate of a fission counter is proportional to the fission rate, which in turn is proportional to the neutron flux. The relationship among these three quantities is similar to the equations given for the BF_3 counter.

For relative measurements, the thickness of the fissile material coating is not very critical except that it should be less than the range of the fission fragments. For absolute measurements, however—i.e., measurements for which every fission should be detected—at least one fission fragment from each fission should produce a recorded pulse. To achieve this, the thickness of the coating must be limited so that fission fragments being produced anywhere in the layer of the fissile material generate a pulse larger than that of alphas, betas, or gammas, which are always present.† Pulses from gammas present a problem only when the counter is used in an extremely intense gamma field. Pulses from alphas, however, are always present and should be discriminated from those of fission fragments. Fortunately, the difference between the ionization produced by alphas and that produced by fission fragments is so large that such discrimination is easy. Figure 14.5 shows a typical integral pulse-height spectrum. Notice that the alpha pulses start at such a low level that a discriminator level can be set to cut them off.

Fission counters may be used for detection of either fast or thermal neutrons. If the counter wall is coated with ^{235}U, essentially only thermal neutrons are detected because the ^{235}U fission cross section for thermal neutrons is about 500 times higher than that of fast neutrons. If the counter is coated with ^{238}U or ^{232}Th, only fast neutrons with kinetic energy greater than 1 MeV are detected because the fission cross section of these isotopes has a threshold at about that energy.

Fission counters are used for detection of the number and not the energy of

†All fissile isotopes emit alpha particles. Betas and gammas may be emitted either by the fissile isotope or by fission fragments.

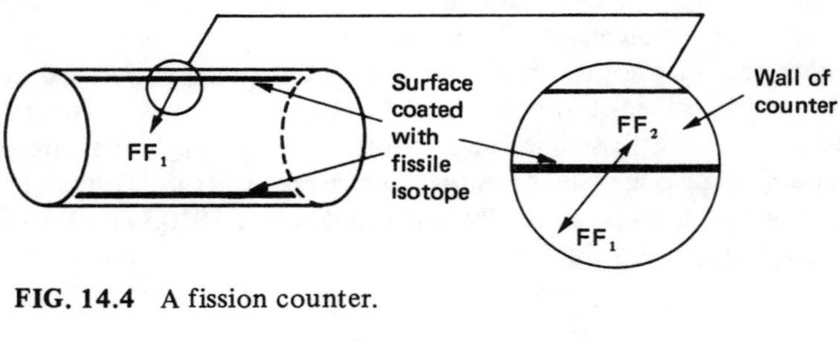

FIG. 14.4 A fission counter.

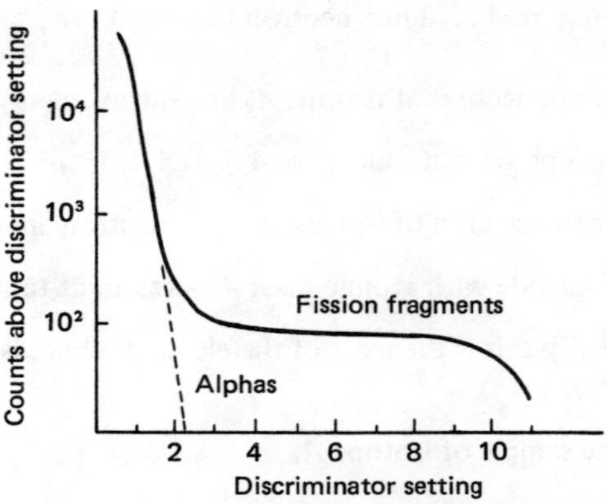

FIG. 14.5 An integral pulse height spectrum taken with a fission counter.

neutrons. They can be used, however, for differentiation of thermal and fast neutron flux by using a combination of ^{235}U- and ^{238}U-coated counters (see also Sec. 14.6).

The sensitivity of a fission counter decreases with exposure because of the depletion of the fissile isotope (the same phenomenon as depletion of boron atoms–see Sec. 14.2.1). The decrease in sensitivity may be halted, to a certain extent, if the counter wall is coated with a mixture of fertile and fissile materials. One such combination is 90 percent ^{234}U and 10 percent ^{235}U. The ^{235}U is partially replenished with new atoms produced by neutron capture in ^{234}U. A ^{235}U–^{238}U combination will have a similar effect, thanks to ^{239}Pu produced as a result of neutron capture in ^{238}U.

Fission counters are used extensively for both out-of-core and in-core measurements of neutron flux in nuclear reactors. In out-of-core situations, they monitor the neutron population during the early stages of power ascension when the neutron flux level is very low. For in-core measurements, fission counters are used for flux mapping (and consequently determination of the core power distribution). They are manufactured as long thin cylindrical probes that can be driven in and out of the core with the reactor in power. Typical commercial fission counters for in-core use have diameters of about 1.5 mm (0.06 in), use uranium enriched to at least 90 percent in ^{235}U as the sensitive material, and can be used to measure neutron fluxes up to 10^{18} neutrons/m^2·s (10^{14} neutrons/cm^2·s).

Another method of measuring fission rates is by using fission track detectors, as discussed in Sec. 17.5.

14.4 NEUTRON DETECTION BY FOIL ACTIVATION

Neutron detection by foil activation is based on the creation of a radioisotope by neutron capture, and subsequent counting of the radiation emitted by that radioisotope. Foil activation is important not only for neutron flux measurements but also for neutron activation analysis, which is the subject of Chap. 15. This section presents the basic equations involved.

Consider a target of mass m being irradiated in a neutron flux $\phi(E)$, where

$\sigma_i(E)$ = neutron absorption cross section of isotope A_i at neutron energy E

λ_{i+1} = decay constant of isotope with atomic mass $A + 1(A_{i+1})$

$\sigma_{i+1}(E)$ = neutron absorption cross section of isotope A_{i+1} at neutron energy E

$N_i(t)$ = number of atoms of nuclide with atomic mass A_i, present at time t

m = mass of target (normally this is the mass of the element whose isotope A_i captures the neutron)

a_i = weight fraction in the sample of isotope A_i

As a result of neutron absorption, the following processes take place.

1 Target atoms of atomic mass A_i are destroyed.
2 Atoms with atomic mass $A + 1$ are produced.
3 Atoms of type A_{i+1} decay.
4 Atoms of type A_{i+1} may be destroyed by absorbing a neutron.

For the target isotope $({}^{A_i}_{Z_i}X)$, the reaction involved is

$${}^{A_i}_{Z_i}X + n \rightarrow {}^{A_{i+1}}_{Z_i}X$$

The destruction of these atoms proceeds according to the equation

$$-\frac{dN_i(t)}{dt} = N_i(t) \int_0^\infty dE\, \sigma_i(E)\phi(E) \tag{14.12}$$

In Eq. 14.12 and all others in this section, it is assumed that the presence of the target does not disturb the flux; i.e., the foil does not cause depression of the flux. Corrections that take into account foil self-absorption can be found in Beckurts and Wirtz (Chap. 11) and in reference 12. The integral over energy in Eq. 14.12 is usually expressed as

$$\int_0^\infty dE\, \sigma_i(E)\phi(E) = \bar{\sigma}_i \int_0^\infty \phi(E)\, dE = \bar{\sigma}_i\phi = \sigma_i\phi \tag{14.13}$$

That is, an average cross section is used, even though the bar that indicates averaging is normally dropped. From now on, Eq. 14.13 will be used without the bar, but the reader should keep in mind that σ is an average over the neutron energy spectrum.

The solution of Eq. 14.12 is, using Eq. 14.13,

$$N_i(t) = N_i(0)e^{-\sigma_i\phi t} \tag{14.14}$$

where

$$N_i(0) = \frac{a_i m N_A}{A_i} = \text{number of atoms of isotope } A_i \text{ at } t = 0$$

The net production of the A_{i+1} isotope is expressed by

$$\frac{dN_{i+1}(t)}{dt} = \text{production-destruction-decay}$$

or

$$\frac{dN_{i+1}(t)}{dt} = N_i(t)\sigma_i\phi - N_{i+1}(t)\sigma_{i+1}\phi - \lambda_{i+1}N_{i+1}(t) \tag{14.15}$$

With initial condition $N_{i+1}(t) = 0$, the solution of Eq. 14.15 is

$$N_{i+1}(t) = \frac{\sigma_i N_i(0)\phi}{\lambda_{i+1} + \sigma_{i+1}\phi - \sigma_i\phi}\{\exp(-\sigma_i\phi t) - \exp[-(\lambda_{i+1} + \sigma_{i+1}\phi)t]\} \tag{14.16}$$

The activity of this target,† $A_{i+1}(t)$, is, after irradiation for time t,

$$A_{i+1}(t) = N_{i+1}(t)\lambda_{i+1} = \frac{\sigma_i N_i(0)\phi}{1 + (\sigma_{i+1} - \sigma_i)\phi/\lambda_{i+1}}\{\exp(-\sigma_i\phi t) - \exp[-(\lambda_{i+1} + \sigma_{i+1}\phi)t]\} \tag{14.17}$$

Equation 14.17 refers to the most general case. In practice, targets are selected in such a way that

1 The fraction of target nuclei destroyed is negligible, i.e., $\sigma_i\phi t \ll 1$.
2 The radioisotope produced has such a neutron absorption cross section that $\lambda_{i+1} \gg \sigma_{i+1}\phi$.

If conditions (1) and (2) are met, Eq. 14.17 takes the form

$$A_{i+1}(t) = \sigma_i N_i(0)\phi[1 - \exp(-\lambda_{i+1}t)] \tag{14.18}$$

which is the more familiar form of the activity or activation equation.

If one plots activity as a function of irradiation time, the result is Fig. 14.6. Two regions are observed.

1 For irradiation times that are short compared to the half-life of the radioisotope produced, the activity increases linearly with time. Indeed, if $\lambda_{i+1}t \ll 1$, then $e^{-\lambda_{i+1}t} \approx 1 - \lambda_{i+1}{}^{+}$ and

$$A_{i+1}(t) \approx \sigma_i N_i(0)\phi\lambda_{i+1}t = \sigma_i N_i(0)\phi\frac{\ln 2}{T_{i+1}}t$$

where T_{i+1} is the half-life of the isotope produced.

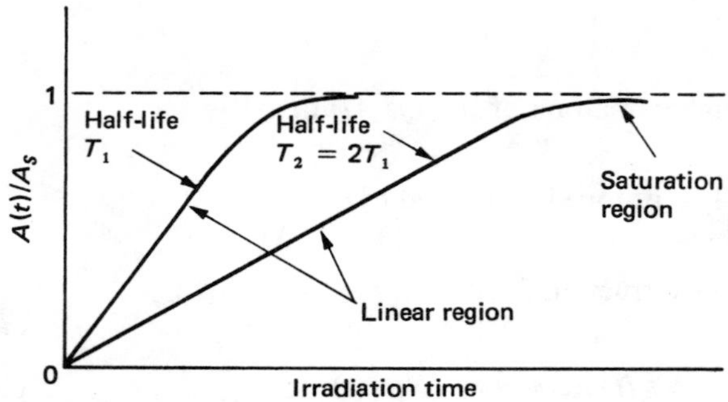

FIG. 14.6 Activity versus irradiation time (shown as a fraction of saturation activity).

2 For irradiation times many times longer than the half-life of the radioisotope, the activity reaches a saturation value (A_s). Theoretically, the value

$$A_s = \sigma_i N_i(0)\phi \tag{14.19}$$

is reached for t equals infinity. In practice, the activity produces is taken as equal to A_s for $t \approx$ 6–7 half-lives.

Table 14.2 gives the fraction of saturation activity produced for several irradiation times.

EXAMPLE 14.3

The isotope ^{197}Au is irradiated in a thermal neutron flux of 10^{18} neutrons/m$^2 \cdot$s. The cross section for neutron capture is 99 b, and the half-life of the radioactive ^{198}Au produced is 2.7 days. (a) How long does the sample have to be irradiated for 0.1 percent of the target atom to be destroyed? (b) What is the irradiation time necessary to produce 95 percent of saturation activity? (c) If the mass of the sample is 4×10^{-6} kg, what is the irradiation time necessary to produce 7×10^4 Bq (2 μCi) of activity?

ANSWER

(a) Using Eq. 14.14:

$$\frac{N(t)}{N(0)} = 0.999 = e^{-\sigma\phi t} \qquad \text{or} \qquad t = \frac{1}{\sigma\phi} \ln \frac{1}{0.999}$$

TABLE 14.2 Fraction of Saturation Activity Produced as a Function of Irradiation Time

Irradiation time (in half-lives)	$A_{i+1}(t)/A_s$
4	0.937
5	0.969
6	0.984
7	0.992
8	0.996

$$t = \frac{1}{(99 \times 10^{-28})10^{18}} \ln \frac{1}{0.999} = 1.01 \times 10^5 \text{ s} = 28 \text{ h}$$

(b) Using Eq. 14.18, the irradiation time t is such that $1 - \exp(-\lambda t) = 0.95$ or

$$t = \frac{T}{\ln 2} \ln \left(\frac{1}{1 - 0.95} \right) = 11.67 \text{ days}$$

(c) Using Eq. 14.18,

$$A(t) = \sigma N(0)\phi(1 - e^{-\lambda t}) = \sigma m \frac{N_A}{A} \phi(1 - e^{-\lambda t})$$

The saturation activity is

$$A_s = \sigma m \frac{N_A}{A} \phi = (99 \times 10^{-28})(4 \times 10^{-6}) \frac{6.023 \times 10^{23}}{197 \times 10^3} (10^{18}) = 12 \times 10^4 \text{ Bq}$$

It is useful to evaluate A_s first, because if A_s is less than the activity desired, it is impossible to obtain such activity under the conditions given. In this example, A_s is greater than $A(t)$ and the required irradiation time t is

$$t = -\frac{T}{\ln 2} \ln \left[1 - \frac{A(t)}{A_s} \right] = -\frac{2.7 \text{ days}}{\ln 2} \ln \left(1 - \frac{7.4 \times 10^4}{12 \times 10^4} \right) = 3.73 \text{ days}$$

14.4.1 Determination of the Neutron Flux by Counting the Activity of the Foils

As shown in Eq. 14.18, the activity of the irradiated foil is proportional to the neutron flux. Determination of the flux requires measurement of the activity, a task accomplished as follows.

Let the irradiation time be t_0. In practice, counting of the foil starts some time after irradiation stops, and it is customary to consider the end of irradiation as time $t = 0$ (Fig. 14.7). At time t after irradiation stops, the activity is, using Eq. 14.18,

$$A_{i+1}(t) = N_i(0)\sigma_i\phi[1 - \exp(-\lambda_{i+1} t_0)] e^{-\lambda_{i+1} t} \tag{14.20}$$

If the sample is counted between t_1 and t_2, the number of disintegrations in that period is

$$D(t_1, t_2) = \int_{t_1}^{t_2} A_{i+1}(t)\, dt = \frac{N_i(0)\sigma_i\phi}{\lambda_{i+1}} [1 - \exp(-\lambda_{i+1} t_0)] (e^{-\lambda_{i+1} t_1} - e^{-\lambda_{i+1} t_2}) \tag{14.21}$$

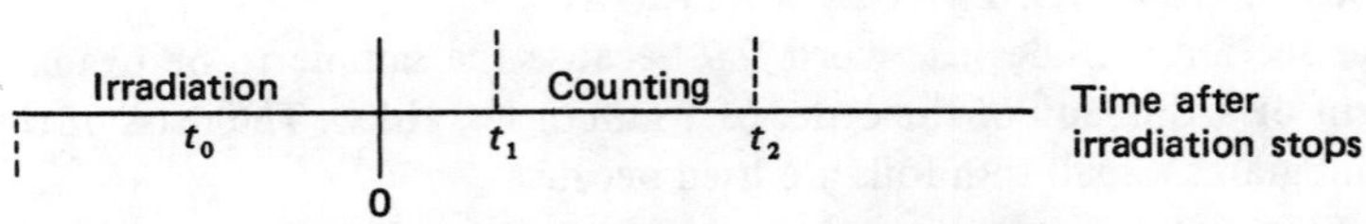

FIG. 14.7 Time scale for counting an irradiated sample. Time $t = 0$ coincides with the end of the irradiation period.

Assuming that one counts particles with energy E_k for which e_k is the probability of emission per decay, and the counting system is such that ϵ_k is the efficiency of the detection of particles with energy E_k, where

Ω = solid angle

B = background counts recorded in time $T = t_2 - t_1$

then the gross counts recorded, G_k, will be

$$G_k = \epsilon_k e_k F\Omega \frac{N_i(0)\sigma_i \phi}{\lambda_{i+1}} [1 - \exp(-\lambda_{i+1} t_0)](e^{-\lambda_{i+1}t_1} - e^{-\lambda_{i+1}t_2}) + B \tag{14.22}$$

The factor F in Eq. 14.22 takes into account any other corrections (i.e., backscattering, foil self-absorption), that may be necessary (see Sec. 8.3). If dead-time correction is necessary, it should be applied to G_k.

The flux ϕ is determined from Eq. 14.22 if all the other factors are known. There are two types of factors in Eq. 14.22:

1 Factors that depend on the sample $[N_i(0), \sigma_i, \lambda_{i+1}]$, which are assumed to be known with negligible error
2 Factors that depend on the counting system (ϵ, F, Ω), which are the main sources of error

To determine the flux distribution only, not the absolute value of the flux, foils are placed at known positions x_j and are irradiated for a time t_0. The foils are then counted using the same detector. At any point x_j, the flux may be written as

$$\phi(x_j) = \frac{L[G(x_j) - B_j]}{m_j C_j} \tag{14.23}$$

where the subscript j indicates position of the foil and where

m_j = mass of foil at position j

$C_j = [\exp(-\lambda_{i+1} t_1) - \exp(-\lambda_{i+1} t_2)]_j$

$$L = \lambda_{i+1} \left\{ \epsilon_k e_k F\Omega \frac{N_A}{A_i} [1 - \exp(-\lambda_{i+1} t_0)] \right\}^{-1}$$

(L includes all the factors that are common to all the foils.)

REMARKS ABOUT THE TARGET MATERIAL

The title of this section includes the word *foil* because the sample to be irradiated is used in the form of a thin foil of the order of 1 mm or less thick. The mass of the foil is only a few milligrams. Small thin foils are used because

1 A thick sample will absorb so many neutrons that the radiation field will be perturbed and the measurement will not give the correct flux.

2 A thick sample will cause a depression of the flux in its interior. In such a case, correction factors will have to be applied to all the equations of this section that contain the flux ϕ.
3 If the radioisotope emits β particles, increased thickness will not necessarily increase the counting rate, because only particles emitted close to the surface within a thickness less than the range will leave the target and have a chance to be recorded.
4 There is no purpose in producing more activity than is necessary.

Foil activation may be used for detection of the number of either fast or thermal neutrons. The use of foils for fast neutron energy measurements is discussed in Sec. 14.6. Foil activation is not used generally for measurement of the energy of thermal neutrons.

14.5 MEASUREMENT OF A NEUTRON ENERGY SPECTRUM BY PROTON RECOIL

Detection of neutrons by proton recoil is based on collisions of neutrons with protons and subsequent detection of the moving proton. Since neutrons and protons have approximately the same mass, a neutron may, in one collision, transfer all its kinetic energy to the proton. However, there is a possibility that the struck proton may have any energy between zero and the maximum possible, as a result of which the relationship between a neutron energy spectrum and a pulse-height distribution of struck protons is not simple. It is the objective of this section to derive a general expression for this relationship. The sections that follow show its application for specific detectors.

Consider the case of a neutron with kinetic energy E_n colliding with a proton at rest (Fig. 14.8). To calculate the proton kinetic energy after the collision, one must apply the equations of conservation of energy and linear momentum (Eqs. 3.81-3.83) using $Q = 0$ and $M_n = M_p$. The result for E_p, the proton kinetic energy as a function of the recoil angle θ, is

$$E_p = E_n \cos^2 \theta \qquad (14.24)$$

In a neutron-proton collision, the maximum value of angle θ is 90°, and the minimum 0°; therefore, the limits of the proton energy are $0 \leqslant E_p \leqslant E_n$. For neutron energies

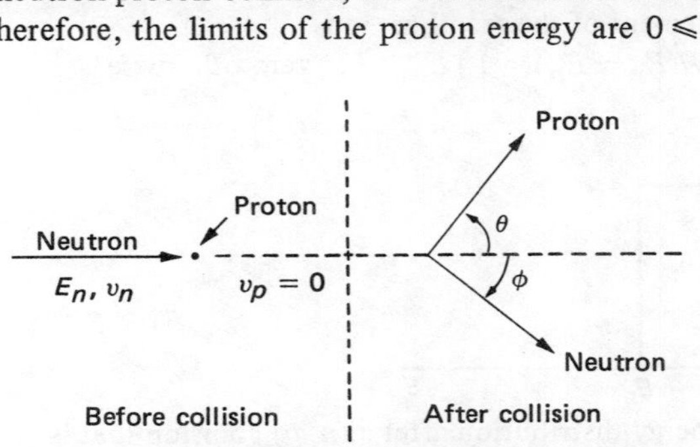

FIG. 14.8 Neutron-proton collision kinematics.

up to about 14 MeV, the (n–p) collision is isotropic in the center-of-mass systems; as a consequence, there is an equal probability for the proton to have any energy between zero and E_n in the laboratory system. That is, if $p(E)\,dE$ is the probability that the proton energy is between E and $E + dE$, after the collision, then

$$p(E)\,dE = \frac{dE}{E_n} \tag{14.25}$$

The function $p(E)$ is shown in Fig. 14.9. What is important for the observer is not $p(E)$ but the proton pulse-height distribution produced by the detector. The relationship between the pulse-height distribution and the neutron spectrum is derived as follows. Let

$\phi(E_n)\,dE_n$ = neutron energy spectrum = flux of neutrons with energy between E_n and $E_n + dE_n$

$N(E_p)\,dE_p$ = proton recoil energy spectrum = number of protons produced (by collisions with neutrons) with energy between E_p and $E_p + dE_p$

$R(E, E_p)\,dE$ = response function of the detector = probability that a proton of energy E_p will be recorded as having energy between E and $E + dE$ (defined before in Sec. 11.5)

$M(E)\,dE$ = measured spectrum = number of protons measured with energy between E and $E + dE$

The measured spectrum $M(E)$ is the pulse-height distribution in energy scale. The response function $R(E, E_p)$ takes into account the finite energy resolution of the detector and the relationship between energy deposition and pulse height.

Assuming isotropic scattering in the center of mass system, the proton energy spectrum is

$$N(E_p)\,dE_p = N_H T \int_0^{E_{\max}} \sigma(E_n)\phi(E_n)\,dE_n \frac{dE_p}{E_n} H(E_n - E_p) \tag{14.26}$$

where N_H = number of hydrogen atoms exposed to the neutron beam
T = time of measurement of the recoil protons
$H(E_n - E_p)$ = step function; $H(E_n - E_p) = 1 \mid E_n \geqslant E_p$, zero otherwise

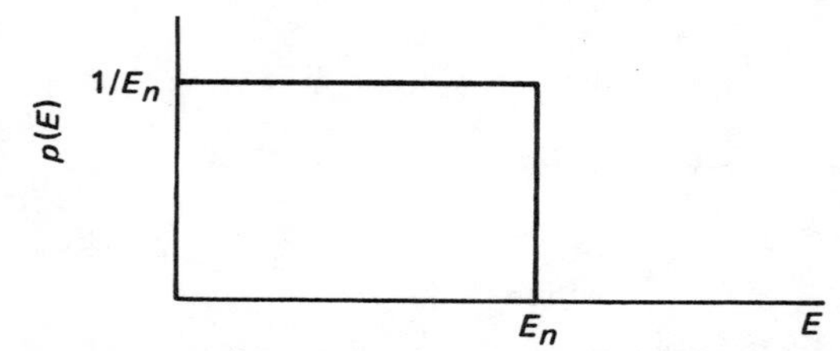

FIG. 14.9 The proton energy distribution after a (n, p) collision that is isotropic in the center of mass system of the two particles.

The measured energy spectrum is then given by

$$M(E)\,dE = \int_0^{E_{max}} dE\,R(E, E_p)N(E_p)\,dE_p \tag{14.27}$$

In Eqs. 14.26 and 14.27, the energy E_{max} is the upper limit of the neutron energy spectrum. Equation 14.27 may be rewritten in the form

$$M(E) = \int_0^{E_{max}} dE_n\,k(E, E_n)\phi(E_n) \tag{14.28}$$

where

$$k(E, E_n) = \int_0^{E_{max}} dE_p\,R(E, E_p)N_H T\,\frac{\sigma(E_n)}{E_n}\,H(E_n - E_p) \tag{14.29}$$

Equation 14.28 has the form of the folding integral (see also Sec. 11.5), while Eq. 14.29 gives the "composite" response function for the proton recoil spectrometer.

EXAMPLE 14.4

As a first application of Eq. 14.28, consider the case of a monoenergetic neutron spectrum and a detector with a Gaussian response function. What is the measured spectrum?

ANSWER

Substituting the Gaussian response function

$$R(E, E_p) = \frac{1}{\sqrt{2\pi}\sigma}\exp\left[-\frac{(E-E_p)^2}{2\sigma^2}\right]$$

into Eq. 14.29 and performing the integration, assuming $E/\sigma \gg 1$, one obtains†

$$k(E, E_n) = \frac{N_H T\sigma(E_n)}{2E_n}\left[1 + \operatorname{erf}\left(\frac{E_n - E}{\sqrt{2}\sigma}\right)\right]$$

where

$$\operatorname{erf}\left(\frac{x}{\sqrt{2}}\right) = \sqrt{\frac{2}{\pi}}\int_0^x e^{-t^2/2}\,dt$$

†There are two σ's involved here: $\sigma(E)$ is the cross section at energy E; σ without an argument is the standard deviation of the Gaussian.

Substituting the value of $k(E, E_n)$ and the monoenergetic flux $\phi(E_n) = S\delta(E_n - E_0)$ into Eq. 14.28 and performing the integration, one obtains

$$M(E) = S \frac{N_H T \sigma(E_0)}{2E_0} \left[1 + \text{erf} \left(\frac{E_0 - E}{\sqrt{2}\sigma} \right) \right] \tag{14.30}$$

The function $M(E)$ given by Eq. 14.30 is shown in Fig. 14.10. It is essentially the same function as that shown in Fig. 14.9, except for the rounding off at the upper energy limit caused by the Gaussian detector response.

The task of neutron spectroscopy is to obtain the neutron energy spectrum $\phi(E)$, which means to unfold Eq. 14.28. Two general methods used to unfold this equation are discussed next.

14.5.1 Differentiation Unfolding of Proton Recoil Spectra

If $R(E, E_p) = \delta(E - E_p)$, then the response function of the proton recoil spectrometer is (using Eq. 14.29)

$$k(E, E_n) = N_H T \frac{\sigma(E_n)}{E_n}$$

and Eq. 14.28 takes the form

$$M(E) = N_H T \int_E^{E_{\max}} dE_n \frac{\sigma(E_n)}{E_n} \phi(E_n) \tag{14.31}$$

The lower limit of the integral is set equal to E because at any energy E only neutrons with energy $E_n > E$ can contribute to $M(E)$. Equation 14.31 may be solved by differentiation to give

$$\phi(E) = - \frac{E}{N_H T \sigma(E)} \left[\frac{dM(E)}{dE} \right] \tag{14.32}$$

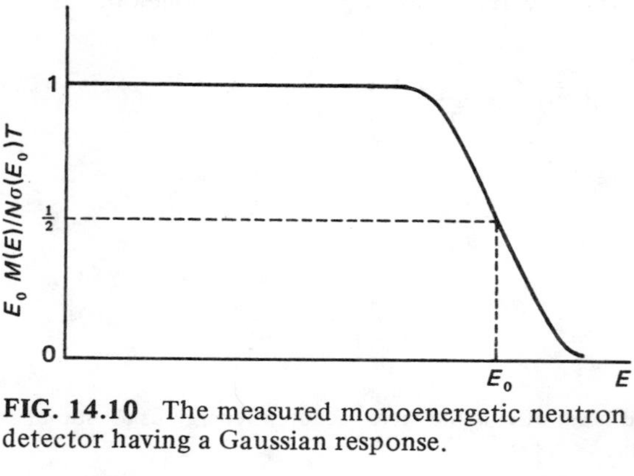

FIG. 14.10 The measured monoenergetic neutron spectrum obtained with a detector having a Gaussian response.

The evaluation of the derivative in Eq. 14.32 is performed by numerical techniques, since it is not the spectrum $M(E)$ that is measured but its "binned" equivalent

$$\int_{E_i}^{E_{i+1}} M(E)\, dE = M_i \tag{14.33}$$

where M_i is the number of counts in channel i of the spectrometer. Several investigators[13-17] applied least-squares fit techniques to numerically perform the differentiation of Eq. 14.32. Usually, each M_i is assigned the energy corresponding to the midpoint of the channel and an Mth order polynomial is least-squares fit to that point and the preceding and following N points [this is an Mth-order, $(2N + 1)$-point fit]. The derivative of the polynomial at the mid-point is used as the derivative $[dM(E)/dE]_i$. This method has the disadvantage that it slightly hardens the unfolded spectrum.

An improved differentiation technique[18] consists of first smoothing the true spectrum $\phi(E)$ to obtain a "smoothed" true spectrum $\phi_s(E)$, given by

$$\phi_s(E) = \int_0^\infty G(E, E')\phi(E')\, dE' \tag{14.34}$$

where $G(E, E')$ is a smoothing function normalized to 1. Substituting Eq. 14.32 into Eq. 14.34, one obtains[18]

$$\phi_s(E) \simeq \sum_{i=1}^{NC} \frac{1}{NT} \left[\frac{E_{i+1}}{\sigma(E_{i+1})} G(E, E_{i+1}) - \frac{E_i}{\sigma(E_i)} G(E - E_i) \right] M_i \tag{14.35}$$

where NC is the number of channels, M_i is given by Eq. 14.33, and the assumption is made that the quantity $[E'/\sigma(E')]\, G(E, E')$ approaches zero at both limits of integration of Eq. 14.34. Results obtained with Eq. 14.35 and a Gaussian smoothing function show no spectrum hardening.

14.5.2 The FERDOR Unfolding Method

The unfolding code FERDOR based on the work of Burrus[19] is described in several articles.[20,21] The measured spectrum given by Eq. 14.28 is written in the form

$$M_i = \int_0^{E_{max}} dE_n\, k_i(E_n)\phi(E_n) \qquad i = 1, NC \tag{14.36}$$

where

$$k_i(E') = \int_{E_i}^{E_{i+1}} dE\, k(E, E') \tag{14.37}$$

$E_{i+1} - E_i = \Delta E_i =$ width of the spectrometer channel at energy E_i

$NC =$ number of spectrometer channels

The functions $k_i(E)$, which represent the detector response, can be measured at selected energies or can be computed.[22]

The code FERDOR, instead of obtaining $\phi(E)$, calculates the quantities S_k given by

$$S_k = \int_0^{E_{max}} W_k(E)\phi(E)\,dE \qquad k = 1, NW \tag{14.38}$$

where $W_k(E) =$ *window function*, chosen by the observer

$NW =$ number of window functions

There are many interpretations of Eq. 14.38, all equally valid.

1 If $W_k(E)$ is taken as the response of an "ideal" detector, then the S_k's are the elements of an "ideal" measured spectrum.
2 The functions $W_k(E)$ are smoothing functions for the true spectrum $\phi(E)$.
3 The S_k's are linear functions of the true spectrum $\phi(E)$ that have to be calculated. For example, if $W_k(E)$ is a cross section, the corresponding S_k is a reaction rate; if $W_k(E)$ is a flux-to-dose conversion factor, then S_k is the dose; or if $W_k(E) = 1$, then S_k is the integral spectrum.

The functions $W_k(E)$ used by FERDOR are Gaussians with a standard deviation given by

$$\sigma_k = \frac{w_k E_k}{2\sqrt{2 \ln 2}} \tag{14.39}$$

where $w_k = \Gamma_k/E_k$, and its value is set by the user. The code calculates the S_k's and their uncertainties based on the values of $k_i(E)$, M_i, and the statistical errors of M_i, using a constrained least-squares method.

Experience has shown that the proper values of w_k depend on both the shape of the neutron spectrum and the statistical errors of M_i, and that small changes in the w_k cause large changes in the errors of the result.[23] This effect has been mitigated in the code FORIST,[24] which is a modification of FERDOR.† In FORIST, the value of w_k is obtained by an iterative process in terms of the desired statistical error of the result S_k. Choosing the widths w_k by this method improves the resolution of the unfolded spectrum, for a fixed desired statistical error.

14.5.3 Proportional Counters Used as Fast-Neutron Spectrometers

Proportional counters filled with hydrogen or methane are used for the measurement of neutron spectra in the energy range 1 keV $< E_n <$ 2 MeV. Neither hydrogen nor

†The codes FERDOR and FORIST are available through the Radiation Shielding Information Center of Oak Ridge National Laboratory.[25,26]

methane are equally useful over the full energy range. Hydrogen-filled counters are used for $E_n < 100$ keV. For higher neutron energy, greater stopping power is needed, and for this reason methane is used instead of hydrogen. Methane-filled counters do not given good results for $E_n < 100$ keV because of spectrum distortion from carbon recoils.

The efficiency of a proportional counter, like that of any other gas counter, depends on its size, the composition and pressure of the gas, and the energy of the incident neutrons. Knowledge of the neutron cross section for interactions with hydrogen and carbon over the energy range of interest is necessary for efficiency calculations. The hydrogen cross section is known to better than 1 percent for neutron energies between 0.2 and 22 MeV. The carbon cross section is less accurately known for $E_n > 2$ MeV. In addition to elastic scattering, other carbon reactions, such as (n, n'), (n, α), and (n, p), are important for $E_n > 4.8$ MeV and should be included in the response function of the counter. Table 14.3 gives hydrogen and carbon cross sections for $0.2 < E_n < 20$ MeV.

The relationship between the neutron spectrum and the measured pulse height distribution is given by Eq. 14.28. The response function $k(E, E')$ (Eq. 14.29) may be measured or calculated. In either case, the following effects have to be taken into account in obtaining $k(E, E')$:[14–17,27,28]

WALL-AND-END EFFECTS

Tracks of protons generated close to the wall or close to the ends of the counter have a high probability for incomplete energy deposition and collection of ionization.

TABLE 14.3 Total Hydrogen and Carbon Cross Sections for $0.2 < E_n < 20$ MeV[22]

Neutron energy (MeV)	$\sigma_t(^1\text{H})$ (b)	$\sigma_t(^{12}\text{C})$ (b)
0.2015	9.523	4.260
0.3220	7.719	3.857
0.4265	6.712	3.583
0.6170	5.532	3.174
0.9035	4.515	2.720
1.205	3.902	2.345
1.613	3.287	1.965
2.234	2.737	1.606
3.329	2.183	1.760
4.236	1.827	1.943
4.919	1.616	1.198
6.017	1.422	1.094
7.038	1.246	1.023
8.029	1.137	1.466
10.00	0.940	1.140
11.98	0.801	1.259
14.01	0.693	1.360
16.06	0.605	1.431
17.81	0.541	1.448
19.91	0.483	1.450
21.81	0.442	1.445

Proton-recoil tracks close to the wall are truncated by collisions with the wall material before the struck proton deposits all its energy in the counter. Protons being produced close to the end of the counter and moving towards either end enter into a region of lower electric field where there is little or no gas multiplication and where there is, consequently, incomplete collection of the charge produced.

As a result of wall-and-end effects, lower energy pulses are generated that tend to increase the response function towards the lower energies (Fig. 14.11). Corrections for this effect have been calculated,[17,29] but unfortunately each calculation applies only to a specific geometry.

The magnitude of wall-and-end effects increases as the size of the counter decreases. It also increases as the neutron energy increases. In fact, for neutrons in the megavolt range, the distortion of the spectrum due to these effects becomes so significant that it sets the upper energy limit (~2 MeV) for the use of a proportional counter as a spectrometer.

ELECTRIC FIELD DISTORTION

The gas multiplication in a proportional counter depends on the intensity of the electric field. Close to the ends of a cylindrical counter, the strength of the electric field becomes gradually less intense than in most of the counter volume. This effect produces lower pulses from proton recoils at the ends of the counter. Detectors with large length-to-diameter ratio are less affected by this problem. Theoretical corrections of this effect have been developed and successfully applied.[17]

EFFECT OF CARBON RECOILS

Neutrons detected by methane-filled counters collide not only with hydrogen nuclei but also with carbon atoms. The ionization produced by carbon recoils is indistinguishable from that produced by protons. However, carbon recoils produce pulses that are smaller than those from protons because of differences in both kinematics and ionization ability. The maximum fraction of neutron energy that can be imparted to a carbon nucleus in one collision is 0.29 (versus 1 for a hydrogen nucleus), and the relative ionization efficiency of a carbon to a proton recoil is about 0.5.[28] Thus, the effect of carbon recoils is to add pulses at the low-energy region of the response function (Fig. 14.11). Carbon-recoil effects are so significant for $E_n < 100$ keV that methane-filled counters are not usable below this energy.

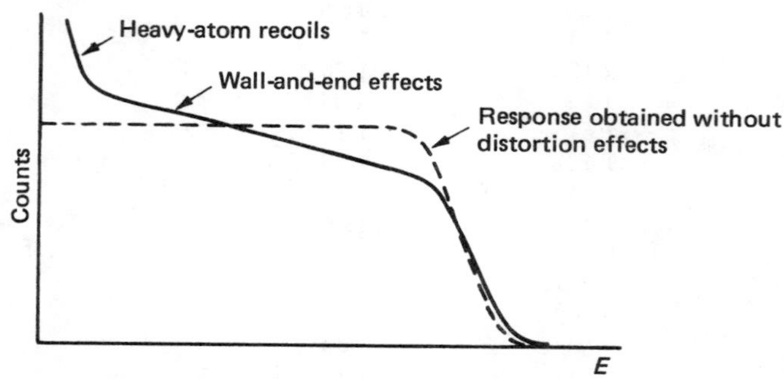

FIG. 14.11 Distortion of the response function because of wall-and-end effects and heavy-atom recoils.

VARIATION OF ENERGY NEEDED TO PRODUCE ONE ION PAIR

As stated in Sec. 14.5, the measured spectrum may differ from the proton-recoil spectrum because of nonlinearity in the relationship between proton energy and ionization produced. That relationship is expressed by the quantity $w(E)$, defined as the energy needed by a proton of energy E to produce one electron-ion pair. To be able to use a counter as a spectrometer, the value of $w(E)$ should be accurately known for the gas of the counter for all energies below the maximum neutron energy measured. Experiments have shown that the value of $w(E)$ is essentially constant for hydrogen for neutron energies above ~20 keV. Below that energy, $w(E)$ changes slightly with energy.[17,28] For methane, $w(E)$ seems to be essentially constant between 100 keV and 1 MeV.[28]

GAMMA-RAY DISCRIMINATION

Proportional counters used in a mixed neutron-gamma field detect both types of radiation. Discrimination of γ-ray pulses has been accomplished by utilizing the fact that the Compton electrons produced by the gammas have longer range than proton recoils. The *time-of-rise method*, which is now almost universally used, takes advantage of the faster risetime of the proton pulse relative to that of the electrons. Proton range is so much shorter than electron range that all the ions produced by the proton arrive at the anode at about the same time and generate a pulse with a fast risetime. On the other hand, ions produced by electrons along their path arrive at the anode over a period of time and generate a pulse with a slower risetime. Thus, using appropriate electronics, the pulses from gammas can be rejected.

FINITE RESOLUTION OF PROTON DETECTOR

The resolution of a proportional counter for monoenergetic protons is derived from two factors. One is a statistical broadening that depends on the number of ion pairs produced. The other is a "mechanical" broadening due to imperfections in the design of the counter and impurities in the filling gas. At an energy of 615 keV, the energy resolution is of the order of 4 percent, but it deteriorates to about 60 percent at 1 keV.

Response functions of proportional counters have been measured and calculated by several people. Verbinski and Giovannini[28] gave a critical study of response functions of gas-filled counters as well as a comparative study of the different codes used to unfold their spectra. Figures 14.12 and 14.13 show measured and calculated response functions for hydrogen and methane-filled proportional counters.

Coarse calibration of proportional counters is achieved by using ^{3}He and N_2 as additives in the gas of the counter and employing the reactions

$$^{14}\mathrm{N}(n,p)^{14}\mathrm{C} \qquad Q = 627 \text{ keV}$$

$$^{3}\mathrm{He}(n,p)^{3}\mathrm{H} \qquad Q = 765 \text{ keV}$$

Fine calibration is obtained by placing the detector inside neutron filters made of aluminum, NaCl, and Teflon.[30] The filters generate dips in the unfolded spectrum, which coincide with the energies of cross-section resonances of the corresponding isotope. Fine calibration is achieved by adjusting the gas gain in the counter until the energies of the dips of the unfolded spectrum coincide with the energies of the resonances.

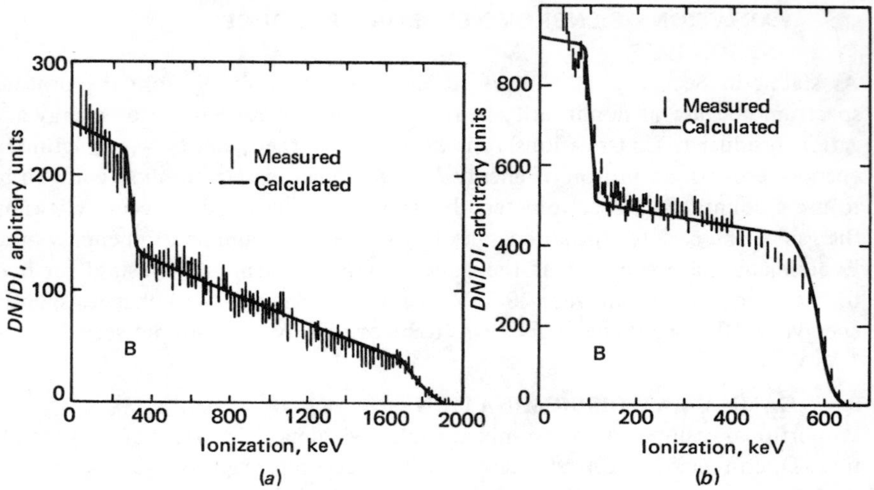

FIG. 14.12 Measured and calculated response functions for a methane-filled proportional counter (1.5 in diameter, 3.5 atm pressure). (*a*) At 75°C, 1772 keV. (*b*) At 75°C, 592 keV. (From ref. 28; *DN/DI* is the proton ionization spectrum.)

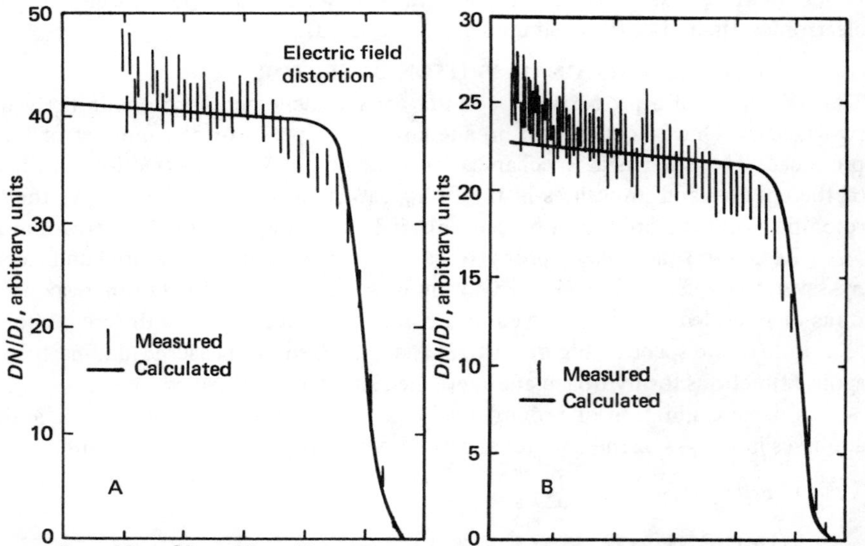

FIG. 14.13 Measured and calculated response functions for a cylindrical hydrogen-filled proportional counter (1.5 in diameter, 2.6 atm pressure, H_2 + 5 percent N_2), (*a*) electrid field distortion, 45°C, 50 keV; (*b*) 45°C, 100 keV. (From ref. 28; *DN/DI* is the proton ionization spectrum.)

14.5.4 Organic Scintillators Used as Fast-Neutron Spectrometers

Organic scintillators have proven to be excellent fast-neutron detectors because they have high and known efficiency, good energy resolution, and low sensitivity to gammas. The high efficiency is due to their hydrogen content (1.1 hydrogen atoms per carbon atom, density about 10^3 kg/m^3 = 1 g/cm^3), the relatively high hydrogen cross section (2.5 b for 2.5-MeV neutrons), and the ability to make and use them in large sizes. Organic scintillators are now (1982) the main detectors used for neutron spectroscopy from ~10 keV to 200 MeV. An excellent review of organic scintillator properties is given in reference 31.

Stilbene scintillators were used as early as 1957. Stilbene as a crystal is very sensitive to mechanical and thermal shock and shows an anisotropic response to neutrons–i.e., neutrons incident from different directions, with respect to the crystal lattice, produce different light output. Liquid organic scintillators have none of these problems; in addition, they have higher *H/C* ratio, and light production from carbon recoils relatively lower than in stilbene. For all these reasons, liquid organic scintillators are almost exclusively used for detecting fast neutrons.

The NE-series† of organic scintillators has been studied in detail and used extensively,[22,28,31] in particular NE-213. The NE-213 scintillator, which is most commonly used, consists of xylene, activators, the organic compound POPOP (as a wavelength shifter), and napthalene, which is added to improve light emission. The density of NE-213 is about 870 kg/m^3 (0.87 g/cm^3), and its composition is taken to be $CH_{1.21}$.

As the size of an organic scintillator increases, the efficiency increases, the energy resolution deteriorates, and the background increases. The optimum size for megavolt neutrons seems to be a scintillator with a volume 10^{-4} m^3 (100 cm^3), i.e., a cylinder 50 mm in diameter and 50 mm tall. The efficiency of the NE-213 scintillator has been determined by Verbinski et al.[22] using a combination of measurements and Monte Carlo calculations for 20 neutron energies between 0.2 and 22 MeV.

The response of an organic scintillator to monoenergetic neutrons depends on effects similar to those discussed in the previous section for proportional counters, with the exception of electric field distortions. The most important cause of a response different from the ideal rectangular distribution shown in Fig. 14.9 is the nonlinear relation between the energy of the proton and the amount of light produced by the scintillation process. For organic scintillators, the light production by protons and heavier ions is essentially proportional to the $\frac{3}{2}$ power of the energy deposited[22] in the energy range $0.3 < E < 4$ MeV, and linear for lower energies.[32] The light production by electrons varies almost linearly with energy[32] (Fig. 14.14).

Response functions for the NE-213 organic scintillator were first obtained by Verbinski et al.[22] These authors measured the NE-213 response for 20 energies between 0.2 and 22 MeV and then normalized the spectra to Monte Carlo calculations. One additional response was constructed at 10 MeV from Monte Carlo calculations and from interpolation of the measured responses at 8 and 12 MeV. These 21 neutron response functions are included in the FERDOR/COOLC package,[25] but they consti-

†Manufactured by Nuclear Enterprises, Winnepeg, Ontario, Canada.

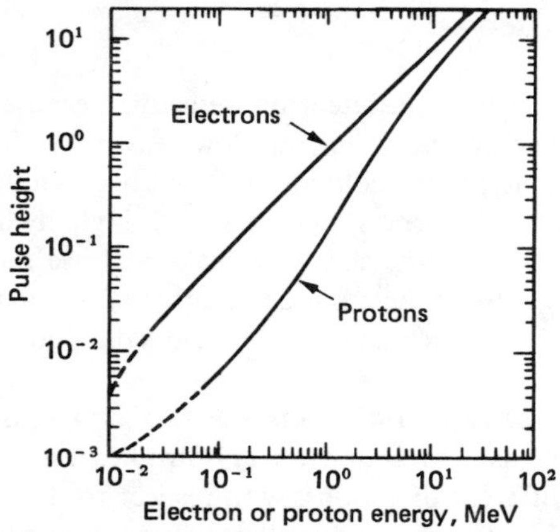

FIG. 14.14 Light response of NE-110 (plastic) and NE-213 scintillator as a function of electron and proton energy (from ref. 34).

tute an insufficient response matrix. Ingersoll and Wehring,[33,35] using an interpolation scheme, expanded this data into an 81-column matrix, and used it successfully to unfold neutron spectra up to 20 MeV. This expanded matrix, along with a measured gamma-ray response function,[33] is included in the FORIST code package.[26] Figure 14.15 shows typical response functions for monoenergetic neutrons up to 8.12 MeV. Figure 14.16 shows a pulse-height distribution and an unfolded spectrum.

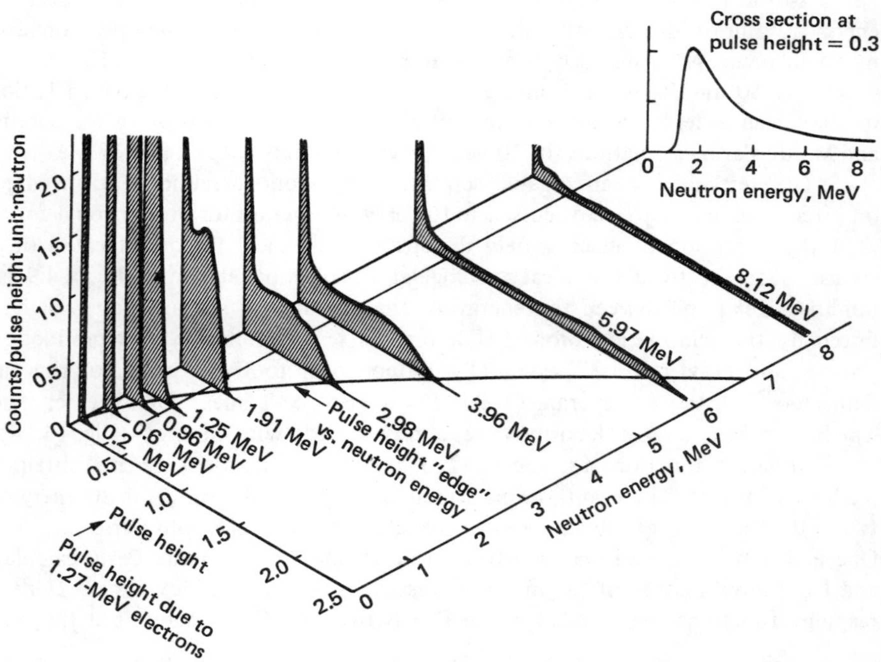

FIG. 14.15 Response functions of NE-213 spectrometer. The curves represent the quantity $k_i(E)$, Eq. 14.37 (from ref. 20).

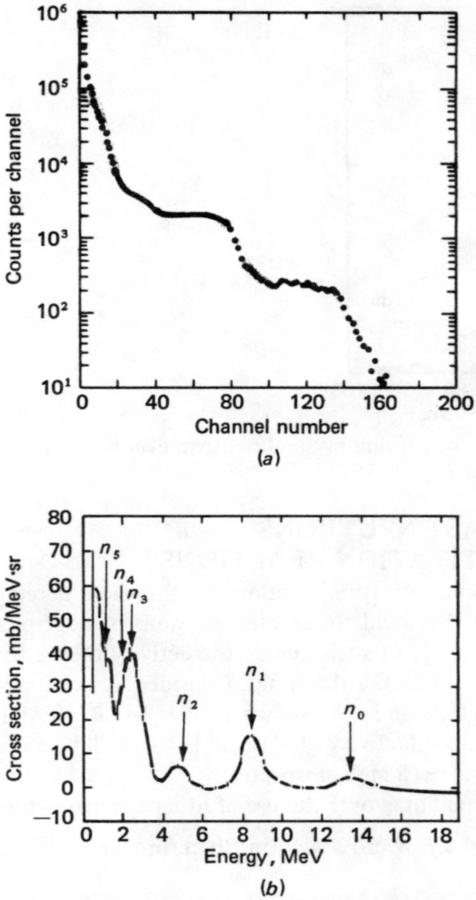

FIG. 14.16 Pulse-height distribution and unfolded neutron spectrum obtained with a 46-mm × 46.5-mm diameter NE-213 organic scintillator (from ref. 20).

Neutron-gamma discrimination is essential for satisfactory performance of an organic scintillator as a neutron spectrometer. Fortunately, rejection of gamma pulses can be achieved by electronic means. The method is called pulse shape discrimination (PSD) and is based on the difference in scintillator response to gamma and neutron associated events. The electrons, which are produced by gammas, cause scintillations at a rate faster than that due to protons produced by neutrons. Thus, the electron pulses, which are associated with photon interactions, have a faster risetime than the proton pulses associated with neutrons. There are many PSD circuits. All of them generate a pulse with amplitude dependent upon the fast and slow comonents of the scintillation. The PSD circuit used by Burrus and Verbinski[20] produces a large positive pulse for neutrons and a small positive or large negative pulse for gammas. Ingersoll et al.[35] used a TAC to generate a signal proportional to the risetime of the pulses produced in the scintillator (Fig. 14.17).

Organic scintillators can be used in time-of-flight measurements because they have shown timing resolution less than 1 ns[31] (see Sec. 14.8).

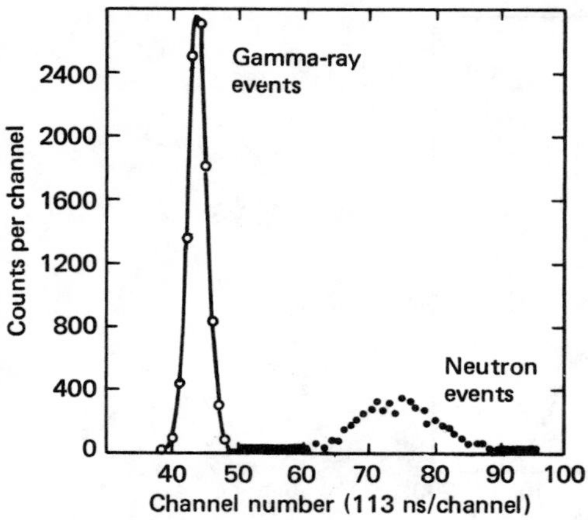

FIG. 14.17 Risetime spectrum of gamma-ray and neutron events recorded by a NE-213 organic scintillator (from ref. 35).

14.6 DETECTION OF FAST NEUTRONS USING THRESHOLD ACTIVATION REACTIONS

Detection of fast neutrons by threshold activation reactions (or threshold detectors) is based on the existence of an energy threshold for certain reactions of neutrons with nuclei. Thus, if one activates a foil made of such nuclei, the activity of the foil will give a measure of the neutron flux above the threshold. Consider, for example, the (n, α) and $(n, 2n)$ cross sections of ^{27}Al and ^{46}Ti shown in Fig. 14.18. If Al and Ti foils are irradiated, the activity produced (activity of ^{24}Na and ^{45}Ti) will be a measure of the neutron flux above ~5 MeV and ~13 MeV, respectively.

The main advantages of this technique, over the use of other spectrometers, are:

1 The foils have a small volume and a low cross section; therefore, they do not disturb the neutron field.
2 The foils are almost insensitive to gammas.
3 Their small size makes the location of foils possible in places where no other spectrometer would fit.
4 The counting equipment does not have to be carried to the radiation area.

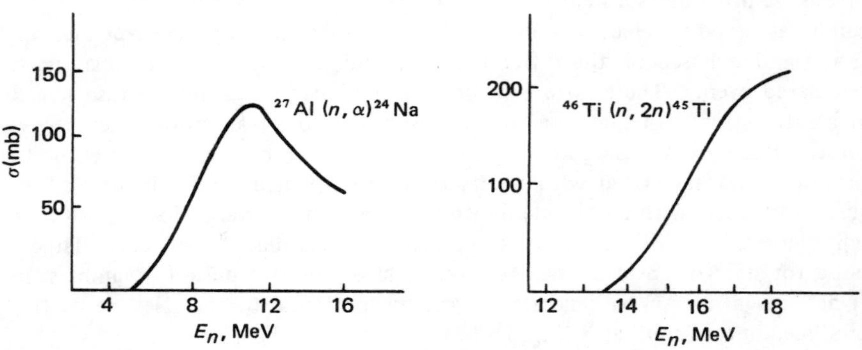

FIG. 14.18 The (n, α) and $(n, 2n)$ cross sections for ^{27}Al and ^{46}Ti.

As shown in Sec. 14.4, the saturation activity A_s of a foil is given by

$$A_s = N \int_0^\infty \sigma(E)\phi(E)\, dE \tag{14.40}$$

where N, $\sigma(E)$, and $\phi(E)$ have been defined in Sec. 14.4. Table 14.4 gives a partial list of the many reactions one can use to cover a given neutron energy range. In general, reactions are selected according to the energy range of the neutron spectrum and the counting equipment available. There are, however, certain criteria that make certain reactions and certain foils more desirable than others:

1 The cross section for the reaction should be well known as a function of energy.
2 The type, energy, and relative intensity of the radiations of the product of the reaction should be well known.
3 The half-life of the radionuclide produced should be well known and should be at least several minutes long.
4 The foil material should be available in high purity, to avoid interference reactions from impurities.

TABLE 14.4 A List of Threshold Reactions

Material	Reaction	Threshold (MeV)	Half-life of product
Teflon	$^{19}F(n, 2n)^{18}F$	11.6	109.7 min
Li	$^{7}Li(n, \alpha n')^{3}H$	3.8	–
Mg	$^{24}Mg(n, p)^{24}Na$	6	15 h
Al	$^{27}Al(n, \alpha)^{24}Na$	4.9	15 h
Al	$^{27}Al(n, p)^{27}Mg$	3.8	9.45 min
Ti	$^{46}Ti(n, p)^{46}Sc$	5.5	83.3 d
Ti	$^{47}Ti(n, p)^{47}Sc$	2.1	3.41 d
Ti	$^{48}Ti(n, p)^{48}Sc$	6.8	43.7 h
Fe	$^{56}Fe(n, p)^{56}Mn$	4.9	2.58 h
Fe	$^{54}Fe(n, p)^{54}Mn$	2.2	312.5 d
Co	$^{59}Co(n, \alpha)^{56}Mn$	5.2	2.58 h
Ni	$^{58}Ni(n, 2n)^{57}Ni$	13	36 h
Ni	$^{58}Ni(n, p)^{58}Co$	2.9	71.3 d
Cu	$^{63}Cu(n, 2n)^{62}Cu$	11.9	9.8 min
Cu	$^{63}Cu(n, \alpha)^{60}Cu$	6.1	5.27 y
Zn	$^{64}Zn(n, p)^{64}Cu$	2	12.7 h
I	$^{127}I(n, 2n)^{126}I$	9.3	13 d
Au	$^{197}Au(n, 2n)^{196}Au$	8.6	6.17 d
Np	$^{237}Np(n, \text{fission})^{140}Ba^{\dagger}$	0.5	12.8 d
U	$^{238}U(n, \text{fission})^{140}Ba^{\dagger}$	1.45	12.8 d

†Various fission products are available for counting, e.g., ^{55}Zn, ^{103}Ru, ^{140}La. ^{140}Ba is shown as typical.

To determine the neutron flux as a function of energy by the threshold reaction technique, one irradiates n foils and obtains n equations for the saturation activity per target nucleus,†

$$A_i = \int_0^\infty \sigma_i(E)\phi(E)\,dE \qquad i = 1, n \tag{14.41}$$

Note that the activity A_i is not the total activity of the foil but only the activity due to the reaction associated with the cross section $\sigma_i(E)$. For example, if one irradiates an aluminum foil, the total activity will be the result of the (n, α) and (n, p) reactions listed in Table 14.4 and the (n, γ) reaction that will also occur. If the user intends to examine the (n, p) reaction, the activity that should be used in Eq. 14.41 is that of ^{27}Mg. Activity due to ^{24}Na [from the (n, α) reaction] and ^{28}Al [from the (n, γ) reaction] should be disregarded.

It is advantageous, but not necessary, to choose reactions that result in the same type of radiation being emitted by all the foils used. Then the same counting equipment can be used with all the foils. The most common choice is gammas, and the detector is a Ge(Li) spectrometer.

The calculation of $\phi(E)$ based on Eq. 14.41 is another case of unfolding. Usually the flux is expressed in terms of a number of energy groups G. If $G < n$, unfolding of Eq. 14.41 is a simple case of least-squares fit. Unfortunately, in most of the cases of practical interest, $G > n$ and the only way to obtain $\phi(E)$ is to assume a certain a priori form for it and then try to improve upon this initial guess. The result depends on the choice of the input spectrum, the set of threshold reactions chosen, the errors of the measured activities, and the uncertainties of the cross sections involved. The several unfolding codes that are used differ mainly in the choice of the input spectrum. A brief description of three such codes, SAND-II,[36] SPECTRA,[37] and RDMM[38] is given next.

14.6.1 The Code SAND-II

The code SAND-II assumes an estimate of the flux $\phi_j^{(0)}|_{j=1,G}$, and for subsequent iterations applies a nonlinear adjustment to the flux. For the kth iteration,

$$\phi_j^{(k+1)} = \phi_j^{(k)} \exp C_j^{(k)} \qquad j = 1, \ldots, G \tag{14.42}$$

where $C_j^{(k)}$ is the kth iteration correction term for the jth flux group. The value of $C_j^{(k)}$ depends on weighting factors, which in turn are calculated from the measured activities. The activity of every foil is written in terms of the flux as shown by Eq. 14.43:

$$A_{ij}^{(k)} = \sigma_{ij}\phi_j^{(k)}(E_{j+1} - E_j) \tag{14.43}$$

†If saturation activity cannot be obtained because of long half-life, an appropriate time-correction factor is used.

where σ_{ij} = ith foil cross section averaged over flux group j

$A_{ij}^{(k)}$ = *the portion of* A_i contributed by neutrons in the jth group, during kth iteration

To avoid spurious variations introduced to the result by the iteration, SAND-II applies smoothing[39,40] to the weighting factors used for the calculation of the flux correction factor $C_j^{(k)}$. Experience has shown that SAND-II is better than either SPECTRA or RDMM.

14.6.2 The Code SPECTRA

The code SPECTRA assumes that the flux is a piecewise-linear function of energy; i.e., for any energy group j, one can write

$$\phi(E) = \frac{E - E_j}{E_{j+1} - E_j}\,\phi_{j+1} + \frac{E_{j+1} - E}{E_{j+1} - E_j}\,\phi_j \tag{14.44}$$

The flux $\phi(E)$ from Eq. 14.44 is introduced into Eq. 14.41, and the integration is performed to give a matrix equation $\mathbf{A} = \mathbf{S}\mathbf{\Phi}$ where $\mathbf{A}$ is an $n \times 1$ vector whose components are the measured activities, $\mathbf{S}$ is an $n \times G$ matrix whose elements are integrals of the cross section of the form

$$S_{ij} = \int_{E_{j-1}}^{E_j} \frac{E - E_{j-1}}{E_j - E_{j-1}}\,\sigma_i(E)\,dE + \int_{E_j}^{E_{j+1}} \frac{E_{j+1} - E}{E_{j+1} - E_j}\,\sigma_i(E)\,dE \tag{14.45}$$

and $\mathbf{\Phi}$ is a $G \times 1$ vector whose components are the desired fluxes. The best flux $\mathbf{\Phi}$ is that flux which minimizes the quantity

$$\sum_{i=1}^{n} \left(\frac{A_i - A_{ci}}{A_i}\right)^2$$

where A_i and A_{ci} are measured and calculated activities, respectively.

14.6.3 The Relative Deviation Minimization Method (RDMM)

The RDMM[38] assumes that the flux can be represented as a series of linearly independent functions,

$$\phi_m(E) = W(E) \sum_{k=1}^{m} a_k \psi_k(E) \tag{14.46}$$

where $W(E)$ = weighting function (e.g., e^{-E})

a_k = coefficients of the expansion

$\psi_k(E)$ = polynomial, simple or orthogonal (e.g., Laguerre or Chebyshev polynomials may be used)

Equation 14.46 represents an iterative procedure for the evaluation of the flux, because a new term is added to the expansion, after each iteration, up to a maximum of n. The best approximation for the flux is that which minimizes the quantity

$$Q(m, a_1, a_2, \ldots, a_m) = \sum_{i=1}^{n} \left[\frac{A_i - \int_0^\infty \sigma_i(E)\phi_m(E)\, dE}{A_i} \right]^2 \quad (14.47)$$

By taking the derivatives

$$\frac{\partial Q}{\partial a_k} = 0 \qquad k = 1, m \quad (14.48)$$

one obtains m equations that can be solved for the constants $a_k|_{k=1,m}$.

14.7 NEUTRON ENERGY MEASUREMENT WITH A CRYSTAL SPECTROMETER

The measurement of neutron energy with a crystal spectrometer is based on the same principle of Bragg diffraction as is the measurement of x-ray energy (see Sec. 12.11).

A neutron with kinetic energy E has a de Broglie wavelength equal to

$$\lambda = \left(\frac{h}{p}\right) = \frac{0.02862}{\sqrt{E\,(\text{eV})}}\ \text{nm} = \frac{0.2862}{\sqrt{E\,(\text{eV})}}\ \text{Å} \quad (14.49)$$

where h = Planck's constant

$p = MV = \sqrt{2ME}$ = linear momentum of the neutron

Neutrons with wavelength λ incident upon a crystal with interplanar distance d are scattered by the atoms of the crystal. As a result of constructive interference, a diffracted neutron beam appears at an angle θ satisfying the Bragg condition (Fig. 14.19),

$$n\lambda = 2d \sin\theta \quad (14.50)$$

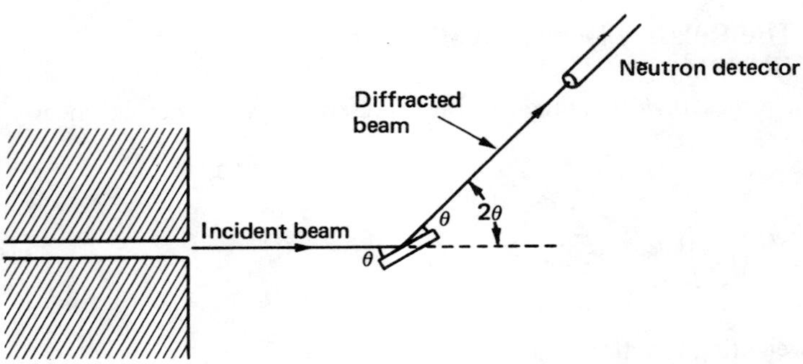

FIG. 14.19 The arrangement of a neutron diffraction spectrometer.

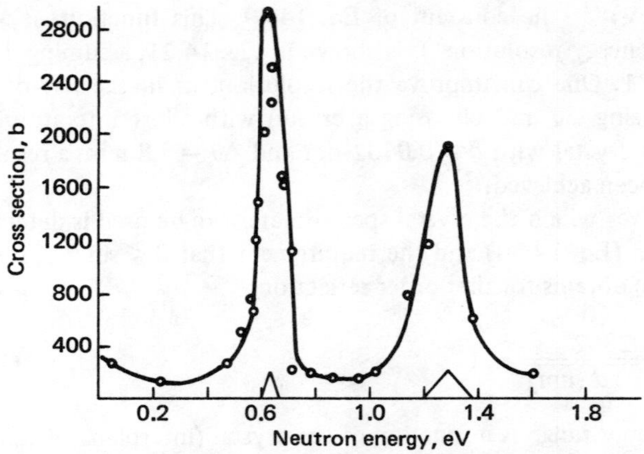

FIG. 14.20 The total neutron cross section of iridium (from ref. 41).

where n is the order of reflection (see Sec. 12.11). If the incident beam is polyenergetic, the neutron detector set at an angle θ will detect neutrons of wavelength λ satisfying Eq. 14.50, i.e., neutrons having kinetic energy E related to λ by

$$E\,(\text{eV}) = \frac{8.191 \times 10^{-4}}{\lambda^2\,(\text{nm})} \tag{14.51}$$

Neutron crystal spectrometers are used either to analyze a polyenergetic neutron source or, more frequently, to provide a source of monoenergetic neutrons. Considering again Fig. 14.19, even though the incident beam may consist of neutrons of many energies, the neutrons diffracted at an angle θ constitute a monoenergetic neutron beam of energy given by Eq. 14.51.

Monoenergetic neutron sources at the energy range provided by crystal spectrometers are necessary for the study of low-energy neutron cross sections with resonances. Consider, as an example, the total neutron cross section of iridium shown in Fig. 14.20. To be able to measure the resonances of this cross section, one needs neutron energy resolution less than 0.1 eV, resolution which can be achieved only with crystal spectrometers[41,42] or time-of-flight measurements (see Sec. 14.8).

Most of the discussion presented in Sec. 12.11 for x-ray crystal spectrometers is also valid for neutron spectrometers (i.e., rocking curve, alignment, higher order reflections, types of spectrometers). There are some differences, however, which are discussed next.

The resolving power of a neutron crystal spectrometer is given (based on Eqs. 14.49 and 14.50) by

$$\frac{E}{\Delta E} = \frac{\lambda}{2\,\Delta\lambda} = \frac{\tan\theta}{2\,\Delta\theta} = \frac{nk}{4(\Delta\theta)d\sqrt{E}}\left(1 - \frac{n^2k^2}{4d^2E}\right)^{-1/2} \tag{14.52}$$

where $k = 0.2862$ nm$\cdot$(eV)$^{1/2}$, the constant of Eq. 14.49. This function, as well as its inverse (which is the energy resolution†), is shown in Fig. 14.21, assuming $d = 0.2$ nm, $\Delta\theta = 0.3°$, and $n = 1$. One can improve the resolution, as in the use of x-ray spectrometers, by decreasing $\Delta\theta$ and choosing a crystal with short interatomic distances. Using a beryllium crystal with $d = 0.0732$ nm and $\Delta\theta = 7.8$ min, a resolution of 2 percent at 1 eV has been achieved.[42]

The energy range over which the crystal spectrometer can be used is determined from the Bragg condition (Eq. 14.50) and the requirement that $0 < \sin\theta < 1$. Using Eqs. 14.49 and 14.50, one obtains for first-order reflection

$$\sin\theta_{min} < \frac{0.02862}{2\sqrt{E\,(\text{eV})}\,d\,(\text{nm})} < 1 \tag{14.53}$$

which shows that the energy range is a function of the crystal (interplanar distance d) and the minimum observable angle θ_{min}. If one assumes $d = 0.2$ nm (LiF crystal) and $\theta_{min} = 0.5°$, the energy range becomes $0.005 \text{ eV} < E < 67$ eV. Both energy limits increase if a crystal with smaller interplanar distance d is used. In practice, the upper limit is determined by the energy resolution that is acceptable for the experiment. As Fig. 14.21 shows, the resolution deteriorates rather rapidly as energy increases. Neutron crystal spectrometers are generally used for $E < 100$ eV. Crystals that have been used include LiF, calcite, mica, beryllium, and copper.

14.8 THE TIME-OF-FLIGHT METHOD

The time-of-flight (TOF) method determines the neutron energy with a resolution that is better than with any other detector. The principle of neutron TOF is the same as

†Better resolution would be obtained with higher order reflections ($n > 1$), but unfortunately the intensity is much lower.

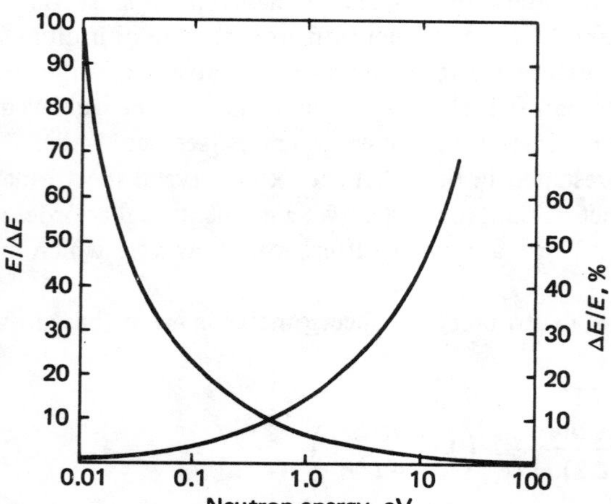

FIG. 14.21 Resolving power and energy resolution of a neutron crystal spectrometer ($n = 1$, $d = 1$ nm, $\Delta\theta = 0.5°$).

for heavy ions (see Sec. 13.6). As it was pointed out in Sec. 13.6, using the TOF technique the particle energy can be measured extremely accurately if the mass of the particle is known. The mass of the neutron is known (to within 1 keV), and energy resolution as good as 0.01 percent has been achieved.

In a TOF measurement, one determines the speed of the neutron v from the time t it takes to travel a flight path of length L. The kinetic energy of the neutron is given by

$$E = Mc^2\left(\frac{1}{\sqrt{1-\beta^2}} - 1\right) = Mc^2\left(\frac{1}{\sqrt{1-L^2/c^2t^2}} - 1\right) \tag{14.54}$$

where

$$Mc^2 = 939.553 \text{ MeV} = \text{rest mass energy of the neutron}$$

The nonrelativistic equation is the familiar one,

$$E_{\text{NR}} = \frac{1}{2}Mv^2 = \frac{1}{2}M\frac{L^2}{t^2} \tag{14.55}$$

Which equation should be used depends on the energy range measured and the resolution of the experiment. At 1 MeV, the nonrelativistic equation, Eq. 14.55, introduces an error of 0.16 percent.

The energy resolution is, using Eqs. 14.54 and 14.55,

Relativistic:

$$\frac{\Delta E}{E} = \frac{E + Mc^2}{E}\frac{\beta^2}{1-\beta^2}\sqrt{\left(\frac{\Delta L}{L}\right)^2 + \left(\frac{\Delta t}{t}\right)^2} \tag{14.56}$$

Nonrelativistic:

$$\frac{\Delta E}{E} = 2\sqrt{\left(\frac{\Delta L}{L}\right)^2 + \left(\frac{\Delta t}{t}\right)^2} \tag{14.57}$$

In neutron TOF experiments, the neutron source is a burst of neutrons generated either by a velocity selector (chopper) or by an ion beam, as explained later in this section. The time-of-flight t is the difference between the time of the neutron burst and the time of neutron detection.

The uncertainty Δt consists of three parts.

1. Δt_s, uncertainty in the time of neutron emission, is essentially equal to the width of the neutron burst and ranges from a few hundred nanoseconds to less than a hundred picoseconds.
2. Δt_d, uncertainty in the time of neutron detection, depends on the pulse risetime, since it is the pulse risetime that signals the time of detection. Neutron detectors used today have a pulse risetime equal to 5 ns or less.[43]
3. Δt_m is the uncertainty in neutron slowing-down time if the source is surrounded by a moderator.

The uncertainty ΔL is due to the finite thicknesses of the neutron-producing target and the neutron detector. The uncertainty in the measurement of L itself can be made negligible. The longer the flight path is, the smaller the uncertainty $\Delta L/L$ becomes. As the length L increases, however, the intensity of the source should increase, and by a greater factor, in order that the counting rate in the detector stay the same.

It is customary to use the quantity t/L as a *figure of merit* for TOF experiments. From Eqs. 14.54 and 14.55, one obtains

$$\left(\frac{t}{L}\right)_{\text{Rel}} = \frac{1}{c}\left[1 - \left(\frac{Mc^2}{E + Mc^2}\right)^2\right]^{-1/2} \tag{14.58}$$

$$\left(\frac{t}{L}\right)_{\text{NR}} = \frac{1}{c}\sqrt{\frac{Mc^2}{2E}} \tag{14.59}$$

Table 14.5 gives typical t/L values.

The requirements for slow-neutron TOF experiments fall in the μs/m range, and those of fast neutrons in the ns/m range. Because of this large difference in timing requirements, it is impossible to span the whole neutron energy range (eV to MeV) with the same TOF spectrometer.

The change of resolution with neutron energy is the same for TOF and crystal spectrometers. In both systems, the energy spread ΔE changes, essentially, as $E^{3/2}$ (compare Eqs. 14.52 and 14.57):

$$\Delta E = 2E\sqrt{\left(\frac{\Delta L}{L}\right)^2 + \left(\frac{\Delta t}{t}\right)^2} \approx 2E\frac{\Delta t}{t} = 2E\frac{L}{t}\frac{\Delta t}{L} \approx Ev \approx E^{3/2}$$

14.8.1 The Neutron Velocity Selector (Neutron Chopper)

The first velocity selector was designed by Fermi and his co-workers in the 1940s and is now known as the Fermi chopper.[44] The Fermi chopper consisted of a multiple sandwich of aluminum and cadmium foils that fit tightly into a steel cylinder about 38 mm (1.5 in) in diameter (Fig. 14.22). The cylinder was rotated at speeds of up to 15,000 r/min, thus allowing only bursts of neutrons to go through the aluminum channels. Based on the geometry of Fig. 14.22, no neutrons from a parallel beam

TABLE 14.5 Typical Values of t/L for Several Neutron Energies

E (eV)	t/L (μs/m)	E (MeV)	t/L (ns/m)
0.01	722	0.1	228
0.1	228.5	1	72.3
1	72.2	2	51.2
10	22.8	5	32.4
100	7.2	10	23
1000	2.3	20	16

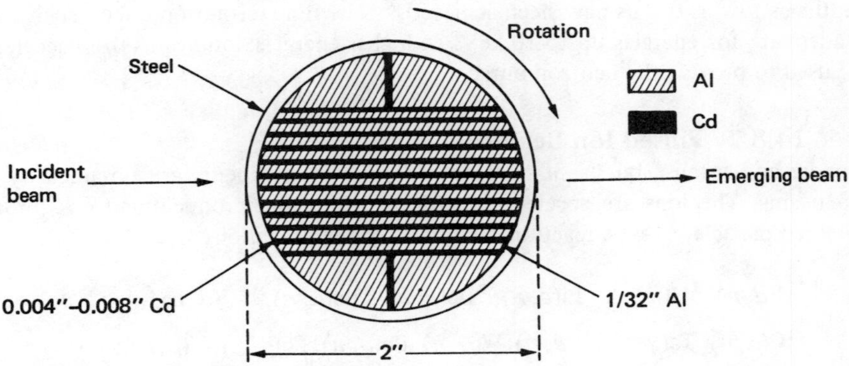

FIG. 14.22 The Fermi chopper.

would go through the channel when the chopper was more than $\Delta\theta/2$ degrees from its fully open position (Fig. 14.23), where $\Delta\theta$ = (width of channel)/(radius of cylinder). The spinning cylinder was viewed with two photocells, one giving a direct measure of the rotation and the other sending to the neutron detector a signal used for the measurement of the time-of-flight of the transmitted neutrons.

Fermi's chopper was a "slow" chopper, the word slow referring to the speed of the neutrons, and was used for neutrons up to 1 eV. Fast choppers have also been developed for use with neutron energy up to keV range,[45–47] with the rotating cylinders of the chopper having different design, depending on the requirements of the measurement.

The most important characteristic of any chopper is the width of the neutron burst. In all choppers, the shape of the pulse is essentially triangular with the base of the triangle being inversely proportional to the rotating speed of the shutter. The shape changes slightly with the neutron speed and the shape of the channel. The width of the channel, which also affects the pulse, is a compromise between acceptable time resolution and adequate counting rate. Using choppers, neutron bursts with

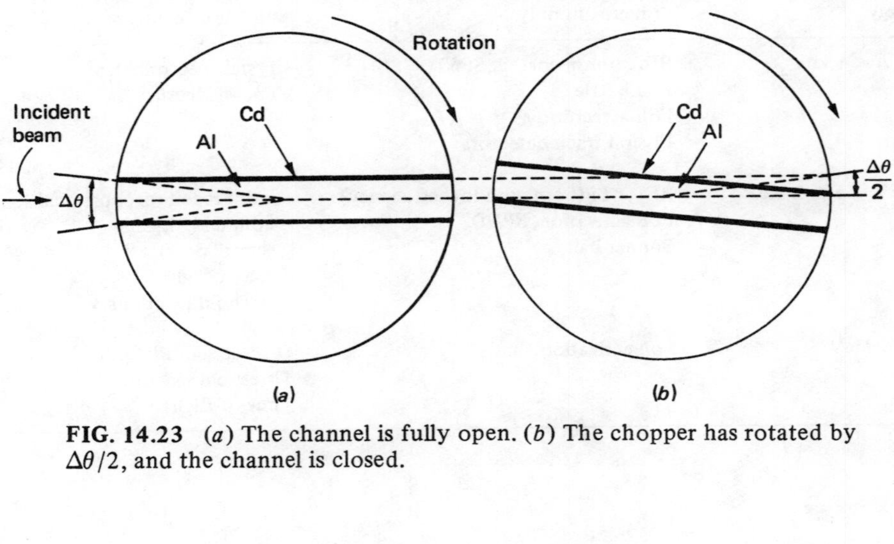

FIG. 14.23 (*a*) The channel is fully open. (*b*) The chopper has rotated by $\Delta\theta/2$, and the channel is closed.

widths as low as 0.5 μs have been achieved.[43] The time resolution due to such a width is adequate for energies up to 10 keV. At higher energies, ion beams from accelerators are used to provide the neutron burst.

14.8.2 Pulsed Ion Beams

Narrow and intense bursts of neutrons for TOF experiments are obtained by using ion beams. The ions are accelerated, strike a target and produce neutrons through a (charged particle, n) type reaction. Examples of such reactions are:

$$^{2}\mathrm{H}(d,n),\ ^{3}\mathrm{H}(p,n),\ ^{7}\mathrm{Li}(p,n),\ ^{9}\mathrm{Be}(p,n),\ ^{9}\mathrm{Be}(\alpha,n),\ ^{12}\mathrm{C}(d,n),$$

$$^{13}\mathrm{C}(\alpha,n),\ \mathrm{Ta}(\gamma,n),\ \mathrm{W}(p,n),\ \mathrm{W}(\gamma,n),\ \mathrm{Pb}(p,n),\ \mathrm{Pb}(\gamma,n),\ \mathrm{U}(p,n),$$

$$\mathrm{U}(\gamma,n),\ \mathrm{U}(\gamma,\text{fission})$$

Neutrons produced by these reactions are in the megavolt range.

The first accelerator to be used for neutron production was the cyclotron.[48] Since that time, other types of accelerators have been utilized in TOF experiments[49–52] –such as electron linear accelerators; the Los Alamos Meson Physics Facility (LAMPF), which accelerates protons to 800 MeV; and the ORELA facility at Oak Ridge, Tenn. The width of neutron bursts produced by accelerators can be lower than 100 ps.[53,54] If the burst becomes that narrow (~50 ps), the resolution is limited by the time response of the neutron detector.

14.9 CONCLUDING REMARKS

Neutron detection is, in general, more complicated and more difficult than detection of either charged particles or photons for two reasons. First, neutrons have no charge

TABLE 14.6 Summary of Neutron Detectors and Their Range of Application

Neutron energy	Measurement of number of neutrons only	Measurement of energy and number of neutrons
$0 < E < 1$ keV	BF_3, boron-coated, SPND,† TLD,† ^{6}Li, ^{3}He Foil activation Fission track detectors†	Crystal spectrometer Time-of-flight
1 keV $< E \lesssim 2$ MeV	BF_3, ^{6}Li (both with low efficiency) Foil activation, SPND† Bonner ball	Proton recoil (proportional counters, organic scintillators) Time-of-flight Threshold reactions
$E \gtrsim 1$ MeV	Foil activation	Organic scintillators Threshold reactions Time-of-flight

†See Chap. 17.

and can only be detected indirectly through photons or charged particles which they generate. Second, the neutron energy range spans at least 10 decades ($10^{-3} < E < 10^8$ eV), over which the type and cross sections for neutron reactions change drastically.

This chapter discussed all the important methods of neutron detection. Some special detectors are also presented in Chap. 17. Table 14.6 gives a summary of all the methods for neutron detection and spectrometry.

PROBLEMS

14.1 Prove that for thermal neutrons, the kinetic energies of the alpha particle and the lithium in the $^{10}B(n, \alpha)^7Li$ reaction are given by

$$T_\alpha = \frac{M_{Li}}{M_\alpha + M_{Li}} Q \qquad T_{Li} = \frac{M_\alpha}{M_\alpha + M_{Li}} Q$$

14.2 If the neutron energy is 1 MeV, what is the maximum energy of the proton in the $^3He(n, p)^3H$ reaction?

14.3 Obtain the efficiency curve as a function of neutron energy for a 1-in-diameter proportional counter filled with ^3He at 10 atm. Assume $1/v$ cross section from 0.01 to 1000 eV. The neutron beam is perpendicular to the counter axis. Compare your result with that of Fig. 14.3.

14.4 Show that the sensitivity of a neutron counter (BF_3 or boron-lined or fission counter) decreases with time as $\exp(-\sigma_a \phi t)$.

14.5 What is the maximum thickness of ^{235}U coating inside a fission counter if it is required that a 60-MeV fission fragment lose no more than 10 percent of its energy as it goes through the uranium deposit? For the fission fragment, assume $Z = 15, A = 100$.

14.6 How long should one irradiate an ^{115}In foil (100 mm^2 area, 1 mm thick) in a thermal neutron flux of 10^{14} neutrons/m^2·s to obtain 1 mCi (3.7×10^7 Bq) of activity? ($\sigma = 194$ b, $\rho = 7.3 \times 10^3$ kg/m^3, $T_{1/2} = 54$ min.)

14.7 An aluminum foil is left in a reactor for 15 s in a flux of 10^{16} neutrons/m^2·s. What is the activity produced? ($\sigma = 0.23$ b, $T_{1/2} = 2.3$ min, $m = 10^{-6}$ kg.)

14.8 The betas from the Al foil of Prob. 14.7 were counted in a 2π detection system with $\epsilon = 0.95$. Counting started 1 min after the end of the irradiation and stopped 2 min later. If the background is 20 counts/min, how many counts will the scaler record?

14.9 Calculate the irradiation time needed to produce ^{198}Au in such a quantity that the gross counting rate is 1000 counts/min using a counter with $\Omega = 10^{-2}$, $\epsilon = 0.90$, and $F = 1$. The background is 100 counts/min. It takes 10 min to get the sample from the reactor, place it under the counter, and start counting. [$\phi = 10^{16}$ neutrons/m^2·s, $T_{1/2} = 2.7$ days, $\sigma = 99$ b, $m(^{197}Au) = 10^{-9}$ kg.]

14.10 In a light-water reactor, how long will it take for the initial amount of ^{235}U to be reduced by 50 percent? ($\phi = 10^{17}$ neutrons/m^2·s, $\sigma_a = 670$ b.)

14.11 Prove Eq. 14.24. Also show that the neutron and the proton directions after collision are 90° apart ($\theta + \phi = 90°$, Fig. 14.8).

14.12 Calculate the measured neutron spectrum obtained by the proton recoil method if the detector response is a δ-function and the source spectrum is the "square" function shown in the figure below. Assume $\sigma(n, p)$ is constant for the range $E_1 \leqslant E \leqslant E_2$.

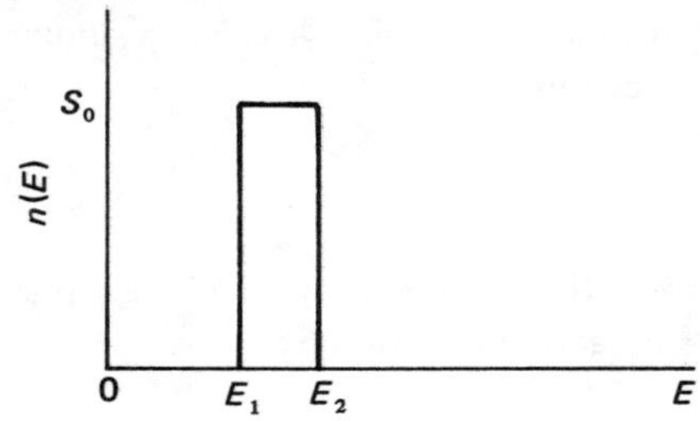

14.13 Assuming that the threshold-reaction cross sections are ideal step functions, as shown in this figure, indicate how the neutron energy spectrum could be obtained. There are N such cross sections with thresholds at $E_i|_{i=1,N}$ and $E_{i+1} - E_i = \Delta E =$ constant.

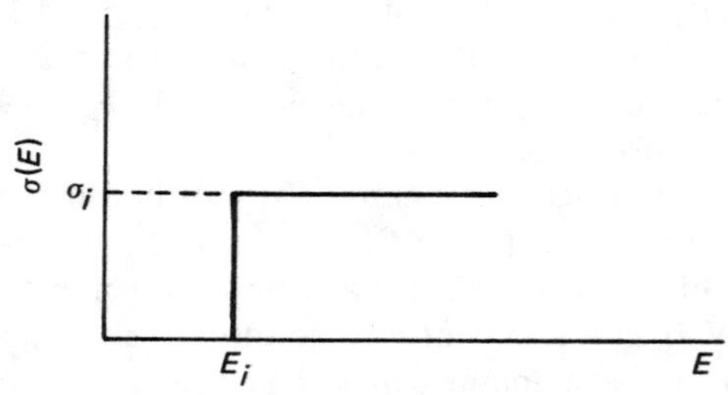

14.14 Prove Eq. 14.49.

14.15 Prove Eq. 14.56 and show that it takes the form of Eq. 14.57 in a nonrelativistic region.

14.16 A neutron TOF experiment will be designed for the measurement of 1-MeV neutrons, with the requirement that the energy resolution is 0.1 percent. What should the length of the flight path be if $\Delta t = 1$ ns and $\Delta L/L$ is negligible?

14.17 The original Fermi chopper consisted of a cylinder 1.5 inches in diameter with a maximum rotational speed of 15,000 revolutions/min. The open channels consisted of aluminum sheets $\frac{1}{32}$ inch thick.

(a) Calculate the angle during which the channel is fully open.

(b) Assuming the maximum rotational speed, what is the minimum neutron speed necessary for a neutron to make it through the channel?

BIBLIOGRAPHY

"A Review of Radiation Energy Spectra Unfolding," Proceedings of a Seminar-Workshop, ORNL/RSIC-40 (1976).

Beckurts, K. H., and Wirtz, K., *Neutron Physics*, Springer-Verlag, Berlin, 1964.

Knoll, G., *Radiation Detection and Measurement*, Wiley, New York, 1979.

REFERENCES

1 Bishop, G. B., *Nucl. Instr. Meth.* **62**:247 (1968).
2 Murray, R. B., *Nucl. Instr. Meth.* **2**:237 (1958).
3 Ginther, R. J., and Schulman, J. H., *IRE Trans. Nucl. Sci.* **N.S-5**(3):92 (1958).
4 Voitovetskii, V. K., Tolmcheva, N. S., and Arsaev, M. I., *Atomn. Energ.* **6**:321 (1959).
5 Spowart, A. R., *Nucl. Instr. Meth.* **135**:441 (1975).
6 Spowart, A. R., *Nucl. Instr. Meth.* **140**:19 (1977).
7 Hill, N. W., Harvey, J. A., Slaughter, G. G., and St. James, A., 147, ORNL-4743 (1972).
8 Haacke, L. C., Hewett, J. S., and McNeill, K. S., *Nucl. Instr. Meth.* **144**:507 (1977).
9 Benjamin, R. W., Ahlfeld, C. E., Harvey, J. A., and Hill, N. W., *Nucl. Sci. Eng.* **55**:440 (1974).
10 Bramblett, R. L., Ewing, R. L., and Bonner, T. W., *Nucl. Instr. Meth.* **9**:1 (1960).
11 Bonner, T. W., *Nucl. Phys.* **23**:116 (1961).
12 Hanna, G. C., *Nucl. Sci. Eng.* **15**:325 (1963).
13 Brunfelter, B., Kockum, J., and Zetterstrom, H. O., *Nucl. Instr. Meth.* **40**:84 (1966).
14 Bennett, E. F., *Nucl. Sci. Eng.* **27**:16 (1967).
15 Bennett, E. F., *Nucl. Sci. Eng.* **27**:28 (1967).
16 Bennett, E. F., Gold, R., and Olson, I. K., ANL-7394 (1968).
17 Bennett, E. F., and Yule, T. J., ANL-7763 (1971).
18 Johnson, R. H., ORNL/RSIC-40, p. 41 (1976).
19 Burrus, W. R., ORNL-3743 (June 1965).
20 Burrus, W. R., and Verbinski, V. V., *Nucl. Instr. Meth.* **67**:181 (1969).
21 Burrus, W. R., ORNL/RSIC-40, p. 2 (1976).
22 Verbinski, V. V., Burrus, W. R., Love, T. A., Zobel, W., Hill, N. W., and Textor, R., *Nucl. Instr. Meth.* **65**:8 (1968).
23 Kendrick, H., and Harris, L., Jr., GULF-RT-A10821 (1971).
24 Johnson, R. H., Wehring, B. W., and Dorning, J. J., *Nucl. Sci. Eng.* **73**:93 (1980).
25 "COOLC and FERDOR Spectra Unfolding Code," RSIC PSR-17, ORNL (19xx).
26 "FORIST Spectra Unfolding Code," RSIC PSR-92, ORNL (1975).
27 Bennett, E. F., and Yule, T. J., *Nucl. Sci. Eng.* **46**:236 (1971).
28 Verbinski, V. V., and Giovannini, R., *Nucl. Instr. Meth.* **114**:205 (1974).
29 Snidow, N. L., and Warren, H. D., *Nucl. Instr. Meth.* **51**:109 (1967).
30 Vehar, D. W., Clikeman, F. M., and Johnson, R. H., *ANS Trans.* **33**:697 (1979).
31 Brooks, F. D., *Nucl. Instr. Meth.* **162**:477 (1979).
32 Furuta, Y., Kinbara, S., and Kaieda, K., *Nucl. Instr. Meth.* **84**:269 (1970).
33 Ingersoll, D. T., and Wehring, B. W., *Nucl. Instr. Meth.* **147**:551 (1977).
34 Harvey, J. A., and Hill, N. W., *Nucl. Instr. Meth.* **162**:501 (1979).
35 Johnson, R. H., Ingersoll, D. T., Wehring, B. W., and Dorning, J. J., *Nucl. Instr. Meth.* **145**:337 (1977).
36 McElroy, W. N., Berg, S., and Crockett, T., AFWL-TR-67-41, vols. I–IV (1967).
37 Greer, C. R., and Walker, J. V., SC-DC-66-1512 (1966).
38 DiCola, G., and Rota, A., *Nucl. Sci. Eng.* **23**:344 (1965).
39 Berg, S., BNWL-855 (1968).
40 Oster, C. A., Simons, R. L., Odette, G. R., McElroy, W. N., and Lippincott, E. P., ORNL/RSIC-40, p. 148 (1976).
41 Sawyer, R. B., Wollan, E. O., Bernstein, S., and Peterson, K. C., *Phys. Rev.* **72**:109 (1947).
42 Borst, L. B., and Sailor, V. L., *Rev. Sci. Instr.* **24**:141 (1953).
43 Firk, F. W. K., *Nucl. Instr. Meth.* **162**:539 (1979).
44 Fermi, E., and Marshall, L., *Phys. Rev.* **72**:193 (1947).
45 Selove, W., *Rev. of Sci. Instr.* **23**:350 (1952).
46 Seidl, F. G. P., Hughes, D. J., Palevsky, H., Levin, J. S., Kato, W. Y., and Sjostrand, N. G., *Phys. Rev.* **95**:476 (1954).
47 Bollinger, L. M., Cote, R. E., and Thomas, G. E., *Proceedings of the 2nd Intern. Conference on Peaceful Uses of Atomic Energy,* **14**:239 (1958).
48 Alvarez, L. W., *Phys. Rev.* **54**:609 (1938).
49 Bareford, C. F., and Kelliker, M. G., *Phillips Tech. Rev.* **15**:1 (1953).

50 Firk, F. W. K., and Bowey, E. M., "Comptes Rendus du Congres Intern. de PHysique Nucleaire," P. Gugenberger (ed.), C.N.R.S., Paris, 1964.
51 Wasson, O. A., and Draper, J. E., *Nucl. Phys.* **73**:499 (1965).
52 Harvey, J. A., "Proc. Int. Conf. on Interactions of Neutrons with Nuclei," E. Sheldon (ed.), US ERDA Tech Info. Center, Oak Ridge, Tenn., p. 144.
53 Norris, N. J., and Hanst, R. K., *Edgerton, Germeshausen and Grier, Prog. Rep. EGG*, 1183-2142 (1967).
54 Laszewski, R. M., Holt, R. J., and Jackson, H. E., *Phys. Rev. Let.* **38**:813 (1977).

15

activation analysis

15.1 INTRODUCTION

Activation analysis achieves a qualitative and quantitative analysis of an unknown sample by irradiating the sample and thus producing radioactive nuclides from stable or unstable isotopes present in the sample. The radioactive nuclides can be identified from properties of the radiations they emit:

1 Type of radiation
2 Energy of radiation
3 Intensity of radiation
4 Half-life

The basic principle of activation analysis is not new. It was applied for the first time in 1936 by Hevesy and Levi,[1] who determined the amount of dysprosium in an yttrium sample. The dysprosium in the sample became radioactive when bombarded with neutrons from a Ra–Be source. Two years later, Seaborg and Livingood[2] determined the gallium content in an iron sample by bombarding it with deuterons. The sensitivity of the method increased considerably with the availability of high neutron fluxes from nuclear reactors. Although charged particles, gamma rays, and 14-MeV neutrons may be used as the bombarding particles, thermal neutrons are, by far, the particles most frequently utilized for the irradiation of the sample.

Activation analysis has become, because of its extremely high sensitivity, an indispensable tool in a wide variety of fields[3] ranging from science and engineering[4] to industry,[5,6] minerals exploration,[7–14] medicine,[15–21] environmental monitoring,[12,22–24] and forensic science.[25–28] The purpose of this chapter is not to present all the aspects, details, and applications of this field, but to discuss the major steps that comprise the method, the interpretation of the results, the errors and sensitivity of the method, and certain representative applications. The reader will find many more details and an extensive list of applications in the bibliography and the references given at the end of the chapter.

The activation analysis method consists of the following major steps, to be discussed next:

1 Selection of the optimum nuclear reaction
2 Preparation of the sample for irradiation
3 Irradiation of the sample
4 Counting of the irradiated sample
5 Analysis of the counting results

15.2 SELECTION OF THE OPTIMUM NUCLEAR REACTION

The optimum nuclear reaction is chosen with these considerations in mind.

1 Production of large activity should occur within a reasonable irradiation time.
2 The radioisotope produced should have a reasonable half-life ($T >$ min).
3 The type and energy of the radiation emitted by the radioisotope should not present great counting difficulties.
4 A minimum number of interfering reactions should be involved.

If the sample is completely unknown, one starts with neutron irradiation because neutrons are absorbed by almost all isotopes. If the composition of the sample is known, then the best reaction for the identification of the isotope of interest should be chosen. Sometimes there is more than one reaction available for the same isotope. For example, aluminum bombarded with fast neutrons may be detected by three different reactions:

$$^{27}\text{Al}(n, p)^{27}\text{Mg}$$

$$^{27}\text{Al}(n, \alpha)^{24}\text{Na}$$

$$^{27}\text{Al}(n, \gamma)^{28}\text{Al}$$

The "optimum nuclear reaction" depends not only on the isotope and the bombarding particles but also on the composition of the sample that is analyzed. For example, the $^{27}\text{Al}(n, \gamma)^{28}\text{Al}$ reaction may be the best for detection of aluminum in a certain sample. However, if the sample contains silicon in addition to aluminum, the reaction $^{28}\text{Si}(n, p)^{28}\text{Al}$ also produces ^{28}Al, and thus causes an interference to the measurement. If silicon is present, it may be better to use the reaction $^{27}\text{Al}(n, p)^{27}\text{Mg}$ or $^{27}\text{Al}(n, \alpha)^{24}\text{Na}$. More details about interfering reactions are given in Sec. 15.9.

The most commonly used neutron reaction is the (n, γ) reaction, which takes place with almost all isotopes (although with different probability) and has no threshold. In general, the (n, γ) cross section is higher for thermal than for fast neutrons. Other neutron interactions are (n, α), (n, p), and $(n, 2n)$ reactions. Except for a few exothermic (n, α) reactions, the others have a threshold; therefore, they can occur with fast neutrons only. Table 15.1 lists neutron reactions for the identification of several elements. Details for many more elements and reactions can be found in the bibliography of this chapter.

Charged-particle reactions are also used in activation analysis. Their disadvantage over neutron reactions is that they are mostly endothermic, i.e., they have a threshold. Table 15.2 gives several examples of such reactions.

Photon activation complements neutron and charged-particle activation. Photons are better than neutrons in certain cases. For example, photons are preferred if the product of the neutron activation is an isotope that has a very short half-life or emits only low-energy betas or low-energy x-rays. The cross-sections for photonuclear reactions are generally smaller than those for neutrons and charged particles. Table 15.3 gives several photonuclear reactions that have been used in activation analysis.

TABLE 15.1 Neutron Activation Reactions

Element	Symbol	Reaction	Threshold energy (MeV)	Half-life of product	Main radiation emitted and its energy (MeV)
Aluminum	Al	$^{27}Al(n, \gamma)^{28}Al$	(–)†	2.3 min	β^- (2.85), γ (1.78)
		$^{27}Al(n, p)^{27}Mg$	1.9	9.46 min	β^- (1.75), γ (0.84, 1.013)
		$^{27}Al(n, \alpha)^{24}Na$	3.27	15 h	β^- (1.389), γ (1.369, 2.754)
Arsenic	As	$^{75}As(n, \gamma)^{76}As$	(–)	26.4 h	β^- (2.97), γ (0.559)
		$^{75}As(n, \alpha)^{72}Ga$	(–)	14.1 h	β^- (3.15), γ (0.835)
		$^{75}As(n, 2n)^{74}As$	8.27	17.9 d	β^- (3.15), γ (0.835)
Cadmium	Cd	$^{110}Cd(n, \gamma)^{111m}(Cd)$	(–)	48.6 min	e^-, γ (0.247)
		$^{110}Cd(n, p)^{110m}Ag$	2.12	235 d	γ (0.658)
Calcium	Ca	$^{48}Ca(n, \gamma)^{49}Ca$	(–)	8.8 min	γ (3.07), β (1.95)
Chlorine	Cl	$^{37}Cl(n, \gamma)^{38}Cl$	(–)	37.2 min	β^- (4.91), γ (1.6, 2.17)
		$^{37}Cl(n, p)^{37}S$	3.6	5 min	β^-, γ (3.09)
Copper	Cu	$^{63}Cu(n, 2n)^{62}Cu$	11.01	9.76 min	γ (0.511)
Fluorine	F	$^{19}F(n, \alpha)^{16}N$	1.57	7.15 s	γ (6.13)
Gold	Au	$^{197}Au(n, \gamma)^{198}Au$	(–)	2.7 d	β^- (0.962), γ (0.412)
		$^{197}Au(n, 2n)^{196}Au$	7.36	6.18 d	β^-, γ (0.356)
Iodine	I	$^{127}I(n, \gamma)^{128}I$	(–)	25 min	β^- (2.12), γ (0.441)
Iron	Fe	$^{58}Fe(n, \gamma)^{59}Fe$	(–)	45.5 d	β^-, γ (1.095, 1.292)
		$^{56}Fe(n, p)^{56}Mn$	2.98	2.57 d	β^-, γ (0.847)
Lead	Pb	$^{208}Pb(n, 2n)^{207m}Pb$	7.45	0.885 s	γ (0.570)
Mercury	Hg	$^{200}Hg(n, 2n)^{199m}Hg$	8.11	43 min	γ (0.158)
		$^{196}Hg(n, \gamma)^{197}Hg$	(–)	65 h	γ (0.077)
		$^{202}Hg(n, \gamma)^{203}Hg$	(–)	46.9 d	γ (0.279)
Nickel	Ni	$^{58}N(n, 2n)^{57}Ni$	12.09	36 h	β^+, γ (0.511, 1.37)
Nitrogen	N	$^{14}N(n, 2n)^{13}N$	11.31	10 min	γ (0.511)
Oxygen	O	$^{16}O(n, p)^{16}N$	10.2	7.1 s	γ (6.13)
Phosphorus	P	$^{31}P(n, \alpha)^{28}Al$	2	2.3 min	β^-, γ (1.78)
Potassium	K	$^{39}K(n, 2n)^{38}K$	13.41	7.7 min	γ (0.511, 2.17)
Silicon	Si	$^{28}Si(n, p)^{28}Al$	3.99	2.3 min	γ (1.78)
Silver	Ag	$^{109}Ag(n, \gamma)^{110}Ag$	(–)	24 s	γ (0.66)

†(–) = No threshold.

TABLE 15.2 Charged-Particle Reactions

Element	Symbol	Reaction	Threshold energy (MeV)	Half-life of product	Main radiation emitted and its energy (MeV)
Boron	B	$^{10}B(\alpha, n)^{13}N$	–	10 min	β^+, γ (0.511)
		$^{10}B(p, \gamma)^{11}C$	0.4	20.4 min	β^+, γ (0.511)
		$^{11}B(p, n)^{11}C$	2.76		
Carbon	C	$^{12}C(p, n)^{11}C$	~19		
Nitrogen	N	$^{14}N(p, \alpha)^{11}C$	~3.5		
Oxygen	O	$^{16}O(\alpha, d)^{18}F$	35	109.8 min	
Sodium	Na	$^{23}Na(\alpha, n)^{26m}Al$	4.62	6.7 s	β^+, γ (0.511)
Aluminum	Al	$^{27}Al(\alpha, n)^{30}P$	3.38	2.55 min	β^+, γ (0.511)
Copper	Cu	$^{63}Cu(\alpha, n)^{66}Ga$	7.69	9.45 h	
Phosphorus	P	$^{31}P(d, p)^{32}P$	–	14.3 d	β^- (E_{max} = 1.17)
Iron	Fe	$^{54}Fe(d, n)^{55}Co$	–	18.2 h	β^+, γ (0.511, 0.93)

TABLE 15.3 Photonuclear Reactions

Element	Symbol	Reaction	Threshold energy (MeV)	Half-life of product	Main radiation emitted and its energy (MeV)
Carbon	C	$^{12}C(\gamma, n)^{11}C$	18.7	20.4 min	β^+, γ (0.511)
Fluorine	F	$^{19}F(\gamma, n)^{18}F$	10.5	109.8 min	β^+, γ (0.511)
Nitrogen	N	$^{14}N(\gamma, n)^{13}N$	10.5	10 min	β^+, γ (0.511)
Oxygen	O	$^{16}O(\gamma, n)^{15}O$	15.7	2.03 min	β^+, γ (0.511)
Copper	Cu	$^{63}Cu(\gamma, n)^{62}Cu$	10.8	9.7 min	β^+, γ (0.511)
		$^{65}Cu(\gamma, n)^{64}Cu$	9.9	12.8 h	β^+, γ (0.511)
Silver	Ag	$^{107}Ag(\gamma, n)^{106}Ag$	9.4	24 min	β^+, γ (0.511)
Sodium	Na	$^{23}Na(\gamma, n)^{22}Na$	12.4	2.62 y	γ (0.511)
Lead	Pb	$^{204}Pb(\gamma, n)^{203}Pb$	8.4	52.1 h	γ (0.279)

15.3 PREPARATION OF THE SAMPLE FOR IRRADIATION

A sample should be prepared properly and then placed in a container before it is irradiated. The person who prepares the sample should be extremely careful not to contaminate it. Activation analysis is so sensitive that it can determine traces of elements undetectable by chemical methods. If the sample is left on a table for a certain period of time, it collects dust that acts as a contaminant. Touch by hand may transfer enough salt to cause the irradiated sample to show the presence of sodium and chlorine. To avoid contamination, samples should be handled in dry boxes or in clean rooms. The person who prepares the sample should use clean instruments (knife, file, tweezers, etc.) and also wear clean plastic gloves.

Solid samples should have their surfaces cleaned with a suitable cleaning fluid to remove any surface contamination. The weight of the sample is determined after cleaning it. For maximum accuracy, the weight is determined again after irradiation and counting are completed.

Liquids and powders cannot be cleaned, so they are handled in clean containers, avoiding contamination from the container wall. For liquid samples, care should be exercised to avoid loss of fluid when the sample is transferred in and out of the container. In the case of powder (or powderized) samples, the observer should be certain that a truly representative sample has been prepared. This is especially important if the main sample under analysis is not homogeneous.

The packaging material or container should

1 have high radiation and thermal resistance (i.e., it should not decompose, melt, or evaporate in the irradiation environment)
2 have low content of elements that become radioactive
3 be inexpensive and easy to handle

Materials that are used most frequently are polyethylene, silica, and aluminum foil. Polyethylene satisfies (2) and (3) of the requirements listed above, but it has low resistance to radiation and temperature. It becomes brittle after exposure to a fluence of 10^{21}-10^{22} neutrons/m^2. Polyethylene tubes of different diameters are routinely used in radiation laboratories. The tubes can be easily cleaned and sealed. Silica containers are not as useful as polyethylene because they are not as pure, they become radioactive, and sealing is more complicated. Aluminum foil is useful for packaging solids, but it becomes radioactive through $^{27}Al(n,\gamma)^{28}Al$ and $^{27}Al(n,\alpha)^{24}Na$ reactions. The second reaction is more troublesome than the first because the half-life of ^{24}Na is 15 h, while the half-life of ^{28}Al is only 2.3 min.

15.4 SOURCES OF RADIATION

Intensities, energies, and special characteristics of the various radiation sources are briefly discussed in this section.

15.4.1 Sources of Neutrons

Neutron sources include reactors, accelerators, and isotopic sources. Nuclear reactors are, by far, the most frequently used irradiation facilities. They provide high fluxes (upper limit 10^{18} neutrons/m$^2\cdot$s) of mostly thermal neutrons ($E < 1$ eV). Fast neutrons in the keV range are also available, but at lower flux levels.

When short-lived isotopes are involved, a higher activity is produced by irradiating the sample in a reactor that can be pulsed (see Lenihan and Thomson). Such a reactor producing a high flux of about 10^{20} neutrons/m$^2\cdot$s for a short period of time (milliseconds) is the TRIGA reactor, marketed by General Atomic.

Accelerators produce fast neutrons as products of charged-particle reactions. The most popular device is the so-called neutron generator, which operates on the reaction

$$^2_1H + ^3_1H \rightarrow ^1_0n + ^4_2He + 17.586 \text{ MeV}$$

The cross section for this exothermic reaction peaks at a deuteron kinetic energy of about 120 keV with a value of about 5 b. The neutrons produced have an energy of about 14 MeV. (The neutron kinetic energy changes slightly with the direction of emission.) The maximum neutron flux provided by a neutron generator is of the order of 10^{12} neutrons/m$^2\cdot$s.

Neutrons with an average energy of about 2.5 MeV are produced by the (d, d) reaction

$$^2_1H + ^2_1H \rightarrow ^1_0n + ^3_2H + 3.266 \text{ MeV}$$

The cross section for this reaction peaks at about 2 MeV bombarding deuteron energy with a value of about 100 mb. At acceleration voltages normally used in neutron generators (~150 keV), the cross section is about 30 mb. The (d, d) reaction offers neutron fluxes of the order of 10^9 neutrons/m$^2\cdot$s. It is important to note that both the (d, t) and the (d, d) reactions produce essentially monoenergetic neutrons.

Isotopic neutron sources are based on (α, n) and (γ, n) reactions, and on spontaneous fission (^{252}Cf). They all produce fast neutrons. The (α, n) and (γ, n) sources produce the neutrons through the reactions

$${}^4_2\text{He} + {}^9_4\text{Be} \rightarrow {}^1_0n + {}^{12}_6\text{C}$$

$$\gamma + {}^9_4\text{Be} \rightarrow {}^1_0n + {}^8_4\text{Be}$$

The isotope ^{252}Cf is the only spontaneous fission (SF) source of neutrons easily available. It provides fission spectrum neutrons with an average of 2.3 MeV. The characteristics of isotopic neutron sources are given in Table 15.4.

15.4.2 Sources of Charged Particles

Apart from certain α-emitting radioisotopes, accelerators are the only practical sources of charged particles. They can provide almost any charged particle or ion for bombardment of the target for a wide range of energies. The particles most commonly used as projectiles are protons, deuterons, alphas, tritons, and ^{3}He nuclei. The beam current of the accelerator is related to particles per second hitting the target by the equation[29]

$$I\,(\text{particles/s}) = 6.2 \times 10^{15}\,\frac{i\,(\text{mA})}{z} \qquad (15.1)$$

where i = the beam current in mA

z = the charge of the accelerated particle in units of the electronic charge

Knowing the number of particles per second hitting the target and the cross section for the reaction, one can calculate the reaction rate.

15.4.3 Source of Photons

The sources of photons include radioisotopes, nuclear reactions, and bremsstrahlung radiation.

There are many radioisotopes that emit gamma rays. The most useful ones are:

1 ^{24}Na with a 15-h half-life emitting γ's with energy 1.37 and 2.75 MeV
2 ^{60}Co with 5.3-year half-life emitting γ's with energy 1.17 and 1.33 MeV
3 ^{124}Sb with a 60-d half-life emitting γ's with energy 1.71, 2.1, and 2.3 MeV
4 ^{22}Na with a 2.6-y half-life emitting a gamma with energy 1.275 MeV (^{22}Na being a positron emitter is also a source of 0.511-MeV gamma rays)

TABLE 15.4 Isotopic Neutron Sources†

Source	Reaction	Yield	Average neutron energy (MeV)	Half-life isotope involved
^{226}Ra–Be	(α, n)	7×10^5 neutrons/s·g (Ra)	~4	1,600 y
^{210}Po–Be	(α, n)	1×10^{10} neutrons/s·g (Po)	~4	138.4 d
^{239}Pu–Be	(α, n)	1.0×10^5 neutrons/s·g (Pu)	~4	24,131 y
^{241}Am–Be	(α, n)	7×10^6 neutrons/s·g (Am)	~4	432 y
^{124}Sb–Be	(γ, n)	1.0×10^{10} neutrons/s·g	0.024	60.2 d
^{242}Cm–^{241}Am–Be	(α, n)	1×10^{10} neutrons/s·g (Cm)	4	162.8 d (^{242}Cm)
^{252}Cf	(SF)	2×10^{12} neutrons/s·g	2.3	2.65 y

†From Lenihan, Thomson, and Guinn.

Photons of extremely high energy may be produced by nuclear reactions. Examples are:

$^3H(p, \gamma)^4He$ $E_\gamma = 19.8$ MeV

$^7Li(p, \gamma)^8Be$ $E_\gamma = 14.8$ and 17.6 MeV

$^{11}B(p, \gamma)^{12}C$ $E_\gamma = 11.7$ and 16.1 MeV

Unfortunately, the gamma fluxes generated by these reactions are very small, relative to neutron fluxes produced by reactors.

Bremsstrahlung is produced with the help of electron accelerators. The electrons are accelerated to a certain energy and then are allowed to hit a solid target. The radiation produced has a continuous energy spectrum, extending from zero energy up to the maximum electron kinetic energy. Large photon fluxes are produced, and may be used for activation of rather large samples.

15.5 IRRADIATION OF THE SAMPLE

Depending on the selected reaction, irradiation of the sample may take place in a reactor, in an accelerator, or with an isotopic source. After the selection of an irradiation facility, the next step is a decision about the irradiation time. If the sample contains known isotopes at approximately known amounts, it is easy to estimate the proper irradiation time. If, on the other hand, the sample is completely unknown, one irradiates the sample for an arbitrary time, checks some of the isotopes present (from the emitted radiations), and then irradiates the sample again for a time that will provide enough activity for proper isotope identification with the desired accuracy.

The equation that gives the activity produced after irradiating the sample for time t_0 is (for derivation see Sec. 14.4)

$$A(t_0) = a_i m \frac{N_A}{A_i} \sigma_i \phi (1 - e^{-\lambda t_0}) \tag{15.2}$$

where m = mass of the element of interest in the sample

a_i = weight fraction (abundance) of isotope with atomic mass A_i (A_i is an isotope of the element with mass m)

λ = decay constant of the radioisotope produced

σ_i = cross section for the reaction that makes the isotope with atomic weight A_i radioactive

ϕ = particle flux (particles/$m^2 \cdot s$)

N_A = Avogadro's number

Equation 15.2 is valid if (see Sec. 14.4)

1 The number of target nuclei stays essentially constant, i.e., $\sigma_i \phi t \ll 1$.
2 The radioisotope produced has such a small neutron-absorption cross section that $\lambda_{i+1} \gg \phi \sigma_{i+1}$.
3 The flux is uniform throughout the target.

If the half-life of the radioisotope is much shorter than t_0 ($t_0 \gtrsim 6\tau_{1/2}$), saturation activity (Asat) is obtained, given by

$$A_{\text{sat}} = a_i m \frac{N_A}{A_i} \sigma_i \phi \tag{15.3}$$

Equation 15.3 indicates that, for a particular isotope, the activity increases by irradiating a larger mass in a higher flux ϕ.

The size of the sample (mass m) is dictated by four factors:

1 The maximum activity that can be safely handled under the conditions of the laboratory (i.e., shielding of source and detector, existence of remote control, automated remote handling of samples).
2 The size of the sample holder.
3 The self-absorption of the radiation emitted by the sample. This is particularly important if the radiation detected is betas or soft x-rays.
4 The size of the detector. Little, if anything, is gained by using a sample much larger than the detector size.

The flux ϕ is determined by the limitations of the irradiation facility. At the present time (1982), the maximum flux is about 10^{18} neutrons/$m^2 \cdot s$ (thermal neutron flux).

15.6 COUNTING OF THE SAMPLE

After irradiation is completed, the sample is counted using an appropriate system. The qualitative and quantitative determination of an isotope is based on the analysis of the energy spectrum of the radiations emitted by the radioisotope of interest. Sometimes it may be necessary to use information about the half-life of the isotope(s). In such a case, counting may have to be repeated several times at specified time intervals.

The counting system depends on the radiation detected. Modern activation analysis systems depend on the detection of gamma rays and x-rays and very seldom on detection of other particles. For this reason, the discussion in the rest of this chapter is based on the assumption that the irradiated sample emits photons.

A basic counting system for activation analysis consists of a detector [Ge(Li) or Si(Li)], electronics (i.e., preamplifier, amplifier), and a multichannel analyzer (MCA). Modern MCAs do much more than record the data. They are minicomputers that analyze the recorded spectra. Examples are Canberra's SCORPIO Computer Analyzer and Tracor Northern's Spectroscopy System TN-4000.

15.7 ANALYSIS OF THE RESULTS

The analysis of an activation analysis spectrum is based on the procedures described in Sec. 12.7.3. It is performed either by the MCA itself, if that instrument has such capability, or by a digital computer. Several computer codes have been written for that purpose.[30-34]

Activation analysis may be qualitative or quantitative. In a qualitative measure-

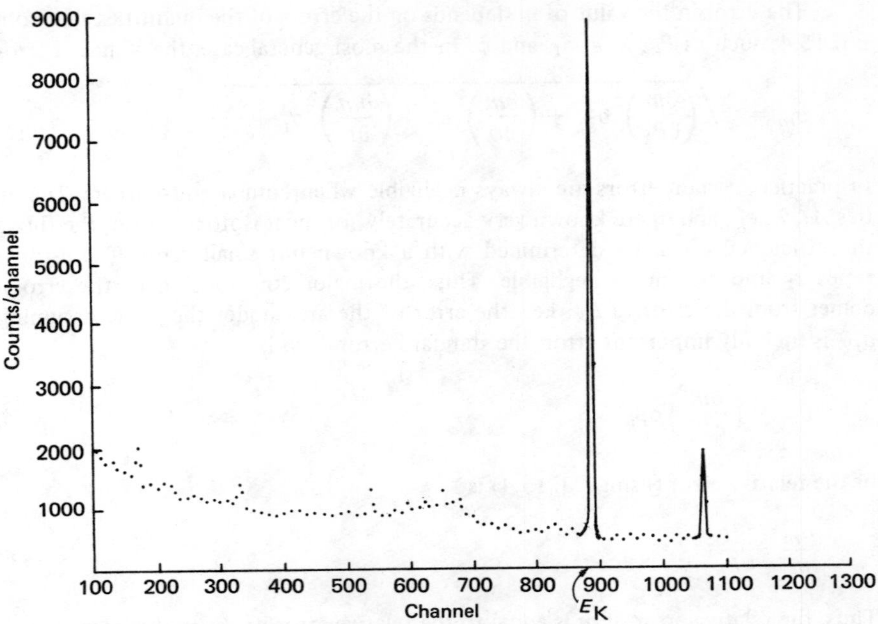

FIG. 15.1 A typical Ge(Li) energy spectrum. The element will be identified using the main peak at channel 888.

ment, only identification of the element is involved. This is accomplished, as shown in Sec. 12.7.3, from the energies and intensities of the peaks of the spectrum. In a quantitative measurement, on the other hand, in addition to identification, the amount of element in the sample is also determined. To illustrate how the mass is determined and what the errors and sensitivity of the method are, consider the energy spectrum of Fig. 15.1 as an example.

Assume that the mass of an element in the sample will be determined from the full-energy peak at E_k. Using the notation of Sec. 14.4, the mass m is given by (see Eq. 14.22)

$$m = \frac{P_k A_i \lambda}{\epsilon(E_k) e_k a_i N_A \sigma_i \phi (1 - e^{-\lambda t_0})(e^{-\lambda t_1} - e^{-\lambda t_2})} \tag{15.4}$$

where P_k = net number of counts under the peak (determined by one of the methods described in Sec. 12.7.3)

$\epsilon(E_k)^\dagger$ = absolute full-energy peak detector efficiency at energy E_k

e_k = probability that a photon of energy E_k is emitted per decay of the isotope (also known as intensity of this gamma)

$t_2 - t_1$ = counting time

†In activation analysis, the efficiency is determined in such a way as to include the solid angle Ω and the other correction factors $F(E_k)$ discussed in Chap. 8.

The error in the value of m depends on the errors of the quantities that comprise Eq. 15.4, such as P_k, λ, ϵ, σ_i, and ϕ. In the most general case, the standard error m is

$$\sigma_m = \sqrt{\left(\frac{\partial m}{\partial P_k}\right)^2 \sigma_{P_k}^2 + \left(\frac{\partial m}{\partial \phi}\right)^2 \sigma_\phi^2 + \left(\frac{\partial m}{\partial t}\right)^2 \sigma_t^2 + \cdots} \tag{15.5}$$

In practice, certain errors are always negligible when compared to others. The quantities A_i, λ, e_k, and σ_i are known very accurately for most isotopes. Also, the flux ϕ and the efficiency ϵ can be determined with a known but small error. The error in the times t_1 and t_2 can be negligible. Thus, the major contribution to the error of m comes from the error of P_k, i.e., the error of the area under the peak. Assuming that σ_{P_k} is the only important error, the standard error of m is

$$\sigma_m = \left(\frac{\partial m}{\partial P_k}\right) \sigma_{P_k} \tag{15.6}$$

or the relative error (using Eq. 15.4) is

$$\frac{\sigma_m}{m} = \frac{\sigma_{P_k}}{P_k} \tag{15.7}$$

Thus, the relative error of m is equal to the relative error of P_k, in this case.

It should be emphasized that only one well identified peak, and not the whole spectrum, is needed for the determination of an element in the sample.

In practice, the objective of the measurement is often to identify the mass of a particular trace element in the sample. Then the unknown mass is determined in a simpler way by irradiating, along with the unknown, a standard sample with a known mass of the trace element and counting both samples with the same counter. If m_s is the mass of the standard and m_x the mass of the unknown, using Eq. 15.4 one obtains

$$m_x = m_s \frac{(P_k)_x (e^{-\lambda t_1} - e^{-\lambda t_2})_s}{(P_k)_s (e^{-\lambda t_1} - e^{-\lambda t_2})_x} \tag{15.8}$$

where the times t_1 and t_2 are different for the standard and the unknown, but in both cases $t_2 - t_1 = T$. Use of Eq. 15.8 constitutes a relative method, in contrast to the use of Eq. 15.4, which represents an absolute method.

15.8 SENSITIVITY OF ACTIVATION ANALYSIS

Sensitivity of the activation analysis method for a particular element refers to the minimum mass of that element that can be reliability detected. The minimum detectable mass is determined from Eq. 15.4 by assuming the most favorable conditions for the measurement and by setting an upper limit for the acceptable error of the result. The process is similar to the determination of the minimum detectable activity discussed in Sec. 2.20.

Assuming that the observer is willing to accept a maximum error σ_m such that

$$\sigma_m \leqslant fm \tag{15.9}$$

($f < 1$) and that the only error in the determination of the mass comes from the error in the number of counts under the peak, a limiting counting rate can be obtained as shown in Sec. 2.20. If one defines a net counting rate r as†

$$r = \frac{G_k}{T} - b_k \tag{15.10}$$

where

$$G_k = P_k + b_k T \tag{15.11}$$

then the result for the minimum acceptable counting rate is

$$r \geqslant \frac{1 + \sqrt{1 + 4f^2 bT + 4f^2 T^2 \sigma_b^2}}{2f^2 T} \tag{15.12}$$

which is, of course, identical to Eq. 2.105.

EXAMPLE 15.1

What is the minimum mass of gold that can be detected by neutron activation analysis under the conditions listed below?

$\phi = 10^{16}$ neutrons/m$^2 \cdot$s $\qquad \epsilon = 0.30$

$t_0 = 2$ h $\qquad t_1 = 5$ min

$f = 0.50 \qquad t_2 = 125$ min

$b = 20 \pm 0.2$ counts/min

For gold, $A = 197, a = 1, e_k = 1, \sigma = 99$ b, and $T_{1/2} = 2.7$ days.

ANSWER

Using Eq. 15.12, the minimum acceptable counting rate is

$$r = \frac{1 + \sqrt{1 + 4(0.5)^2 20(120) + 4(0.5)^2 (120)^2 (0.2)^2}}{4(0.5)^2 120} = 0.5 \text{ counts/min}$$

Using Eqs. 15.10 and 15.11,

$$P_k = rT = 0.5(120) = 60 \text{ counts}$$

Since the half-life of ^{198}Au is 2.7 days,

$$\lambda = \frac{\ln 2}{2.7 \text{ d}} = 1.78 \times 10^{-4} \text{ s}^{-1}$$

†Note that both G_k and b_k refer to gross counts and background of the peak k, and not the whole spectrum.

and the exponential factors in Eq. 15.4 become

$$\exp(-\lambda t_0) = \exp(-1.78 \times 10^{-4} \times 2 \times 3600) = 0.277$$

$$\exp(-\lambda t_1) = \exp(-1.78 \times 10^{-4} \times 300) = 0.948$$

$$\exp(-\lambda t_2) = \exp(-1.78 \times 10^{-4} \times 125 \times 60) = 0.263$$

Equation 15.4 gives:

$$m = \frac{60(197)(1.78 \times 10^{-4})\ \text{s}^{-1}}{(0.30)(6.023 \times 10^{23})(99 \times 10^{-28}\ \text{m}^2)(10^{16}\ \text{neutrons/m}^2 \cdot \text{s})(1 - 0.277)(0.948 - 0.263)}$$
$$= 2.37 \times 10^{-13}\ \text{g} = 2.37 \times 10^{-16}\ \text{kg}$$

■ *EXAMPLE 15.2*

What is the absolute minimum mass of an element that can be detected under the most favorable conditions?

ANSWER

The absolute minimum mass will be determined if one assumes:

Efficiency 100 percent ($\epsilon = 1$)
Intensity 100 percent ($e_k = 1$)
Saturation activity ($1 - e^{-\lambda t_0} = 1$)
Maximum thermal neutron flux ($\sim 10^{18}$ neutrons/m$^2 \cdot$s)

Then Eq. 15.4 takes the form

$$m = \frac{P_k A_i \lambda}{a_i N_A \sigma_i \phi (e^{-\lambda t_1} - e^{-\lambda t_2})} \qquad (15.13)$$

Factors that may further affect the result given by Eq. 15.13 depend on the background of the counting system and the maximum acceptable error (both background and acceptable error affect the minimum acceptable value of P_k).

15.9 INTERFERENCE REACTIONS

One source of error in activation analysis is interference reactions. These are reactions that produce the same isotope as the one being counted, through bombardment of a different isotope in the sample. As an example, assume that a sample is analyzed for magnesium by using fast-neutron activation. The reaction of interest is ^{24}Mg$(n,p)^{24}$Na. Therefore, the activity of ^{24}Na will be recorded, and from that the amount of ^{24}Mg can be determined. If the sample contains ^{23}Na and ^{27}Al, two other reactions may take place which also lead to ^{24}Na. They are

$$^{23}\text{Na}(n,\gamma)^{24}\text{Na}$$

$$^{27}\text{Al}(n,\alpha)^{24}\text{Na}$$

If this is the case and the investigator does not consider these last two reactions, the mass of ^{24}Mg will be determined to be higher than it is.

Interference reactions are discussed in detail in many activation analysis books (see Racovic, and Nargolwalla and Przybylowicz). A few representative examples are given below:

$^{68}Zn(n, \gamma)^{69m}Zn$ and $^{69}Ge(n, p)^{69m}Zn$ and $^{72}Ge(n, \alpha)^{69m}Zn$
$^{32}S(n, p)^{32}P$ and $^{31}P(n, \gamma)^{32}P$ and $^{35}Cl(n, \alpha)^{32}P$
$^{17}O(n, \alpha)^{14}C$ and $^{13}C(n, \gamma)^{14}C$ and $^{14}N(n, p)^{14}C$
$^{98}Tc(n, \gamma)^{99m}Tc$ and $^{98}Mo(n, \gamma)^{99}Mo \xrightarrow{\beta^-} {}^{99m}Tc$
$^{31}P(n, \gamma)^{32}P$ and $^{30}Si(n, \gamma)^{31}Si \xrightarrow{\beta^-} {}^{31}P(n, \gamma)^{32}P$
$^{55}Mn(n, \gamma)^{56}Mn$ and $^{54}Cr(n, \gamma)^{55}Cr \xrightarrow{\beta^-} {}^{55}Mn(n, \gamma)^{56}Mn$
$^{17}F(n, 2n)^{18}F$ and $^{17}O(p, \gamma)^{18}F$ and $^{18}O(p, n)^{18}F$

In the last two reactions, the proton is produced by fast neutrons interacting with target nuclei.

15.10 ADVANTAGES AND DISADVANTAGES OF THE ACTIVATION ANALYSIS METHOD

One of the greatest advantages of activation analysis is its ability to detect most of the isotopes with an extremely high sensitivity. Other advantages are that the method

1 Is nondestructive (in most cases)
2 Needs a sample with a very small mass
3 Can detect more than one element at a time
4 Identifies different isotopes of the same element
5 Provides results rapidly
6 Is not affected by the chemical form of the element of interest

The greatest disadvantage of the method is that it requires expensive equipment, and the analysis of the results is not trivial. Also, activation analysis does not provide information about the chemical compound in which the nuclide of interest belongs.

Overall, activation analysis is a very powerful technique, as demonstrated by its wide use in so many different fields–i.e., chemistry, biology, medicine, forensic medicine, industry, archaeology, and environmental research. For details regarding these special applications, the reader is referred to references 3-29 and to the bibliography of this chapter, in particular to the books by Nargolwalla and Przybylowicz, and by Rakovic.

PROBLEMS

15.1 Traces of manganese are suspected in an unknown sample that has been irradiated for 30 min in a flux of 10^{16} neutrons/m$^2\cdot$s. Counting started 5 min after the irradiation ended. The 0.8 MeV gamma of ^{56}Mn was detected by a counter with a 4 percent efficiency ($F.\epsilon.\Omega$). The sample gave 500 counts in 5 min while the background was 30 ± 1 counts/min. Based on this information calculate the mass of manganese in the sample and the standard error of this measurement. For ^{55}Mn $\sigma = 14$ b. For ^{56}Mn $T_{1/2} = 2.58$ h.

15.2 What should the minimum activity of a sample be if it is required that in the worst case $r = 0.8\sigma_r$, and the sample is counted in a system for which $\epsilon = 0.50$, and the background is 100 ± 5 counts/min. The sample can only be counted for 1 min.

15.3 In a neutron activation analysis experiment, a 10-percent-efficient Ge(Li) detector with 12 ± 0.1 counts/min background is used for the measurement of 0.6-MeV γ's emitted by the sample. If the counting time is 5 min,

(a) What is the minimum acceptable counting rate if the maximum acceptable error is 60 percent?

(b) What is the minimum mass that can be detected if the isotope under investigation has $A = 75$, $\sigma = 0.21$ b, $T_{1/2}(^{76}\text{X}) = 20$ h, irradiation time = 8 h, and $\phi = 10^{16}$ neutrons/m^2·s?

15.4 What is the absolute minimum mass of phosphorus that can be detected using the reaction (n, α) under irradiation with 14-MeV neutrons? Assume $\phi = 10^{12}$ neutrons/m^2·s, $\sigma = 0.150$ b, counting system background = 15 ± 0.5 counts/min, counting time = 2 min, and maximum acceptable error is 30 percent.

BIBLIOGRAPHY

Bowen, H. J. M., and Gibbons, D., *Radioactivation Analysis*, Oxford University Press, London, 1963.

Coomber, D. I. (ed.), *Radiochemical Methods in Analysis*, Plenum Press, New York, 1975.

Lederer, C. M., and Shirley, V. S. (eds.), *Table of Isotopes*, 7th ed., Wiley, New York, 1978.

Lenihan, M. M. A., Thomson, S. J., and Guinn, V. P., *Advances in Activation Analysis*, Academic, New York, 1972, vol. 2.

Lutz, G. J., Boreni, R. J., Maddock, R. S., and Wing, J. (eds.), "NBS Technical Note 467 Activation Analysis: A Bibliography Through 1971," National Bureau of Standards, 1972.

Lyon, W. S. (ed.), *Guide to Activation Analysis*, Van Nostrand, Princeton, N.J., 1964.

Nargolwalla, S. S., and Przybylowicz, E. P., *Activation Analysis with Neutron Generators*, Wiley, New York, 1973.

Radiochemical Methods of Analysis, Proceedings of a Symposium on Radiochemical Methods of Analysis, IAEA, Salzburg, Austria, 1964.

Rakovic, M., *Activation Analysis* (English translation), CRC Press, Cleveland, Ohio, 1970.

REFERENCES

1 Hevesy, G., and Levi, H., *K. Danske Vidensk. Sels. Mat.-fys. Medd.* **14**:5 (1936).
2 Seaborg, G. T., and Livingood, J. J., *J. Am. Chem. Soc.* **60**:1784 (1938).
3 "International Nuclear and Atomic Activation Analysis Conference," October 1975, Gatlinburg, TN, Suppl. *L3 to ANS Trans.* **21**:1–45 (1975).
4 Ruf, H., and Fredrich, M., *Nucl. Tech.* **37**:79 (1978).
5 Wood, D. E., *Nucl. Instr. Meth.* **92**:511 (1971).
6 Marsh, R. H., and Bechnell, D. E., *ANS Trans.* **14**:105 (1971).
7 Wogman, N. A., Rieck, H. G., Jr., Kosorok, J. R., and Perkins, R. W., BNWL-SA-4434 (1972).
8 Nargolwalla, S. S., *ANS Trans.* **21**:107 (1975).
9 Ruch, R. R., Cahil, R. A., Frost, J. K., Camp, L. R., and Gluskoter, H. J., *ANS Trans.* **21**:107 (1975).
10 Millard, H. T., and Swanson, V. E., *ANS Trans.* **21**:108 (1975).
11 Profio, A. E., and Pouraghabagher, A. R., *ANS Trans.* **21**:109 (1975).
12 Harrison, S. H., *ANS Trans.* **21**:110 (1975).
13 Rhodes, J. R., *ANS Trans.* **21**:113 (1976).

14 Darnley, A. G., *ANS Trans.* **21**:113 (1976).
15 Marshall, T. D., and Knight, A., *Nucl. Instr. Meth.* **92**:595 (1971).
16 Battye, C. K., Knight, V., Marshall, T. O., Knight, A., and Godfrey, B. E., *Nucl. Instr. Meth.* **92**:601 (1971).
17 Livingston, H. D., *ANS Trans.* **14**:98 (1971).
18 Podolsky, S., Cardarelli, J. A., and Burrows, B. A., *ANS Trans.* **14**:99 (1971).
19 McNeill, K. A., Cabeza, L., and Harrison, J. E., *ANS Trans.* **14**:100 (1971).
20 Comar, D., *ANS Trans.* **14**:101 (1971).
21 Bird, E. D., *ANS Trans.* **14**:102 (1971).
22 Friedman, M. H., and Tanner, J. T., *ANS Trans.* **21**:97 (1975).
23 Hankins, D. A., Babb, A. L., and Scribner, B. H., *ANS Trans.* **21**:98 (1975).
24 Ricci, E., *ANS Trans.* **21**:99 (1975).
25 Ruch, R. R., Guinn, V. P., and Pinker, R. H., *Nucl. Sci. Eng.* **20**:381 (1964).
26 Guinn, V. P., and Rumos, J. C., *ANS Trans.* **14**:105 (1971).
27 Scheringer, H. L., and Lukens, H. R., *ANS Trans.* **14**:106 (1971).
28 Williamson, T. G., and Harrison, W. W., *ANS Trans.* **14**:107 (1971).
29 Ricci, E., "Charged Particle Activation Analysis," in J. M. A. Lenihan, S. J. Thomson, and V. P. Guinn (eds.), *Advances in Activation Analysis*, Academic, London, 1972, vol. 2.
30 Cline, J. E., Putnam, M. H., and Hermer, R. G., "GAUSS VI, A Computer Program for the Automatic Analysis of Gamma-Ray Spectra from Ge(Li) Spectrometers," ANCR-1113 (1973).
31 Heath, R. L., Helmer, R. G., Schmittroth, L. A., and Cazier, G. A., *Nucl. Instr. Meth.* **47**:281 (1967).
32 Helmer, R. G., Heath, R. L., Schmittroth, L. A., Jayne, G. A., and Wagner, L. M., *Nucl. Instr. Meth.* **47**:305 (1967).
33 Mariscotti, M. A., *Nucl. Instr. Meth.* **50**:309 (1967).
34 Phillips, G. W., and Marlow, K. W., *Nucl. Instr. Meth.* **137**:525 (1976).

16

■ health physics fundamentals

16.1 INTRODUCTION

Health physics is the discipline that consists of all the activities related to the protection of individuals and the general public from potentially harmful effects of ionizing radiation. Ionizing radiation comes from two sources:

1 Natural or background radiation is radiation emitted by radioisotopes that exist on or inside the earth, as well as radiation incident upon the earth from outer space. Humans have been exposed to this natural radiation for as long as they have existed on this planet.

2 Human-made radiation is radiation emitted by all the radioisotopes that have been produced through nuclear reactions (mainly fission), as well as radiation produced by machines used in medical installations (e.g., x-ray machines) or in scientific laboratories (e.g., accelerators).

Health physics is concerned with protection of people from radiation. Since the background radiation has been, is, and will always be on our planet at about the same level everywhere, there is not much a health physicist can do to protect individuals or populations from background radiation. Hence, health physics is, essentially, concerned with human-made radiation.

A health physicist performs many tasks. He or she, most importantly

1 Is responsible for the detection and measurement of radiation in areas of work and and in the environment
2 Is responsible for the proper operation and calibration of detection instruments
3 Inspects at regular intervals the facilities where radiation sources are used
4 Enforces federal and state regulations dealing with proper handling of radiation sources and establishment of acceptable levels of radiation fields at places of work
5 Keeps records of exposure for all individuals under his or her jurisdiction
6 Knows how to clean areas which have been contaminated with radioactive materials
7 Acts as the liaison representative between the regulatory agencies and his organization

Although the term health physics was coined after 1940, and a health physics society was established in 1955, the concern about the harmful effects or radiation had been born much earlier–but probably not early enough. The first recorded radiation damage case occurred in 1896, only a year after the discovery of x-rays, yet the first limits concerning x-ray exposure were set in the 1920s. Today, both national and international groups exist that act as advisory bodies† to the appropriate regulatory agencies.

†International Commission on Radiological Units and Measurements (ICRU); International Commission on Radiological Protection (ICRP); in the United States, the National Council on Radiation Protection and Measurements (NCRP).

Since improperly handled radiation may produce deleterious effects to humans, it is important that individuals who use radiation sources learn the fundamentals of dosimetry, definition of dose units, biological effects of radiation, federal exposure limits, and operation of health physics instruments. This chapter briefly discusses all these items. If more detailed treatment of these topics is needed, consult the bibliography and references given at the end of the chapter.

16.2 RADIATION UNITS

Protection of individuals against radiation necessitates the completion of two tasks:

1 Development of safe radiation exposure limits
2 Construction of instruments that measure the intensity of radiation

Neither of these tasks can be accomplished without the means of quantitative description of radiation, i.e., without defining radiation units.

The radiation effect is measured in terms of exposure or dose. Exposure is defined as charge released per unit mass of air. Dose is defined as energy absorbed per unit mass of material. The first radiation unit to be defined was the ***roentgen*** (symbol R):

1 R = exposure due to x-rays or gamma rays of such intensity that the electrons produced by this radiation in 1 cm^3 of dry air, at standard temperature and pressure, generate along their tracks electron-ion pairs carrying a total charge of 1 esu of either sign

The SI unit of exposure is defined as 1 C/kg air, without any new name proposed for it. Numerically,

$$1 \text{ R} = 2.58 \times 10^{-4} \text{ C/kg air}$$

The roentgen suffers from two limitations:

1 It was defined in terms of electromagnetic radiation only.
2 It was defined in terms of air only.

Radiation protection may involve other types of radiation, and media other than air. For this reason, another unit was defined called the ***radiation absorbed dose*** or ***rad***, defined as

$$1 \text{ rad} = 100 \text{ erg/g}$$

In the SI system a new unit, the *Gray* (Gy), has been proposed to replace the rad:

$$1 \text{ Gy} = 1 \text{ J/kg} = 100 \text{ rad}$$

The rad (or the Gy) has a simple definition and is a unit independent of both type of radiation and material. But the measurement of absorbed dose in terms of rad (or Gy) is neither simple nor straightforward, because it is very difficult to measure energy deposited in a certain mass of tissue. Fortunately, one can bypass this difficulty by measuring energy deposited in air, which is proportional to the exposure, and then relate it to the absorbed dose.

The measurement of exposure is achieved by using ionization chambers, and the result is given in roentgens. Based on the definition of the roentgen, the following relationship can be established between roentgens and rads.

$$1\ \mathrm{R} = \frac{1\ \mathrm{esu}}{1.293 \times 10^3\ \mathrm{g}} (2.082 \times 10^9)\ \text{ion pairs/esu}$$
$$\times (34\ \mathrm{eV/pair})(1.062 \times 10^{-12}\ \mathrm{ergs/eV})$$
$$= 88\ \mathrm{erg/g} = 0.88\ \mathrm{rad} = 8.8\ \mathrm{mGy}$$

If D_a is the absorbed dose in air, and D_e is the exposure in air, the relationship between the two is

$$D_a = 0.88 D_e \tag{16.1}$$

For media other than air, the relationship is obtained as follows.

The absorbed dose (energy per unit mass) in material X is, apart from constants,

$$D_a(\mathrm{X}) \approx I\ (\mathrm{particles/m^2 \cdot s}) \times \mu_a(\mathrm{X})\ (\mathrm{m^2/kg}) \times E\ (\mathrm{J/particle}) \tag{16.2}$$

The absorbed dose in air is

$$D_a(\mathrm{air}) \approx I\ (\mathrm{particles/m^2 \cdot s}) \times \mu_a(\mathrm{air})\ (\mathrm{m^2/kg}) \times E\ (\mathrm{J/particle}) \tag{16.3}$$

The ratio of Eq. 16.2 to 16.3 gives†

$$D_a(\mathrm{X}) \approx \frac{\mu_a(\mathrm{X})}{\mu_a(\mathrm{air})} D_a(\mathrm{air}) = \frac{\mu_a(\mathrm{X})}{\mu_a(\mathrm{air})} 0.88 D_e(\mathrm{air}) \tag{16.4}$$

Equations 16.1 and 16.4 express the fact that the measurement of absorbed dose is a two-step process:

1 Exposure (or exposure rate) is measured.
2 Absorbed dose (or dose rate) is calculated from the measured exposure using Eq. 16.1 or 16.4.

In practice, the instruments that measure radiation dose are, usually, properly calibrated to read rad or Gy.

†Equations 16.2 and 16.3 give dose rate, not dose; but the meaning of Eq. 16.4 is the same in either case.

16.3 THE RELATIVE BIOLOGICAL EFFECTIVENESS–THE DOSE EQUIVALENT

The units of absorbed dose defined in the previous section are quite adequate for the quantitative assessment of the effects of radiation to inanimate objects, like irradiated transistors or reactor fuel. For protection of people, however, the important thing is not the measurement of energy deposited–i.e., the absorbed dose–but the biological effects due to radiation exposure. Unfortunately, biological effects and absorbed dose do not always have one-to-one correspondence, and for this reason a new unit had to be defined: a unit that takes into account the biological effects of radiation.

The ideal unit for the measurement of biological effect should be such that a given dose produces a certain biological effect regardless of the *type and energy of radiation* and also regardless of the *biological effect* chosen as the criterion. Such a unit cannot be defined because of the complexity of the biological organisms and the different ways in which particles deposit their energy in matter.

An ideal unit may not exist, but some unit that "equalizes" biological effects had to be defined. This task has been accomplished by introducing a factor called the *relative biological effectiveness* (RBE), defined as:

$$\text{RBE}_i = \frac{\begin{array}{c}[\text{dose of x- or } \gamma \text{ radiation (200-300 keV)}\\ \text{producing a certain biological effect]}\end{array}}{\begin{array}{c}(\text{dose of radiation of type } i \text{ producing}\\ \text{the same biological effect})\end{array}} \tag{16.5}$$

In understanding the meaning of RBE, note that

1 RBE is defined in terms of photons; therefore, it should be expected thar RBE = 1 for electromagnetic radiation.
2 A given type of radiation does not have a single RBE, but many. RBE values depend on the energy of the radiation, the cell, the biological effect being studied, the total dose, dose rate, and other factors.
3 In practice, one RBE value is used for all biological effects. However, RBE may change with the energy of the particle (see Table 16.1).
4 It is a well-known fact that the biological damage increases as the energy deposited per unit distance, the *linear energy transfer* (LET), increases. Thus, heavier particles (alphas, heavy ions, fission fragments) are more biologically damaging than photons, electrons, and positrons.

In 1963, the International Commission on Radiological Units and Measurements (ICRU) proposed the replacement of RBE by a new factor named the *quality factor* (QF). Here is an excerpt from their recommendation.

> In radiation protection it is necessary to provide a factor that denotes the modification of the effectiveness of a given absorbed dose by LET (Linear Energy Transfer). Unlike RBE, which is always experimentally determined, this factor *must* be *assigned* on the basis of a number of considerations and it is recommended that it be termed the *quality factor* (QF). Provisions for other factors are also made. Thus a *distribution factor*, (DF), may be used

to express the modification of biological effect due to nonuniform distribution of internally deposited radionuclides. The product of absorbed dose and modifying factors is termed the dose equivalent, (DE).

Thus DE = (QF)(DF) · · · (D_a). In practice, only the quality factor is used, and the dose equivalent is expressed by

$$\text{DE} = \text{QF}(D_a)$$

In 1973, the ICRU[1] recommended dropping the "F" from QF, and this practice will be followed here from now on.

Today, RBE is used only in radiobiology, and Q is used in radiation protection. A detailed discussion of similarities and differences between the two factors is given in reference 2. For the radiations and energy ranges considered in this book, RBE and Q are practically the same, and from this point on only the factor Q will be mentioned. Table 16.1 gives Q values for various radiations commonly encountered.

When the unit of absorbed dose is multiplied by the corresponding Q value, the unit of dose equivalent (DE) is obtained. The DE units are

$$1 \text{ rem} = \text{Q} \times \text{rad}$$

and the SI unit

$$1 \text{ Sievert (Sv)} = \text{Q} \times \text{Gy}$$

Thus:

$$1 \text{ Sv} = 100 \text{ rem}$$

Since the DE equalizes biological effects among different types and energy of radiation, one can add rem (or Sv) but not rad (or Gy).

EXAMPLE 16.1

At the open beam port of a research reactor, the absorbed dose rate consists of 10 mrad/h due to gammas, 10 mrad/h due to fast neutrons, and 6 mrad/h due to thermal

TABLE 16.1 Quality Factors for Several Types of Radiation[3,4]

Radiation type	Q	Radiation type	Q
Gamma rays	1	Neutrons:	
X-rays	1	Thermal	2.3
Beta particles:	1	0.005 MeV	2
Electrons	1	0.02 MeV	5
Positrons	1	0.10 MeV	7.4
Protons ($E < 14$ MeV)	10	0.50 MeV	11
Alpha particles		1.00 MeV	10.6
($E < 10$ MeV)	20	5.0 MeV	7.8
Recoil nuclei ($A > 4$)	20	10 MeV	6.8

neutrons. What is the total dose a person will receive by standing in front of the beam for 5 s?

ANSWER
Calculate the DE, as shown below.

From	Abs. dose rate (mrad/h)	Q	DE (mrem/h)	DE (mSv/h)
Gammas	10	1	10	0.1
Fast neutrons	10	10	100	1.0
Thermal neutrons	6	2.3	13.8	0.14
Total			123.8	1.24

The total dose received by the individual is

$$DE = 123.8 \text{ mrem/h} \left(\frac{5}{3600}\right) = 0.17 \text{ mrem} = 1.7 \times 10^{-6} \text{ Sv}$$

16.4 DOSIMETRY

The general *dosimetry problem* may be defined as: Given the intensity of the radiation field at a certain point, calculate the dose rate at that point.

The radiation field is assumed to be known completely, i.e., type and energy of particles should be given as well as their number. The calculation, which is not trivial and is different for charged particles, gammas, and neutrons, is beyond the scope of this brief review. The interested reader should find details of dosimetry calculations in the bibliography given at the end of this chapter.

In practice, the user of radioisotopes seldom does dosimetry calculations (unless of course the subject of the research is related to dosimetry). Instead, the user calculates the dose rate at a certain location based on conversion tables or graphs that relate particle flux to dose rate. Let $N(E)\,dE$ be the number of particles/m^2·s with kinetic energy between E and $E + dE$, and $C(E)$ be the conversion factor for particles with energy E having units (Sv/s)/(particles/m^2·s). A person exposed to the radiation field $N(E)$ will receive a dose equivalent per unit time given by

$$\text{DE (Sv/s)} = \int_0^\infty C(E)N(E)\,dE \tag{16.7}$$

Most of the time, both the radiation field and the conversion factor are not analytic functions but are known as discrete energy group distributions. Thus, one has available the quantities

$$N_g = \int_{E_{g-1}}^{E_g} N(E)\,dE$$ = number of particles/m^2·s with energy between E_{g-1} and E_g (the energy interval $E_g - E_{g-1}$ is called an *energy group*)

and

$$C_g = \frac{1}{E_g - E_{g-1}} \int_{E_{g-1}}^{E_g} C(E)\, dE$$

and Eq. 16.7 takes the form

$$\text{DE (Sv/s)} = \sum_{g-1}^{G} C_g N_g \qquad (16.8)$$

Equation 16.8 will be applied now to dose calculations from sources of charged particles, gammas, and neutrons. In all cases it will be assumed that the source is external to the body.

16.4.1 Dose Due to Charged Particles

If the radiation field is due to an alpha source external to the body, the dose to an exposed individual will be zero, because the alphas emitted by radioisotopes do not penetrate the skin-forming cells.

If the source emits electrons, positrons, or betas, a part of the tissue deeper than the skin might be exposed depending on the energy of the particle. The flux-to-dose rate conversion factors for electrons and betas are given in Table 16.2. Since these particles have a definite range, the dose to a human exposed to an external electron source affects mostly the skin tissue. For example, electrons of 1 MeV will penetrate the body up to a depth of about 4 mm.

TABLE 16.2 Flux-to-Dose Rate Conversion Factors for Electrons and Betas[5]

Electrons			Betas		
E (MeV)$_{max}$	(Sv/s)/ (particles/m² · s)	(mrem/h)/ (particles/cm² · s)	E (MeV)$_{max}$	(Sv/s)/ (particles/m² · s)	(mrem/h)/ (particles/cm² · s)
0.02	2.104 – 13†	0.758	0.2	1.160 – 13	0.417
0.100	0.649 – 13	0.234	0.4	0.772 – 13	0.278
0.200	0.440 – 13	0.158	0.6	0.578 – 13	0.208
0.300	0.369 – 13	0.133	0.8	0.433 – 13	0.156
0.400	0.337 – 13	0.121	1	0.386 – 13	0.119
0.600	0.309 – 13	0.111	1.5	0.330 – 13	0.119
0.800	0.297 – 13	0.107	2	0.317 – 13	0.114
1	0.293 – 13	0.105	3	0.303 – 13	0.109
2	0.297 – 13	0.107			
3	0.303 – 13	0.109			
5	0.320 – 13	0.115			
7	0.324 – 13	0.118			
10	0.342 – 13	0.123			

†Read 2.104×10^{-13}.

If the source of charged particles is inside the body, the dose is obtained from

$$\mathrm{DE} = \frac{\text{energy released/s}}{\text{mass}} Q$$

because the range of charged particles in tissue is very short and all the energy is deposited within a very small volume. If the source strength is

$S(E)\,dE$ = particles emitted per second with kinetic energy between E and $E + dE$

and $S_g = \int_{E_{g-1}}^{E_g} S(E)\,dE$ = particles emitted per second with energy between E_{g-1} and E_g

the equation for the dose rate becomes

$$\mathrm{DE} = \frac{\left(\int_0^\infty dE\, S(E)\right) EQ(E)}{M} = \frac{\left(\sum_{g=1}^{G} S_g \bar{E}_g\right) Q_g}{M} \tag{16.9}$$

where $\bar{E}_g = \frac{1}{2}(E_{g-1} + E_g)$.

The mass to be used in the denominator of Eq. 16.9 depends on the distribution of the source. If the source is distributed in one organ only, then the mass of that organ should be used. If the source is localized, then the mass affected is contained in a sphere of radius equal to the range of the most energetic particle.

For monoenergetic sources, Eq. 16.9 consists of one term, with the energy E_g being the energy of the source particles. In the case of a source of betas, the energy E_g is the average energy of the beta spectrum, which to a first approximation is equal to $E_{max}/3$.†

EXAMPLE 16.2

What is the dose rate due to 3.7×10^{-2} Bq (1 pCi) of a localized 6-MeV alpha source imbedded in tissue?

ANSWER

The range of this alpha in tissue is calculated using Eqs. 4.12 and 4.13.

$$R = [(0.05)(6) + 2.85]6^{3/2} = 46.3 \text{ mm in air}$$

Using the Bragg-Kleeman rule (Eq. 4.13), the range in tissue is

$$R = 3.2 \times 10^{-4} \frac{(A_e)^{1/2}}{\rho}(46.3) = 3.2 \times 10^{-4}\left(\frac{3}{1}\right)(46.3) = 4.44 \times 10^{-5} \text{ m}$$

†A more accurate equation is given in reference 6.

The dose rate is, using Eq. 16.9,

$$DE = \frac{(3.7 \times 10^{-2}\ \alpha/s)(6)(1.602 \times 10^{-13}\ J/\alpha)(10)}{(4/3)\pi(4.44 \times 10^{-5})^3(10^3\ kg/m^3)}$$

$$= 9.7 \times 10^{-4}\ Sv/s = 349\ rem/h$$

This is an extremely large dose rate, because a large amount of energy is deposited in a very small volume.

EXAMPLE 16.3

What is the dose rate due to the alphas of Example 16.2 if it is known that the source is uniformly distributed in the lungs?

ANSWER

In this case, the mass affected is that of the lungs, which is (for a 70 kg person) about 1 kg. The dose rate is

$$DE = \frac{(3.7 \times 10^{-2})(6)(1.602 \times 10^{-13})(10)}{1}$$

$$= 3.56 \times 10^{-13}\ Sv/s = 1.28 \times 10^{-7}\ rem/h$$

EXAMPLE 16.4

What is the dose rate to the thyroid gland due to the betas emitted by 1 mCi of ^{131}I?

ANSWER

The isotope ^{131}I emits two betas, one with $E^{(1)}_{max} = 0.608$ MeV, 85 percent of the time, and a second one with $E^{(2)}_{max} = 0.315$ MeV, 15 percent of the time. The range of these betas in tissue is about 2 mm and 0.9 mm, respectively.

The thyroid gland has a mass of about 0.025 kg (i.e., a volume of about 25 cm^3); therefore, all the beta energy will be deposited in it.

The dose rate is obtained using Eq. 16.9:

$$DE = \frac{(3.7 \times 10^7)[0.85(0.608/3) + 0.15(0.315/3)](1.602 \times 10^{-13})}{25 \times 10^{-3}}$$

$$= 4.53 \times 10^{-5}\ Sv/s = 16.3\ rem/h$$

16.4.2 Dose Due to Photons

A person exposed to a gamma-ray source will receive a whole-body dose because photons, having no definite range, penetrate deep into the body. The distribution of the dose depends, of course, on the energy of the gammas. The flux-to-dose rate conversion factors as a function of photon energy, developed by the American Nuclear Society as an American National Standard,[7] are given in Table 16.3 and Fig. 16.1.

The calculation of the dose rate from a gamma source inside the human body is more complicated and is not presented here. The reader is referred to the bibliography of this chapter (see Cember, Chap. 6).

TABLE 16.3 Gamma-Ray Flux-to-Dose Rate Conversion Factors[7]

Photon energy (MeV)	(Sv/s)/ (particles/m² • s)	(rem/h)/ (particles/cm² • s)	Photon energy (MeV)	(Sv/s)/ (particles/m² • s)	(rem/h)/ (particles/cm² • s)
0.01	1.10 − 15†	3.96 − 06	1.4	6.97 − 16	2.51 − 06
0.03	1.62 − 16	5.82 − 07	1.8	8.30 − 16	2.99 − 06
0.05	8.05 − 17	2.90 − 07	2.2	9.50 − 16	3.42 − 06
0.07	7.17 − 17	2.58 − 07	2.6	1.06 − 15	3.82 − 06
0.1	7.86 − 17	2.83 − 07	2.8	1.11 − 15	4.01 − 06
0.15	1.05 − 16	3.79 − 07	3.25	1.22 − 15	4.41 − 06
0.2	1.39 − 16	5.01 − 07	3.75	1.34 − 15	4.83 − 06
0.25	1.75 − 16	6.31 − 07	4.25	1.47 − 15	5.23 − 06
0.3	2.11 − 16	7.59 − 07	4.75	1.55 − 15	5.60 − 06
0.35	2.44 − 16	8.78 − 07	5.0	1.61 − 15	5.80 − 06
0.4	2.74 − 16	9.85 − 07	5.25	1.67 − 15	6.01 − 06
0.45	3.00 − 16	1.08 − 06	5.75	1.77 − 15	6.37 − 06
0.5	3.25 − 16	1.17 − 06	6.25	1.87 − 15	6.74 − 06
0.55	3.53 − 16	1.27 − 06	6.75	1.97 − 15	7.11 − 06
0.6	3.78 − 16	1.36 − 06	7.5	2.13 − 15	7.66 − 06
0.65	4.00 − 16	1.44 − 06	9.0	2.44 − 15	8.77 − 06
0.7	4.22 − 16	1.52 − 06	11.0	2.86 − 15	1.03 − 05
0.8	4.67 − 16	1.68 − 06	13.0	3.28 − 15	1.18 − 05
1.0	5.50 − 16	1.98 − 06	15.0	3.69 − 15	1.33 − 05

†read 1.10×10^{-15}.

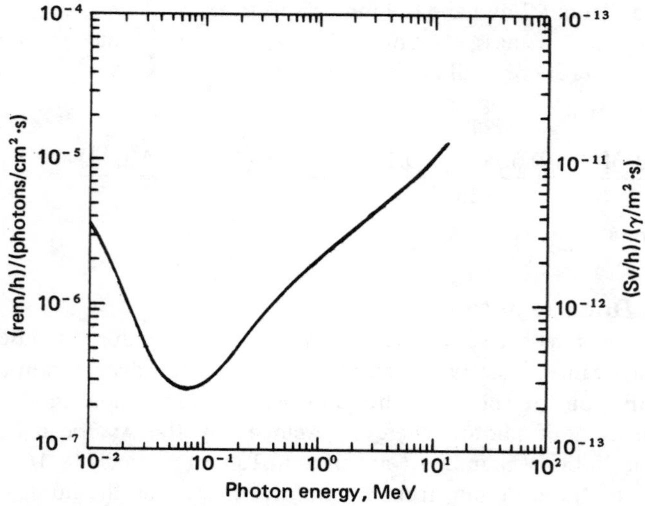

FIG. 16.1 Photon flux-to-dose rate factors for energies from 10^{-2} to 15 MeV (ref. 7).

EXAMPLE 16.5

What is the dose rate at 1 m away from 3.7×10^{10} Bq (1 Ci) of ^{137}Cs, if (a) the attenuating medium is water or (b) the attenuating medium is air?

ANSWER

In both cases the dose rate is calculated by using Eq. 16.8 with a dose-to-flux conversion factor obtained from Table 16.3 for 0.662-MeV gammas. Using linear interpolation, that factor is 4.05×10^{-16} (Sv/s)/(γ/m$^2\cdot$s).

The flux at 1 m from the source is given by $BSe^{-\mu r}/4\pi r^2$, where

B = buildup factor (from App. E)

S = source strength

μ = total attenuation coefficient for 0.662-MeV gammas in air or water

From App. E, the buildup factor is

$$B = 1 + a\mu r[\exp(b\mu r)]$$

where (using linear interpolation) $a = 1.96$ and $b = 0.054$.

(a) For water, the value of the total attenuation coefficient is (from App. D, using linear interpolation)

$$\mu = 0.00861 \text{ m}^2/\text{kg} = 0.0861 \text{ cm}^2/\text{g}$$

The number of mean free paths (mfp) in water is

$$\mu r = (0.00861 \text{ m}^2/\text{kg})(10^3 \text{ kg/m}^3)(1 \text{ m}) = 8.61 \text{ mfp}$$

Thus,

$$B\text{ (water)} = 1 + (1.96)(8.61)[\exp(8.61 \times 0.054)] = 27.86$$

The dose rate in water is

$$D\text{ (water)} = (27.86)(3.7 \times 10^{10})\left[\frac{e^{-8.61}}{4\pi(1^2)}\ \gamma/\text{m}^2\cdot\text{s}\right]$$

$$\times [4.05 \times 10^{-16} \text{ (Sv/s)/}(\gamma/\text{m}^2\cdot\text{s})]$$

$$= 6.04 \times 10^{-9} \text{ Sv/s} = 2.17 \text{ mrem/h}$$

(b) For air, the value of the total attenuation coefficient is (from App. D, using linear interpolation)

$$\mu = 0.0082 \text{ m}^2/\text{kg} = 0.082 \text{ cm}^2/\text{g}$$

The number of mean free paths in air is

$$\mu r = (0.0082 \text{ m}^2/\text{kg})(1.29 \text{ kg/m}^3)(1 \text{ m}) = 0.01 \text{ mfp}$$

The buildup factor is

$$B(\text{air}) = 1 + (1.96)(0.01)[\exp(0.01 \times 0.054)] = 1.02$$

The dose rate in air is

$$D(\text{air}) = (1.02)(3.7 \times 10^{10})\left[\frac{e^{-0.01}}{4\pi(1)^2}\ \gamma/\text{m}^2\cdot\text{s}\right][4.05 \times 10^{-16}\ (\text{Sv/s})/(\gamma/\text{m}^2\cdot\text{s})]$$

$$= 1.2 \times 10^{-6} \text{ Sv/s} = 0.433 \text{ rem/h}$$

It should be pointed out that the value of the buildup factor is significantly greater than 1 if the distance in mean free paths is significantly greater than 1. In Example 16.5, the distance in water is 8.61 mfp and $B = 27.86$, while in air the distance is 0.01 mfp and $B = 1.02$.

The build-up factor constants given in App. E apply to a point isotropic source in an infinite medium (Example 16.5 is such a case). The same constants can be used, however, in other geometries if no better values are available. For example, one could use the constants given in App. E to calculate the dose rate from a point isotropic source in a semiinfinite medium or from a point isotropic source located behind a slab shield. In such cases, the use of the constants from App. E results in an overestimate of the buildup factor. Buildup factors for many different geometries are given in reference 5.

16.4.3 Dose Due to Neutrons

Neutrons hitting the human body deliver energy to it through elastic and inelastic collisions with nuclei, and through secondary radiation emitted by the radioisotopes produced after neutrons are captured.

If an individual is exposed to fast neutrons, most of the energy transfer takes place through elastic collisions with hydrogen (~90 percent) and to a lesser extent through collisions with oxygen and carbon nuclei. [The average neutron energy lost per collision with hydrogen (proton) is 50 percent of the incident neutron energy; the corresponding fractions for carbon and oxygen are 14 percent and 11 percent.] These "recoil" nuclei are charged particles, which lose their energy as they move and slow down in tissue. This is true for neutron energies down to about 20 keV. When the neutron energy reaches or becomes lower than a few keV, the importance of elastic collisions decreases and the reaction $^{14}N(n, p)^{14}C$ produces more significant effects. As discussed in Chap. 14, this is an exothermic reaction producing protons with kinetic energy of 626 keV. Radioactive ^{14}C is also produced, emitting betas with a maximum energy of 156 keV. The biological damage comes mainly from the protons, not from the betas of ^{14}C.

Thermal neutrons are absorbed in the body mainly through the reaction $^1H(n, \gamma)^2H$, which results in the emission of a 2.2-MeV gamma. A reaction of secondary importance is $^{23}Na(n, \gamma)^{24}Na$. The isotope ^{24}Na has a 15-h half-life and emits two energetic gammas with energy 1.37 and 2.75 MeV. Thus, when thermal neutrons are absorbed, damage is caused by the energetic gammas which are produced as a result of the neutron capture.

The neutron flux-to-dose rate conversion factors are given in Table 16.4 and Figs. 16.2 and 16.3.

EXAMPLE 16.6

At the open beam port of a research reactor the neutron flux at a certain power level consists of 1.6×10^8 neutrons/m² · s with energy 100 keV, and 3.5×10^8 neutrons/m² · s with an average energy of 0.025 eV. What is the total dose rate at that point?

ANSWER

Using Table 16.4, the dose rate is

$$DE = (1.6 \times 10^8)(6.03 \times 10^{-15}) + (3.5 \times 10^8)(1.02 \times 10^{-15})$$

$$= 1.32 \times 10^{-6}\ \text{Sv/s} = 0.475\ \text{rem/h}$$

16.5 EXTERNAL AND INTERNAL EXPOSURE: THE BIOLOGICAL HALF-LIFE

Radioisotopes may enter the body by inhalation, drinking, eating, injection, or through broken skin (wound). If the radiation source is inside the body, the exposure is internal and more damaging. No attenuation is provided by skin or clothes, and the person cannot walk away from the source. The exposure continues until the radioisotope decays completely or is excreted by the body.

TABLE 16.4 Neutron Flux-to-Dose Rate Conversion Factors[7]

Neutron energy (MeV)	(Sv/s)/(neutrons/m² ·s)	(rem/h)/(neutrons/cm² ·s)
2.5 – 08†	1.02 – 15	3.67 – 06
1.0 – 07	1.02 – 15	3.67 – 06
1.0 – 06	1.24 – 15	4.46 – 06
1.0 – 05	1.26 – 15	4.54 – 06
1.0 – 04	1.16 – 15	4.18 – 06
1.0 – 03	1.04 – 15	3.76 – 06
1.0 – 02	9.89 – 16	3.56 – 06
1.0 – 01	6.03 – 15	2.17 – 05
5.0 – 01	2.57 – 14	9.26 – 05
1.0	3.67 – 14	1.32 – 04
2.5	3.47 – 14	1.25 – 04
5.0	4.33 – 14	1.56 – 04
7.0	4.08 – 14	1.47 – 04
10.0	4.08 – 14	1.47 – 04
14.0	5.78 – 14	2.08 – 04
20.0	6.30 – 14	2.27 – 04

†Read as 2.5×10^{-8}.

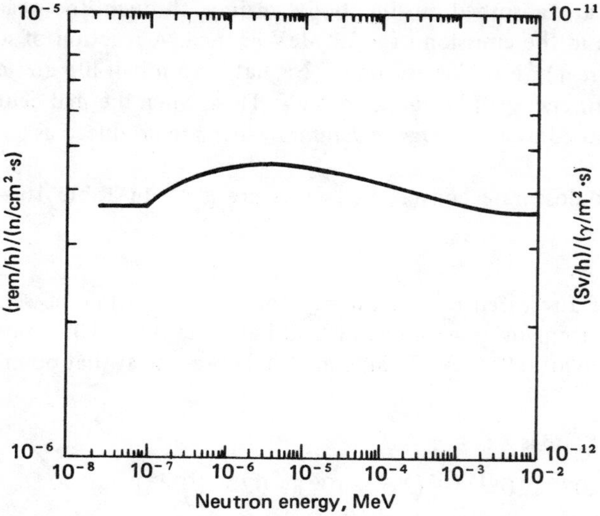

FIG. 16.2 Neutron flux to-dose rate factors for energies 10^{-8} to 10^{-2} MeV (ref. 7).

FIG. 16.3 Neutron flux-to-dose rate factors for energies 10^{-2} to 20 MeV (ref. 7).

A radioisotope is rejected by the body at a rate that depends upon the chemical properties of the elements. All isotopes of the same element are rejected at the same rate, whether they are stable or not. For most radioisotopes, the rate of rejection is proportional to the amount of the isotope in the body. This leads to an exponential elimination law as a result of the combination of decay and rejection.† Let

$N(t)$ = number of radioactive atoms at time t

λ_R = radiological decay constant

λ_B = biological decay constant

= probability of rejection (by the body) per atom per unit time

The rate of change of $N(t)$ is $dN(t)/dt = -\lambda_R N(t) - \lambda_g N(t)$, with solution

$$N(t) = N(0)e^{-(\lambda_R + \lambda_B)T} = N(0)e^{-\lambda_e t} \tag{16.10}$$

where $\lambda_e = \lambda_R + \lambda_B$ = effective decay constant.

A *biological half-life* is defined in terms of λ_B:

$$T_B = \frac{\ln 2}{\lambda_B} \tag{16.11}$$

and an *effective half-life* is defined by the equation

$$T_e = \frac{T_B T_R}{T_B + T_R} = \frac{\ln 2}{\lambda_e} \tag{16.12}$$

The biological excretion rate of an element from the human body is not necessarily the same for the whole body and for a particular organ. In fact, in most cases, the biological elimination rates are different for different organs and for the body as a whole. For example, the biological half-life of iodine is 138 days for rejection from the thyroid, 7 days for the kidneys, 14 days for the bones, and 138 days for the whole body. For this reason, a table of biological and effective half-lives ought to include the organ of reference. Table 16.5 gives biological half-lives for certain common isotopes. The reader should remember that the biological half-life is the same for all isotopes of the same element, but the effective half-life is not.

If $T_B \gg T_R$, the decay removes the material much faster than the body rejects it. An example of such a case is ^{131}I, with $T_B = 138$ days (thyroid), $T_R = 8$ days, and $T_e = 7.6$ days. If $T_B \ll T_R$, the biological elimination is mainly responsible for the removal of the isotope. An example of such a case is tritium, with $T_B = 12$ days, $T_R = 12$ years, and $T_e = 12$ days.

As a result of the combined radioactive and biological elimination of a radioisotope from the whole body or from an organ, the dose rate to the body or the organ is not constant over time. Consider an amount of a certain radioisotope that delivers

†Other rejection laws have been proposed (see p. 35 of reference 8, and reference 9).

TABLE 16.5 Radiological, Biological, and Effective Half-Lives of Certain Common Isotopes[8]

Isotope	Organ of reference	Radiological half-life	Biological half-life	Effective half-life
^{3}H	Total body	12.3 y	12 d	12 d
^{14}C	Total body	5,700 y	10 d	10 d
	Fat		12 d	
	Bone		40 d	
^{32}P	Total body	14.3 d	257 d	13.5 d
	Liver		18 d	8 d
	Bone		1,115 d	14.1 d
	Brain		257 d	13.5 d
^{40}K	Total body	1.28×10^{9} y	58 d	58 d
^{55}Fe	Total body	1,100 d	800 d	463 d
	Spleen		600 d	388 d
	Lungs		3,200 d	819 d
	Liver		554 d	368 d
	Bone		1,680 d	665 d
^{57}Fe		45.1 d		
^{99m}Tc	Total body	0.25 d	1 d	0.2 d
	Kidneys		20 d	0.25 d
	Lungs		5 d	0.24 d
	Skin		10 d	0.24 d
	Liver		30 d	0.25 d
	Bone		25 d	0.25 d
^{129}I	Total body	1.726×10^{7} y	138 d	138 d
	Thyroid		138 d	138 d
	Kidneys		7 d	7 d
	Liver		7 d	7 d
	Spleen		7 d	7 d
	Testes		7 d	7 d
	Bone		14 d	14 d
^{131}I	Thyroid	8 d	138 d	7.6 d
^{235}U	Total body	7.12×10^{8} y	100 d	100 d
	Kidneys		15 d	15 d
	Bone		300 d	300 d
^{238}U		4.66×10^{9} y		
^{239}Pu	Total body	24,000	175 y	175 y
	Liver		82 y	82 y
	Kidneys		87.7 y	87.7 y

a dose rate equal to DE(0) at the time of ingestion. If the effective half-life of the isotope is T_e, the total dose delivered to the body, or the organ, over a period of time T is

$$\mathrm{DE}_T = \int_0^T \mathrm{DE}(0) e^{-\lambda_e t}\, dt = \frac{\mathrm{DE}(0)}{\lambda_e}\left(1 - e^{-\lambda_e T}\right)$$

$$= \frac{\mathrm{DE}(0)}{\ln 2} T_e \left[1 - \exp\left(-\ln 2 \frac{T}{T_e}\right)\right] \tag{16.13}$$

If the time period $T \gg T_e$, then

$$DE_{TOT} = \frac{DE(0)}{\ln 2} T_e \tag{16.14}$$

EXAMPLE 16.7
What is the total dose received by an individual who drank, accidentially, 10^{-7} kg of ^3H_2O?

ANSWER
Assuming that the ^3H_2O is uniformly distributed, the dose rate at the time of the accident ($t = 0$) is given by Eq. 16.9. Tritium is a beta emitter with $E_{max} = 18.6$ keV, $T_R = 12$ years, and $T_B = 12$ days. At $t = 0$, the source strength (i.e., the activity) is

$$S = N\lambda = (10^{-7} \text{ kg})(2 \text{ at/molecule}) \frac{6.022 \times 10^{23} \text{ molecules/mol}}{22 \times 10^{-3} \text{ kg/mol}}$$

$$\times \left[\frac{\ln 2}{12(3.15 \times 10^7}\right] = 1.00 \times 10^{10} \text{ Bq} = 2.71 \text{ mCi}$$

Considering an average-size person (70 kg), the dose rate at $t = 0$ is

$$DE(0) = \frac{[10^{10}(0.018/3) \text{ MeV/s}](1.602 \times 10^{-13} \text{ J/MeV})(1)}{70 \text{ kg}}$$

$$= 1.38 \times 10^{-7} \text{ Sv/s} = 50 \text{ mrem/h}$$

The total dose is obtained by using Eq. 16.14:

$$DE_{TOT} = \frac{1.38 \times 10^{-7} \text{ Sv/s}}{\ln 2} (12 \text{ days})(86{,}400 \text{ s/day})$$

$$= 0.206 \text{ Sv} = 20.6 \text{ rem}$$

16.6 BIOLOGICAL EFFECTS OF RADIATION

The study of the biological effects of radiation is a very complex and difficult one for two main reasons:

1 The human body is a very complicated entity with many organs of different sizes, functions, and sensitivities.
2 Pertinent experiments are practically impossible with humans. The existing human data on the biological effects of radiation come from accidents and through extrapolation from animal studies.

Why does radiation produce damage to biological material? Consider the constituents and some of the functions of the human body. In terms of chemical compounds, the predominant one is water (about 61 percent of the total weight). Other compounds include proteins, nucleic acids, fats, and enzymes.

The basic unit of every living organism is the cell, which consists of a semipermeable membrane enclosing an aqueous suspension of a liquid substance called the cytoplasm. The cell exchanges material with the rest of the organism through the membrane. A typical cell size is about 10^{-5} m (size of a typical atom 10^{-10} m). At the center of the cell there is another region called the nucleus, also enclosed by a semipermeable membrane. The nucleus is the most important part of the cell because it controls its activities. Nucleic acids and chromosomes are the cell's most significant contents.

The two nucleic acids found in the nucleus of a cell are ribonucleic acid (RNA) and deoxyribonucleic acid (DNA). The RNA and DNA molecules are very large (10^6–10^7 μm). The RND molecule controls the synthesis of proteins, which are necessary for the life of the cell. The DNA molecule contains the genetic code of the species.

The chromosomes are threadlike assemblies that are extremely important because they contain the genes that transmit the hereditary information. The genes are parts of the DNA molecules. Every species has a definite number of chromosomes. The human species has 23 pairs, one chromosome of each pair being contributed by each parent. Every cell has 23 pairs of chromosomes with the exception of the egg and the sperm, which have 23 chromosomes each. When fertilization occurs, the first cell of the new organism contains 23 pairs of chromosomes, equally contributed by the two parents.

Cells multiply by a dividing process called mitosis. Just before mitosis is to take place, each chromosome of the cell splits in two. Thus, each of the two new cells has exactly the same number of chromosomes as the parent cell. There are some human cells that do not divide, such as the blood cells and the nerve cells. The blood cells are regenerated by the blood-forming organs, primarily by the bone marrow. The nerve cells, when destroyed, are not supplied again.

Radiation may damage the cell when it delivers extra energy to it, because that energy may be used to destroy parts or functions of the cell. For example, as a result of irradiation, chromosomes or DNA molecules may break. The break may occur either by direct collision with an incoming fast particle (e.g., fast neutron) or as the result of chemical activity initiated by the radiation. It has been determined experimentally that the energy imparted by the radiation may be used to break chemical bonds and create free radicals, which are always chemically active and which may produce new chemical compounds unhealthy for the organism. For example, a water molecule may break into two radicals that in turn may form hydrogen peroxide (H_2O_2)†:

$$H_2O^{\bullet} \rightarrow HO^{\bullet} + H^{\bullet}$$

$$HO^{\bullet} + HO^{\bullet} \rightarrow H_2O_2$$

A damaged cell may react in different ways. It may recover or it may die, or grow out of control if the radiation has damaged the RNA and DNA molecules that provide the instructions to feed and divide the cell. Obviously, the net result of the damage to the organism depends on many factors such as the number and the type

†Chemical poisoning by H_2O_2 shows many of the radiation sickness symptoms.

of cells destroyed. Another effect of irradiation may be damage to the DNA molecules that carry the genetic code. The result of such impairment does not manifest itself in the irradiated cell or organism but in the descendants of the organism.

The biological effects of radiation are divided into two groups:

1 *Somatic effects*, which affect the irradiated person
2 *Genetic or hereditary effects*, which affect the descendants of the irradiated individual

The effects are also divided into two categories in terms of the period between irradiation and appearance:

1 Short-term effects, which appear within a matter of minutes, days, or weeks
2 Long-term effects, which appear years, decades, and sometimes generations later

The somatic effects may be short- or long-term, while the genetic ones are long-term effects.

One question that created, and will always create, a great deal of discussion about biological radiation damage is this:

> Is there a threshold below which the radiation dose received does not produce any harmful effects?

The *threshold* and *nonthreshold* hypotheses, have been proposed as answers to this question. They are illustrated in Fig. 16.4.

According to the threshold hypothesis, there is a dose threshold below which irradiation will not produce observable effects. According to the nonthreshold hypothesis, any dose, no matter how small, will produce some harmful effects.

Which hypothesis is correct? They both are, depending on the effect. If the effect considered is death of a person, or fever, or nausea, or erythema, there is a threshold and Fig. 16.4*a* applies. If the concern is about genetic effects, there is no conclusive evidence for either hypothesis. The radiation protection guides for the general public (see Sec. 16.7) are based on the nonthreshold hypothesis. This is done not because there is a definite proof that this hypothesis is the correct one but because in the absence of definite proof that it is wrong, it is best to follow the conservative approach.

The effects of an acute whole-body irradiation could be summarized as follows. At 0.50 Sv (50 rem) or less whole-body one-time dose, the individual will feel nothing

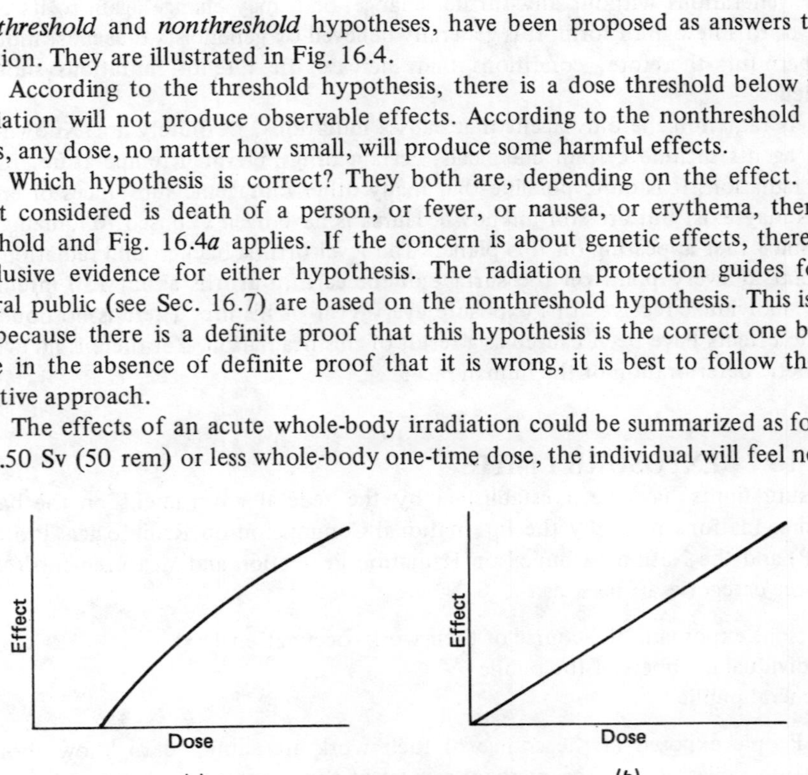

FIG. 16.4 (*a*) The threshold hypothesis. (*b*) The nonthreshold hypothesis.

physically, and clinical tests will not show symptoms of any illness or injury. Depending on the individual, a clinical test may show changes in the blood. A dose of 4–5 Sv (400–500 rem) is indicated as LD-50, meaning that it is lethal to about 50 percent of the people so exposed (death will occur in about 2 months). A dose above 10 Sv (1000 rem) is considered lethal with any treatment. Some of the symptoms of radiation sickness are nausea, vomiting, fever, diarrhea, loss of hair, inability of the organism to fight infection, and destruction of the nervous system.

Long-term effects may appear as a result of an acute irradiation or of a chronic low-level exposure over a long period. Although it is never certain that a particular effect or disease will appear as a long-term effect, it is believed that the probability of occurrence increases as a result of exposure to radiation. Some of the long-term effects are loss of hair, eye cataracts, carcinogenesis, and leukemia.

Genetic effects are those related to the transmission of harmful hereditary information from one generation to the next. It is known today that the carriers of the hereditary code are the genes, which are parts of DNA molecules and are contained in the chromosomes. The gene is an extremely stable entity. Its structure is transmitted from generation to generation without any changes, which means that it transmits identical information from generation to generation. But sometimes a gene may change and become a *mutation.* The mutated gene may be transmitted through many generations without any further change, or it may change again to its original form or to a new third form. It is generally believed by geneticists that most mutations are harmful; therefore, conditions that increase the rate of mutations should be avoided.

Is radiation the only agent that causes mutations? Definitely not. Known mutagenic agents include certain chemicals, certain drugs, elevated temperature, and ionizing radiation. It is quite possible that many other mutagenic substances or environments may exist but are still unknown. Humans have been exposed to ionizing radiation since first appearing on this planet. The level of this background radiation is not constant at every point on the surface of the earth, but it is about 130 mrem/year. Every individual receives this exposure every year of his life. There is no doubt that genetic effects have been caused as a result of this irradiation. Yet there is no evidence of genetic deterioration of the human race.

16.7 EXPOSURE LIMITS

Exposure limits have been established by the federal government on the basis of recommendations made by the International Commission on Radiological Protection (ICRP) and the National Council on Radiation Protection and Measurements (NCRP) for three categories of persons:

1 Persons exposed in the course of their work (occupational exposure)
2 Individual members of the public
3 General public

People exposed in the course of their work are supposed to know about the risks from radiation and accept them as part of the hazards associated with their job. Individual members of the public constitute a relatively small number of people who may be exposed to man-made ionizing radiation without knowing about it and without the free choice of avoiding it. For example, people living close to a nuclear power plant belong to this category.

The philosophy behind the recommended exposure limits for people in groups (1) and (2) is expressed in the following passage from reference 3.

> The permissible dose for an *individual* is that dose, accumulated over a long period of time or resulting from a single exposure, which, in the light of present knowledge, carries a negligible probability of severe somatic or genetic injuries; furthermore, it is such a dose that any effects that ensue more frequently are limited to those of a minor nature that would not be considered unacceptable by the exposed individual and by competent medical authorities.

In addition to this philosophy, the ICRP recommends a dose limitation based on these principles[9]:

1 No practice shall be adopted unless its introduction produces a net benefit.
2 All exposures shall be kept as low as reasonably achievable, economic and social factors being taken into account
3 The dose equivalent to individuals shall not exceed the limits recommended for the appropriate circumstances by the Commission (i.e., ICRP)

In general, the exposure limits for individual members of the public are set 10 times less than for occupational exposure, and the limits for the general public are set still lower. The exposure limits for all these categories are shown in Table 16.6.

The cumulative whole-body dose equivalent (assessed in the gonads, lens of the eye, or red bone marrow) is not allowed to exceed $5(N-18)$ rem,† where N is the

†This is still in effect (1982), but there is a recommendation to abolish it.

TABLE 16.6 Exposure Limits[10]

Category	Limit (per calender quarter)†
A. Occupational exposure	
1 Whole body; head and trunk; active blood forming organs; lens of eyes; or gonads	1.25 rem
2 Hands and forearms; feet and ankles	18.75 rem
3 Skin of whole body	7.5 rem
4 Fertile women (dose to fetus)	0.5 rem (during entire gestation period)
B. Individual members of the public	1/10 of limits in (A)
C. General public	
Whole body	170 mrem/year

†These limits do not include exposure due to natural radiation and medical x-ray examinations.

age of individual in years. This implies that people younger than 18 years should not be involved in activities that result in occupational exposure. New guidelines issued in 1977 by the Environmental Protection Agency[11] attempt to control exposure to any member of the public as a result of activities anywhere in the nuclear fuel cycle.

As mentioned earlier, in addition to exposure from man-made radioisotopes, humans are exposed to natural radiation and also to radiation from several consumer items. Natural radiation delivers, on the average, 130 mrem/year = 1.3 mSv/year per person in the United States. The components of natural radiation are the following:

Cosmic rays (sea level)	35 mrem/year
Ground (from U, Th, and their daughters)	50 mrem/year
Air (mainly from radon isotopes and ^{14}C)	15 mrem/year
H_2O, food, body elements (Rn, ^{40}K, ^{14}C)	30 mrem/year

In addition to natural radiation, people receive the following estimated average doses from man-made radiation:

Medical x-rays	50-70 mrem/year per person†
Smoking 1.5 packs of cigarettes per day	8000 mrem/year[12] (to small portions of the lungs)
Luminous wristwatches	3 mrem/year[12] (if they contain ^{226}Ra)
Color television	0.5 mrem/year[12] (average population exposure)
Nuclear power plants	<1 mrem/year (population average)

Limits have also been set for the maximum permissible concentration (MPC) of radioisotopes in air and water.[8,10] For a particular isotope, the limit is calculated on the basis of the dose to an individual resulting from the maximum amount of that isotope accumulated in the body after continuous intake of that isotope.

16.8 HEALTH PHYSICS INSTRUMENTS

Health physics instruments are detectors like those discussed in Chaps. 5-7. They have some unique requirements, however, because their purpose is not to measure radiation but the effects of radiation on human tissue. Since the material of which the detector is made is not tissue, its response to radiation may be different from that of tissue. Also, the biological damage resulting from a given absorbed energy may be quite different for different tissues. For these reasons, it is necessary to measure the radiation with an instrument and then translate it into response of tissue. In practice, the instruments are properly calibrated to read directly in rad (or Gy) or rem (or Sv). Since different detectors do not have the same efficiency or sensitivity for all types of

†This is the so-called genetically significant dose (GSD). The GSD is the dose to the gonads that if received by every member of the population, would be expected to produce the same total genetic effect to the population as do the actual doses received by the various individuals.

of radiation and at all energies, there is no single instrument that can be used for alphas, betas, gammas, and neutrons.

Health physics instruments are divided into three groups according to the way they are used

1 Personnel portable monitoring instruments
2 Portable survey instruments
3 Nonportable instruments

Personnel monitoring is achieved by requiring individuals working in radiation areas to wear a *pocket dosimeter* and/or a *film badge.* The pocket dosimeter is an ionization chamber or an electrometer and is sensitive to gammas and energetic betas. The film badge responds to betas, gammas, and fast and thermal neutrons. It consists of two or three films packed together, wrapped in a light-tight paper and placed in a plastic or metal container. Using films with different sensitivities, one can differentiate between beta and gamma, and neutron radiation.

There are many portable survey instruments, but the most commonly used are Geiger-Müller counters and the so-called cutie-pie survey meter. The cutie-pie is an ionization chamber normally detecting x-rays and γ rays and some high-energy betas. The nonportable instruments are continuous monitors (e.g., air monitors) or personnel monitors (e.g., hand and foot monitor). A list of several health physics instruments with their characteristics is given in Table 16.7.

In recent years the *thermoluminescent dosimeter* (TLD) has been used frequently. The operation of the TLD is based on the fact that certain crystals, such as CaF_2 and Lif, emit light if they are heated after they have been exposed to radiation. More details are given in Sec. 17.6. Digital dosimeters available in the market today provide either dose rate or integrated dose. Two such models are shown in Fig. 16.5.

16.9 PROPER USE OF RADIOISOTOPES

Since radiation may be hazardous, it is important that individuals who handle ionizing radiation follow certain rules to avoid accidents. The official rules to be followed by all persons licensed to handle radioactive materials have been developed and dictated by the federal government through the Nuclear Regulatory Commission (NRC) and the Environmental Protection Agency (EPA). The NRC standards for protection against radiation are contained in Chap. 1, Title 10 of the Code of Federal Regulations, Part 20 (10 CFR 20).[10] The EPA standards are found in Part VII, Title 40.[11] The exposure limits, based on these guidelines, were discussed in Sec. 16.7.

To protect personnel, areas where radiation sources are used are marked with certain signs. The definitions of "radiation areas" and the corresponding signs are[10]:

> "Restricted area" means any area access to which is controlled by the licensee for purposes of protection of individuals from exposure to radiation and radioactive materials. "Restricted area" shall not include any areas used as residential quarters, although a separate room or rooms in a residential building may be set apart as a restricted area (p. 6.3.2);

TABLE 16.7 A List of Common Health Physics Instruments

Instrument	Detector	Radiation detected	Range (nominal)	Remarks
		Portable		
Film badge	Photographic emulsions	$\gamma, \beta, n_f, n_{th}$	0.1–1,000 rad of mixed radiations	Measurement of integrated dose
Pocket chamber (direct reading)	Ionization chamber (air)	γ	5–200 mR; available with higher ranges	Visual check on gamma exposure
Personnel radiation monitor	G–M tube	γ, x-ray, high-level β	Maximum audible warning at ~0.5 R/h	Audible warning of radiation field
Cutie-pie survey meter	Ionization chamber (air)	γ, x-ray, high-energy β	5–10,000 mrad/h	Dose-rate meter for γ and x-rays (0.008–2 MeV within 10 percent)
		Nonportable		
Continuous β–γ particulate air monitor	G–M tube (shielded)	β, γ	0–5,000 cpm	Light and audible alarms for preset levels
Continuous α particulate air monitor	ZnS(Ag) for α	α, β	0–1,000 cpm	Light and audible alarms for preset levels
Hand and foot monitor	Halogen-quenched G–M tube	β, γ	Low-level	Simultaneous detection of β and γ contamination of hands and shoes

(a)

(b)

FIG. 16.5 (*a*) A pocket-size beta-gamma survey meter with a range from 0.1 mR/h to 100 R/h. (*b*) A pocket dosimeter that displays accumulated dose (from Reactor Experiments Inc.).

"Radiation area" means any area, accessible to personnel, in which there exists radiation, originating in whole or in part within licensed material, at such levels that a major portion of the body could receive in any one hour a dose in excess of 5 millirem, or in any 5 consecutive days a dose in excess of 100 millirems;

"High radiation area" means any area, accessible to personnel, in which there exists radiation originating in whole or in part within licensed material at such levels that a major portion of the body could receive in any one hour a dose in excess of 100 millirem (p. 6.3.5).

Radiation areas should be marked with the radiation symbol shown in Fig. 16.6 and with cautionary signs. If necessary, the radiation area should be roped off or, if it is a room, should be locked to keep people out.

People who work in radiation areas or use radioisotopes should keep in mind the following simple principle:

> Radiation exposure should be avoided, if at all possible. If exposure is necessary, the risk from the exposure should be balanced against the expected benefit. The exposure is justified if the benefit outweighs the risk.

If exposure is justified, the individual to be exposed should take every possible step to minimize the dose received, such as:

1 Try to avoid internal exposure. Substances enter the body by mouth (eating, drinking), by breathing, through wounds, and by injection. Therefore, in places where radioactive materials are handled, do not eat, do not drink, and cover all wounds. If the air is contaminated, wear a mask. Hands should be washed after the operation is over, especially if no protective gloves were used.
2 Stay close to the source of radiation for as short a time interval as possible.
3 Use protective covers, if this is the suggestion of the health physicist.
4 Place the source behind a shield or in a proper container.
5 If practical, wait for the radiation to decay to a safer level before handling it. The exponential decay law is a helpful ally.

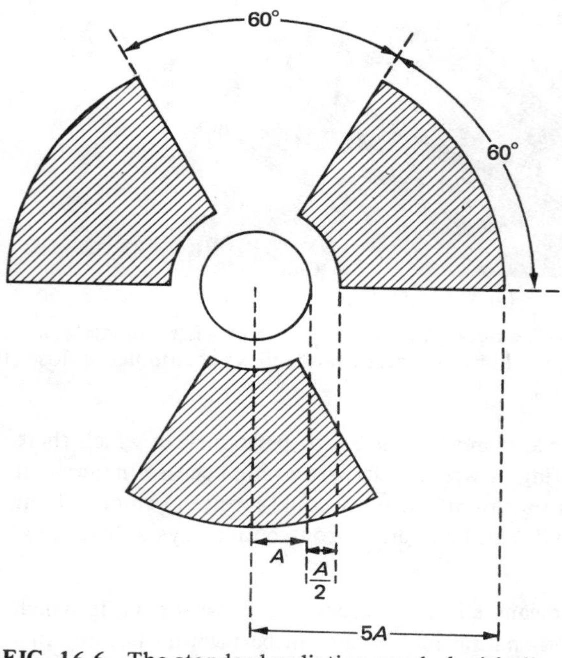

FIG. 16.6 The standard radiation symbol with dimensions as shown has a yellow background, with the hatched area being magenta or purple.

6 Stay as far away from the source as practical. The strength of a point isotropic source decreases as $1/r^2$ (r = distance away from the source).

The shielding medium that should be used is not the same for all types of radiation. Here are simple suggestions for the three types of radiation considered in this book.

CHARGED PARTICLES

These have a definite range. Therefore, to stop them completely, a shielding material with thickness at least equal to the range (in that material) of the most penetrating particle should be placed between the source and the worker. A few millimeters of metal will definitely stop all charged particles emitted by radioisotopic sources. Some bremsstrahlung may get through, though.

GAMMAS

A beam of gammas going through a material of thickness t is attenuated by a factor $\exp(-\mu t)$, where μ is the total linear attenuation coefficient of the gamma in that medium. The higher the value of μ is, the smaller the thickness t that reduces the intensity of the beam by a desired factor. In theory, the beam cannot be attenuated to zero level. In practice, the attenuation is considered complete if the radiation level equals the background.

The attenuation coefficient μ increases with the atomic number of the material. The most useful practical element for gamma shielding is lead ($Z = 82$). Lead is relatively inexpensive; and it is easy to melt it and make shields with it having the desired shape, size, and thickness.

NEUTRONS

Shielding against neutrons is more difficult than either against charged particles or photons. If the source emits fast neutrons, the first step is to provide a material that will thermalize the neutrons. Such materials are water, wax, or paraffin.

Thermalized neutrons are easily absorbed by many isotopes. Examples are ^{115}In, ^{113}Cd, and ^{10}B. Of these the most practical to use is boron. It can be used in powder form or be dissolved in water or liquid wax. A very simple but effective shield for a source of fast neutrons is 0.15 to 0.30 m of wax or paraffin to which boron has been added. The thickness of this borated material may change depending on the strength of the source and the amount of boron added. Cadmium is very useful in sheet form. A cadmium sheet 3 mm to 6 mm ($\frac{1}{8}$ to $\frac{1}{4}$ in) will stop a thermal neutron beam almost completely. Finally, shields have been manufactured which are flexible, like rubber, yet are excellent neutron attenuators.[13]

PROBLEMS

16.1 What is the total dose received by an individual standing in front of the open beam port of a research reactor for 10 s under the radiation levels listed in the figure below?

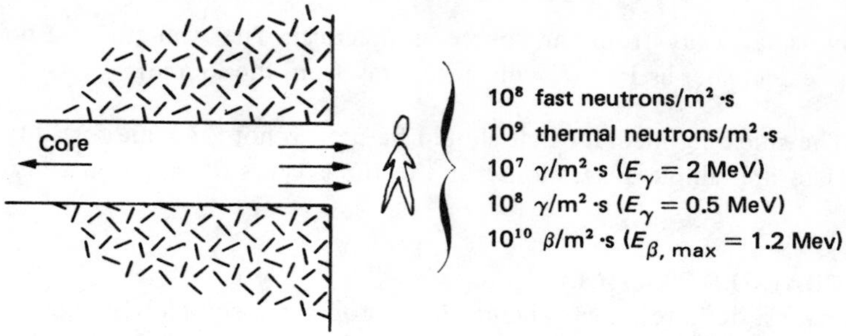

16.2 What is the dose rate per curie of ^{24}Na if it is shielded by 0.025 m of lead as shown below?

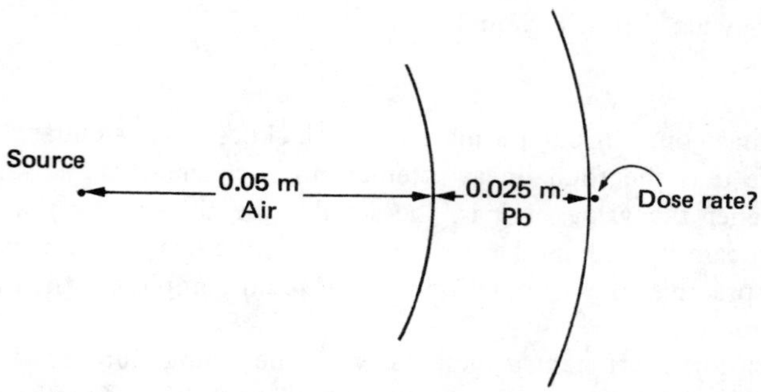

^{24}Na emits a 1.37-MeV gamma and a 2.75-MeV gamma 100 percent of the time, betas with E_{max} = 4.17 MeV 0.003 percent of the time, and betas with E_{max} = 1.389 MeV 100 percent of the time.

16.3 As a result of carelessness, a worker inhaled 1 μg of ^{241}Am. Considering alpha particles only and assuming that the americium is spread uniformly in the bones, (a) what is the dose rate at the time of the accident and (b) what is the total dose to that individual? For ^{241}Am, T_R = 433 years and T_B = 200 years, mass of bones = 10 kg.

16.4 What is the dose rate 0.30 m away from 1 Ci of ^{60}Co if (a) the attenuating medium is air, and (b) if the source is shielded by 0.01 m of aluminum?

16.5 What is the thickness of a lead container that will result in a dose rate at its surface of 2.5 mrem/h = 2.5×10^{-5} Sv/h, if it is used to store 1 Ci of ^{124}Sb? ^{124}Sb emits the following gammas:

Energy (MeV)	Intensity (%)	Energy (MeV)	Intensity (%)	Energy (MeV)	Intensity (%)
0.603	97	0.967	2.4	1.37	5
0.644	7	1.048	2.4	1.45	2
0.720	14	1.31	3	1.692	50
				2.089	7

16.6 What is the dose rate due to the ^{124}Sb of Prob. 16.5 at a distance equal to the thickness of the container, if the attenuating medium is air?

16.7 A 1-Ci sample of ^{60}Co is stored behind a concrete and lead shield as shown below. What is the dose rate at point P?

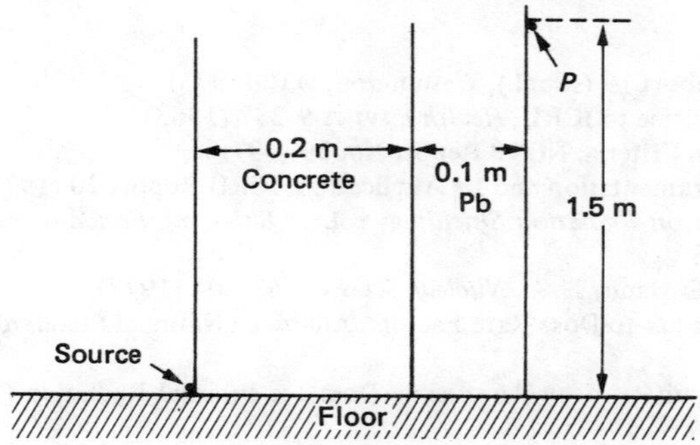

16.8 The isotope ^{99m}Tc is used *in vivo* for diagnostic purposes in humans. It emits x-rays with energy 140 keV and betas with E_{max} = 0.119 MeV. If a person is injected with 1 μCi of this isotope, what is the total dose to the brain, assuming that all the isotope is uniformly distributed there? (Mass of the brain = 1.5 kg.)

16.9 If all the water in the human body were suddenly changed to T_2O ($T = {}^3H$), what would be the total dose to that individual? Assume 61 percent of the body is water.

16.10 If 1 pCi of ^{239}Pu is inhaled by breathing and gets into the lungs, what is going to be the total dose to that individual? (T_R = 24,000 years, T_B = 200 years, mass of lungs = 1 kg.)

16.11 Calculate the total dose rate at the center of a spherical submarine submerged in contaminated water with activity 1 Ci/m^3 of ^{137}Cs. The submarine is made of steel 0.025 m thick. Its radius is 1.5 m and it is filled with air at 1 atm.

16.12 What is the annual dose to a person due to the ^{40}K found in every human body? The isotopic abundance of ^{40}K in potassium is 0.0119 percent. The human body contains 1.7×10^{-3} kg of potassium per kg. ^{40}K emits the following radiations:

Particle	Energy (MeV)	Intensity (%)
β^-	1.31	89
β^+	0.49	10^{-3}
γ	1.46	11
γ	0.511	2×10^{-3}

BIBLIOGRAPHY

Attix, F. H., and Tochilin, E. (eds.), *Radiation Dosimetry* I, II, III, Academic, New York, 1969.
Cember, H., *Introduction to Health Physics*, Pergamon, 1969.
Fitzgerald, J. J., Brownell, G. L., and Mahoney, F. J., *Mathematical Theory of Radiation Dosimetry*, Gordon and Breach, 1967.
Morgan, K. Z., and Turner, J. E. (eds.), *Principles of Radiation Protection*, Wiley, 1967.
Report of the ICRP Committee II on Permissible Dose for Internal Radiation, Health Physics, 3 June 1960.

"The Effects on Populations of Exposure to Low Levels of Ionizing Radiation," BEIR Report, National Academy of Sciences National Research Council (1972).

REFERENCES

1 Dose Equivalent ICRU Report 19 (suppl.), Washington, D.C. (1973).
2 Report of the RBE Committee to ICRU, *Health Physics* **9**:357 (1963).
3 Basic Radiation Protection Criteria, NCRP Report No. 39 (1971).
4 Radiation Protection Instrumentation and Its Application, ICRU Report 20 (1971).
5 *Engineering Compendium on Radiation Shielding*, vol. 1: *Shielding Fundamentals and Methods*, Springer-Verlag, 1968.
6 Stamatelatos, M. G., and England, T. R., *Nuclear Sci. Eng.* **63**:204 (1977).
7 Neutron and Gamma-ray Flux-to-Dose Rate Factors, American National Standard, ANSI/ANS-6.11-1977.
8 Report of the ICRP Committee II on Permissible Dose for Internal Radiation (1959), *Health Physics* **3**:146 (June 1960).
9 Recommendations of the International Commission on Radiological Protection, *Nuclear Safety* **20**(3):330 (1979).
10 Nuclear Regulatory Commission, Title 10, Chap. 1, Part 20, Code of Federal Regulations (10 CFR 20), 1978.
11 Environmental Protection Agency–Environmental Radiation Protection Standards for Nuclear Power Operations, Title 40, Chap. 1, Part 190, Code of Federal Regulations (1977).
12 Radiation Exposure from Consumer Products and Miscellaneous Sources, NCRP Report No. 56 (1978).
13 Reactor Experiments Inc., San Carlos, CA (1980).

17

special detectors and spectrometers

17.1 INTRODUCTION

This chapter discusses detectors that do not quite fit in any of the categories presented earlier. These are detectors that were developed for special applications but, in most cases, proved to be useful in more than one area of radiation detection. Examples are the self-powered detectors developed for in-core flux measurements in power reactors, and materials applied to radiation dosimetry, e.g., thermoluminescent materials and photographic emulsions.

In keeping with the purpose of this book—which is to deal with instruments and methods for the detection of low-energy particles (see Table 1.1)—spark chambers, bubble chambers, and Cerenkov counters are not discussed because they are used with high-energy particles. Electrostatic and magnetic spectrometers are discussed, however, because they can be used with high- as well as low-energy particles.

For every detector type, the principle of operation is presented here first with limitations, advantages and disadvantages, and then with specific applications. The material is intended only as an introduction into this subject. For details, the reader should consult the bibliography and references listed at the end of the chapter.

17.2 SELF-POWERED DETECTORS

Self-powered detectors, as their name implies, operate without an externally applied voltage. The incident radiation (neutrons or gammas or both) generates a signal in the form of a current proportional to to the bombarding flux. The detectors are usually constructed in a coaxial configuration (Fig. 17.1). The central conductor is called the *emitter* and is the material responsible for the generation of the signal. The outer conductor, called the *collector*, is separated from the emitter by an insulator. The collector is made of inconel alloy, and has the form of a metallic sheath encasing the insulator and the emitter.

The principle of signal generation in a self-powered detector is simple. As a result of bombardment by radiation, the emitter releases electrons (betas) that escape to the insulator and leave the emitter positively charged. If the emitter is connected to the collector through a resistor (Fig. 17.1), current flows, which when measured gives an estimate of the incident flux. Note that this is not an emitter-collector system: any beta particle escaping from the emitter contributes to the current, regardless of whether or not it reaches the collector.

Because self-powered detectors have been developed for use inside the core of power reactors, they are designed to have small size (a few millimeters in diameter), to be able to operate for rather long periods of time (years) in the intense radiation

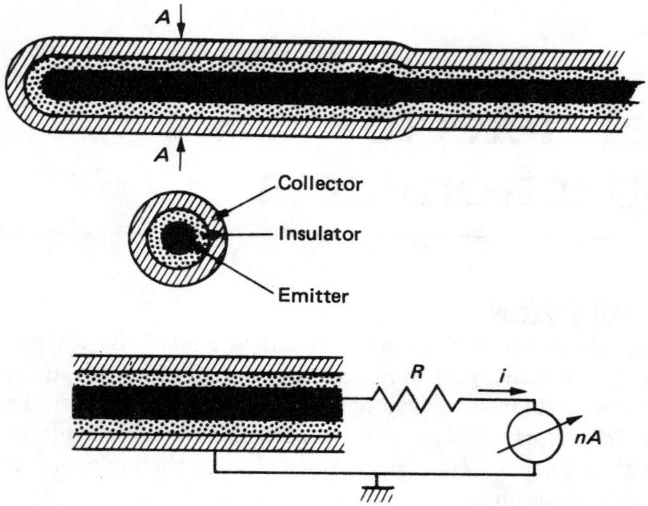

FIG. 17.1 Configuration of a self-powered detector.

field of the reactor core without appreciable deterioration in performance, and finally, to operate without an external power supply.

The performance of a self-powered detector is given in terms of its ***sensitivity S***, defined by the equation

$$S(t) = \frac{\Delta I(t)}{\Delta \phi} \tag{17.1}$$

where $I(t)$ = detector current after exposure to the flux ϕ for time t

ϕ = neutron flux

Thus, the sensitivity represents the change in detector current per unit change in the flux.

Many elements have been considered as emitters for self-powered detectors.[1–4] The ideal emitter should be such that the detector has

1 High sensitivity
2 Low burnup rate
3 Prompt response
4 Sensitivity to neutrons only or gammas only

The material properties that determine these characteristics are discussed in Secs. 17.2.1 and 17.2.2, after the equations for the detector current and sensitivity are derived.

The properties of the insulator are also important. The insulator must have a resistance of about 10^{12} ohms at room temperature and 10^9 ohms at reactor operating temperature. The two insulators commonly used are magnesium oxide (MgO) and aluminum oxide (Al_2O_3). Experiments have shown[5] that the resistance of MgO decreases with exposure to radiation, while that of Al_2O_3 does not change. For this reason, Al_2O_3 is gradually replacing MgO as an insulator for self-powered detectors.

The self-powered detectors are divided into neutron-sensitive and gamma-sensitive, and further subdivided into those with delayed response and those with prompt response. The characteristics of various types of self-powered detectors are presented in Secs. 17.2.1 and 17.2.2.

17.2.1 Self-Powered Neutron Detectors (SPND) with Delayed Response

Rhodium, vanadium, cobalt, and molybdenum have been used as emitters for self-powered neutron detectors (SPNDs). Since rhodium SPNDs are, at this time (1982), the main in-core instruments for the determination of power distribution in pressurized-water reactors (PWR), their discussion is first, in greater detail than the others.

The signal of rhodium SPNDs is produced as a result of activation of the emitter (^{103}Rh) by the incident neutrons, and subsequent decay of the isotope ^{104}Rh that is produced. The decay scheme of ^{104}Rh is shown in Fig. 17.2. An isomeric state with a 4.4-min half-life is produced with a 12-b cross section. The ground state of ^{104}Rh has a 43-s half-life and is formed with a cross section of 138 b (with thermal neutrons). It decays to ^{104}Pd with a maximum β^- energy of 2.5 MeV. This decay, which takes place 98.5 percent of the time, is primarily responsible for the signal of the rhodium detector. The isomeric state with the 4.4-min half-life contributes very little to the signal, but is responsible for a residual current after reactor shutdown.

To identify the factors that improve sensitivity of the detector and lengthen its life, one should look at the processes responsible for the generation of the detector signal. This is done below, and equations for current and sensitivity are derived for any emitter material.

Consider an emitter with an average neutron absorption cross section σ exposed to a total neutron flux† ϕ, and upon absorption of a neutron becoming radioactive

†As explained in Chap. 14, the average absorption cross section is given by $\sigma = \int \sigma(E)\phi \times (e)\, dE/\phi$ and the total flux by $\phi = \int \phi(E)\, dE$.

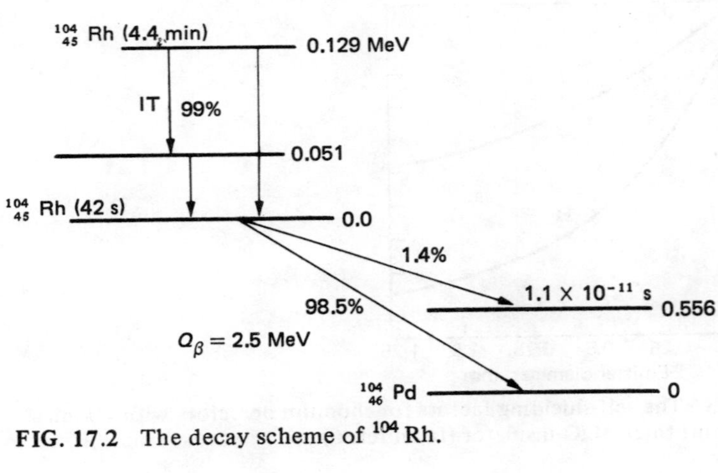

FIG. 17.2 The decay scheme of ^{104}Rh.

with a half-life T [or decay constant $\lambda = (\ln 2)/T$]. The number of radioactive atoms $N(t)$ present after exposure for time t is (see Eq. 14.16)

$$N(t) = \frac{s\sigma_e \phi N_0}{\lambda - \sigma_a \phi} (e^{-s\sigma_a \phi t} - e^{-\lambda t}) \tag{17.2}$$

where N_0 = number of emitter atoms at $t = 0$
σ_a = absorption cross section of emitter
σ_e = cross section that leads to the state that contributes to the signal
s = self-shielding factor ($s < 1$)

The ***self-shielding factor s*** corrects for the fact that the target (emitter) is thick, as a result of which the flux in the emitter is depressed. Thus, interior atoms are "shielded" from exposure to the full flux by the atoms close to the surface. The shielding factor is less than 1 and decreases as the diameter of the emitter increases (see Fig. 17.3). If every decay of the radioisotope releases a particle with charge q, the current at time t is equal to

$$I(t) = kqN(t)\lambda = kq \frac{s\sigma_e \phi N_0}{1 - \sigma_a \phi/\lambda} (e^{-s\sigma_a \phi t} - e^{-\lambda t}) \tag{17.3}$$

where k is a constant that takes into account such effects as self-absorption of betas in the emitter or loss of betas in the insulator (see Fig. 17.3). The factor $\sigma_a \phi/\lambda$ in the denominator can be neglected because, for all emitters of interest, $\sigma_a \phi/\lambda \ll 1$. The exponential factors of Eq. 17.3 have the following meaning.

The factor $\exp(-\lambda t)$ gives the response of the detector. If the flux undergoes a step increase, as shown in Fig. 17.4, the signal will rise exponentially to its saturated value. If the flux goes down suddenly, the signal will decay again exponentially. The speed of response is determined by the half-life of the isotope involved. Rhodium, with a half-life of 42 s, reaches saturation after about 5 min. Vanadium (^{52}V), with a half-life equal to 3.76 min, reaches saturation after about 25 min.

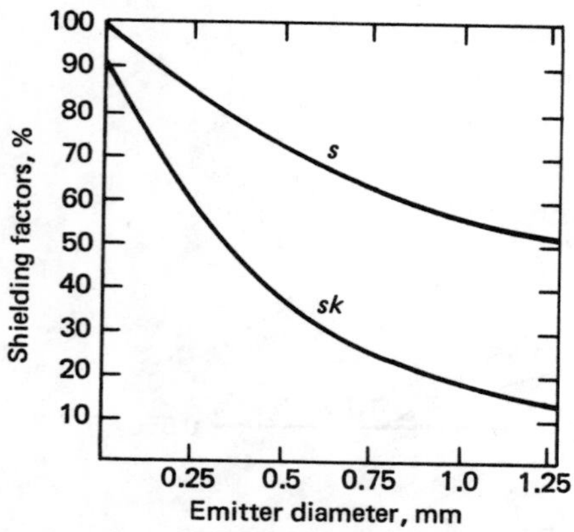

FIG. 17.3 The self-shielding factors for rhodium detectors with 10-mil (0.254-mm) thick MgO insulator (from ref. 6).

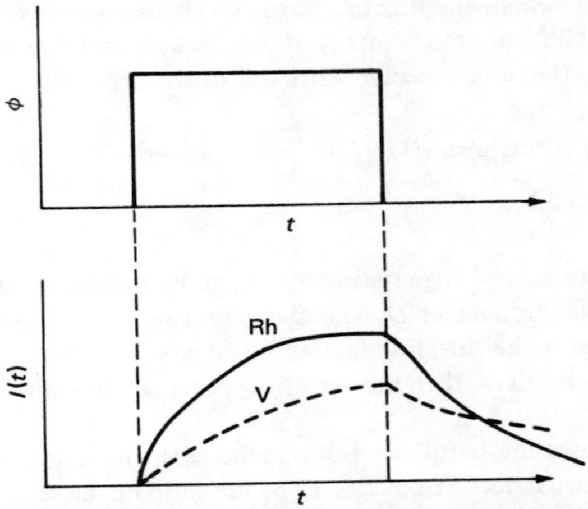

FIG. 17.4 The response of Rh and V SPNDs to a step change in neutron flux.

The factor $\exp(-s\sigma_a\phi t)$ gives the burnup rate of the emitter. It is the factor that determines the lifetime of the detector, because, as seen below, the decrease in sensitivity with time is essentially given by this same factor.

Assuming saturation, the sensitivity of the detector is given (using Eqs. 17.1 and 17.3) by

$$S(t) = \frac{\Delta I(t)}{\Delta\phi} = ks\sigma_e N_0 q(1 - s\sigma_a\phi t)e^{-s\sigma_a\phi t} \text{ A/(neutrons/m}^2\cdot\text{s)} \tag{17.4}$$

If the emitter diameter is D, its length is L, its density is ρ, and its atomic weight is A, then the number of atoms $N_0 = (\rho\pi D^2/4)(LN_A/A)$ where N_A is Avogadro's number.

Substituting into Eq. 17.4, one obtains an equation for the sensitivity per unit length

$$\frac{S(t)}{L} = ksq\sigma_e\rho\frac{\pi D^2}{4}\frac{N_A}{A}(1 - s\sigma_a\phi t)e^{-s\sigma_a\phi t} \text{ [(A/m)/(neutrons/m}^2\cdot\text{s)]} \tag{17.5}$$

EXAMPLE 17.1

What is the sensitivity of Rh detectors 0.5 mm in diameter, per unit detector length under saturation conditions, for a new detector?

ANSWER

For rhodium, $\sigma_e = 139$ b, $\sigma_a = 150$ b, $\rho = 12.4 \times 10^3$ kg/m^3, $A = 103$, and $q = 1.602 \times 10^{-19}$ C. At the beginning of life ($t = 0$),

$$\frac{S(0)}{L} = ks(1.602 \times 10^{-19}\text{ C})(139 \times 10^{-28}\text{ m}^2)(12.4 \times 10^3\text{ kg/m}^3)$$

$$\times\,\pi\frac{(5 \times 10^{-4})^2\text{ m}^2}{4}\left(\frac{6.023 \times 10^{26}}{103}\text{ atoms/kg}\right)$$

$$= 3.17 \times 10^{-23}\,ks \text{ [(A/m)/(neutrons/m}^2\cdot\text{s)]}$$

Typical values of k are about 0.8.[6] Since the flux in a large power reactor (1000 MWe) is about 10^{17} neutrons/m$^2\cdot$s (10^{13} neutrons/cm$^2\cdot$s) at full power, and the typical detector has a length of about 0.10 m, the expected current is of the order of

$$(0.8)(0.8 \times 10^{-23})\ [(\text{A/m})/(\text{neutrons/m}^2\cdot\text{s})]\,(10^{17}\ \text{neutrons/m}^2\cdot\text{s})(0.1\ \text{m})$$
$$= 1.92 \times 10^{-7}\ \text{A} \approx 200\ \text{nA}$$

Equation 17.5 shows that to achieve high sensitivity one should select an emitter with high cross section σ and large diameter D. The diameter affects the sensitivity through the factor D^2 and through the shielding factors k and s, which decrease† as the diameter increases. The net result is that the sensitivity changes roughly as the first power of the diameter.[6]

A high cross section increases the sensitivity but also increases the rate at which the sensitivity decreases with time. Indeed, from Eq. 17.5, the ratio of the sensitivity after exposure for time t to its value at time $t = 0$ is

$$\frac{S(t)}{S(0)} = (1 - s\sigma_a\phi t)e^{-s\sigma_a\phi t} \tag{17.6}$$

Table 17.1 gives the characteristics of several self-powered detectors. Rhodium detectors have not only the best sensitivity but also the largest burnup rate. Their change of sensitivity with time is important and necessitates a correction before the signal is used for the determination of power. The correction is not trivial because of changes in self-absorption effects in the emitter, and may introduce errors unless the detector is calibrated properly.

Despite the drawback of large burnup, rhodium SPNDs are used extensively in nuclear power plants, especially in pressurized-water reactors (PWRs) for the determination of power distribution, fuel burnup, and other information related to the performance of the core. The detectors are inserted into a certain number of "instrumented" fuel assemblies through guide tubes. Every instrumented assembly has seven equally spaced SPNDs, (a background detector and a thermocouple are also included

†The factor s is not constant over prolonged exposure. It tends to increase as the emitter burnup continues, because a smaller number of emitter atoms is left for self-shielding.[6]

TABLE 17.1 Characteristics of Self-Powered Detectors with 0.5-mm Emitter Diameter[3,6]

Emitter	Sensitivity (A/m)/(2×10^{17} neutrons/m$^2\cdot$s)	Burnup†/year (%) for $\phi = 2 \times 10^{17}$ neutrons/m$^2\cdot$s	Response
Rh	2.4×10^{-6}	5	Delayed
V	1.5×10^{-7}	0.3	Delayed
Co	3.4×10^{-8}	2.3	Prompt
Mo	1.7×10^{-8}	0.9	Prompt
Pt	2.6×10^{-7}	0.2	Prompt

†Burnup $\approx 1 - \exp(\sigma_a\phi t)$.

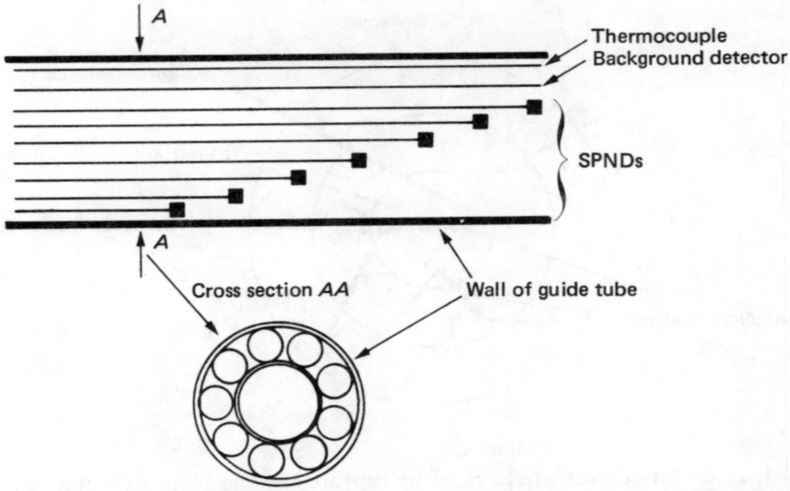

FIG. 17.5 Each detector tube contains seven SPNDs, one background detector, and one thermocouple.

in the package; see Fig. 17.5) for the measurement of the flux at seven axial locations. The outputs of the detectors, corrected for background, are transmitted to the plant computer where, after appropriate corrections are applied, the power, fuel burnup, plutonium production, etc., are calculated. Every PWR has at least 50 instrumented assemblies, which means that the flux is monitored at more than 350 locations.

17.2.2 Self-Powered Neutron Detectors with Prompt Response

Neutron-sensitive self-powered detectors with prompt response operate on a different principle than rhodium and vanadium SPNDs. The emitter, in this case, absorbs a neutron and emits gammas at the time of capture. It is these capture gammas which are responsible for the signal, and since they are only emitted at the time of the neutron capture the detector response is instantaneous. Cobalt (^{59}Co) and molybdenum (^{95}Mo) are two elements seriously considered as emitters. The subsequent discussion is based on cobalt, but the processes involved in the signal generation are the same for molybdenum.

Consider, then, a cobalt SPND (Fig. 17.6). Most of the capture gammas traverse the emitter, the insulator, and the collector without an interaction. Those that do interact produce electrons through the photoelectric or Compton reactions. As these fast electrons travel, they produce an outward flow of charge which generates a current. Because relatively few gammas interact, the sensitivity of a cobalt detector is lower than that of either rhodium or vanadium detectors (see Table 17.1).

The Co detector has one undesirable characteristic due to the product of the neutron capture. That product is ^{60}Co, a β^- emitter with a 5.3-year half-life. The betas from ^{60}Co produce a background signal that builds up with exposure. A way to suppress this background, using platinum shields, has been reported by Goldstein and Todt.[7]

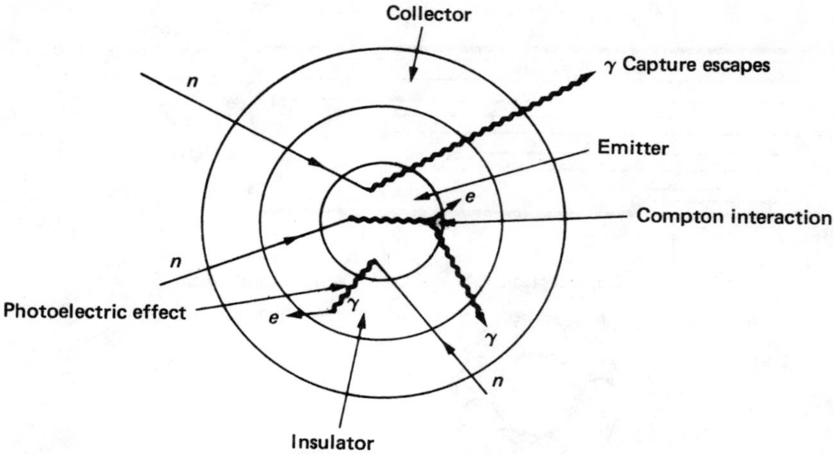

FIG. 17.6 Production of electrons by capture gammas from a Co emitter.

The molybdenum SPND is at the research and development stage. As is shown in Table 17.1, its response is half that of cobalt detectors. Molybdenum detectors, however, do not have any background signal.

17.2.3 Self-Powered Gamma-Sensitive Detectors (SPGD)

In a reactor there is an intense gamma field that could be utilized in a self-powered gamma-sensitive detector (SPGD). Such a device was first described by Shields,[8,9] with platinum as the emitter. Platinum is a good material for a SPGD because it has

1 Low neutron capture cross section
2 High atomic number
3 High density

Low neutron cross section decreases the sensitivity of the device to neutrons. High density and atomic number increase the probability of interaction for gammas (remember, the photoelectric coefficient goes as Z^m, $m = 3$-5).

In a SPGD, the photons generate electrons through photoelectric and Compton interaction (Fig. 17.7), and the electrons by their motion produce the current that is measured. So far, this process looks the same as that in a cobalt SPND. There is a difference, however, between the two types of detectors. In the cobalt detector the gammas interact, mainly, in the emitter. By contrast, in the SPGD the gammas interact in all three regions. Some of the electrons produced in the collector travel inwards and thus cancel some of the current produced by electrons from the emitter. Electrons born in the insulator may contribute to current of either sign. Therefore, the SPGD is a device that produces a current equal to the difference between the outward current from the emitter and insulator and the inward current from the collector and the insulator.

The sensitivity of the platinum SPGD is larger than that of the cobalt one but smaller than that of the rhodium detector (see Table 17.1). The rather large sensitivity, coupled with the prompt response, makes the platinum SPGD an attractive device. The main drawback of this detector is that it is sensitive to neutrons too. In a new detector, about 20 percent of the signal is due to thermal neutrons, mostly through the (n, γ) reaction with ^{195}Pt. This (n, γ) reaction with a cross section of 27 b is responsible for about 90 percent of the burnup of Pt detectors.

When the platinum detector was first developed, it was described as a device that gives a signal directly related to power and not to neutron flux, because it responds to gamma flux, which is proportional to reactor power. Since the device is also sensitive to neutrons, the neutron-to-gamma flux ratio should be accurately known in the core as a function of operational history and fuel burnup. One other factor that should be considered is the contribution to the signal from fission-product gammas, which are present after the reactor is shut down. More research is needed, therefore, before platinum SPGDs replace rhodium detectors as the main instruments of in-core flux determination.

A self-powered gamma detector using a central electrode made of lead and carbon (Pb–C) has also been constructed.[10,11] It has been applied as a threshold gamma-ray detector for prompt fission gammas. In a gamma field involving photons from fission (prompt) and from fission products, this detector can discriminate against low-energy gammas (from fission products) and detect only the prompt fission gammas.

17.2.4 Gamma Thermometers (GTs)

Gamma thermometers (GTs) are self-powered gamma-sensitive detectors that, despite their name, do not measure temperature but reactor power. Gamma thermometers were developed in the early 1960s[12] and have been used in a small number of reactors (Savannah River, Halden, Otto Hahn). The rhodium SPNDs replaced them in commercial light-water reactors, but there has recently (1979) been renewed interest in them because they have certain advantages.

Determination of reactor power with GTs is based on the measurement of the temperature drop along a piece of metal heated by energy deposited mainly from gammas and, to a lesser extent, from neutrons. Figure 17.8 shows the GT used in the Savannah River reactors. The heat generated by the radiation in the iron pin creates a

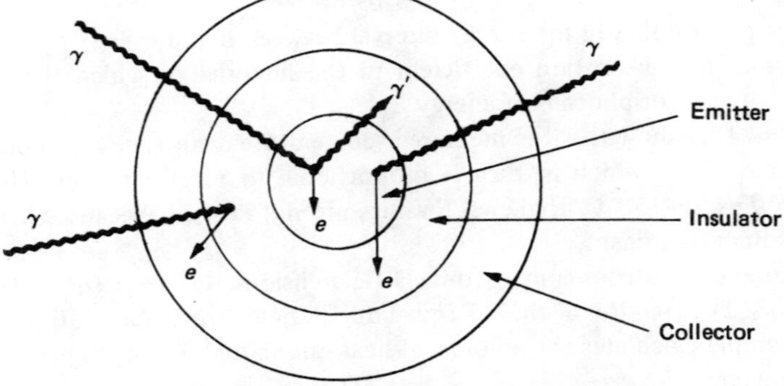

FIG. 17.7 Processes taking place in a gamma-sensitive self-powered detector.

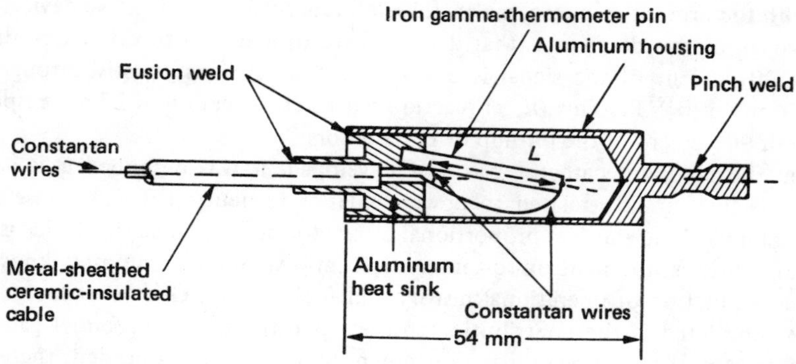

FIG. 17.8 The gamma thermometer used in the Savannah reactors (from ref. 12).

temperature difference as it flows through the metal and escapes into the heat sink. If all the heat generated in the pin goes to the heat sink, the temperature difference between two points a distance L apart is[12]

$$\Delta T = \frac{qL^2}{2k} \tag{17.7}$$

where q = heat generated in the pin per unit volume per second
k = thermal conductivity of the material of which the pin is made
ΔT = temperature difference measured by two thermocouples

In practice, only part of the heat is going to the heat sink. Corrections have to be made for heat lost by conduction through the insulator and by radiation. Both types of losses tend to decrease ΔT.

The heat q is produced by gammas and neutrons. The heat due to photons is

$$q\ (\mathrm{W/m^3}) = \int_0^\infty dE\ \phi_\gamma(E)\mu_a(E)E \tag{17.8}$$

where $\phi_\gamma(E)\,dE$ = photon flux in the energy interval between E and $E + \mathrm{dE}$
$\mu_a(E)$ = energy absorption coefficient in the material of which the pin is made, for photons of energy E

Equations 17.7 and 17.8 show that the measured temperature difference ΔT is proportional to the gamma flux, which in turn is proportional to reactor power. The expected value of ΔT is 15–150°C. Higher ΔT values are not desired, because then the radiation loss becomes significant.

Heat generated by neutrons comes from elastic collisions, and from (n, γ), (n, α), and (n, p) reactions. The fraction of the GT signal due to neutrons is about 10 percent. The following example calculates the amount of heat one should expect to measure in a gamma thermometer.

EXAMPLE 17.2

What is the heat expected to be deposited from gammas in the iron pin of a GT exposed to a flux of 10^{18} $\gamma/m^2 \cdot s$? Assume that the average gamma energy is 1.5 MeV.

ANSWER

Using average energy and total flux, Eq. 17.8 gives

$$q = \phi_\gamma \mu_a(E)E$$
$$= (10^{18}\ \gamma/m^2 \cdot s)(2.48 \times 10^{-3}\ m^2/kg)(1.5\ MeV)(1.602)(10^{-13}\ J/MeV)$$
$$= 596\ W/kg = 0.596\ W/g$$

The response of GTs is not instantaneous, for two reasons. First, the response of the instrument to a sudden change in power (i.e., to the heat rate q) depends on the thermal properties (heat capacity and thermal conductivity) and the square of the length of the metallic pin.[12] If the thermal time constant of the pin is τ, a change of temperature difference from ΔT_1 to ΔT_2 follows the equation

$$\frac{\Delta T_2}{\Delta T_1} = 1 - e^{-t/\tau} \tag{17.9}$$

The time constants for gamma thermometers under development range from a few seconds to about 150 s.

The second reason for the delayed response is the fraction of the GT signal that comes from delayed gammas emitted by fission products. Fission products build up at a certain rate after reactor startup, and also decay at a certain rate after reactor shutdown. It is estimated that about 25 percent of the signal comes from delayed gammas. Thus, a step change in power will create a signal as shown in Fig. 17.9. It takes about 15 min to reach about 95 percent of the saturation fission product activity.[13] The

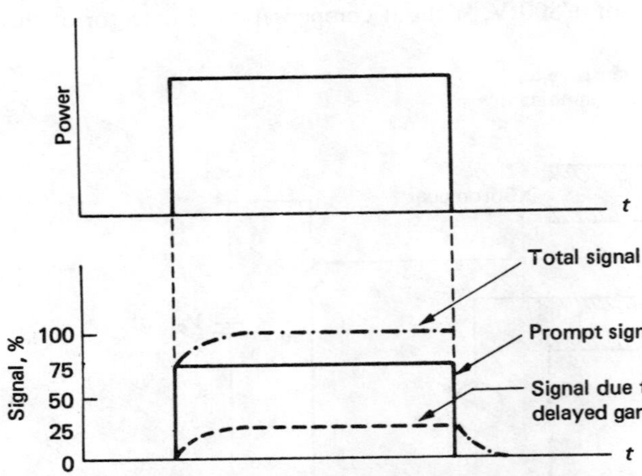

FIG. 17.9 Prompt and delayed GT signals as a result of a step change in power.

presence of a signal after reactor shutdown makes the GTs useful for measurements of decay heat (SPNDs cannot be used after shutdown).

The main advantages of GTs seem to be their almost zero burnup rate–i.e., very long lifetime–and their ability to measure power directly rather than neutron flux. The disadvantages are their delayed response and the possible dependence of the signal on the history of the fuel surrounding them. Research continues on the factors that determine the signal of a GT and their relationship to reactor power.

17.3 COMPENSATED ION CHAMBERS

Neutron counters located close to a reactor core are subjected to both neutron and gamma bombardment. Although a neutron counter–e.g., a ^{10}B counter–is mainly sensitive to neutrons, it responds to gammas too. At low reactor power, when the neutron flux is small, the neutron signal is overshadowed by a signal due to gammas emitted from fission products that had been accumulated from earlier reactor operation. To eliminate the effect of the gammas, a ***compensated ion chamber*** is used.

Compensated ion chambers operate in such a way that the gamma signal is subtracted and the output is proportional to the neutron signal only. Figure 17.10 shows the basic principle of a compensated ion chamber. The counter consists of two compartments. One, coated with boron, is sensitive to both neutron and gammas and produces a signal proportional to the total radiation field. The other is sensitive to gammas only and produces a signal proportional to γ radiation only. As Fig. 17.10 shows, the circuitry is such that the γ signal is subtracted from the $(\gamma + n)$ signal, thus giving a signal proportional to the neutron field only. The signal, in the form of a current, is measured by a picoammeter.

Correct compensation is achieved when the signal is zero in a pure gamma field. This is accomplished by using the proper combination of volumes for the two compartments or by changing the voltages or by a combination of voltage and volume change. Typical compensation voltages (V_c) are of the order of -25 V; the positive voltage V is of the order of $+800$ V. Without compensation, a detector of this type

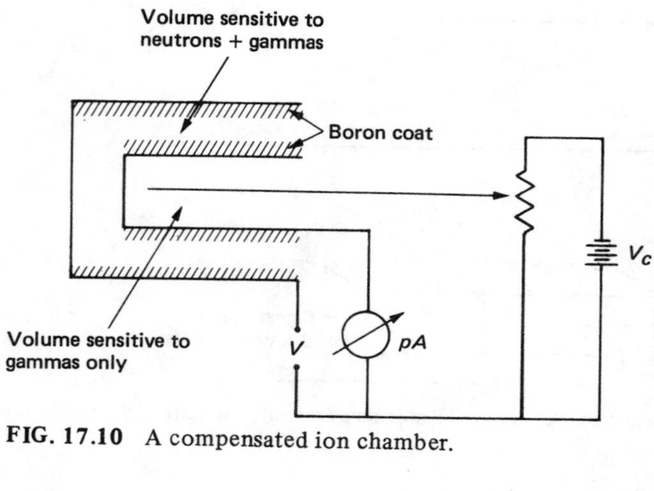

FIG. 17.10 A compensated ion chamber.

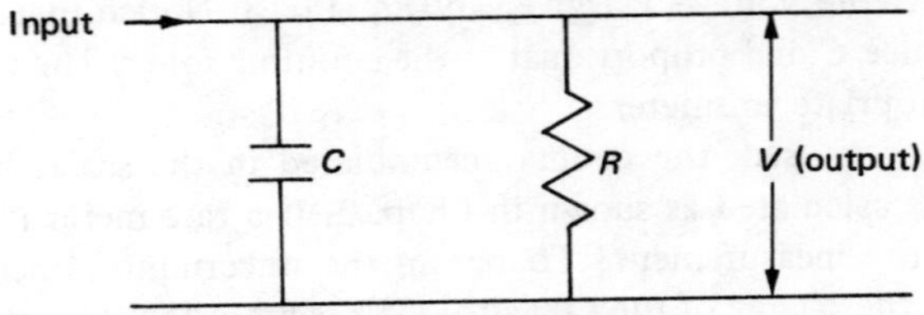

FIG. 17.11 The circuit of a rate meter.

has a useful range from 2×10^8 to 2×10^4 neutrons/cm$^2\cdot$s. With compensation, the useful range is extended downwards by about two orders of magnitude. The sensitivity of compensated ion chambers is of the order of 10^{-18} A/(neutrons/m$^2\cdot$s).

17.4 RATE METERS

A rate meter is a device that measures the average rate of incoming pulses. Rate meters are used for continuous monitoring of an event, where the average counting rate versus time rather than the instantaneous counting rate is needed.

The basic operation of a rate meter is to feed a known charge per pulse into a capacitor that is shunted by a resistor (Fig. 17.11). Let

r = counting rate (pulses/s)

q = charge per pulse

V = voltage across capacitor

R = resistance

Q = capacitor charge

The net rate of change of Q with respect to time is given by the equation:

$$\frac{dQ}{dt} = \text{(charge fed by pulses/s)} - \text{(charge flowing through resistor)}$$

or

$$\frac{dQ}{dt} = rq - \frac{Q}{RC} \tag{17.10}$$

The solution of this differential equation with the initial condition $Q(0) = 0$ is

$$Q(t) = rqRC(1 - e^{-t/RC}) \tag{17.11}$$

or, if one writes the result in terms of the output voltage,

$$V(t) = \frac{Q(t)}{C} = rqR(1 - e^{-t/RC}) \tag{17.12}$$

For time $t \gg RC$, equilibrium is reached and the value of the voltage is

$$V_\infty = rqR \tag{17.13}$$

The signal of a rate meter is the voltage V_∞ given by Eq. 17.13. Notice that V_∞ is independent of the capacitance C and proportional to the counting rate r. The voltage V_∞ is measured with an appropriate voltmeter.

If a pulse-type detector is used, the counts accumulated in the scaler have a statistical uncertainty that is calculated as shown in Chap. 2. If a rate meter is used, what is the uncertainty of the measurement? To obtain the uncertainty, one starts with Eq. 17.11, which gives the charge of the capacitor C. It is important to note that the charge changes exponentially with time. Thus, the contribution of the charge from a pulse arriving at $t = 0$ is not instantaneous but continues for a period of time.

Consider an observation point t_0 (Fig. 17.12). The standard deviation σ_Q of the charge collected at $t = t_0$ is the result of contributions from pulses having arrived earlier. If the counting rate is r, the number of pulses in a time interval Δt is, on the average, $r\,\Delta t$. The statistical uncertainty of this number is $\pm\sqrt{r\,\Delta t}$, or the uncertainty of the charge is $\pm q\sqrt{r\,\Delta t}$. One can show that a single pulse arriving at time t contributes to the signal at time $t = t_0$ an amount of charge equal to $q \exp[-(t_0 - t)/RC]$. Therefore, the variance of the charge at time $t = t_0$ is

$$\sigma_Q^2 = \int_0^{t_0} [q\sqrt{r\,dt}\, e^{-(t_0-t)/RC}]^2 \tag{17.14}$$

Integration of Eq. 17.14 gives the result

$$\sigma_Q^2 = 0.5q^2 rRC(1 - e^{-2t_0/RC}) \tag{17.15}$$

For $t_0 \gg RC$, Eq. 17.15 takes the form

$$\sigma_Q = q\sqrt{\frac{r(RC)}{2}} \tag{17.16}$$

At equilibrium, $Q = rqRC$ (from Eq. 17.11); therefore

$$\sigma_r = \frac{\sigma_Q}{qRC} = \sqrt{\frac{r}{2RC}} \tag{17.17}$$

The quantity RC is the time constant of the circuit shown in Fig. 17.11. Equation 17.17 states that any instantaneous reading on a rate meter has a standard deviation equal to that of the counting rate that is obtained by counting for a time equal to $2RC$ (assuming the background is negligible).

17.5 SOLID–STATE TRACK RECORDERS (SSTRs)

The technique of measuring the number of particles by observing their tracks in certain organic and inorganic materials has been used for the study of phenomena in such diverse fields as geology, archeology, astrophysics, nuclear physics, and dosimetry. The

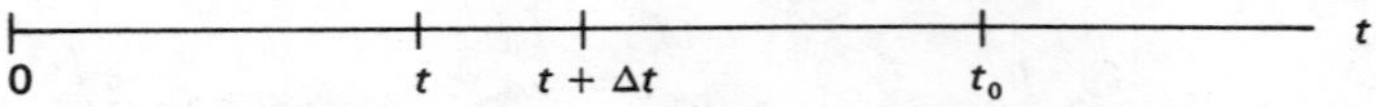

FIG. 17.12 Pulses arriving during Δt, and at t, contribute to σ_Q at $t = t_0$.

book by Fleischer, Price, and Walker is an excellent review of materials used as SSTRs, of the theories for track formation, and of the techniques used for track etching. In the field of radiation measurements, SSTRs are used for fission rate measurements[14,15] and as alpha[16] and neutron dosimeters.[17–19]

The SSTR technique is based on the damage created in a solid along the path of a heavily ionizing particle such as an alpha particle or a fission fragment. The damage along the path, called a track, may become visible under an ordinary optical microscope after etching with suitable chemicals. The visible tracks are counted either by direct observation by a human or with the help of automated instruments.[20–22]

Many theories have been proposed for the production of tracks by ionizing particles in solids, but none explains all the phenomena involved for both organic and inorganic materials. The basic mechanisms of energy loss are known (see Chap. 4). A fast-moving charged particle loses energy by excitation and ionization. Ionization creates charge centers in any solid. The ejected electrons, also called δ rays, can produce further excitation and ionization. In organic materials, such as polymers, the deexcitation may break the long molecular chains and produce free radicals. As the ion slows down, it starts picking up electrons, and thus its charge decreases. Close to the end of its path, atomic rather than electronic collisions are the dominant mode of energy loss. The result of atomic collisions is atom displacement and creation of a vacancy. Pictorially, the formation of the track in a crystal and a polymer is believed to be taking place as shown in Fig. 17.13.

The best means of observing the tracks is by etching the SSTR material with a chemical which preferentially attacks the damaged material and enlarges the original track. It is believed that the damaged region is preferentially attacked because it becomes more active chemically than the surrounding undamaged region. This is a result of the free energy associated with the disorders created along the trajectory of the particle. The particular method of etching depends on the material used. Mica is etched with a 27-*M* hydrofluoric acid for 2-3 h at room temperature. Lexan,† a polymer, is etched with 6-*N* NaOH for 30 min at 70°C. No matter what the chemical or the material is, the track is enlarged because the chemical attacks the surface of the SSTR and the interior surface of the original track and creates a conical hole as shown in Fig. 17.14. Thus, tracks that originally intersected the surface will appear as circles or ellipses under the microscope. If the etching continues further, some tracks may be "etched away"; therefore the level of etching is important, especially in absolute measurements.

Figure 17.15 shows a photograph–a fission fragment etched tracks in three different materials.

An SSTR used for fission rate measurements consists of a thin, uniform deposit of a fissionable material in contact with a material such as mica or Lexan. When the SSTR is exposed to neutrons, fissions take place, and the fission fragments are imbedded into the SSTR material and produce the tracks. For absolute measurement of

†Lexan is a polymer developed by General Electric, with atomic composition $C_{16}H_{14}O_3$.

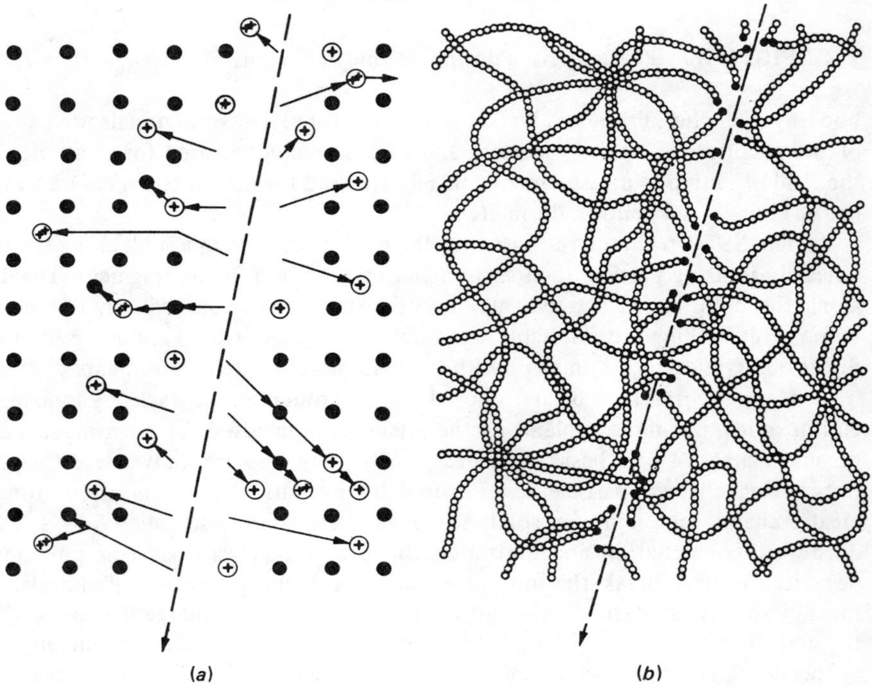

FIG. 17.13 The formation of a particle track in (*a*) a crystal and (*b*) a polymer. (From Fleischer-Price-Walker.)

fission rate, the observer should be certain that all the fission fragments enter the SSTR and create visible tracks; or, if some fragments are lost, a correction should be applied to the observed fission rate. Some fission fragments will not escape no matter how thin the fissionable deposit is (Fig. 17.16). As Fig. 17.16 shows, a fission fragment emitted toward the SSTR will not reach it unless it is emitted within a cone defined by the angle $2\theta = 2 \arccos (x/R)$, where x is the depth at which the fission took place and R is the range of that fragment. If $t < R$, where t is the thickness of the deposit, the fraction of fragments reaching the SSTR is equal to $1 - t/2R$. If one

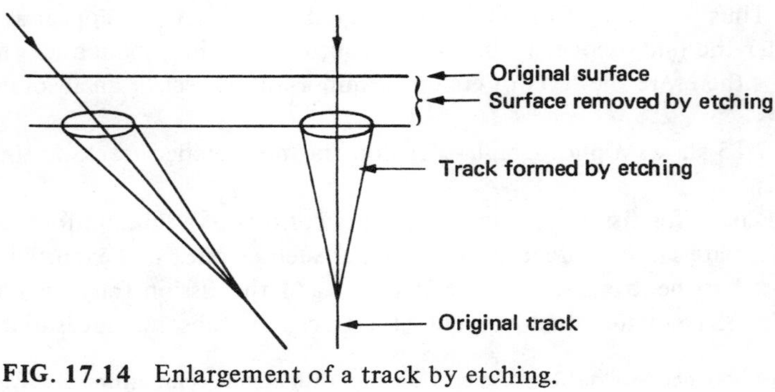

FIG. 17.14 Enlargement of a track by etching.

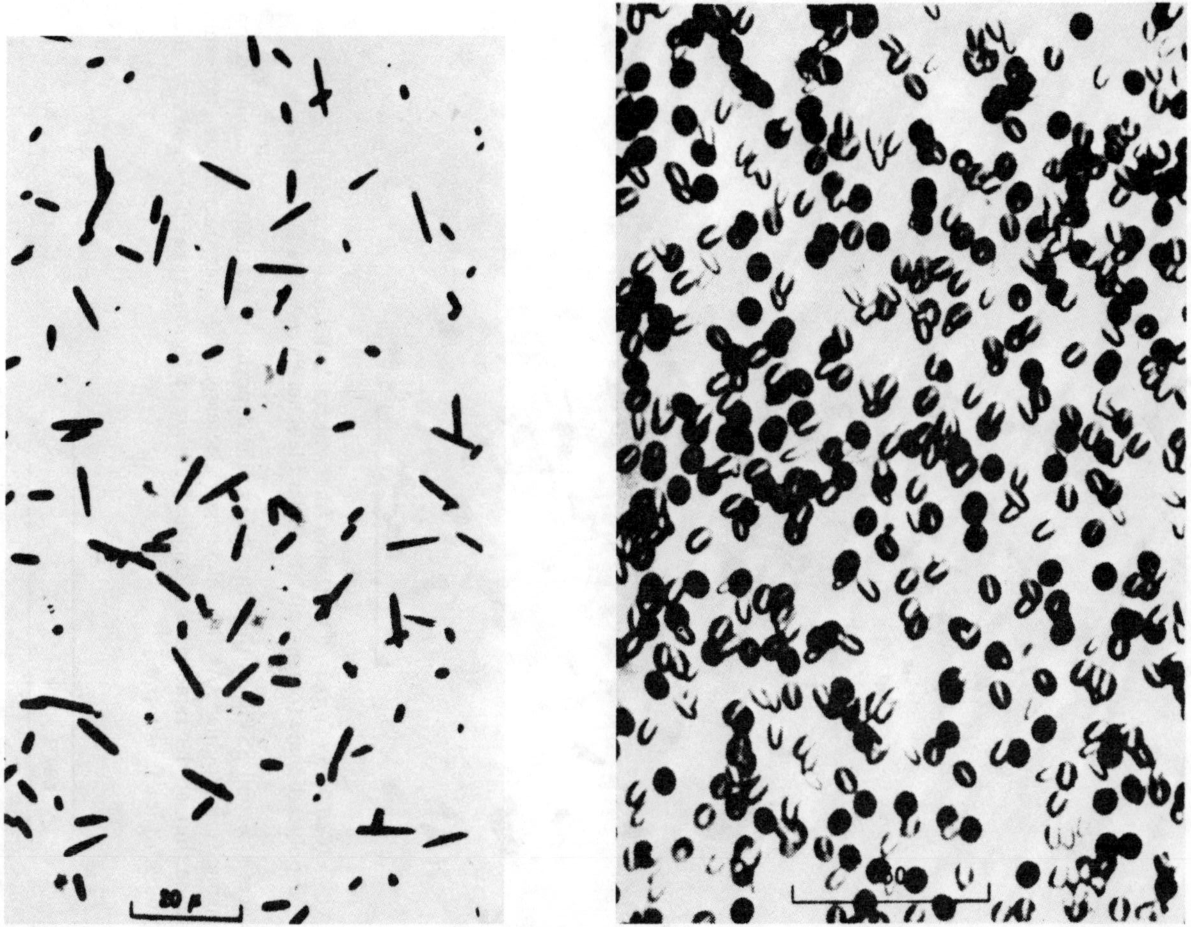

(a) (b)

FIG. 17.15 ^{25}Cf fission fragment tracks after etching in (*a*) crystal (orthoclase), (*b*) soda-lime glass. (From Fleischer-Price-Walker.)

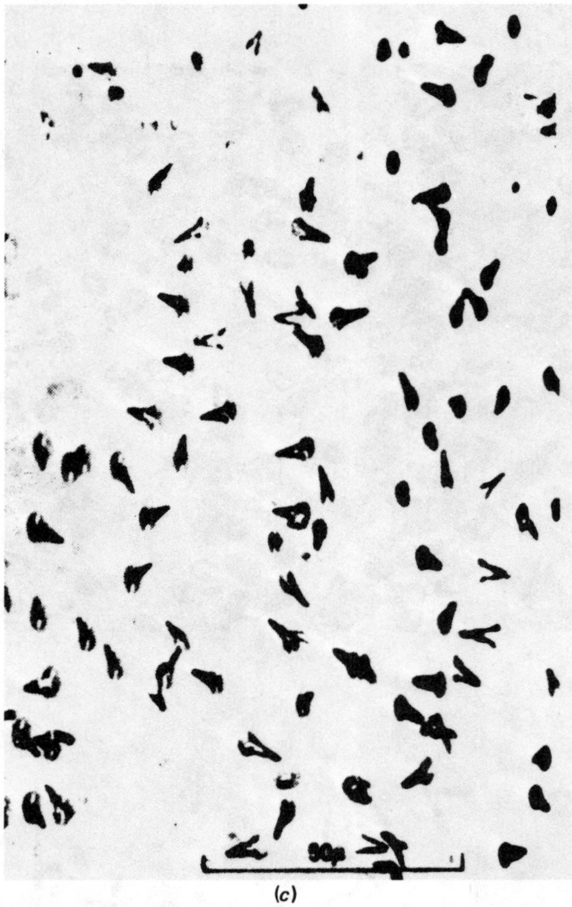

FIG. 17.15 (*Cont.*) ^{25}Cf fission fragment tracks after etching in (*c*) Lexan polycarbonate (organic glass). (From Fleischer-Price-Walker.)

places a second SSTR on the other side of the deposit, the fraction of escaping fission fragments is doubled. As the thickness t decreases, a greater fraction of fragments escape, but at the same time a smaller number of fissions takes place and a smaller number of tracks is formed.

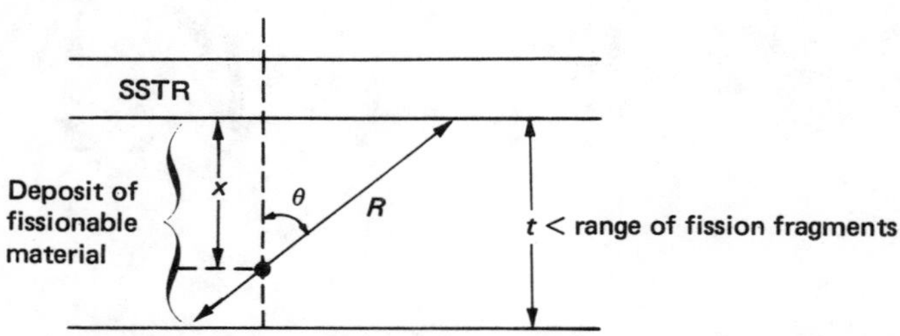

FIG. 17.16 A fission fragment emitted at an angle greater than θ ($\theta = \arccos x/R$) will not reach SSTR.

A neutron dosimeter based on the SSTR technique consists of a foil with a known amount of a fissionable isotope placed in contact with an SSTR material. Problems associated with such dosimeters are discussed in a review article by Becker.[23] A useful relationship between track density and neutron fluence, valid to within 1 percent for neutron energies present in a reactor environment, is (Fleischer, Price, Walker, Chap. 9):

$$\rho\,(\text{tracks/m}^2) = 1.6 \times 10^{-5}\,(\phi t) \times \sigma_f\,(\text{b}) \tag{17.18}$$

In addition to fission-fragment track measurements, thermal neutrons can also be detected by alpha tracks from (n, α) reactions with ^{6}Li or ^{10}B. Finally, SSTRs have been used as monitors of alpha particles,[16] especially alphas emitted by radon and its daughters.

17.6 THERMOLUMINESCENT DOSIMETERS (TLDs)

Thermoluminescent dosimeters (TLDs) are based on the property of thermoluminescence, which can be understood if one refers to the electronic energy-band diagram of crystals (see also Chap. 7). When ionizing radiation bombards a crystal, the energy given to the electrons may bring about several results (Fig. 17.17). The electron may acquire enough energy to move from the valence to the conduction band, in which case the event is called ionization. Or, the electron acquires enough energy to move to an excited state (to the exciton band) and form an exciton. An exciton, consisting of an electron and a hole bound electrostatically, can migrate through the crystal. Electrons, holes, and excitons may be caught in many "traps" that exist in the solid. Traps are formed in a variety of ways. Foreign atoms (impurities), interstitial atoms, dislocations, vacancies, and imperfections may act as traps. The trapped carriers remain in place for long periods of time if the temperature of the crystal stays constant or decreases. If the temperature is raised, however, the probability of escape increases. As electrons and holes are freed and return to the ground state, they emit light (Fig. 17.18).

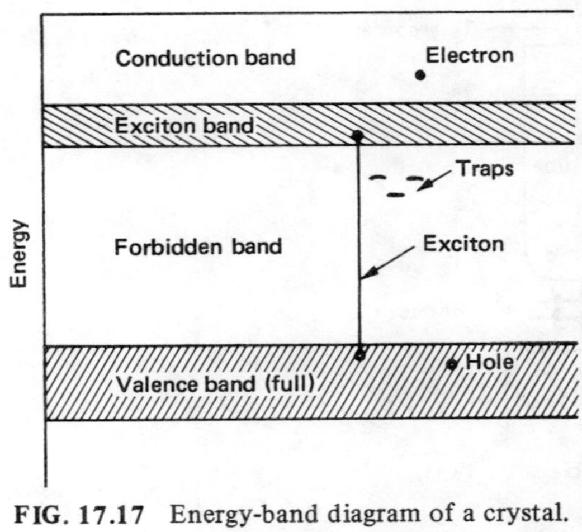

FIG. 17.17 Energy-band diagram of a crystal.

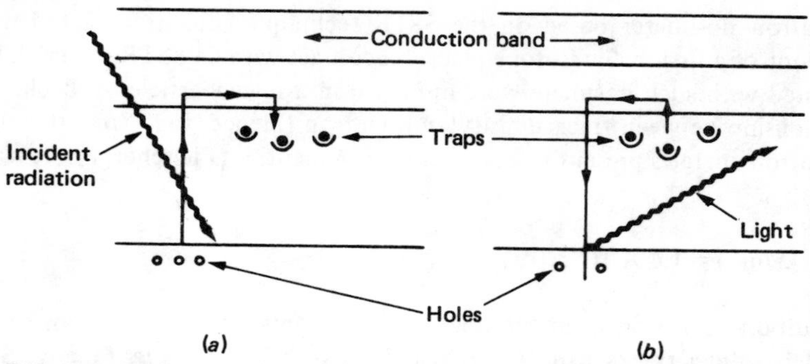

FIG. 17.18 (*a*) As a result of irradiation, some carriers fall into traps. (*b*) Upon heating, the carriers are given enough energy to escape from the traps and return to the valence band, with the emission of light.

The emission of this light is called thermoluminescence and is the property upon which the operation of TLDs is based.

A TLD is essentially a piece of a thermoluminescent material, exposed to the radiation being measured. After irradiation stops, the TLD is heated under controlled conditions (Fig. 17.19), and the light intensity is measured either as a function of temperature or as a function of the time during which the temperature is raised. The result of such a measurement is a graph called the *glow curve* (Fig. 17.20). Glow curves have more than one peak, corresponding to traps at various energy levels. The amplitudes of the peaks are proportional to the number of carriers trapped in the corresponding energy traps. The absorbed dose may be measured either from the total light emitted by the glow curve or from the height of one or more peaks of the glow curve. The TLD is annealed–i.e., it returns to its original condition–and is ready to be used again after being heated long enough so that all the traps have been emptied; then it is left to cool down to room temperature. Measurement of the light from the glow curve, and subsequent annealing, are performed by instruments generally called *readers*, which are available commercially (e.g., from Harshaw Co.).

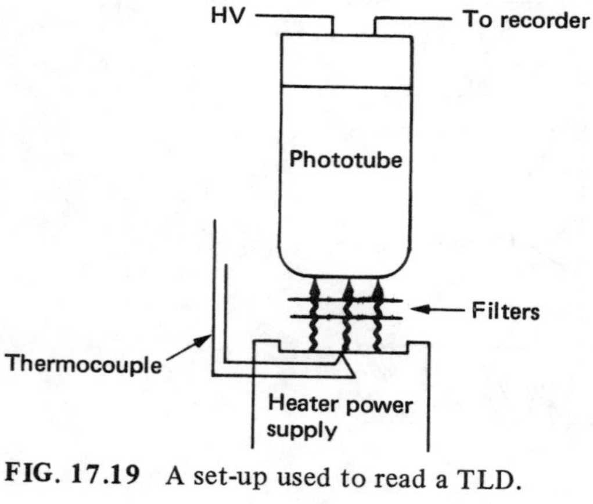

FIG. 17.19 A set-up used to read a TLD.

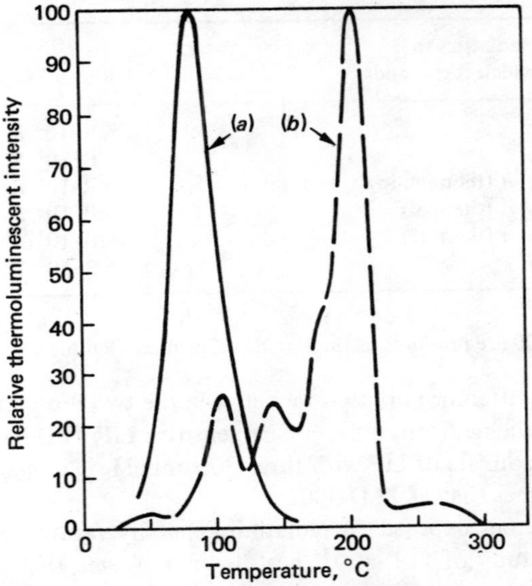

FIG. 17.20 Typical thermoluminescent glow curves. (*a*) Glow curve of $CuSO_4$:Mn heated at 6°C/min. (*b*) LiF (TLD-100) exposed to 10^4 R and heated at 20°C/min. (From ref. 24.)

There are many thermoluminescent materials, but those useful for dosimetry should have the following characteristics.

1 Retention of trapped carriers for long periods of time at temperatures encountered during the exposure
2 Large amount of light output
3 Linear response over a large dose range
4 Perfect annealing to enable repetitive use

Materials commercially available that satisfy most of these requirements are $CaSO_4$:Mn, CaF_2 (natural), CaF_2:Mn, $Li_2B_4O_7$:Mn, and LiF. $CaSO_4$ and CaF_2 are used for gammas only. The lithium compounds can be used for gammas or neutrons. The main characteristics of these materials are given in Table 17.2. Various uses are described in references 25–33.

The three LiF TLDs listed in Table 17.2 have found wide use for measurements in mixed neutron-gamma fields.[31] The TLD-100 containing natural lithium (92.6 percent ^{7}Li, 7.4 percent ^{6}Li) responds to gammas and thermal neutrons. Thermal neutrons are detected through the (n, α) reaction with ^{6}Li, which has a cross section equal to 950 b for thermal energies. If a TLD-100 is exposed to about 3×10^{11} neutrons/m^2 (thermal), its light output is equivalent to that from 1 R of gamma radiation. The TLD-600, containing lithium enriched to 95.62 percent in ^{6}Li, is extremely sensitive to thermal neutrons and also to gammas. The TLD-700, containing 99.993 percent ^{7}Li, is sensitive to gammas only, because the neutron cross section for ^{7}Li is very small (about 0.033 b for thermal neutrons). In a mixed neutron-gamma field, one can achieve γ-n discrimination by exposing to the radiation a TLD-100 plus a TLD-700, or

TABLE 17.2 Properties of Certain TLDs†

Material	Radiation to which it responds	Range
$CaSO_4$ (Mn)	γ	μR–10^3 R
CaF_2 (Natural)	γ	mR–10^3 R
CaF_2 (Mn)	γ, n (thermal, low response)	mR–10^5 R
LiF (TLD-100)‡	γ, n (thermal)	mR–10^4 R
LiF (TLD-600)	γ, n (thermal)	mR–10^5 R
LiF (TLD-700)	γ	mR–10^5 R

†From reference 24.
‡TLD-100, TLD-600, TLD-700 are products of the Harshaw Chemical Company.

a TLD-600 plus a TLD-700. The difference in response between the two dosimeters of either pair gives the dose due to the neutrons only. A very sensitive LiF TLD has been reported by Nakajima et al.[32] It consists of LiF with three dopants: Mg, Cu, and P. Its sensitivity is supposed to be 23 times that of TLD-600.

Thermoluminescent dosimeters exposed to radiation generally return to their original state after annealing. A study of TLD-600[33] has shown, however, that the annealing erases the effects of past exposure only partially. A zero dose reading, which increases with exposure, is always present.

17.7 NUCLEAR-TRACK AND PHOTOGRAPHIC EMULSIONS

Emulsions used for radiation detection consist of silver halide microcrystals or grains—usually AgBr—dispersed in gelatin. The function of the gelatin is to support the grains while allowing the chemicals of the developer to attack them. After exposure to radiation, the emulsion is developed and fixed like a regular photographic film. The developer attacks preferentially the AgBr of the grains that were hit by the radiation and reduces it to metallic Ag, which then becomes visible.

The process by which the incident radiation makes the exposed grains more sensitive to the developer than the nonexposed ones is not yet fully understood. It is believed that the particles hit the grain and create "latent-image" centers which consist of up to a few hundred Ag atoms. Under the influence of the developer, these latent-image centers stimulate the reduction of the whole grain to metallic Ag. More details on this subject are given in references 34–36.

Nuclear-track emulsions consist of grains with linear dimensions of about 30 μm randomly dispersed in gelatin with a mean distance between grains many times greater than their diameter. As the particle traverses the emulsion, it leaves behind a track of latent-image centers located at the struck grains. The subsequent action of the developer makes the track visible. The mass and the charge of the particle can be determined from the characteristics of the track, mainly from the length of the track and its grain density.

The advantages of nuclear emulsions are simplicity, reliability, compactness, stability, continuous sensitivity over a long period of time, ability to identify the type of particle, ability to record the path (i.e., position) of a particle, large recording range (from very low to almost unlimited flux), and capability to "remember" (the tracks, once formed, become a permanent record). The disadvantages are delay in getting the information (the emulsions have to be developed), impossibility of coincidence measurements, difficulty in reading the information (if the grain density is low and there are many tracks from background radiation), and inability to vary emulsion density (the composition of the emulsions cannot be changed arbitrarily; thus, one is limited to studies with nuclei normally present in the emulsion and a few others that can be added in limited concentrations).†

Nuclear emulsions are used today for charged-particle detection in high-energy physics and in neutron dosimetry. Fast neutrons are detected through proton recoils produced by collisions with the hydrogen of the gelatin. Personnel film badges based on this principle are widely used around reactors and accelerators. The absorbed dose is proportional to the number of tracks per unit area. Automated scanning devices have been built to count the number of tracks.[20]

Using photographic emulsions, detection is based on the density of developed grains. The size of the grains is now about 1 μm and individual particle tracks cannot be resolved. Photographic emulsions are mainly used in dosimetry, and that function is discussed next in some detail.

The quantitative interpretation of a developed photographic film depends on the relationship between the density of developed grains–from now on called the film density D–and the particle fluence F (particles/m^2). As reported by Dudley,[24] if one makes the assumptions that

1 A single hit is sufficient to make a grain developable.
2 All grains have the same projection area a, which stays the same after development.
3 There is no absorption or scattering of the particles in the emulsion.
4 The incident particles hit the emulsion along the normal.

then this relationship between D and F is[37]

$$D = (\log e)aN(1 - e^{-aF}) \tag{17.19}$$

where a = average area of a developed grain (m^2)
N = number of grains/m^2 in the emulsion

For low fluences, Eq. 17.19 takes the form

$$D = (\log e)a^2 NF \tag{17.20}$$

which indicates a linear relationship between film density and fluence. Although no real film satisfies Eq. 17.20 exactly, the linear relationship is valid over the relatively narrow range of densities encountered in practice.[24]

The sensitivity of photographic emulsions is defined as the reciprocal of the absorbed dose required to give a density $D = 0.3$. The density D is measured by

†Two companies that make nuclear emulsions are Kodak and Ilford. Emulsion composition is given in reference 36.

allowing light of initial intensity I_0 to pass through the film. If I is the intensity of the transmitted light,

$$D = \log \frac{I_0}{I} \tag{17.21}$$

Thus, D is a quantitative expression of the degree of blackening of the film. The effect produced by the radiation fluence is equal to the total density minus the background density, which is the density exhibited by a film not yet exposed. The background density is of the order of 0.1 to 0.2. Measurable densities may range from 0.1 to 6. The human eye is capable of detecting differences of about 0.01, depending on the sharpness of the demarkation lines.

Photographic emulsion dosimeters are used to monitor electrons, photons, and neutrons. The response of an emulsion is essentially the same for either electrons or photons, which is not surprising since photons deposit their energy through electron-generating interactions (photoelectric effect, Compton effect, pair production). The sensitivity of emulsions as a function of particle energy is not constant either for electrons or for photons. The photon sensitivity may become practically constant with energy, if the film is covered with a thin plastic scintillator. Experiments have shown that the sensitivity also depends on the angular distribution of the incident radiation (parallel beam or isotropic). As a result of these dependencies of the sensitivity, the user of photographic emulsions should observe very carefully the geometry of the measurement and the calibration of the measuring instrument.

Fast neutron detection with photographic emulsions is accomplished in the same way as in nuclear track emulsions—namely, through proton recoils as a result of collisions with the hydrogen of the gelatin. Thermal neutrons are detected through the $^{14}N(n,p)^{14}C$ reaction,† which produces a 0.615-MeV proton, and through the β particles emitted by the neutron-activated isotopes of silver (^{108}Ag and ^{110}Ag). An increase in sensitivity is accomplished if the film is "loaded" with either lithium or boron. Both of these elements produce alphas through the reactions $^6Li(n,\alpha)^3H$ and $^{10}B(n,\alpha)^7Li$. Another way to increase sensitivity is by covering the film with a cadmium foil. Cadmium absorbs essentially all the thermal neutrons. Gammas, and especially betas, emitted after the neutron capture hit the film and produce an intensified result.

17.8 MAGNETIC SPECTROMETERS

Magnetic spectrometers (or analyzers) are instruments that exhibit high-energy resolution. They are used mainly in experiments involving high-energy particles, but they can be used with low-energy particles as well. Energy resolutions achieved with this type of spectrometer are of the order of $\Delta E/E \approx 10^{-4}$ for high-energy electrons[38] and $\Delta E/E \approx 10^{-3}$ for heavy ions.[39]

Many types of magnetic spectrometers have been developed, differing mainly in the shape of the magnetic field used.[38–43] The basic principle common to all of them is the action of the force from a magnetic field on a moving charged particle. If a

†This reaction is responsible for absorbed dose in tissue.

particle with charge Ze moves with velocity $\mathbf{v}$ in a magnetic field of strength $\mathbf{B}$, the force $\mathbf{F}$ on the particle is given by

$$\mathbf{F} = Ze\mathbf{v} \times \mathbf{B} \tag{17.22}$$

Since the magnetic force is always perpendicular to the direction of motion of the particle, it does not change the energy but only the direction of motion of the particle. If the velocity is perpendicular to the magnetic field, the particle trajectory will form a circle of radius ρ given by the equation

$$\frac{M'v^2}{\rho} = ZevB \tag{17.23}$$

which after rearrangement becomes

$$M'v = ZeB\rho \tag{17.24}$$

where $M' = M/\sqrt{1-\beta^2}$, the total mass of the particle.

Equation 17.24 is the basis of magnetic spectroscopy. It expresses the fact that the momentum of the particle is proportional to the quantity $B\rho$, called the ***magnetic rigidity*** of the particle. The relationship between kinetic energy and rigidity is obtained from the equation linking energy and momentum (see Chap. 3). The relativistic equation for the kinetic energy T is

$$T = \sqrt{(Mc^2)^2 + Z^2e^2(B\rho)^2} - Mc^2 \tag{17.25}$$

The nonrelativistic equation is

$$T = \frac{Z^2e^2(B\rho)^2}{2M} \tag{17.26}$$

To understand how a magnetic spectrometer works, consider two particles with momenta m_1v_1 and m_2v_2 coming into a space with a constant magnetic field B (Fig. 17.21). Assuming that the magnetic field is perpendicular to the velocity and also

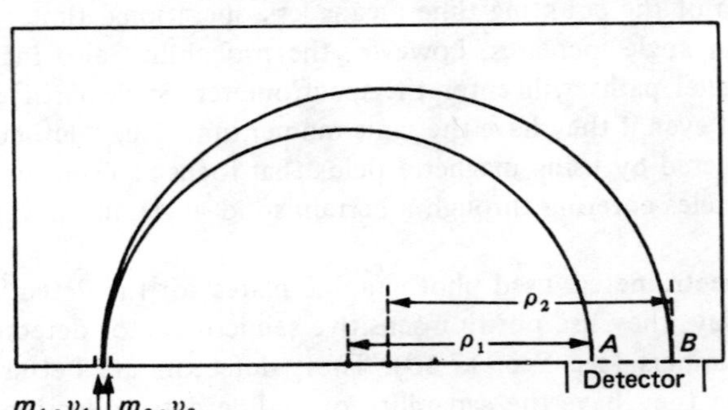

FIG. 17.21 Two particles with velocities v_1 and v_2 follow trajectories with radius ρ_1 and ρ_2 respectively, and are separated spatially (points A and B).

perpendicular to the plane of the figure, the two particles will move along circular paths of radii ρ_1 and ρ_2 given by Eq. 17.24. If the particles are allowed to complete half a circle, they will hit the "focal" plane at points A and B. Thus, if one determines ρ_1 and ρ_2 and the charge of the particles is known, the momentum can be determined from Eq. 17.24. If the mass of the particle is also known, the energy can be determined from Eq. 17.25 or 17.26. Obviously, for a given rigidity $B\rho$, the energy E depends on the factor Z^2/M. If one determines the energy by other means–e.g., by using a semiconductor detector–the magnetic spectrometer may give the Z^2/M value.

In practice, the value of B is accurately known; therefore, the real task is the accurate determination of the radius ρ. The radius ρ (or the distance 2ρ) is measured with the help of position-sensitive detectors placed at the focal plane. Any uncertainty in the value of ρ, or in general in the value of $B\rho$, introduces an uncertainty in the value of energy. If all the particles move in parallel paths when they enter the spectrometer, the energy resolution obtained from Eq. 17.25 or Eq. 17.26 is given by

Relativistic

$$\frac{\Delta T}{T} = \left(1 + \frac{MC^2}{MC^2 + T}\right) \frac{\Delta (B\rho)}{B\rho} \tag{17.27}$$

Nonrelativistic

$$\frac{\Delta T}{T} = 2 \frac{\Delta (Be)}{B\rho} \tag{17.28}$$

As explained in Sec. 17.10, the position can be determined with an accuracy of less than a millimeter. Since a typical value of ρ is 0.80 m, the expected resolution is

$$\frac{\Delta T}{T} \approx \frac{\Delta \rho}{\rho} \approx \frac{1}{800} \approx 10^{-3}$$

As shown in Fig. 17.21, the particles enter the spectrometer through a slit, whose size defines the solid angle accepting particles from the source. This solid angle should be as large as possible, because the more particles enter the spectrometer, the less the necessary counting time is. Since magnetic spectrometers are commonly used with accelerators, reduction of the counting time means less operational time for the accelerator. When the solid angle increases, however, the probability also increases that particles with nonparallel paths will enter the spectrometer. Such particles will not focus at the same point even if they have the same momentum. This "defocusing" is a difficult problem, corrected by using magnetic fields that focus at a spot or a line of the detector all the particles entering through a certain solid angle and having the same momentum.[38–40]

The first magnetic spectrometers used photographic plates for the detection of the deflected particles. Today, they use position-sensitive semiconductor detectors or multiwire proportional chambers (see Sec. 17.10). These detectors are better than photographic plates because they have the capability of on-line data collection, provide time information for coincidence experiments, and measure the energy of the particle.

17.9 ELECTROSTATIC SPECTROMETERS

Electrostatic spectrometers (or analyzers) do not have such extensive use as magnetic spectrometers. They are only used for certain types of experiments where an electrostatic field is more convenient than a magnetic one.

An electrostatic spectrometer uses a radial electric field formed between two cylindrical electrodes, as shown in Fig. 17.22. If there is a potential difference V between the electrodes EF and $E'F'$, the electric field in the space between the two electrodes is equal to

$$\left[\frac{V}{\ln(r_o/r_i)}\right]\frac{1}{r}$$

where r is the distance from the center of the electrodes. A particle entering the spectrometer at EE' with velocity $\mathbf{v}$ perpendicular to EE' will follow under the influence of the radial force

$$F = (Ze)\left[\frac{V}{\ln(r_o/r_i)}\right]\frac{1}{r}$$

a circular path of radius r_f. It can be shown[44] that particles entering at EE' with the same speed but traveling along a direction slightly different than the normal to EE' will focus at the point F_0, provided the angle $\phi = \pi/\sqrt{2} = 127°17'$ (Fig. 17.22). All the particles that arrive at F_0 have a kinetic energy that can be calculated in terms of the radius r_f and the strength of the electric field.[45] More details about the focusing properties and energy resolution of electrostatic spectrometers can be found in references 46–48. Electrostatic analyzers with parallel plates have also been constructed.

17.10 POSITION-SENSITIVE DETECTORS

Detectors that in addition to the measurement of the energy also indicate the position of the particle have been developed for application with magnetic or electrostatic

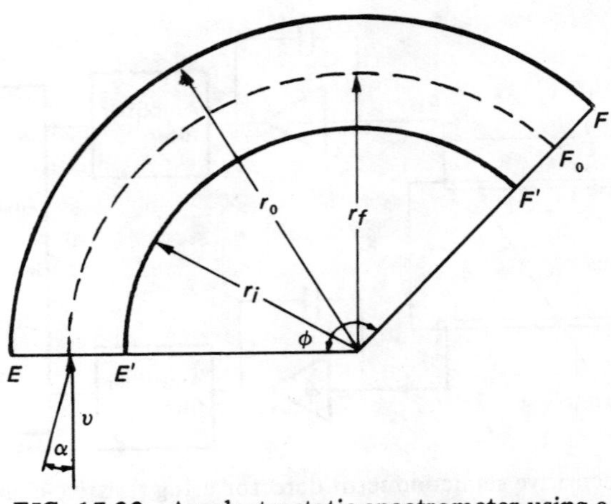

FIG. 17.22 An electrostatic spectrometer using a radial field.

spectrometers and for imaging devices used in biological and medical research. As discussed in Secs. 17.8 and 17.9, magnetic or electrostatic spectrometers are mostly used with high energy particles. Biological and medical imaging devices, on the other hand, involve mostly x-rays and in general low-energy radiation. It is because of the latter use that position-sensitive detectors are briefly discussed here.

17.10.1 Position-Sensitive Semiconductor Detectors

Most of the position-sensitive semiconductor detectors determine the position of the incident particle by employing the method of ***resistive-charge division.***[49] To illustrate the method, consider the detector in Fig. 17.23. The detector is a reverse-biased p–n junction with electrodes on both front and back. The front electrode with considerable resistivity has two electrical contacts a distance L apart. The back electrode has low resistivity and provides a good electrical contact to the base material. When a particle enters the detector, electrons and holes are created that move under the influence of the electric field. If the resistivity of the front electrode is homogeneous, and charge-sensitive low-impedance amplifiers are used, the charge collected at one of the two contacts of the front electrode is proportional to the distance between the point of impact and the other contact. The total charge q_0 collected through the single contact of the back side is, of course, proportional to the energy deposited in the sensitive region of the junction. This technique of determining the position by comparing the signals from q_1 and q_0 is called the ***amplitude method.***

The signal q_1 changes with time, as shown in Fig. 17.24. The time scale is in units of the time constant $\tau_D = R_D C_D$, where R_D and C_D are the resistance and capacitance of the detector, respectively. Figure 17.24 shows that the risetime of the signal depends on the position. This property is the basis of a second technique for determination of position, called the *time method.* The position is now determined from the difference in arrival times of the signals from the charges q_1 and q_2 (Fig. 17.23).

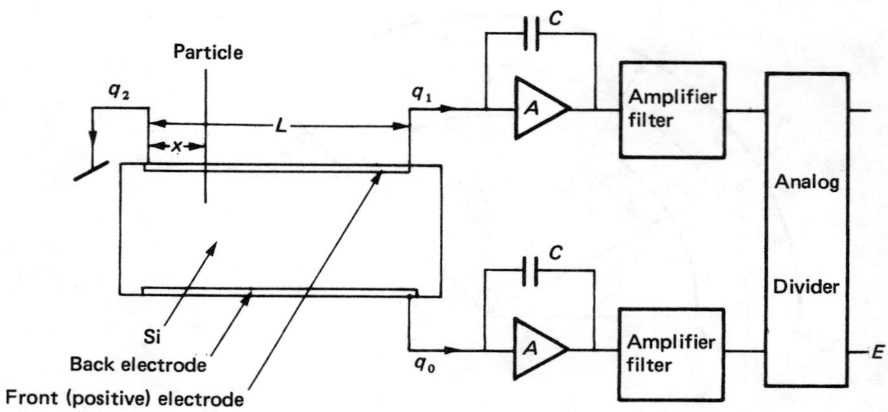

FIG. 17.23 A position-sensitive semiconductor detector using resistive-charge division (from ref. 49).

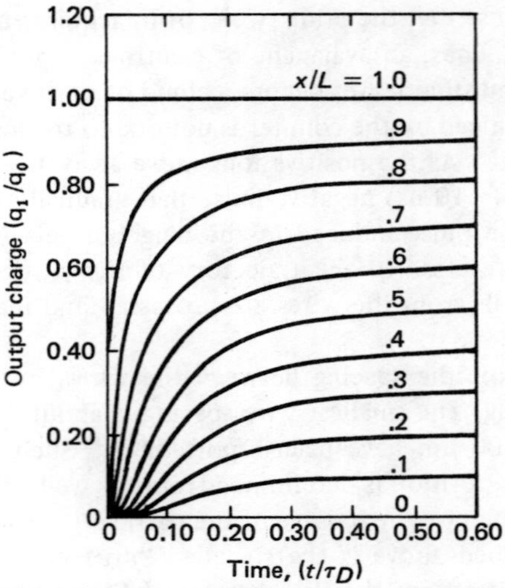

FIG. 17.24 Time dependence of the position signal for different positions of incidence (from ref. 49).

The detector described above is of the "continuous" type. Position-sensitive detectors of the "discrete" type have also been developed.[50] They consist of individual semiconductor elements all placed on the same base material, with each element connected to its own preamplifier-amplifier system. Two-dimensional detectors of the continuous type[51] as well as of the discrete type[52] have also been tried.

17.10.2 Multiwire Proportional Chambers (MWPC)

Multiwire proportional chambers (MWPC)† have been developed for use as position-sensitive focal detectors for magnetic spectrometers.[54–61] They can provide excellent position resolution, operate with counting rates as high as 10^6 counts/s, and provide a large solid angle at the focal plane of the spectrometer.

The basic design of a MWPC is shown in Fig. 17.25. A series of thin, equally spaced anode wires is positioned between two parallel plates serving as cathodes. A noble gas mixed with an organic component fills the space between the cathodes. A positive voltage is applied to the anode wires. When a particle goes through the counter, electrons and ions are created by the processes explained in Chap. 5. The

†Single-wire position-sensitive proportional counters operating by charge-division (as described in Sec. 17.10.1) have also been used.[53]

FIG. 17.25 The basic design of an MWPC.

electric field close to the wires is so intense that the primary electrons acquire enough energy to produce secondary ionization. Thus, an avalanche of electrons is produced that is collected within a time of 1 ns, but that leaves behind a cloud of positive ions.

It is significant that the pulse produced by the counter is not due to the motion of the electrons, but to that of the ions.[54] As the positive ions move away from the anode wires, they generate a fast-rising (~10 ns) negative pulse that gradually slows down and lasts a few microseconds. The pulse induced in the neighboring wires is positive. Thus, the active wire (the wire close to the trajectory of the particle) is distinguished from the others. The signals from the wires are processed and read by either digital or analog systems.[55]

The position resolution depends on the spacing between the wires, but it is better than the actual size of the spacing. The smallest wire spacing is about 1 mm, while position resolutions better than 100 μm have been reported.[62,63] Such space resolution has been achieved because the position is determined from the well-defined centroid of the charge distribution generated by the passage of the particle.

A variation of the detector described above is the so-called "drift chamber." The drift chamber determines the position from the time it takes the electrons produced by the incoming particle to drift to the nearest anode wire.[64] A two-dimensional MWPC has also been constructed for detection of neutrons scattered from biological samples.[65] It is a ^{3}He gas-filled counter that detects neutrons through the (n, p) reaction.

PROBLEMS

17.1 What is the burnup rate per month of an SPND using ^{235}U as the emitter and being exposed to a thermal flux of 2×10^{17} neutrons/m$^2\cdot$s?

17.2 How long will it take for the sensitivity of a rhodium SPND to decrease to 50 percent of its initial value? Assume a thermal flux of 10^{18} neutrons/m$^2\cdot$s.

17.3 Using semiquantitative arguments, show that the sensitivity of a Co detector increases as D^m, where D is the emitter diameter and $2 < m < 3$.

17.4 What is the heat expected to be deposited from neutrons in the iron pin of a gamma thermometer exposed to a thermal flux of 10^{18} neutrons/m$^2\cdot$s?

17.5 Prove that the probability that a fission fragment will escape from one side of a fissile deposit is $\frac{1}{2}(1 - x/R)$, where R is the average range of fission fragments and x is the depth in the deposit where the fission took place.

17.6 Prove that the total number of fission fragments escaping from one side of a fissile deposit of thickness t ($t <$ range) is equal to $F(t - t^2/2R)$, where F is the number of fissions per unit volume.

17.7 Prove that the energy resolution of a magnetic spectrometer is given (relativistically) by the equation

$$\frac{\Delta T}{T} = \left(1 + \frac{MC^2}{MC^2 + T}\right)\frac{\Delta (B\rho)}{B\rho}$$

(Definition of the symbols used is given in Sec. 17.8.)

BIBLIOGRAPHY

Attix, F. H., and Roesch, W. C. (eds.), *Radiation Dosimetry*, vol. II, 2nd ed., Academic Press, New York, 1966.

Fleischer, R. L., Price, P. B., and Walker, R. M., *Nuclear Tracks in Solids*, University of California Press, Berkeley, Calif., 1975.

REFERENCES

1 Hilborn, J. W., *Nucleonics* **22**(2):69 (1964).
2 Joslin, C. W., *Nucl. Eng. Intern.* (May 1972).
3 Kroon, J. C., Smith, F. M., and Taylor, R. I., *ANS Trans.* **23**:459 (1976).
4 Balcar, E., Bock, H., and Hahn, F., *Nucl. Instr. Meth.* **153**:429 (1978).
5 Bozarth, D. P., and Warren, H. D., *ANS Trans.* **23**:517 (1976).
6 Hawer, J. M., and Beckerley, J. G., TID-25952-P1, vol. 1 (1973).
7 Goldstein, N. R., and Todt, W. H., *IEEE Trans.* **NS-26**:1 (1979).
8 Shields, R. B., *IEEE Trans. Nucl. Sci.* **NS-20**:603 (1973).
9 Lynch, G. F., *IEEE Trans. Nucl. Sci.* **NS-25**:1094 (1978).
10 Levert, F. E., Beyerlein, R. A., and Cox, S. A., *Nucl. Instr. Meth.* **160**:61 (1979).
11 Levert, F. E., *Nucl. Tech.* **44**:188 (1979).
12 Stutheit, J. S., *Nucl. Instr. Meth.* **63**:300 (1968).
13 Tsoulfanidis, N., Wehring, B. W., and Wyman, M. R., *Nucl. Sci. Eng.* **43**:42 (1979).
14 Draper, E. L., Jr., *Nucl. Sci. Eng.* **46**:31 (1971).
15 Gilliam, D. M., Dudey, N. D., and Popek, R. J., *Nucl. Tech.* **25**:237 (1975).
16 Lovett, D. B., *Health Phys.* **16**:623 (1969).
17 Seitz, M. G., Walker, R. M., and Carpenter, B. S., *J. Appl. Phys.* **44**:510 (1973).
18 Gomaa, M. A., and Hamdan, M. A., *Nucl. Instr. Meth.* **146**:527 (1977).
19 Aframian, A., *Nucl. Tech.* **42**:343 (1979).
20 Paretti, C., and Ricci, A., *Nucl. Instr. Meth.* **122**:389 (1974).
21 Cohn, C. E., and Gold, R., *Rev. Sci. Instr.* **43**:13 (1972).
22 Chou, H. P., Johnson, R. H., and Clikeman, F. M., *Proceedings Intern. Conf. on Nucl. Cross Sections for Technology*, Knoxville, Tenn., Oct. 1979.
23 Becker, K., in F. H. Attix (ed.), *Topics in Radiation Dosimetry*, Suppl. 2, pp. 79–142, Academic Press, New York, 1972.
24 Dudley, R. A., in F. H. Attix and W. C. Roesch (eds.), *Radiation Dosimetry*, II, Academic Press, New York, 1966.
25 Charalambous, S., and Petridou, C., *Nucl. Instr. Meth.* **137**:441 (1976).
26 Simons, G. G., and Emmons, L. L., *Nucl. Instr. Meth.* **160**:79 (1979).
27 Ohno, A., and Matsuura, S., *Nucl. Tech. Meth.* **47**:485 (1980).
28 Bacci, C., Calicchia, A., Pugliani, L., Salvadori, P., and Furetta, C., *Health Phys.* **38**:21 (1980).
29 Vohra, K. G., Bhatt, R. C., Chandra, B., Pradhan, A. S., Lakshmanan, A. R., and Shastry, S. S., *Health Phys.* **38**:113 (1980).
30 Liu, N. H., Gilliam, J. D., and Anderson, D. W., *Health Phys.* **38**:359 (1980).
31 McKlveen, J. W., and Schwenk, M., *Nucl. Tech.* **31**:257 (1976).
32 Nakajima, T., Murayama, Y., Matsuzawa, T., and Kayano, A., *Nucl. Instr. Meth.* **157**:155 (1978).
33 Piesch, E., Burgkhardt, X., and Sayed, A. M., *Nucl. Instr. Meth.* **157**:179 (1978).
34 Powell, C. F., Fowler, P. H., and Perskins. D. H., *The Study of Elementary Particles by the Photographic Method*, Pergamon Press, Bristol, 1959.
35 Welb, J. H., *Phys. Rev.* **74**:511 (1948).
36 Barkas, W. H., *Nuclear Research Emulsions*, vol. 1, (1963), vol. 2 (1973), Academic Press, New York.
37 Silberstein, L., *Phil. Mag.* **44**:257 (1929).
38 Bertoxzi, W., Hynes, M. V., Sargent, C. P., Turchinetz, W., and Williamson, C., *Nucl. Instr. Meth.* **162**:211 (1979).

39 Fulbright, H. W., *Nucl. Instr. Meth.* **162**:355 (1979).
40 Enge, H. A., *Nucl. Instr. Meth.* **162**:161 (1979).
41 Laurent, H., and Schapira, J. P., *Nucl. Instr. Meth.* **162**:181 (1979).
42 Mladjenovic, M. S., *Nucl. Instr. Meth.* **162**:193 (1979).
43 Bartholomew, G. A., and Lee-Whiting, G. E., *Nucl. Instr. Meth.* **162**:239 (1979).
44 Hughes, A. L., and Rojansky, V., *Phys. Rev.* **34**:284 (1929).
45 Rogers, F. T., Jr., *Rev. Sci. Instr.* **8**:22 (1937).
46 Rogers, F. T., Jr., *Rev. Sci. Instr.* **11**:19 (1939).
47 Rogers, F. T., Jr., *Rev. Sci. Instr.* **14**:216 (1943).
48 Rogers, F. T., Jr., *Rev. Sci. Instr.* **22**:723 (1951).
49 Laegsgaard, E., *Nucl. Instr. Meth.* **162**:93 (1979).
50 Haase, E. L., Fawzi, M. A., Saylor, D. P., and Vellen, E., *Nucl. Instr. Meth.* **97**:465 (1971).
51 Kalbitzer, S., Barer, R., Melzer, W., and Stumpfi, W., *Nucl. Instr. Meth.* **54**:323 (1967).
52 Hofker, W. K., Ooosthock, D. P., Hoeberechts, A. M. E., Van Dantzig, X., Mulder, K., Obserski, J. E. J., Koerts, L. A., Dieperink, J. H., Kok, E., and Rumphorst, R. F., *IEEE Trans.* **NS-13**(3): 208 (1966).
53 Ford, J. L. C., Jr., *Nucl. Instr. Meth.* **162**:277 (1979).
54 Charpak, G., *Nature* **270**:479 (1977).
55 Ball, G. C., *Nucl. Instr. Meth.* **162**:263 (1979).
56 Allemond, R., Bourdel, J., Roudant, E., Convert, P., Ibel, K., Jacobe, J., Cotton, J. P., and Farnoux, B., *Nucl. Instr. Meth.* **126**:29 (1975).
57 Beardsworth, E., Fischer, J., Iwata, S., Lavine, M. J., Rudeka, V., and Thorn, C. E., *Nucl. Instr. Meth.* **127**:29 (1975).
58 Glussel, P., *Nucl. Instr. Meth.* **140**:61 (1977).
59 Kitahara, T., and Isozumi, Y., *Nucl. Instr. Meth.* **140**:263 (1977).
60 Fischer, J., Okuno, H., and Walenta, A. H., *Nucl. Instr. Meth.* **151**:451 (1978).
61 Gustaldi, V., et al., *Nucl. Instr. Meth.* **156**:297 (1978).
62 Lacy, J. L., and Lindsey, R. S., *Nucl. Instr. Meth.* **119**:483 (1974).
63 Bertozzi, W., Hynes, M. V., Sargent, C. P., Creswell, C., Dunn, P. C., Hirsch, A., Leitch, M., Norum, B., Rad, F. N., and Susanuma, T., *Nucl. Instr. Meth.* **141**:457 (1977).
64 Breskin, A., Charpak, G., Gabioud, B., Sauli, F., Trantner, N., Duinker, W., and Schultz, G., *Nucl. Instr. Meth.* **119**:9 (1974).
65 Alberi, J., Fischer, J., Radeka, V., Rogers, L. C., and Schoenborn, B., *Nucl. Instr. Meth.* **127**: 507 (1975).

appendix a

■ useful constants and conversion factors

TABLE A1 Useful Constants

Constant	Symbol or definition	Value
Avogadro's number	N_A	0.6022169×10^{23} at/mol
Elementary charge	e	$1.6021917 \times 10^{-19}$ C = 4.803250 esu
Atomic mass unit	u	1.660531×10^{-27} kg = 931.478 MeV
Atomic mass unit	$\frac{1}{12}$ of mass $^{12}_{6}C$	
Electron rest mass	m	9.109558×10^{-31} kg = 0.511 MeV
Proton rest mass	M_p	1.672614×10^{-27} kg = 938.256 MeV
Neutron rest mass	M_n	1.674919×10^{-27} kg = 939.549 MeV
Planck constant	h	6.626196×10^{-34} J·s
Boltzmann constant	k	1.380622×10^{-23} J/K
Standard atmosphere		101,325 Pa = 14.696 lb/in²
Fine-structure constant	$\alpha = e^2/(\hbar \cdot c)$	1/137.14
Classical electron radius	$r_0 = e^2/(m \cdot c^2)$	2.818042×10^{-15} m
Bohr radius	$\hbar^2/(m \cdot e^2)$	0.529177×10^{-10} m
Compton wavelength	$h/(m \cdot c)$	2.424631×10^{-12} m

TABLE A2 Conversion Table

To convert	Multiply by	To obtain (symbol)
MeV	1.602×10^{-13}	Joules (J)
Pounds (lb)	2.2046	Kilograms (kg)
Inches	2.54×10^{-2}	Meters (m)
Lb/in²	1.4504×10^{-4}	Pascal (Pa)
Btu/h	0.293	Watts (W)
Pounds force	4.4482	Newtons (N)
Flux (particles/cm² · s)	10^4	Flux (particles/m² · s)
Density (g/cm³)	10^3	Density (kg/m³)
μ (cm²/g)	10^{-1}	μ (m²/kg)
μ (cm⁻¹)	100	μ (m⁻¹)
Σ (cm⁻¹)	100	Σ (m⁻¹)
Range (g/cm²)	10	Range (kg/m²)
Curies	3.7×10^{10}	Becquerels (Bq)
Rads	10^{-2}	Grays (Gy)
Rems	10^{-2}	Sieverts (Sv)

TABLE A3 Prefix and SI Symbols of Multiplication Factors

10^{12} tera (T)	10^{-6} micro (μ)
10^{9} giga (G)	10^{-9} nano (n)
10^{6} mega (M)	10^{-12} pico (p)
10^{3} kilo (k)	10^{-15} femto (f)
10^{-3} milli (m)	10^{-18} atto (a)

appendix b

■ atomic masses and other properties of isotopes

Name	Symbol	Z	A	Isotopic mass (u)†	Natural abundance (%)	Density of element (in 10^3 kg/m^3)	σ_a (b) for (n, γ) (0.0253-eV neutrons)
Aluminum	Al	13	27	26.98153	100	2.7‡	0.235
Antimony	Sb	51	121	120.9038	57.25	6.62	5.9
			123	122.9041	42.75		4.1
Argon	Ar	18	36	35.96755	0.337	Gas	6.0
			38	37.96272	0.063		0.8
			40	39.96238	99.60		0.61
Arsenic	As	33	75	74.9216	100	5.73	4.5
Beryllium	Be	4	9	9.01218	100	1.85	0.0095
Bismuth	Bi	83	209	208.9804	100	9.8	0.034
Boron	B	5	10	10.01294	19.78	2.3	3,837
			11	11.00931	80.22		0.04
Cadmium	Cd	48	106	105.9070	1.22	8.65	2,450.0
			108	107.9040	0.88		2.0
			110	109.9039	12.39		0.1
			111	110.9042	12.75		–
			112	111.9028	24.07		0.03
			113	112.9046	12.26		20,000.0
			114	113.9036	28.86		0.14
			116	115.9050	7.58		1.4
Carbon	C	6	12	12.0000	98.89	1.60	0.0034
			13	13.00335	1.11		0.0009
Cesium	Cs	55	133	132.9051	100	1.9	29.0
Cobalt	Co	27	59	59.93344	100	8.8	37.2
Copper	Cu	29	63	62.9298	60.09	8.96	4.5
			65	64.9278	30.91		2.2
Gadolinium	Gd	64	152	151.9195	0.20	7.95	< 180
			154	153.9207	2.15		100.0
			155	154.9226	14.73		61,000.0
			156	155.9221	20.47		11.5
			157	156.9339	15.68		254,000.0
			158	157.9241	24.87		3.5
Germanium	Ge	32	70	69.9243	20.52	5.36	3.68
			72	71.9217	27.43		0.98
			73	72.9234	7.76		14.0
			74	72.9212	36.54		0.45
			76	75.9214	7.76		0.2
Gold	Au	79	197	196.9666	100	19.32	98.8

(*See footnotes on page 549.*)

Name	Symbol	Z	A	Isotopic mass (u)†	Natural abundance (%)	Density of element (in 10^3 kg/m^3)	σ_a (b) for (n, γ) (0.0253-eV neutrons)
Helium	He	2	3	3.01603	0.00013	Gas	5,327.0
			4	4.00260	99.99987		–
Hydrogen	H	1	1	1.007825	99.985	Gas	0.332
			2	2.01410	0.015		0.0005
Indium	In	49	113	112.9043	4.28	7.31	11.1
			115	114.9041	95.72		193.2
Iodine	I	53	127	126.9044	100	4.93	6.2
Iron	Fe	26	54	53.9396	5.82	7.87	2.3
			56	55.9349	91.66		2.7
			57	56.9354	2.19		2.5
			58	57.9333	0.33		1.2
Lead	Pb	82	204	203.9730	1.48	11.34	0.17
			206	205.9745	23.6		0.0305
			207	206.9759	22.6		0.709
			208	207.9766	52.3		< 0.03
Lithium	Li	3	6	6.01512	7.42	0.53	245 (n, α)
			7	7.01600	92.58		0.037
Mercury	Hg	80	196	195.9650	0.146	13.55	3,100.0
			198	197.9668	10.02		0.018
			199	198.9683	16.84		2,500.0
			200	199.9683	23.13		50.0
			201	200.9703	13.22		50.0
			202	201.9706	29.80		4.5
			204	203.9735	6.85		0.4
Nickel	Ni	28	58	57.9353	67.88	8.90	4.4
			60	59.9308	26.23		2.6
			61	60.9310	1.19		2.0
			62	61.9283	3.66		14.2
			64	63.9280	1.08		1.5
Nitrogen	N	7	14	14.00307	99.63	Gas	1.81
			15	15.00011	0.37		0.00004
Oxygen	O	8	16	15.99491	99.759	Gas	0.000178
			17	16.99914	0.037		0.04
			18	17.99915	0.204		0.00016
Phosphorus	P	15	31	30.97376	100	1.82	0.19
Platinum	Pt	78	190	189.9600	0.0127	21.45	150.0
			192	191.9614	0.78		8.0
			194	193.9628	32.19		1.2
			195	194.9648	33.8		27.0
			196	195.9650	25.3		1.0
			198	197.9675	7.21		4.0
Rhodium	Rh	45	103	102.9048	100	12.41	150.0
Silicon	Si	14	28	27.97693	92.21	2.33	0.080
			29	28.97649	4.70		0.28
			30	29.97376	3.09		0.40
Silver	Ag	47	107	106.9041	51.82	10.49	37.0
			109	108.9047	45.18		92.0

(*See footnotes on page 549.*)

Name	Symbol	Z	A	Isotopic mass (u)†	Natural abundance (%)	Density of element (in 10^3 kg/m^3)	σ_a (b) for (n, γ) (0.0253-eV neutrons)
Sodium	Na	11	23	22.98977	100	0.97	0.534
Uranium	U	92	234	234.0409	0.0057	19.1	95.0
			235	235.0439	0.710		678.0
			238	238.0508	99.284		2.73

†Isotopic masses from *Nuclear Heat Transfer* by M. M. El-Wakil, International Textbook Co., N.Y. (1971).

‡Same number gives density in g/cm^3 or 10^3 kg/m^3.

appendix c

■ alpha, beta, and gamma sources commonly used

TABLE C1 Alpha Sources

Isotope	Half-life	Alpha energy (MeV)	Relative intensity† (%)
^{210}Po	138.38 d	5.304	100
^{234}U	2.446×10^5 y	4.774	72
		4.723	28
^{235}U	7.038×10^8 y	4.397	57
		4.367	18
^{238}U	4.468×10^9 y	4.196	77
		4.149	23
^{239}Pu	2.413×10^4 y	5.155	73.3
		5.143	15.1
		5.105	11.5
^{241}Am	432.02 y	5.486	86
		5.443	12.7
^{252}Cf	2.646 y	6.118	84
		6.076	15.8

†Only intensities greater than 10 percent are listed.

TABLE C2 Electron and Beta Sources

Isotope	Half-life	Type of particle	Energy (MeV)
^{3}H	12.33 y	β^-	0.0186
^{14}C	5730 y	β^-	0.1565
^{32}P	14.28 d	β^-	1.7104
^{35}S	87.4 d	β^-	0.1675
^{36}Cl	3.0×10^5 y	β^-	0.7095
^{89}Sr	50.55 d	β^-	1.463
^{90}Y	64 h	β^-	2.282
^{99}Tc	2.14×10^5 y	β^-	0.292
^{63}Ni	100 y	β^-	0.0659
^{113}Sn	115.1 d	IC electron	$E_K = 0.3625$
			$E_L = 0.3875$
^{137}Cs†	30.17 y	IC electron	$E_K = 0.626$
			$E_L = 0.656$
^{207}Bi	38 y	IC electron	$E_K = 0.4816, 0.5558$
			$E_L = 0.9754, 1.0496$
			$E_M = 1.060$

†^{137}Cs is also a β^- emitter.

TABLE C3 Gamma Rays Used as Reference Standards for Energy Calibration†

Parent nuclide	Gamma-ray energy (keV)	Parent nuclide	Gamma-ray energy (keV)	Parent nuclide	Gamma-ray energy (keV)	Parent nuclide	Gamma-ray energy (keV)
^{183}Ta	52.596 ± 0.001	^{183}Ta	264.072 ± 0.006	^{94}Nb	702.627 ± 0.019	^{56}Co	1238.256 ± 0.027
^{241}Am	59.537 ± 0.001	^{75}Se	264.651 ± 0.008	^{110m}Ag	706.669 ± 0.020	^{182}Ta	1257.390 ± 0.028
^{75}Se	66.055 ± 0.009	^{133}Ba	276.397 ± 0.012	^{124}Sb	713.761 ± 0.018	^{160}Tb	1271.850 ± 0.026
^{182}Ta	67.750 ± 0.001	^{203}Hg	279.188 ± 0.006	^{185}Os	717.424 ± 0.018	^{182}Ta	1273.703 ± 0.028
^{153}Gd	69.676 ± 0.002	^{203}Pb	279.188 ± 0.006	^{124}Sb	722.767 ± 0.018	^{22}Na	1274.511 ± 0.028
^{133}Ba	80.998 ± 0.008	^{75}Se	279.528 ± 0.008	^{95}Zr–^{95}Nb	724.184 ± 0.018	^{182}Ta	1289.126 ± 0.029
^{183}Ta	82.919 ± 0.001	^{183}Ta	291.724 ± 0.006	^{110m}Ag	744.254 ± 0.020	^{58}Fe	1291.564 ± 0.028
^{170}Tm	84.254 ± 0.003	^{192}Ir	295.949 ± 0.006	^{95}Zr–^{95}Nb	756.715 ± 0.019	^{82}Br	1317.440 ± 0.027
^{182}Ta	84.680 ± 0.002	^{160}Tb	298.572 ± 0.006	^{110m}Ag	763.928 ± 0.019	^{132}Cs	1317.890 ± 0.028
^{183}Ta	84.712 ± 0.002	^{133}Ba	302.851 ± 0.015	^{95}Zr–^{95}Nb	765.786 ± 0.019	^{124}Sb	1325.478 ± 0.029
^{160}Tb	86.786 ± 0.002	^{75}Se	303.913 ± 0.007	^{82}Br	776.502 ± 0.016	^{60}Co	1332.464 ± 0.028
^{109}Cd	88.037 ± 0.005	^{192}Ir	308.445 ± 0.007	^{95m}Tc	786.184 ± 0.017	^{56}Co	1360.175 ± 0.029
^{160}Tb	93.917 ± 0.006	^{192}Ir	316.497 ± 0.007	^{58}Co	810.757 ± 0.021	^{124}Sb	1368.130 ± 0.029
^{75}Se	96.733 ± 0.002	^{51}Cr	320.078 ± 0.008	^{110m}Ag	818.018 ± 0.022	^{24}Na	1368.599 ± 0.029
^{153}Gd	97.432 ± 0.003	^{183}Ta	353.999 ± 0.004	^{95m}Tc	820.608 ± 0.019	^{182}Ta	1373.807 ± 0.030
^{183}Ta	99.080 ± 0.002	^{133}Ba	356.005 ± 0.017	^{82}Br	827.809 ± 0.019	^{110m}Ag	1384.267 ± 0.029
^{182}Ta	100.105 ± 0.001	^{183}Ta	365.615 ± 0.007	^{54}Mn	834.827 ± 0.021	^{182}Ta	1387.376 ± 0.030
^{153}Gd	103.180 ± 0.002	^{133}Ba	383.851 ± 0.020	^{95m}Tc	835.132 ± 0.018	^{82}Br	1474.853 ± 0.030
^{183}Ta	107.932 ± 0.002	^{113}Sn	391.688 ± 0.010	^{56}Co	846.751 ± 0.019	^{110m}Ag	1475.757 ± 0.034
^{177}Lu	112.954 ± 0.003	^{75}Se	400.646 ± 0.009	^{58}Co	863.935 ± 0.018	^{144}Ce	1489.124 ± 0.032
^{182}Ta	113.673 ± 0.002	^{203}Pb	401.315 ± 0.013	^{94}Nb	871.099 ± 0.018	^{110m}Ag	1505.006 ± 0.032
^{182}Ta	116.418 ± 0.002	^{183}Ta	406.589 ± 0.006	^{185}Os	874.814 ± 0.019	^{110m}Ag	1562.264 ± 0.033
^{75}Se	121.115 ± 0.003	^{198}Au	411.794 ± 0.008	^{160}Tb	879.364 ± 0.018	^{82}Br	1650.296 ± 0.034
^{57}Co	122.063 ± 0.004	^{110m}Ag	446.791 ± 0.010	^{185}Os	880.272 ± 0.019	^{58}Co	1674.679 ± 0.036
^{185}Os	125.358 ± 0.004	^{192}Ir	468.062 ± 0.010	^{84}Rb	881.595 ± 0.018	^{124}Sb	1690.942 ± 0.036
^{75}Se	136.000 ± 0.005	^{7}Be	477.593 ± 0.012	^{192}Ir	884.523 ± 0.018	^{207}Bi	1770.188 ± 0.037
^{57}Co	136.473 ± 0.004	^{192}Ir	484.570 ± 0.011	^{110m}Ag	884.667 ± 0.018	^{56}Co	1771.303 ± 0.037
^{59}Fe	142.648 ± 0.004	^{85}Sr	513.996 ± 0.016	^{46}Sc	889.258 ± 0.018	^{125}Sn	1806.652 ± 0.038
^{183}Ta	144.127 ± 0.002	^{82}Br	554.334 ± 0.012	^{88}Y	898.021 ± 0.019	^{56}Co	1810.672 ± 0.039
^{141}Ce	145.440 ± 0.003	^{207}Bi	569.689 ± 0.013	^{110m}Ag	937.483 ± 0.020	^{88}Y	1836.014 ± 0.037

^{182}Ta	152.434 ± 0.002	^{95m}Tc	582.068 ± 0.013	^{160}Tb	962.295 ± 0.020	^{125}Sn	1889.844 ± 0.039
^{182}Ta	156.387 ± 0.002	^{228}Th	583.174 ± 0.013	^{160}Tb	966.151 ± 0.020	^{84}Rb	1897.727 ± 0.038
^{199}Au	158.370 ± 0.003	^{192}Ir	588.572 ± 0.012	^{124}Sb	968.188 ± 0.022	^{56}Co	1963.669 ± 0.040
^{185}Os	162.854 ± 0.008	^{185}Os	592.066 ± 0.014	^{84}Rb	1016.143 ± 0.021	^{132}Cs	1985.581 ± 0.041
^{139}Ce	165.853 ± 0.007	^{124}Sb	602.715 ± 0.013	^{56}Co	1037.815 ± 0.022	^{125}Sn	2002.089 ± 0.043
^{182}Ta	179.393 ± 0.003	^{192}Ir	604.401 ± 0.012	^{95m}Tc	1039.247 ± 0.022	^{56}Co	2015.133 ± 0.040
^{59}Fe	192.344 ± 0.006	^{192}Ir	612.450 ± 0.013	^{82}Br	1043.973 ± 0.022	^{56}Co	2034.706 ± 0.041
^{183}Ta	192.646 ± 0.005	^{82}Br	619.088 ± 0.013	^{124}Sb	1045.106 ± 0.022	^{124}Sb	2090.889 ± 0.044
^{160}Tb	197.030 ± 0.004	^{110m}Ag	620.342 ± 0.016	^{207}Bi	1063.635 ± 0.024	^{56}Co	2113.049 ± 0.045
^{182}Ta	198.356 ± 0.004	^{124}Sb	645.835 ± 0.017	^{198}Au	1087.633 ± 0.024	^{144}Ce	2185.608 ± 0.046
^{75}Se	198.596 ± 0.007	^{185}Os	646.111 ± 0.017	^{59}Fe	1099.224 ± 0.025	^{125}Sn	2200.965 ± 0.047
^{95m}Tc	204.117 ± 0.005	^{110m}Ag	657.744 ± 0.017	^{65}Zn	1115.518 ± 0.025	^{56}Co	2212.862 ± 0.049
^{199}Au	208.196 ± 0.005	^{137}Cs	661.638 ± 0.019	^{46}Sc	1120.516 ± 0.025	^{125}Sn	2275.710 ± 0.048
^{177}Lu	208.362 ± 0.010	^{132}Cs	667.698 ± 0.017	^{182}Ta	1121.272 ± 0.026	^{56}Co	2598.400 ± 0.053
^{160}Tb	215.641 ± 0.004	^{198}Au	675.871 ± 0.018	^{60}Co	1173.208 ± 0.025	^{228}Th	2614.471 ± 0.054
^{82}Br	221.476 ± 0.005	^{110m}Ag	677.601 ± 0.018	^{56}Co	1175.067 ± 0.026	^{24}Na	2753.965 ± 0.056
^{182}Ta	222.110 ± 0.003	^{203}Pb	680.495 ± 0.017	^{160}Tb	1177.934 ± 0.024	^{56}Co	3201.878 ± 0.064
^{182}Ta	229.322 ± 0.006	^{110m}Ag	686.998 ± 0.019	^{182}Ta	1189.022 ± 0.027	^{56}Co	3253.341 ± 0.065
^{185}Os	234.158 ± 0.010	^{144}Ce	696.492 ± 0.019	^{182}Ta	1221.376 ± 0.027	^{56}Co	3272.912 ± 0.065
^{95m}Tc	253.066 ± 0.006	^{82}Br	698.358 ± 0.016	^{182}Ta	1230.989 ± 0.028	^{56}Co	3451.064 ± 0.069

†From R. L. Heath, "Gamma Ray Spectrum Catalogue, Ge(Li) and Si(Li) Spectrometry Vol. 2," Aerojet Nuclear Co. Report ANCR-1000-2, USAEC (1974).

appendix d

tables of photon attenuation coefficients

TABLE D1 Total Mass Attenuation Coefficients in cm^2/g[†] for Gamma Rays[‡]

Photon energy (MeV)	H	Be	C	N	O	Na
1.00 - 02[§]	3.85 - 01	5.36 - 01	2.17 + 00	3.57 + 00	5.58 + 00	1.51 + 01
1.50 - 02	3.76 - 01	2.68 - 01	7.22 - 01	1.09 + 00	1.62 + 00	4.37 + 00
2.00 - 02	3.69 - 01	2.06 - 01	3.88 - 01	5.41 - 01	7.54 - 01	1.88 + 00
3.00 - 02	3.57 - 01	1.71 - 01	2.30 - 01	2.76 - 01	3.35 - 01	6.39 - 01
4.00 - 02	3.46 - 01	1.59 - 01	1.93 - 01	2.12 - 01	2.36 - 01	3.55 - 01
5.00 - 02	3.35 - 01	1.52 - 01	1.79 - 01	1.87 - 01	1.99 - 01	2.54 - 01
6.00 - 02	3.26 - 01	1.47 - 01	1.70 - 01	1.74 - 01	1.81 - 01	2.09 - 01
8.00 - 02	3.09 - 01	1.39 - 01	1.58 - 01	1.60 - 01	1.62 - 01	1.70 - 01
1.00 - 01	2.94 - 01	1.32 - 01	1.50 - 01	1.50 - 01	1.52 - 01	1.52 - 01
1.50 - 01	2.65 - 01	1.19 - 01	1.34 - 01	1.34 - 01	1.34 - 01	1.31 - 01
2.00 - 01	2.43 - 01	1.09 - 01	1.23 - 01	1.23 - 01	1.23 - 01	1.18 - 01
3.00 - 01	2.11 - 01	9.45 - 02	1.07 - 01	1.06 - 01	1.07 - 01	1.02 - 01
4.00 - 01	1.89 - 01	8.47 - 02	9.55 - 02	9.54 - 02	9.54 - 02	9.14 - 02
5.00 - 01	1.73 - 01	7.73 - 02	8.72 - 02	8.71 - 02	8.71 - 02	8.34 - 02
6.00 - 01	1.60 - 01	7.15 - 02	8.07 - 02	8.05 - 02	8.06 - 02	7.72 - 02
8.00 - 01	1.40 - 01	6.29 - 02	7.09 - 02	7.08 - 02	7.08 - 02	6.78 - 02
1.00 + 00	1.26 - 01	5.65 - 02	6.37 - 02	6.36 - 02	6.37 - 02	6.09 - 02
1.50 + 00	1.03 - 01	4.60 - 02	5.19 - 02	5.18 - 02	5.18 - 02	4.97 - 02
2.00 + 00	8.75 - 02	3.94 - 02	4.45 - 02	4.45 - 02	4.46 - 02	4.28 - 02
3.00 + 00	6.91 - 02	3.14 - 02	3.57 - 02	3.58 - 02	3.60 - 02	3.49 - 02
4.00 + 00	5.81 - 02	2.66 - 02	3.05 - 02	3.07 - 02	3.10 - 02	3.04 - 02
5.00 + 00	5.05 - 02	2.35 - 02	2.71 - 02	2.74 - 02	2.78 - 02	2.76 - 02
6.00 + 00	4.50 - 02	2.12 - 02	2.47 - 02	2.51 - 02	2.55 - 02	2.56 - 02
8.00 + 00	3.75 - 02	1.82 - 02	2.16 - 02	2.21 - 02	2.26 - 02	2.32 - 02
1.00 + 01	3.25 - 02	1.63 - 02	1.96 - 02	2.02 - 02	2.09 - 02	2.18 - 02

	Al	Si	Fe	Cu	Pb	NaI
1.00 - 02	2.58 + 01	3.36 + 01	1.72 + 02	2.23 + 02	1.28 + 02	1.36 + 02
1.50 - 02	7.66 + 00	9.97 + 00	5.57 + 01	7.33 + 01	1.12 + 02	4.59 + 01
2.00 - 02	3.24 + 00	4.19 + 00	2.51 + 01	3.30 + 01	8.34 + 01	2.12 + 01
3.00 - 02	1.03 + 00	1.31 + 00	7.88 + 00	1.06 + 01	2.84 + 01	6.86 + 00

(*See footnotes on page 557.*)

TABLE D1 Total Mass Attenuation Coefficients in cm^2/g† for Gamma Rays‡ (*Continued*)

Photon energy (MeV)	Al	Si	Fe	Cu	Pb	NaI
4.00 − 02	5.14 − 01	6.35 − 01	3.46 + 00	4.71 + 00	1.31 + 01	1.89 + 01
5.00 − 02	3.34 − 01	3.96 − 01	1.84 + 00	2.50 + 00	7.22 + 00	1.05 + 01
6.00 − 02	2.55 − 01	2.92 − 01	1.13 + 00	1.52 + 00	4.43 + 00	6.42 + 00
8.00 − 02	1.89 − 01	2.07 − 01	5.50 − 01	7.18 − 01	2.07 + 00	3.00 + 00
1.00 − 01	1.62 − 01	1.73 − 01	3.42 − 01	4.27 − 01	5.23 + 00	1.64 + 00
1.50 − 01	1.34 − 01	1.40 − 01	1.84 − 01	2.08 − 01	1.89 + 00	5.90 − 01
2.00 − 01	1.20 − 01	1.25 − 01	1.39 − 01	1.48 − 01	9.45 − 01	3.14 − 01
3.00 − 01	1.03 − 01	1.07 − 01	1.07 − 01	1.08 − 01	3.83 − 01	1.58 − 01
4.00 − 01	9.22 − 02	9.54 − 02	9.21 − 02	9.19 − 02	2.20 − 01	1.12 − 01
5.00 − 01	8.41 − 02	8.70 − 02	8.29 − 02	8.22 − 02	1.54 − 01	9.21 − 02
6.00 − 01	7.77 − 02	8.05 − 02	7.62 − 02	7.52 − 02	1.20 − 01	8.02 − 02
8.00 − 01	6.83 − 02	7.06 − 02	6.65 − 02	6.55 − 02	8.56 − 02	6.63 − 02
1.00 + 00	6.14 − 02	6.35 − 02	5.96 − 02	5.86 − 02	6.90 − 02	5.80 − 02
1.50 + 00	5.00 − 02	5.18 − 02	4.87 − 02	4.79 − 02	5.10 − 02	4.66 − 02
2.00 + 00	4.32 − 02	4.48 − 02	4.25 − 02	4.19 − 02	4.50 − 02	4.12 − 02
3.00 + 00	3.54 − 02	3.68 − 02	3.62 − 02	3.59 − 02	4.16 − 02	3.66 − 02
4.00 + 00	3.11 − 02	3.24 − 02	3.31 − 02	3.32 − 02	4.14 − 02	3.50 − 02
5.00 + 00	2.84 − 02	2.97 − 02	3.14 − 02	3.18 − 02	4.24 − 02	3.46 − 02
6.00 + 00	2.66 − 02	2.79 − 02	3.05 − 02	3.10 − 02	4.34 − 02	3.47 − 02
8.00 + 00	2.44 − 02	2.57 − 02	2.98 − 02	3.06 − 02	4.59 − 02	3.55 − 02
1.00 + 01	2.31 − 02	2.46 − 02	2.98 − 02	3.08 − 02	4.84 − 02	3.68 − 02

	Air	H_2O	Compact bone	Muscle, striated	concrete	Pyrex glass	Lucite $(C_5H_8O_2)_n$
1.00 − 02	4.82 + 00	4.99 + 00	2.00 + 01	5.09 + 00	2.65 + 01	1.67 + 01	3.11 + 00
1.50 − 02	1.45 + 00	1.48 + 00	6.15 + 00	1.53 + 00	8.01 + 00	4.95 + 00	9.82 − 01
2.00 − 02	6.91 − 01	7.11 − 01	2.68 + 00	7.31 − 01	3.45 + 00	2.13 + 00	5.03 − 01
3.00 − 02	3.18 − 01	3.38 − 01	9.07 − 01	3.42 − 01	1.12 + 00	7.24 − 01	2.74 − 01
4.00 − 02	2.29 − 01	2.48 − 01	4.78 − 01	2.49 − 01	5.59 − 01	3.94 − 01	2.19 − 01
5.00 − 02	1.96 − 01	2.14 − 01	3.27 − 01	2.14 − 01	3.61 − 01	2.77 − 01	1.98 − 01
6.00 − 02	1.79 − 01	1.97 − 01	2.58 − 01	1.97 − 01	2.73 − 01	2.25 − 01	1.86 − 01
8.00 − 02	1.62 − 01	1.79 − 01	2.00 − 01	1.78 − 01	2.00 − 01	1.80 − 01	1.72 − 01
1.00 − 01	1.51 − 01	1.68 − 01	1.74 − 01	1.66 − 01	1.70 − 01	1.60 − 01	1.62 − 01
1.50 − 01	1.34 − 01	1.49 − 01	1.47 − 01	1.48 − 01	1.40 − 01	1.36 − 01	1.45 − 01
2.00 − 01	1.23 − 01	1.36 − 01	1.32 − 01	1.35 − 01	1.25 − 01	1.23 − 01	1.32 − 01
3.00 − 01	1.06 − 01	1.18 − 01	1.14 − 01	1.17 − 01	1.07 − 01	1.06 − 01	1.15 − 01
4.00 − 01	9.53 − 02	1.06 − 01	1.01 − 01	1.05 − 01	9.58 − 02	9.50 − 02	1.03 − 01
5.00 − 01	8.70 − 02	9.67 − 02	9.25 − 02	9.59 − 02	8.73 − 02	8.67 − 02	9.41 − 02
6.00 − 01	8.05 − 02	8.95 − 02	8.56 − 02	8.87 − 02	8.07 − 02	8.02 − 02	8.70 − 02
8.00 − 01	7.07 − 02	7.86 − 02	7.51 − 02	7.79 − 02	7.09 − 02	7.04 − 02	7.65 − 02

(*See footnotes on page 557.*)

TABLE D1 Total Mass Attenuation Coefficients in cm^2/g† for Gamma Rays‡ *(Continued)*

Photon energy (MeV)	Air	H_2O	Compact bone	Muscle, striated	concrete	Pyrex glass	Lucite $(C_5H_8O_2)n$
1.00 + 00	6.36 − 02	7.07 − 02	6.75 − 02	7.00 − 02	6.37 − 02	6.33 − 02	6.87 − 02
1.50 + 00	5.18 − 02	5.75 − 02	5.50 − 02	5.70 − 02	5.19 − 02	5.16 − 02	5.59 − 02
2.00 + 00	4.45 − 02	4.94 − 02	4.73 − 02	4.89 − 02	4.48 − 02	4.44 − 02	4.80 − 02
3.00 + 00	3.58 − 02	3.97 − 02	3.83 − 02	3.93 − 02	3.65 − 02	3.61 − 02	3.85 − 02
4.00 + 00	3.08 − 02	3.40 − 02	3.31 − 02	3.37 − 02	3.19 − 02	3.14 − 02	3.29 − 02
5.00 + 00	2.75 − 02	3.03 − 02	2.97 − 02	3.00 − 02	2.90 − 02	2.84 − 02	2.92 − 02
6.00 + 00	2.52 − 02	2.77 − 02	2.74 − 02	2.74 − 02	2.70 − 02	2.63 − 02	2.66 − 02
8.00 + 00	2.23 − 02	2.43 − 02	2.44 − 02	2.40 − 02	2.45 − 02	2.37 − 02	2.32 − 02
1.00 + 01	2.04 − 02	2.22 − 02	2.26 − 02	2.19 − 02	2.31 − 02	2.22 − 02	2.11 − 02

†Multiply by 10^{-1} to obtain m^2/kg.

‡From J. H. Hubbell, "Photon Cross Sections, Attenuation Coefficients, and Energy Absorption Coefficients from 10 keV to 100 GeV," NSRDS-NBS 29 C (1969).

§Read × 10^{-2}

TABLE D2 Mass Energy Absorption Coefficients in cm^2/g† for Gamma Rays‡

Photon energy (MeV)	H	Be	C	N	O	Na
0.01	0.00986	0.368	1.97	3.38	5.39	14.9
0.015	0.0110	0.104	0.536	0.908	1.44	4.20
0.02	0.0135	0.0469	0.208	0.362	0.575	1.70
0.03	0.0185	0.0195	0.0594	0.105	0.165	0.475
0.04	0.0231	0.0146	0.0306	0.0493	0.0733	0.199
0.05	0.0271	0.0142	0.0233	0.0319	0.0437	0.106
0.06	0.0306	0.0147	0.0211	0.0256	0.0322	0.0668
0.08	0.0362	0.0166	0.0205	0.0223	0.0249	0.0382
0.10	0.0406	0.0184	0.0215	0.0224	0.0237	0.0297
0.15	0.0481	0.0216	0.0245	0.0247	0.0251	0.0260
0.2	0.0525	0.0235	0.0265	0.0267	0.0268	0.0264
0.3	0.0569	0.0255	0.0287	0.0287	0.0288	0.0277
0.4	0.0586	0.0262	0.0295	0.0295	0.0295	0.0284
0.5	0.0593	0.0265	0.0297	0.0296	0.0297	0.0285
0.6	0.0587	0.0263	0.0295	0.0295	0.0296	0.0284
0.8	0.0574	0.0256	0.0288	0.0289	0.0289	0.0275
1.0	0.0555	0.0248	0.0279	0.0279	0.0278	0.0266
1.5	0.0507	0.0227	0.0255	0.0255	0.0254	0.0243
2	0.0464	0.0208	0.0234	0.0234	0.0234	0.0225
3	0.0398	0.0180	0.0204	0.0205	0.0206	0.0199

(See footnotes on page 559.)

TABLE D2 Mass Energy Absorption Coefficients in cm^2/g† for Gamma Rays‡ (*Continued*)

Photon energy (MeV)	H	Be	C	N	O	Na
4	0.0352	0.0161	0.0185	0.0186	0.0188	0.0184
5	0.0317	0.0148	0.0171	0.0173	0.0175	0.0174
6	0.0290	0.0138	0.0161	0.0163	0.0166	0.0161
8	0.0252	0.0123	0.0147	0.0151	0.0155	0.0159
10	0.0225	0.0114	0.0138	0.0143	0.0148	0.0155

	Al	Si	Fe	Cu	Pb	Air
0.01	25.5	33.3	142	160	127	4.61
0.015	7.47	9.75	49.3	59.4	91.7	1.27
0.02	3.06	4.01	22.8	28.2	69.1	0.511
0.03	0.868	1.14	7.28	9.50	24.6	0.148
0.04	0.357	0.472	3.17	4.24	11.8	0.0668
0.05	0.184	0.241	1.64	2.22	6.54	0.0406
0.06	0.111	0.144	0.961	1.32	4.08	0.0305
0.08	0.0562	0.0700	0.414	0.573	1.86	0.0243
0.10	0.0386	0.0459	0.219	0.302	2.28	0.0234
0.15	0.0285	0.0312	0.0814	0.106	1.15	0.0250
0.2	0.0276	0.0292	0.0495	0.0597	0.629	0.0268
0.3	0.0282	0.0294	0.0335	0.0370	0.259	0.0287
0.4	0.0287	0.0298	0.0308	0.0318	0.143	0.0295
0.5	0.0286	0.0298	0.0295	0.0298	0.0951	0.0296
0.6	0.0286	0.0295	0.0286	0.0286	0.0710	0.0295
0.8	0.0277	0.0288	0.0273	0.0271	0.0481	0.0289
1.0	0.0269	0.0277	0.0262	0.0258	0.0377	0.0278
1.5	0.0245	0.0253	0.0237	0.0233	0.0271	0.0254
2	0.0226	0.0234	0.0220	0.0217	0.0240	0.0234
3	0.0202	0.0210	0.0204	0.0202	0.0234	0.0205
4	0.0188	0.0196	0.0199	0.0200	0.0245	0.0186
5	0.0179	0.0187	0.0198	0.0200	0.0259	0.0174
6	0.0172	0.0182	0.0199	0.0202	0.0272	0.0164
8	0.0168	0.0177	0.0204	0.0209	0.0294	0.0152
10	0.0165	0.0175	0.0209	0.0215	0.0310	0.0145

	H_2O	Compact bone	Muscle, striated	Concrete	Pyrex glass	Lucite $(C_5H_8O_2)_n$
0.01	4.79	19.2	4.87	25.5	16.5	2.91
0.015	1.28	5.84	1.32	7.66	4.75	0.783
0.02	0.512	2.46	0.533	3.22	1.94	0.310
0.03	0.149	0.720	0.154	0.936	0.554	0.0899

(*See footnotes on page 559.*)

TABLE D2 Mass Energy Absorption Coefficients in cm^2/g† for Gamma Rays‡
(Continued)

Photon energy (MeV)	H_2O	Compact bone	Muscle, striated	Concrete	Pyrex glass	Lucite $(C_5H_8O_2)_n$
0.04	0.0677	0.304	0.0701	0.393	0.232	0.0437
0.05	0.0418	0.161	0.0431	0.204	0.122	0.0301
0.06	0.0320	0.0998	0.0328	0.124	0.0768	0.0254
0.08	0.0262	0.0537	0.0264	0.0625	0.0428	0.0232
0.10	0.0256	0.0387	0.0256	0.0424	0.0325	0.0238
0.15	0.0277	0.0305	0.0275	0.0290	0.0274	0.0266
0.2	0.0297	0.0301	0.0294	0.0290	0.0276	0.0287
0.3	0.0319	0.0310	0.0317	0.0295	0.0289	0.0310
0.4	0.0328	0.0315	0.0325	0.0298	0.0295	0.0318
0.5	0.0330	0.0317	0.0328	0.0300	0.0297	0.0322
0.6	0.0329	0.0314	0.0325	0.0297	0.0294	0.0319
0.8	0.0321	0.0306	0.0318	0.0289	0.0287	0.0311
1.0	0.0309	0.0295	0.0306	0.0279	0.0277	0.0301
1.5	0.0282	0.0270	0.0280	0.0254	0.0252	0.0275
2	0.0260	0.0249	0.0257	0.0235	0.0233	0.0253
3	0.0227	0.0219	0.0225	0.0209	0.0207	0.0220
4	0.0206	0.0200	0.0204	0.0193	0.0190	0.0199
5	0.0191	0.0187	0.0189	0.0182	0.0179	0.0184
6	0.0180	0.0178	0.0178	0.0176	0.0171	0.0173
8	0.0166	0.0167	0.0164	0.0168	0.0163	0.0158
10	0.0157	0.0159	0.0155	0.0163	0.0157	0.0148

†Multiply by 10^{-1} to obtain m^2/kg.

‡From J. H. Hubbell, "Photon Cross Sections, Attenuation Coefficients, and Energy Absorption Coefficients from 10 keV to 100 GeV," NSRDS-NBS 29 C (1969).

appendix e

■ table of buildup factor constants

Coefficients of the Berger Equation for Dose Buildup Factors

$B = 1 + a\mu r \exp(b\mu r)$†

	Concrete‡		Air§		Water§	
E (MeV)	a	b	a	b	a	b
0.015	0.01	0.029	0.08	−0.034	0.09	−0.036
0.02	0.03	0.041	0.23	−0.032	0.26	−0.032
0.03	0.10	−0.036	0.93	−0.009	1.01	−0.006
0.04	0.26	0.035	2.40	0.018	2.58	0.024
0.05	0.52	−0.026	4.05	0.050	4.36	0.057
0.06	0.78	−0.008	5.27	0.075	5.59	0.082
0.08	1.42	0.007	6.11	0.102	6.47	0.108
0.1	1.83	0.028	5.93	0.113	6.11	0.120
0.15	2.19	0.054	4.70	0.121	4.88	0.125
0.2	2.20	0.065	3.94	0.113	4.13	0.118
0.3	2.03	0.067	3.10	0.094	3.18	0.096
0.4	1.87	0.061	2.61	0.079	2.67	0.080
0.5	1.73	0.055	2.29	0.067	2.32	0.068
0.6	1.60	0.049	2.05	0.058	2.07	0.059
0.8	1.41	0.040	1.71	0.045	1.74	0.045
1	1.27	0.032	1.50	0.035	1.50	0.036
1.5	1.02	0.021	1.16	0.021	1.16	0.021
2	0.89	0.014	0.97	0.013	0.97	0.013
3	0.71	0.007	0.75	0.005	0.74	0.005
4	0.59	0.004	0.61	0.001	0.62	0.000
5	0.49	0.004	0.53	−0.002	0.52	−0.002
6	0.45	0.002	0.47	−0.004	0.47	−0.005
8	0.36	0.001	0.37	−0.004	0.38	−0.006
10	0.30	0.003	0.31	−0.004	0.31	−0.005
15	0.21	0.004	0.23	−0.006	0.23	−0.008

†Good up to 40 mean free paths, for point isotropic source in infinite medium.
‡From A. B. Chilton, *Nuc. Sci. Eng.* **69**:436 (1979).
§From A. B. Chilton, C. M. Eisenhauer, and G. L. Simmons, *Nuc. Sci. Eng.* **73**:97 (1980).

■ index